Control and Optimization Technologies in Renewable Energy and Integrated Energy Systems

Control and Optimization Technologies in Renewable Energy and Integrated Energy Systems

Guest Editors

Temitayo Olowu
Raghav Khanna

MDPI

Basel • Beijing • Wuhan • Barcelona • Belgrade • Novi Sad • Cluj • Manchester

Guest Editors

Temitayo Olowu
Energy and Environment,
Science & Technology
Idaho National Laboratory
Idaho Falls, ID
USA

Raghav Khanna
Electrical Engineering and
Computer Science
University of Toledo
Toledo, OH
USA

Editorial Office
MDPI AG
Grosspeteranlage 5
4052 Basel, Switzerland

This is a reprint of the Special Issue, published open access by the journal *Electronics* (ISSN 2079-9292), freely accessible at: https://www.mdpi.com/journal/electronics/special_issues/137012H33M.

For citation purposes, cite each article independently as indicated on the article page online and as indicated below:

Lastname, A.A.; Lastname, B.B. Article Title. *Journal Name* **Year**, *Volume Number*, Page Range.

ISBN 978-3-7258-6270-2 (Hbk)
ISBN 978-3-7258-6271-9 (PDF)
https://doi.org/10.3390/books978-3-7258-6271-9

Contents

About the Editors

Temitayo Olowu

Temitayo Olowu (Member, IEEE) received his OND (2001) and HND (2005) from the Federal Polytechnic Ado-Ekiti and his BS (2010) and MS (2014) from the University of Benin, Nigeria, all in Electrical Engineering (Power Systems and Machines). He was a recipient of the Petroleum Technology Development Fund PhD Scholarship in Nigeria. He obtained his PhD degree from Florida International University (FIU), Miami, Florida, in 2021. He is currently working as a Research Engineer at the Energy and Environment Science and Technology Directorate, Idaho National Laboratory. His research focus includes power systems optimization and control, integrated energy systems, and resiliency.

Raghav Khanna

Raghav Khanna (Senior Member, IEEE) received the B.S., M.S., and Ph.D. degrees in electrical engineering from the University of Pittsburgh, Pittsburgh, PA, USA, in 2007, 2010, and 2014, respectively. He has worked for several industries, including Lockheed Martin, Philadelphia, PA; PPG Industries, Pittsburgh; and HRL Laboratories, Malibu, CA, USA. He is currently an Associate Professor with the Electrical Engineering and Computer Science Department, The University of Toledo, Toledo, OH, USA, where he holds the position of Leidich Family Endowed Professor of power and energy systems. His research interests include the characterization and modeling of wide-bandgap semiconductors for applications in next-generation power electronics, including renewable energy, electric vehicles, aerospace and maritime systems, and low-power consumer electronics.

electronics MDPI

Editorial

Control and Optimization Technologies in Renewable Energy and Integrated Energy Systems

Temitayo O. Olowu

Idaho National Laboratory, Idaho Falls, ID 83415, USA; temitayo.olowu@inl.gov

The rapid deployment of inverter-dominated generation, hybrid storage fleets, and cyber-connected automation is reshaping renewable and integrated energy systems. This Special Issue assembles contributions that advance control and optimization technologies for reliable, low-carbon operation across grid-tied and islanded contexts. These works collectively confront enduring challenges, including frequency stability with low inertia, voltage regulation under uncertainty, multi-energy coordination, forecasting for scheduling, and the security of increasingly digitalized assets. This issue also highlights gaps in interoperability, scalability, and resilience-by-design.

A consistent theme is the orchestration of converter-interfaced resources to provide grid services alongside legacy assets. Authors of contribution 1 demonstrate a multi-objective coordination of smart inverters with conventional devices to balance voltage support and ancillary services while respecting device limits. Complementing coordinated services, a survey and evaluation of virtual inertia implementations for autonomous AC/DC microgrids, clarifying design trade-offs between emulated inertia, damping, and stability margins as grid-forming/-following roles evolve, is conducted in contribution 2. Extending to hybrid storage, contribution 3 present a coordinated strategy that couples adaptive virtual inertia with virtual droop while optimizing state-of-charge trajectories, highlighting how control layers can be co-designed with energy management objectives .

Optimization-driven management under uncertainty has become highly important with the penetration of variable energy sources such as solar and wind systems. Authors of contribution 4 integrate unscented transformation with enhanced metaheuristics to improve microgrid scheduling quality and robustness under renewable variability. At the system scale, contribution 5 cast low-carbon economic dispatch for regional integrated energy systems with wind/solar uncertainty as a robust optimization, quantifying the emissions–cost trade-space and demonstrating the operational value of uncertainty sets. On the demand side, Laitsos et al. propose an enhanced sequence-to-sequence deep transfer learning framework for day-ahead load forecasting that strengthens generalization across domains—a prerequisite for dependable model-predictive scheduling contribution 6.

Some of the papers in this Special Issue target the control of energy conversion hardware and power quality. The methodology in contribution 7 develops a fast repetitive control strategy that improves tracking performance for power conversion systems with periodic disturbances, while contribution 8 introduce a fractional-delay PI multi-resonant-type repetitive controller based on a Farrow-structure filter to enhance harmonic compensation in grid-tied inverters. For hardware-in-the-loop validation, authors of contribution 9 validate flywheel energy storage control for wind-farm power smoothing, emphasizing the importance of high-fidelity testbeds for controller verification. In hydrogen-coupled systems, contribution 10 design fast frequency response strategies for electrolysis-driven plants, evidencing the flexibility value of sector-coupled resources.

Cyber–physical integration and infrastructure interfacing add further dimensions. Contribution 11 propose a distributed-ledger-enabled smart contract for total power factor management in grids with customer-owned wind farms, pointing to auditable, automated compliance mechanisms in prosumer-rich networks. Authors of contribution 12 study mitigation strategies for DC neutral-point potentials on transformers induced by metro stray currents, underscoring the need to consider transportation electrification externalities in distribution design and protection. Authors of contribution 13 envision a prosumer hydro plant network as a sustainable distributed energy depot, illustrating community-scale flexibility that complements utility-scale resources. At the converter level, contribution 14 show how modular multilevel converter control can maximize AC fault currents while maintaining grid code compliance, clarifying the dynamic limits for protection coordination in converter-rich grids.

Beyond the specific advances, the aggregated findings reveal persistent gaps. Firstly, interoperable controls that seamlessly coordinate grid-forming and grid-following resources across vendors and voltage levels remain immature; standardized interfaces for services (frequency, voltage, inertia, and fault response) are needed to reduce integration cost and commissioning risk. Secondly, scalable uncertainty-aware optimization must move from case-specific tuning toward reusable formulations with certified robustness and computational tractability for real-time deployment. Thirdly, experimental validation at scale is still scarce; controller designs that perform in simulations can falter in the field without hardware-in-the-loop, power-hardware-in-the-loop, and staged commissioning protocols. Fourthly, resilience-by-design, including cyber hardening and fail-operational behavior under contingencies, must be co-engineered with control. Finally, human-in-the-loop operability (explainable settings, safe defaults, anomaly detection, and self-tuning policies) is crucial as more distributed energy resources are integrated into the existing grid infrastructure.

Looking ahead, I see several promising directions. (i) Unifying grid-forming control with protection-aware objectives to safeguard dynamic stability amid low inertia and high fault-ride-through demands; (ii) hierarchical and distributed optimization that couples stochastic commitment with fast control synthesis, leveraging surrogate models and certified learning; (iii) hybrid storage dispatch that explicitly values ramp equity and lifecycle-aware control; (iv) forecasting–control co-design where uncertainty quantification flows directly into constraint tightening and reserve activation; (v) cyber–physical trust that enables verifiable services from prosumers and virtual power plants; and (vi) reproducible validation pathways from simulations to HIL/PHIL tests, which will shorten the time-to-field. Together, these research thrusts can deliver flexible, resilient, and low-emission energy systems that are as dependable as they are sustainable.

Conflicts of Interest: The author declares no conflict of interest.

List of Contributions

1. Olowu, T.O.; Odeyomi, O. Multi-Objective Coordinated Control of Smart Inverters and Legacy Devices. *Electronics* **2025**, *14*, 297. https://doi.org/10.3390/electronics14020297.
2. Hardan, F.; Norman, R.; Tricoli, P. Virtual Inertia Methods for Supporting Frequency Stabilisation in Autonomous AC/DC Microgrids. *Electronics* **2025**, *14*, 91. https://doi.org/10.3390/electronics14010091.
3. Xing, C.; Xiao, J.; Li, P.; Xi, X.; Chen, Y.; Guo, Q. Adaptive Virtual Inertial Control and Virtual Droop Control Coordinated Control Strategy for Hybrid Energy Storage Taking into Account State of Charge Optimization. *Electronics* **2024**, *13*, 1228. https://doi.org/10.3390/electronics13071228.

4. Alghamdi, A.S. Optimizing Microgrid Performance: Integrating Unscented Transformation and Enhanced Cheetah Optimization for Renewable Energy Management. *Electronics* **2024**, *13*, 4563. https://doi.org/10.3390/electronics13224563.
5. Zhang, M.; Wang, B.; Wei, J. The Robust Optimization of Low-Carbon Economic Dispatching for Regional Integrated Energy Systems Considering Wind and Solar Uncertainty. *Electronics* **2024**, *13*, 3480. https://doi.org/10.3390/electronics13173480.
6. Laitsos, V.; Vontzos, G.; Tsiovoulos, A.; Bargiotas, D.; Tsoukalas, L.H. Enhanced Sequence-to-Sequence Deep Transfer Learning for Day-Ahead Electricity Load Forecasting. *Electronics* **2024**, *13*, 1996. https://doi.org/10.3390/electronics13101996.
7. Zhou, J.; Sun, Y.; Chen, S.; Lan, T. A Fast Repetitive Control Strategy for a Power Conversion System. *Electronics* **2024**, *13*, 1186. https://doi.org/10.3390/electronics13071186.
8. Liang, F.; Lee, H.J.; Zhao, Q. A Novel Fractional Delay Proportional–Integral Multi-Resonant-Type Repetitive Control Based on a Farrow-Structure Filter for Grid-Tied Inverters. *Electronics* **2023**, *12*, 4010. https://doi.org/10.3390/electronics12194010.
9. Yang, L.; Zhao, Q. Hardware-in-the-Loop Simulation of Flywheel Energy Storage Systems for Power Control in Wind Farms. *Electronics* **2024**, *13*, 3610. https://doi.org/10.3390/electronics13183610.
10. Shi, T.; Xu, Z.; Gu, L.; Zhou, H. Research on Fast Frequency Response Control Strategy of Hydrogen Production Systems. *Electronics* **2024**, *13*, 956. https://doi.org/10.3390/electronics13050956.
11. Piesciorovsky, E.C.; Hahn, G.; Borges Hink, R.; Werth, A. Total Power Factor Smart Contract with Cyber Grid Guard Using Distributed Ledger Technology for Electrical Utility Grid with Customer-Owned Wind Farm. *Electronics* **2024**, *13*, 4055. https://doi.org/10.3390/electronics13204055.
12. Wang, A.; Lin, S.; Wu, G.; Li, X. Mitigation Strategy of Neutral-Point DC for Transformer Caused by Metro Stray Currents. *Electronics* **2024**, *13*, 2467. https://doi.org/10.3390/electronics13132467.
13. Morawski, M.; Ignaciuk, P. A Prosumer Hydro Plant Network as a Sustainable Distributed Energy Depot. *Electronics* **2024**, *13*, 3043. https://doi.org/10.3390/electronics13153043.
14. Vidal-Albalate, R.; Belenguer, E.; Magraner, F. Modular Multilevel Converter Control Strategy for AC Fault Current Maximization and Grid Code Compliance. *Electronics* **2025**, *14*, 1763. https://doi.org/10.3390/electronics14091763.

electronics

MDPI

Article

Multi-Objective Coordinated Control of Smart Inverters and Legacy Devices

Temitayo O. Olowu [1,*,†] and Olusola Odeyomi [2,†]

1 Idaho National Laboratory, Idaho Falls, ID 83415, USA
2 Department of Computer Science, North Carolina Agricultural and Technical State University, Greensboro, NC 27411, USA; otodeyomi@ncat.edu
* Correspondence: temitayo.olowu@inl.gov
† These authors contributed equally to this work.

Abstract: This work proposes multi-objective two-stage distribution optimal power flow (D-OPF) to coordinate the use of smart inverters (SIs) and existing voltage control legacy devices. The first stage of multi-objective D-OPF aims to solve a mixed-integer nonlinear programming (MINLP) formulation that minimizes both voltage variation and active power loss, with SI modes, SI settings, voltage regulator (VR) taps, and capacitor bank (CB) status as control variables. The Pareto Optimal Solutions obtained from the first-stage MINLP are used to determine the optimal active–reactive power dispatch from the SIs by solving a nonlinear programming formulation in the second stage of the proposed D-OPF. This model guarantees that the setpoints for active–reactive power align with the droop characteristics of the SIs, ensuring practicability and the autonomous dispatch of active–reactive power by the SIs according to IEEE 1547-2018. The effectiveness of the proposed method is tested on the IEEE 123 distribution network by contrasting the two proposed D-OPF models, with one prioritizing SIs for voltage control and power loss minimization and the other not prioritizing SIs. The simulation results demonstrate that prioritizing SIs with optimal mode and droop settings can improve voltage control and power loss minimization. The proposed model (with SI prioritization) also reduces the usage of traditional grid control devices and optimizes the dispatch of active–reactive power. The POS also shows that the SI modes, droops, and legacy device settings can be effectively obtained based on the desired objective priority.

Keywords: multi-objective optimization; smart inverters; legacy voltage control devices; voltage control

1. Introduction

As inverter-based distributed energy resources (DERs) become more prevalent, smart inverters (SIs) are emerging as feasible alternatives for supplying voltage and reactive power support within electric power transmission and distribution grids.

In traditional control strategies, legacy grid components like on-load tap changers (OLTCs), voltage regulators (VRs), and capacitor banks (CAPs) are employed for volt-VAr (VV) control (VVC). Capacitor banks, widely used for reactive power compensation, face limitations primarily due to mechanical and electrical stress. The dielectric materials inside capacitors degrade over time, leading to reduced capacitance and potential failure. Transient overvoltages, harmonics, and temperature fluctuations also accelerate aging and can cause damage. Additionally, the switching mechanisms, such as contactors,

 https://doi.org/10.3390/electronics14020297

experience mechanical wear, which reduces reliability. Capacitor banks are also susceptible to harmonic resonance, which can amplify specific frequencies, leading to overheating or other failures. These limitations make regular maintenance crucial to ensuring their effectiveness and longevity [1,2]

On-load tap changers (OLTCs) and voltage regulators (VRs), essential for voltage regulation in transformers, are constrained by their mechanical nature and slower response times. Frequent tap changes result in the mechanical wear of moving parts, such as diverter switches and contacts, leading to degradation over time. Arcing during operation erodes contacts, increasing resistance and the risk of overheating. Oil-immersed OLTCs face additional challenges, as arcing degrades the insulating properties of the oil, requiring periodic replacement. Furthermore, OLTCs operate on a timescale of seconds, making them unsuitable for responding to fast transients in the power system. Despite their robust design and longer lifespan compared with other devices, they require regular maintenance to mitigate mechanical and operational limitations [3,4]. However, due to mechanical switching and limitations on daily switching operations, these legacy devices may struggle to fully counteract rapid voltage fluctuations caused by photovoltaics (PVs).

The operational limitations of smart inverters are primarily tied to their reliance on advanced power electronics and control algorithms. While they offer features like voltage regulation and frequency support, their performance can be constrained by response times and thermal management. Prolonged operation under high loads or at extreme temperatures can cause overheating, reducing efficiency and lifespan. Frequent switching in power electronics, such as IGBTs or MOSFETs, accelerates wear and eventual failure. Moreover, smart inverters depend on communication protocols for grid coordination, introducing vulnerabilities such as latency, data loss, or cybersecurity risks, which can compromise reliability [5,6].

As PV systems proliferate within distribution networks, smart inverters (SIs) will become crucial assets for VVC both now and in the future. Inverter-based PV systems offer swift, adaptable, and precisely controlled active–reactive power support, making them suitable for participating in voltage control [7,8]. Hence, as responsive active–reactive power sources, PV-based inverters can be synchronized with legacy grid devices to regulate voltage and reactive power across different operational timescales. According to [7], SIs can switch between various modes of operation. The strategic selection of SI modes (such as volt-Watt (VW), VV, and constant power factor (CPF)) proves highly advantageous due to the fluctuating power output of PVs, particularly at high penetration levels. This flexibility also enables PV systems to offer ancillary services to the power grid based on current grid dynamics. However, integrating SI constraints into distribution optimal power flow (D-OPF) formulations poses a significant mathematical challenge due to the diverse options for SI mode selection and the complex characteristics of SI droop curves as outlined in IEEE 1547. Another main hurdle in integrating SIs and legacy devices into D-OPF models is the sub-optimal use of SI voltage control resources. When the SI variables are co-optimized with legacy devices, this often leads to SI droop settings that sometimes do not maximize the voltage control capabilities of the SIs. To address some of these issues, this paper aims to formulate two D-OPF models that assign the SI modes as integer control variables. In addition to the SI modes, we also present a simplified SI droop formulation and legacy devices for D-OPF. One of the proposed D-OPF models solves multi-objective MINLP without prioritizing the use of SIs and the second D-OPF prioritizes the use of SIs. This work is an extension to the conference paper [9], and its main technical contributions include the following:

- A novel approach to investigate the multi-mode and multi-droop configurations of smart inverters (SIs) for coordinated control alongside legacy devices across two

distinct operational timeframes. Current D-OPF models lack the capability to optimize the selection of SI modes within their formulations. This is often desirable since power generation from inverter-based resources such as PV systems often varies intermittently, and the grid dynamics could require the SIs to operate in different modes throughout the day at high PV penetration.
- The prioritization of the use of SIs in order to maximize their voltage control and/or active power loss minimization contribution. Subsequently, the status of the capacitor banks (CAPs) and the tap position of the on-load tap changer (OLTC) are individually optimized. This approach is contrasted with an alternative D-OPF formulation that does not emphasize the utilization of smart inverters (SIs).
- The use of multi-objective optimization in the slow-timescale D-OPF formulation gives the flexibility of obtaining the SI modes and droop settings. The Pareto Optimal Solutions obtained using this approach allow for the flexible selection of the SI modes and droop settings based on the objective functions with higher priority. This also allows users (such as power utility companies) of the proposed formulation to observe the different trade-offs in optimization objectives in selecting SI modes, SI droop settings, CAP status, and VR tap position.

The rest of this paper is structured as follows: related works in Section 2; Section 3 details the mathematical formulation of SI modes and SI droops; Section 4 presents the D-OPF formulation; Section 5 presents the coordinated control algorithms; Section 6 presents the simulation validation of the proposed D-OPF models. The paper is concluded with future work in Section 7.

2. Related Works

To determine the optimal setpoints of voltage-regulating devices on the grid, an efficient D-OPF formulation is required. This allows for the determination of the optimal setpoints for devices such as VR, CAPs, and SIs. Numerous studies in the literature have introduced D-OPF models aimed at determining the optimal setpoints for SIs. In [10], a method to optimally select the active and reactive power setpoints for SIs with the aim of effective voltage regulation is proposed. To convexify the D-OPF problem, the authors adopted semidefinite relaxation and sparsity-promoting regularization methods. Similar to [10], the authors of [11,12] developed a multi-objective approach to solving the D-OPF problem using weighted sum. Their efforts were aimed at determining the optimal active–reactive power setpoints of the SIs to improve both the voltage and power loss performance of the distribution networks used in their studies. To leverage recent developments in machine learning techniques, the authors of [13,14] proposed coordinated VVC using deep reinforcement learning (DRL) soft actor–critic. The proposed DRL-based method achieved fast control of the SIs under uncertainties in power generation from solar PVs. In the proposed methods in these works [10–14], though the VVC was modeled, other inverter voltage control modes (such as VW and CPF) that might be more effective depending on the location of the PVs and the time of the day were not considered. Also, the inverter reactive dispatch solutions obtained in [10–12,15–17], which need to be sent from time to time to the SIs, disallow autonomous reactive power injection from the SIs based on the voltage at their point of interconnection.

Since the SIs are not solely responsible for voltage regulation and reactive power control in the distribution feeders, VV optimization (VVO) models that integrate the use of CBs, and VRs with SIs were proposed in [15–17]. In [15], a bi-level VV optimization (VVO) framework to enhance grid performance was proposed. At the first level, a linear approximate three-phase power flow model is employed to optimize the control of CBs, VRs, and SIs. At the second level, the control setpoints for SIs are modified to achieve an optimal solution. The first-level optimization is formulated as a mixed-integer linear program (MILP), while the second-level optimization is formulated as a nonlinear program. The results obtained from the proposed framework showed its effectiveness in achieving Conservative Voltage Reduction (CVR). The main drawback in the above work (though the legacy devices were considered) is the absence of the SI droop formulation and the use of reactive power control only. Similarly, reference [17] also coordinated SIs, CBs, and VRs to achieve energy savings. The first optimization layer is formulated to determine the optimal settings of the load tap changers (LTCs) and the CBs to achieve energy savings. The droops of the SIs are subsequently optimized to improve the energy savings within the network and achieve improved CVR. Though the authors of [17] modeled the SI VV droop, a single SI mode was considered in their formulation. To integrate the SI mode and droop mathematical formulation in D-OPF, a recent research study in [18] modeled the VV and VW droop constraints. This was performed by using a mixed-integer linear programming formulation, and these droop constraints were integrated into a linear-based D-OPF model through first-order voltage approximation methods. Similarly, the authors of [19] also formulated a two-stage VV control framework that dispatches the CB and the VRs on a slow timescale and the SI droop on a fast timescale. While the authors in [18–25] incorporated droop settings as variables in the D-OPF formulation, a significant limitation of these studies lies in their assumption of predefined SI mode selection. This assumption may result in sub-optimal solutions, as the sensitivity of the feeder voltage to various SI modes differs because of the variations in X/R, as seen from each node of the feeder [26,27]. The variation in the feeder voltage sensitivity to changes in reactive or active power across its nodes means that allowing the SI to take up a single mode during operation will limit its ability to effectively carry out voltage regulation and manage the systems' power loss. Also, for the effective operation of the distribution grids, SIs need to be coordinated with existing legacy control devices at different timescales to facilitate efficient voltage control and regulation. Thus, configuring the D-OPF to maximize the utilization of SI resources while minimizing the reliance on legacy devices becomes essential. Optimizing the use of relatively faster SIs can reduce the wear and tear of other legacy voltage control devices, thereby making distribution feeder management cost-effective. Addressing the prioritization of the use of SIs, modeling all the various SI modes and droops, and multi-objective D-OPF are the motivations for this work.

3. Mathematical Modeling of SI Modes and Droop Settings

The mathematical model for SI functionalities considering their various modes and droop settings, following the guidelines set by the IEEE 1547-2018 standard [7], is presented in this section.

3.1. VW Mode

In VW mode, the SI dynamically changes its active power injection in response to changes in the voltage at the point of interconnection (POI). A steep VW curve slope (illustrated in Figure 1a) indicates a more pronounced change in active power dispatch in response to voltage variations. This relationship is mathematically represented by Equation (1) [9,28].

$$P_i^G = \begin{cases} P_i^{pv} & V_i^L \le v_i \le V_i^a \\ P_i^{min} + \psi_{i,2}(v_i - V_i^b) & V_i^a \le v_i \le V_i^b \\ P_i^{min} & V_i^b \le v_i \le V_i^H \end{cases} \quad (1)$$

In VW mode, the active power constraint is defined as

$$P_i^{min} \leq P_i^G \leq P_i^{pv}, \quad \forall i \in \mathcal{N}_{pv}.$$

Additionally, the reactive power is constrained as expressed in Equations (2) and (3):

$$Q_i^G \leq \sqrt{(S_i^{SI})^2 - (P_i^G)^2}, \quad \forall i \in \mathcal{N}_{pv}. \tag{2}$$

$$-Q_i^{pv} \leq Q_i^G \leq Q_i^{pv}, \quad \forall i \in \mathcal{N}_{pv}. \tag{3}$$

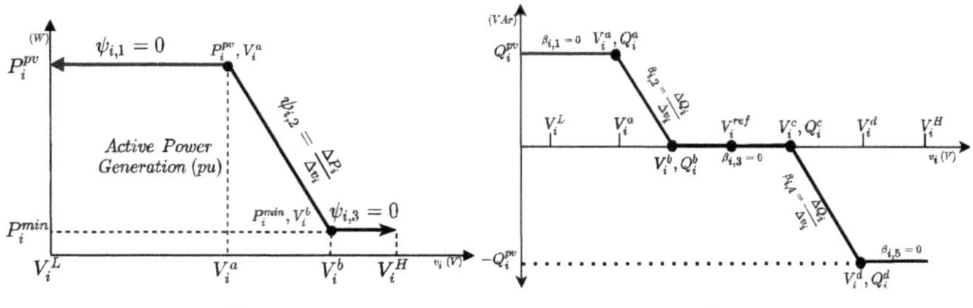

Figure 1. (**a**) VW curve. (**b**) VV curve.

3.2. VV Mode

In VV mode, the reactive power injection or absorption by the SI is modulated in response to the changes in voltage at the POI. As shown in Figure 1b, this is governed by the slope of the VV curve. This mode can be mathematically modeled as expressed in Equation (4).

$$Q_i^G = \begin{cases} Q_i^{pv} & V_i^L \leq v_i \leq V_i^a \\ (v_i - V_i^b)\beta_{i,2} & V_i^a \leq v_i \leq V_i^b \\ 0 & V_i^b \leq v_i \leq V_i^c \\ (v_i - V_i^c)\beta_{i,4} & V_i^c \leq v_i \leq V_i^d \\ -Q_i^{pv} & V_i^d \leq v_i \leq V_i^H \end{cases} \tag{4}$$

Here, $\beta_{i,2} = \frac{Q_i^{pv}}{V_i^a - V_i^b}$ and $\beta_{i,4} = \frac{Q_i^{pv}}{V_i^c - V_i^d}$ are the VV curve gradients within their respective voltage ranges. Reactive power injection or absorption in this mode can be constrained as expressed in Equation (5):

$$-Q_i^{pv} \leq Q_i^G \leq Q_i^{pv}, \quad \forall i \in \mathcal{N}_{pv}. \tag{5}$$

The active power is constrained as expressed Equations (6) and (7):

$$P_i^G \leq \sqrt{(S_i^{SI})^2 - (Q_i^G)^2}, \quad \forall i \in \mathcal{N}_{pv}. \tag{6}$$

$$P_i^{min} \leq P_i^G \leq P_i^{pv}, \quad \forall i \in \mathcal{N}_{pv}. \tag{7}$$

In VV mode, the SI can operate in either VAr-priority (Q-priority) or Watt-priority (P-priority), as illustrated in Figure 2a.

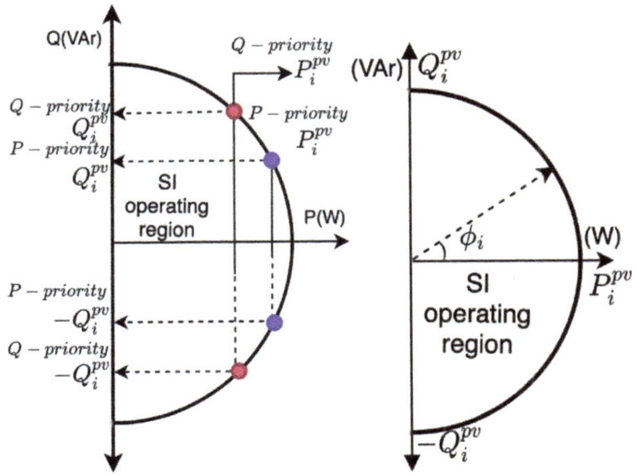

(a) (b)

Figure 2. (**a**) VV (P-/Q-priority) operating region [29]. (**b**) CPF mode.

3.2.1. VV Q-Priority

When SI is set to VV Q-priority mode, its reactive power injection or absorption is prioritized over active power. In this mode, the active power is constrained as expressed in Equation (8):

$$P_i^G \leq \sqrt{(S_i^{SI})^2 - (Q_i^G)^2}, \quad \forall i \in \mathcal{N}_{pv} \tag{8}$$

where $Q_i^G \leq S_i^{SI}$, $P_i^{min} \leq P_i^G \leq P_i^{pv}$, and the curtailed active power is defined as $P_i^{curt} = P_i^{pv} - P_i^G$.

3.2.2. VV P-Priority

When the smart inverter (SI) operates in VV mode with P-priority, the entire active power generated by the photovoltaic (PV) system is dispatched. Depending on the apparent power rating of the SI, the reactive power available for voltage control is as expressed in Equation (9).

$$Q_i^G \leq \sqrt{(S_i^{SI})^2 - (P_i^{pv})^2}, \quad \forall i \in \mathcal{N}_{pv}. \tag{9}$$

3.3. *CPF Mode*

In CPF mode, the smart inverter is designed to absorb or inject reactive or active power based on a set of power factor settings as expressed in Equation (10). For given power factor settings ϕ_i (constrained between ϕ_i^{min} and ϕ_i^{max}) and assuming a two-quadrant operation for the SIs, its reactive and active power dispatch is as shown in Figure 2b.

The reactive power that can be injected or absorbed in this mode is given by

$$Q_i^G = P_i^{pv} \tan\phi_i, \quad \forall i \in \mathcal{N}_{pv}. \tag{10}$$

4. OPF Mathematical Formulation

In this paper, the objective function for the OPF problem involves minimizing active power loss and the total voltage deviation. The first objective function, which is the total voltage deviations caused by the voltage control actions of SIs, CBs, and VRs, is expressed in Equation (11).

$$OF_1 = \min \sum_{i \in \mathcal{N}} \left| \frac{R_i^{eq} \Delta P_i^G + X_i^{eq} \Delta Q_i^G}{v_i} + \frac{1}{v_i} X_i^{eq} \Delta Q_i^c \right.$$
$$\left. + \Delta Z(tp_i) \cdot I_0 + Z_0 \cdot \Delta I(tp_i) \right| \quad (11)$$

Minimizing the active power loss set as the second objective function is expressed in Equation (12):

$$OF_2 = \sum_{(i,j) \in \mathrm{L}} r_{ij} \times \frac{P_{ij}^2 + Q_{ij}^2}{v_i} \quad (12)$$

The distribution grid model is incorporated into the optimization as constraints by using power flow equations, ensuring adherence to physical limitations, as shown in Equations (13)–(17) [30].

$$\Delta P_i(v_i, \delta_i) = P_i^G - P_i^L, \quad \forall i \in \mathcal{N} \quad (13)$$

$$\Delta Q_i(v_i, \delta_i) = Q_i^G - Q_i^L, \quad \forall i \in \mathcal{N} \quad (14)$$

where

$$\Delta Q_i(v_i, \delta_i) = v_i \sum_{k \in \mathcal{N}} v_k (\mathcal{G}_{ik} \cos(\delta_{ik}) + \mathcal{B}_{ik} \sin(\delta_{ik})), \quad (15)$$

$$\Delta P_i(v_i, \delta_i) = v_i \sum_{k \in \mathcal{N}} v_k (\mathcal{G}_{ik} \sin(\delta_{ik}) - \mathcal{B}_{ik} \cos(\delta_{ik})). \quad (16)$$

The nodal voltage constraint in the network is enforced as

$$v^{min} \leq v_i \leq v^{max}, \quad \forall i \in \mathcal{N}. \quad (17)$$

The use of discrete-controlled VRs and CBs renders this formulation an MINLP problem for the first stage of the D-OPF. This involves solving Equations (11) and (12), subject to constraints in Equation (18).

$$\text{Stage-1:} \begin{cases} SIM_i \in \{m_1, \ldots, m_5\}, & \forall i \in \mathcal{N}_{pv} \\ SIS_i \in \{V_i^L, \ldots, V_i^H, \phi_i\}, & \forall i \in \mathcal{N}_{pv} \\ tp_i \in \{-16, \ldots, +16\}, & \forall i \in \mathcal{N}_{tp} \\ Q_i^c = q_i^c \, tc_i, \quad \forall i \in \mathcal{N}_C, & tc_i \in [0, 1] \end{cases} \quad (18)$$

Conversely, the subsequent stage does not involve dispatching integer or binary variables for legacy devices or optimizing droop settings and modes for SIs. Thus, it can be structured as a nonlinear programming (NLP) problem, involving either Equation (11) or (12), subject to Equation (19).

$$\text{Stage-2:} \begin{cases} SIM_i = SIM_i^{opt}, & \forall i \in \mathcal{N}_{pv} \\ SIS_i = [V_i^L, \ldots, V_i^H, \phi_i]^{opt}, & \forall i \in \mathcal{N}_{pv} \\ tp_i = tp_i^{opt}, & \forall i \in \mathcal{N}_{tp} \\ Q_i^c = q_i^c \, tc_i^{opt}, & \forall i \in \mathcal{N}_C \\ P_i^{min} \leq P_i^G \leq P_i^{pv}, & \forall i \in \mathcal{N}_{pv} \\ -Q_i^{pv} \leq Q_i^G \leq Q_i^{pv}, & \forall i \in \mathcal{N}_{pv} \end{cases} \quad (19)$$

5. Optimization and Coordinated Control of VRs, CBs, and SIs

As described earlier, to minimize voltage deviation and active power loss, alongside the control of VRs and CBs, we integrated five SI modes: VW, VV P-priority, VV Q-priority, CPF leading, and CPF lagging. We present two OPF methods, referred to as D-OPF-1 and D-OPF-2. D-OPF-1 co-optimizes the legacy devices and the SIs in a single multi-objective D-OPF, while D-OPF-2 prioritizes the use of the SIs and subsequently adds the legacy devices if they can be used to improve the performance of the objective functions.

5.1. D-OPF-1

This algorithm tackles power flow optimization in a two-stage approach:

- Stage 1: This stage focuses on broader grid management decisions made at hourly intervals. It considers five control modes for the SIs, breakpoint settings for their droop based on those modes, CPF values for the SI CPF mode, settings for the VRs, and the status of the CAPs. Here, the SIs are not prioritized over traditional grid devices for control purposes. The stage is as follows:
  - The algorithm starts by solving a multi-objective optimization problem considering these control variables. The objective functions for this stage are defined in Equations (11) and (12).
  - The minimum POS values of the objective functions from this stage (denoted by SIM_i^{opt}, SIS_i^{opt}, tc_i^{opt}, and tp_i^{opt}) provide the settings for the second stage.
- Stage 2: This stage focuses on fine-tuning power delivery by the SIs at a 1-minute resolution. It uses the optimal values from stage 1 and optimizes the actual active and reactive power dispatch setpoints for the SIs. The stage is as follows:
  - This stage is solved 60 times within each hour using the hourly optimal values from stage 1.
  - After each hour, the algorithm returns to stage 1 to determine new optimal settings for the next hour.

A detailed breakdown of the algorithm can be found in Figure 3.

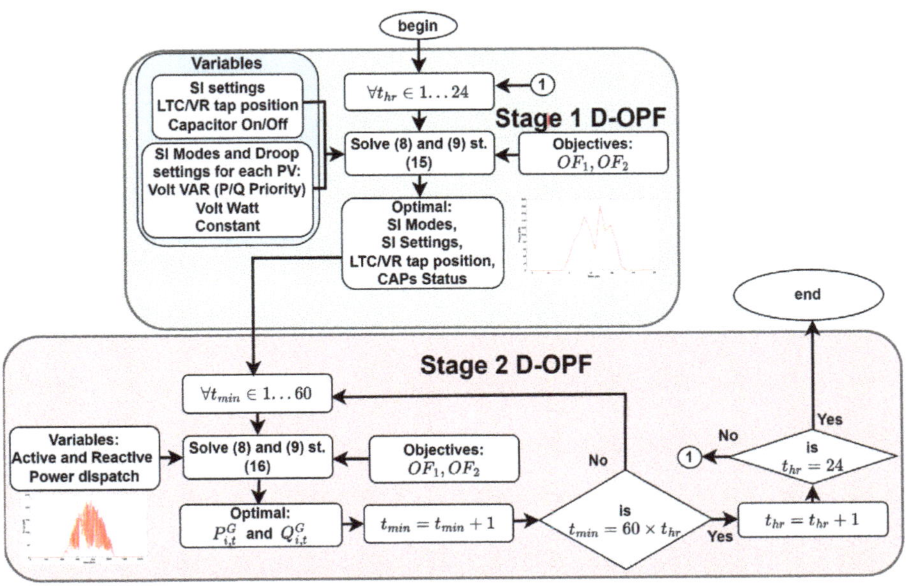

Figure 3. Without SI priority (D-OPF-1).

5.2. D-OPF-2

The D-OPF-2 method addresses computational efficiency and prioritizes SI control over legacy grid devices. It achieves this by solving control variables sequentially in three steps within the first stage of D-OPF. In the first step in stage 1 of D-OPF-2, the optimal modes and settings for the SIs are solved.

The flow charts of the proposed D-OPF-2 are presented in Figures 4 and 5.

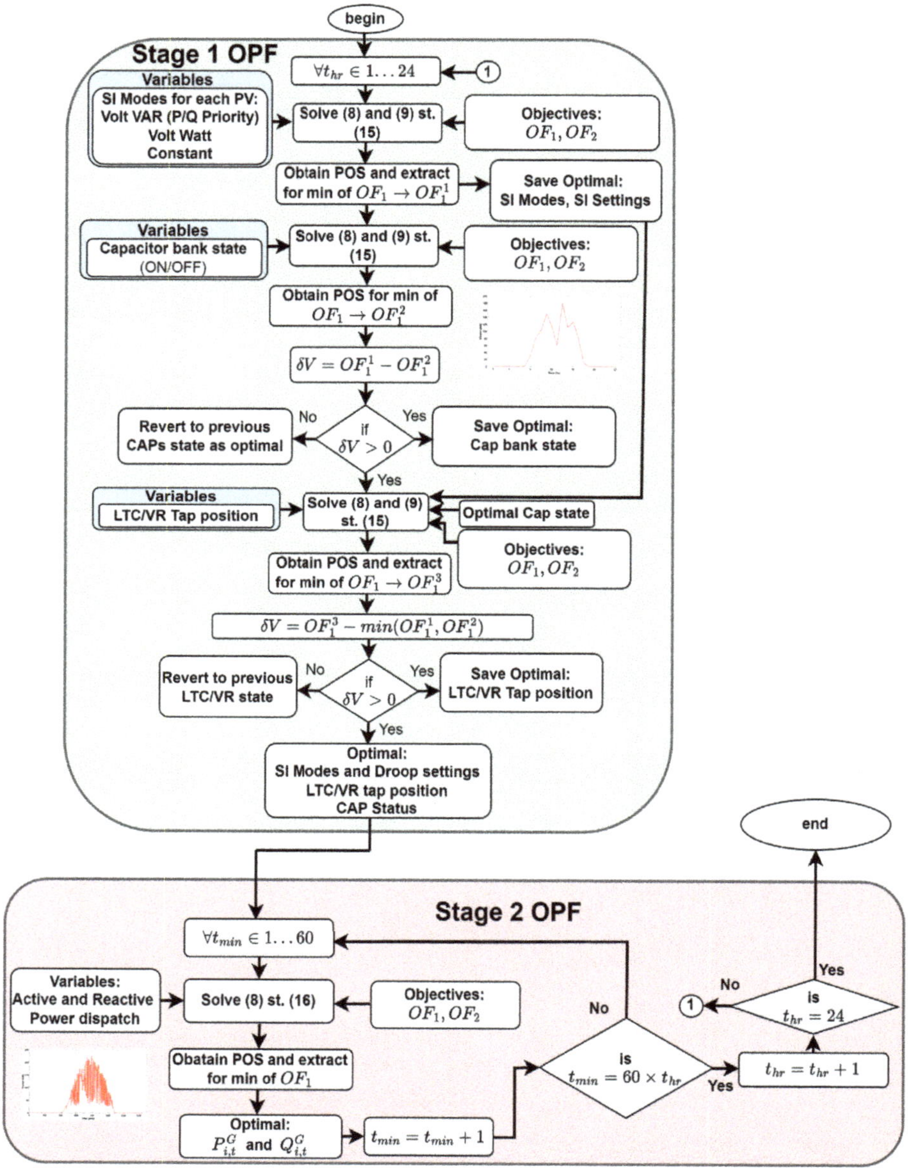

Figure 4. With SI priority (D-OPF-2) for improved voltage deviation minimization.

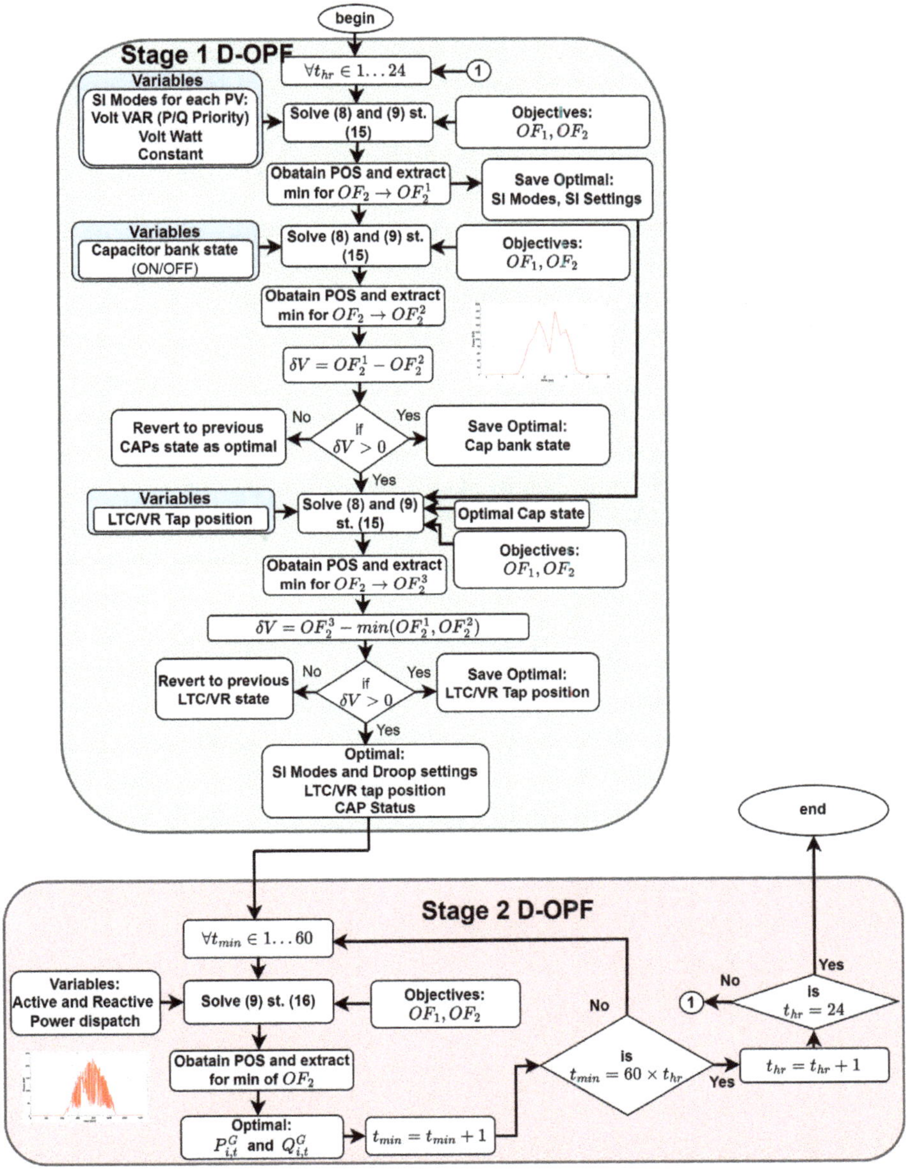

Figure 5. With SI Priority (D-OPF-2) for improved power loss minimization.

- Stage 1: Here, the SIs are prioritized over traditional grid devices for control purposes. The stage is as follows:
  - The algorithm starts by solving a multi-objective optimization problem considering the SI modes and droops as the only control variables. The objective functions for this stage are defined in Equations (11) and (12).
  - The minimum POS values of the objective functions from this stage (denoted by SIM_i^{opt}, SIS_i^{opt}, tc_i^{opt}, and tp_i^{opt}) provide the settings for the second D-OPF problem in stage 1. This prioritization allows the SIs to reach their full potential in regulating voltage.

  - By using the SI modes and settings in the first solution, a second multi-objective D-OPF problem is solved by using the CAP status as the only control variable. The second set of POS in stage 1 can either be selected to minimize the voltage deviation OF_1 or reduce the power loss OF_2. If any improvement is achieved, the new CAP status is used; otherwise, the previous state of the CAPs is used. The SI modes, SI settings, and CAP status are passed on to the third D-OPF problem in stage 1.
  - By using the SI modes and settings, and CAP status from the second solution, a third multi-objective D-OPF problem is solved using the VR Tap settings as the only control variables. The third set of POS in stage 1 can either be selected to improve the voltage deviation OF_1 or reduce the power loss OF_2. If any improvement is achieved, the new VR Tap settings are used; otherwise, the previous VR Tap settings are used. The final set of optimal control variables for the SIs, CAPs and VRs are passed on to the second stage of D-OPF.
- Stage 2: This stage focuses on fine-tuning power delivery at a 1-minute resolution. It uses the optimal values from stage 1 and optimizes the actual active and reactive power setpoints for the SIs. The stage is as follows:
  - This stage is solved 60 times within each hour by using the hourly optimal values from stage 1.
  - After each hour, the algorithm returns to stage 1 to determine new optimal settings for the next hour.

6. Simulation Results and Analysis

Figure 6 shows the IEEE 123 bus distribution network used to evaluate the effectiveness of the proposed D-OPF models.

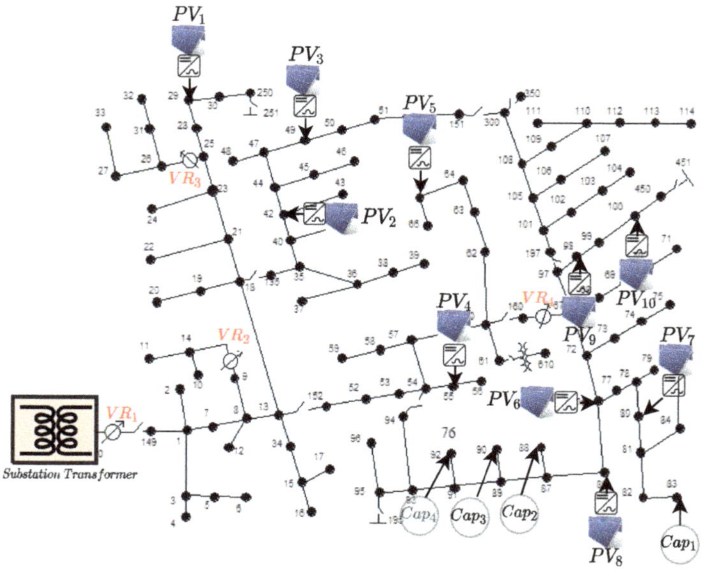

Figure 6. IEEE 123 distribution network with ten solar PVs.

The nominal voltage for the IEEE 123 test feeder is 4.16 kV. The network incorporates four voltage regulators, denoted by VR_1, VR_2, VR_3, and VR_4, to maintain voltage levels. Additionally, four capacitor banks (Cap_1, Cap_2, Cap_3, and Cap_4) are installed to improve power quality. These capacitor banks consist of one three-phase unit rated at 600 kVAr and three single-phase units rated at 50 kVAr each. This feeder integrates ten photovoltaic (PV) systems, each with a capacity of 100 kW. To manage reactive power, the apparent power (kVA) rating of the inverter is set to 125% of the kW rating of the PVs. The maximum reactive power of the SIs is limited to $Q_i^G = \frac{\sqrt{1.25^2-1^2}}{1.25} = 0.6$ in VV (P-priority) [31] to prioritize active power delivery. In VV (Q-priority), the SIs can curtail active power generation to meet the reactive power setpoint ($Q_i^G \leq S_i^{SI}$ [32]). This mode allows for more flexibility in voltage regulation. The SIs have five operational modes: VW, VV (P-priority), VV (Q-priority), and CPF leading and lagging. Simulations are performed on a cloudy day with PV generation data. Hourly data are used to optimize the settings of voltage regulators (VRs), capacitor banks (CAPs), SI modes, and SI droop settings. However, for dispatching the actual active and reactive power output of the SIs, a higher resolution (1-minute) PV generation profile is employed.

6.1. Optimal SI Modes and Settings

The optimal SI modes and droop settings obtained from the two proposed D-OPF algorithms are presented.

6.1.1. D-OPF-1 Optimal SI Modes and Settings

The Pareto optimal SI modes and settings for minimum values of OF_1 and OF_2 for stage 1 of D-OPF-1 are as shown in Figure 7.

The plots for 9 h between 8 am and 4 pm are presented due to space constraints. This time frame was chosen as it captures most of the solar generation period based on the solar power generation profile used. As shown in Figure 7a,b, the algorithm effectively chooses the optimal operating mode for each SI every hour. Notably, all the considered SI modes (VW, VV (P-priority), VV (Q-priority), and CPF leading and lagging) are utilized by the SIs to achieve effective voltage regulation and minimized power loss. Table 1 shows the number of times within a 24 h period each of the SI modes is selected by all the 10 PVs.

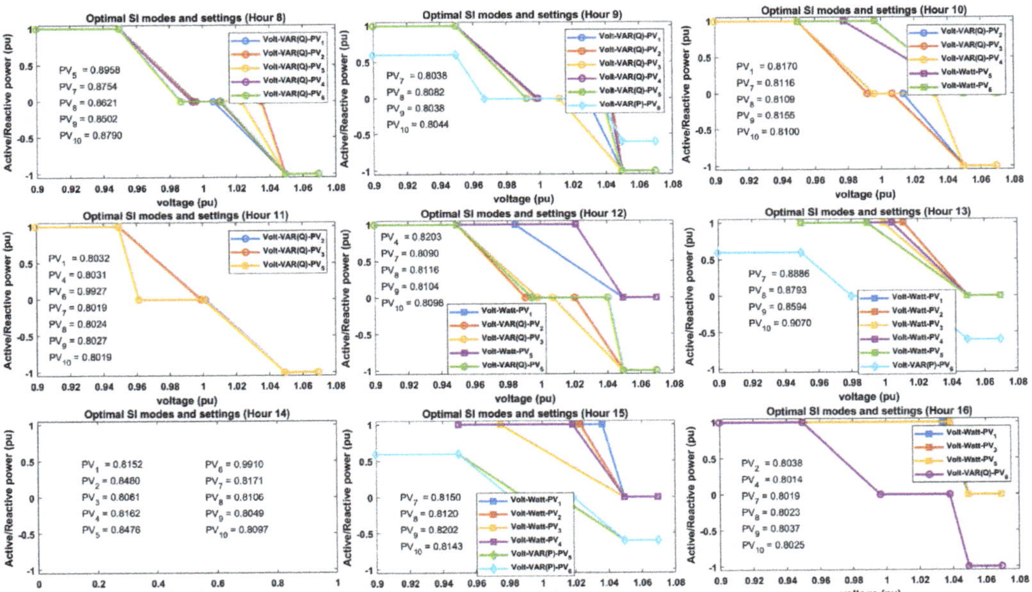

(a)

Figure 7. *Cont.*

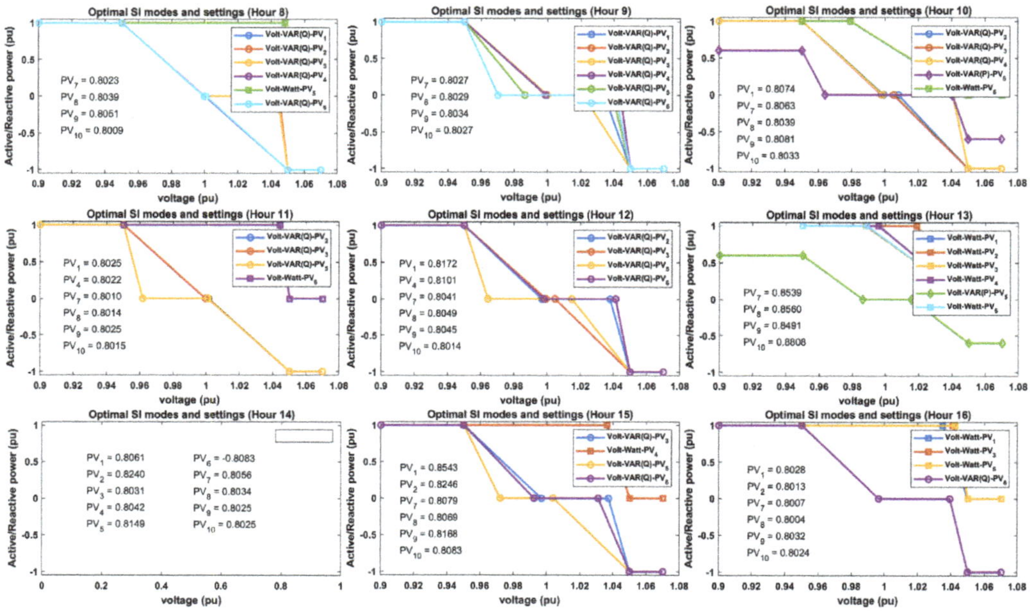

(b)

Figure 7. Optimal SI modes and settings using D-OPF-1: (**a**) min OF_1; (**b**) min OF_2.

Table 1. Hours 8-16 SI modes using D-OPF-1.

SI Modes	min $\lvert OF_1 \rvert$	min $\lvert OF_2 \rvert$
CPF (leading and lagging)	50	51
VW	16	12
VV (P-priority)	4	2
VV (Q-priority)	20	25

6.1.2. D-OPF-2 Optimal SI Modes and Settings

The Pareto optimal SI modes and settings for minimum values of OF_1 and OF_2 for stage-one of D-OPF-2 are as shown in Figure 8. The number of different SI modes and settings used using this approach within a 24 h period is presented in Table 2. Comparing the SI modes and settings for D-OPF-1 and D-OPF-2, the following differences can be observed: Firstly, D-OPF-2 uses more VV (Q-priority) compared with D-OPF-1. This is due to the effectiveness of this mode in carrying out voltage regulation since it curtails some active power in order to allow for more reactive power injection/absorption. The effectiveness of active or reactive power control for voltage regulation also depends on the X/R ratio of the feeder. secondly, it can also be seen from Figure 8a,b that the VV droop plots for D-OPF-2 has narrower dead-band compared to that of D-OPF-1. This allows the SIs to aggressively carry out voltage control and possible active power loss minimization. These plots highlight the benefits of prioritizing the use of SIs in D-OPF-2 compared with D-OPF-1.

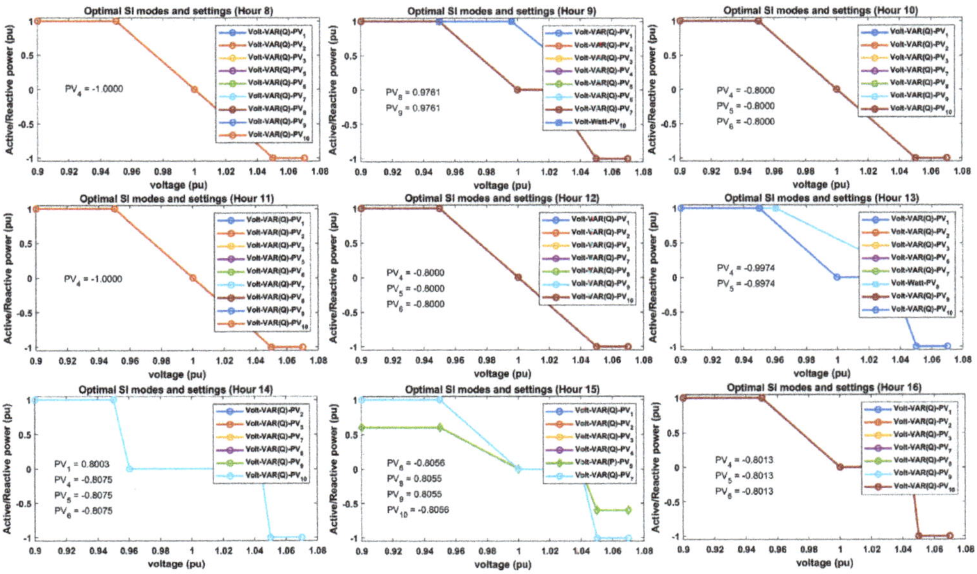

(a)

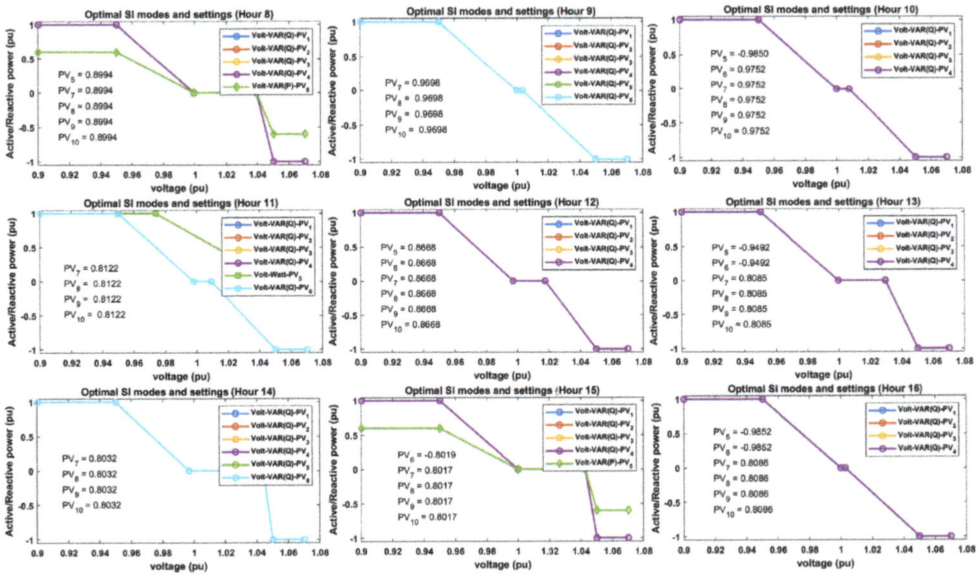

(b)

Figure 8. Optimal SI modes and settings using D-OPF-2: (**a**) min OF_1; (**b**) min OF_2.

The results in Figures 7 and 8 show that the algorithm dynamically selects the most suitable operational mode for each smart inverter (SI) on an hourly basis. This ensures adaptability to changing grid conditions, such as varying loads and solar generation throughout the day. Each mode is selected based on its effectiveness in addressing specific voltage regulation needs and minimizing power losses at a given time, thereby achieving both grid stability and energy efficiency. The algorithm ensures that droop settings align with grid requirements in real time by switching between different control modes.

Table 2. Hours 8-16 SI modes using D-OPF-2.

SI Modes	min $\lvert OF_1 \rvert$	min $\lvert OF_2 \rvert$
CPF (leading and lagging)	23	46
VW	2	1
VV (P-priority)	1	2
VV (Q-priority)	64	41

6.2. Optimal Tap Positions and CAP Status

Figures 9 and 10 show the optimal tap positions obtained by the proposed D-OPF-1 and D-OPF-2 algorithms for VR_1, VR_2, VR_3, and VR_4. The total number of tap changes from both algorithms' POS is summarized in Tables 3 and 4.

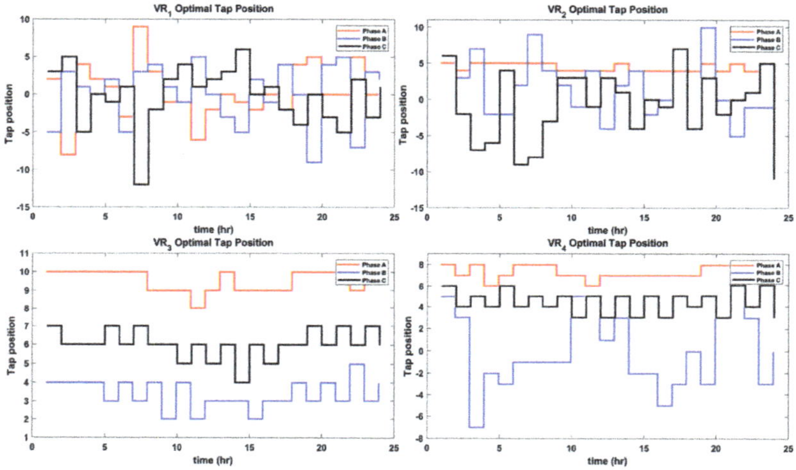

(a)

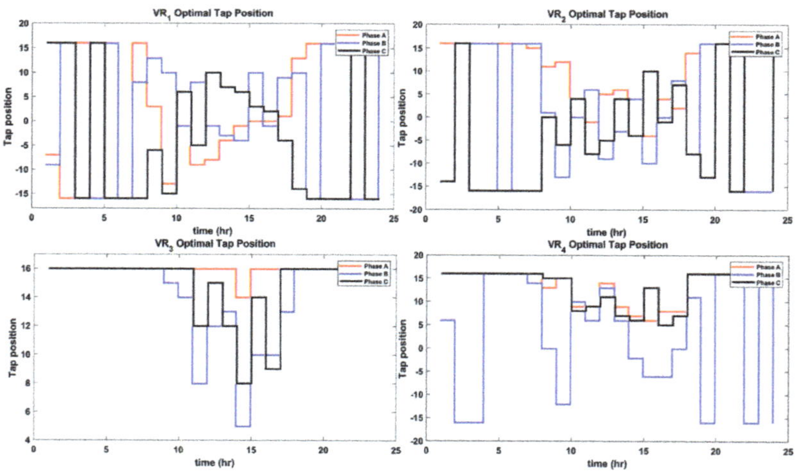

(b)

Figure 9. D-OPF-1 Pareto optimal tap positions of VR_1, VR_2, VR_3 and VR_4 for (**a**) min OF_1 and (**b**) min OF_2.

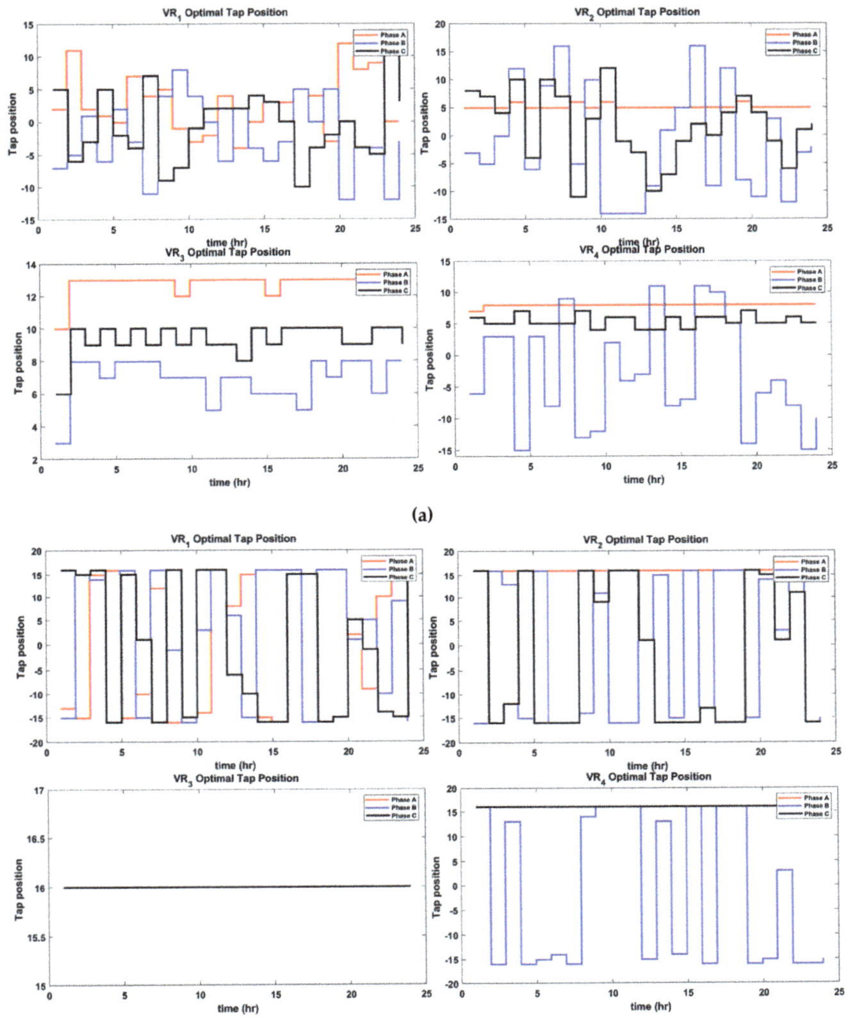

Figure 10. D-OPF-2 Pareto optimal tap positions of VR_1, VR_2, VR_3 and VR_4 for (**a**) min OF_1 and (**b**) min OF_2.

Table 3. Tap changes for D-OPF-1.

	min $\lvert OF_1\rvert$			min $\lvert OF_2\rvert$		
	Ph A	**Ph B**	**Ph C**	**Ph A**	**Ph B**	**Ph C**
VR_1	21	23	23	20	21	17
VR_2	11	21	22	13	16	18
VR_3	8	17	19	2	9	7
VR_4	11	18	23	9	18	10
Total		217			160	

These values represent the Pareto optimal tap position that corresponds to the minimum values of OF_1 and OF_2. As seen from the figures and tables, the total number of optimal tap changes made by using D-OPF-2 is much less than that of D-OPF-1. There is a 12.9% reduction in tap changes for minimizing OF_1 by using D-OPF-2 compared with D-OPF-1 and a 27.5% reduction in tap changes for minimizing OF_2 by using D-OPF-2 compared with D-OPF-1. It is also worthy of note that both algorithms made a higher number of tap changes to minimize OF_1 compared with OF_2. This is expected since changes in the VR tap position have more influence on regulating the network voltage. The total numbers of the ON/OFF statuses of the four CAPs are presented in Tables 5 and 6. The results show that the D-OPF-2 has fewer CAP ON statuses compared with D-OPF-1. This can be attributed to more reactive power control achieved by prioritizing the use of the SIs in D-OPF-2.

Table 4. Tap changes for D-OPF-2.

	min $\lvert OF_1 \rvert$			min $\lvert OF_2 \rvert$		
	Ph A	**Ph B**	**Ph C**	**Ph A**	**Ph B**	**Ph C**
VR_1	21	22	21	21	20	20
VR_2	8	21	23	0	20	16
VR_3	5	13	17	0	0	0
VR_4	1	22	15	06	19	0
Total		189			116	

Table 5. Total number of CAP ON/OFF statuses for D-OPF-1.

	Cap_1, Cap_2, Cap_3, and Cap_4, $\sum_1^{24} tc_{i,T}^{opt}$	
	min OF_1	**min OF_2**
ON	96	96
OFF	0	0

Table 6. Total number of CAP ON/OFF statuses for D-OPF-2.

	Cap_1, Cap_2, Cap_3 and Cap_4, $\sum_1^{24} tc_{i,T}^{opt}$	
	min OF_1	**min OF_2**
ON	9	64
OFF	87	32

As seen in Figures 9 and 10, tap changes are critical to adjusting the voltage on the grid. Fewer changes imply reduced wear of equipment, lower maintenance costs, and smoother voltage regulation, enhancing grid reliability. By leveraging reactive power from the SIs with D-OPF-2, the grid benefits from decentralized voltage support. This reduces the need for the frequent mechanical switching of CAPs or VRs, improving operational efficiency. D-OPF-2 demonstrates an advanced control mechanism by balancing reactive power contributions from the SIs and mechanical devices (VRs and CAPs). This synergy minimizes voltage deviations and power losses with fewer interventions, aligning with the goals of modern smart grid management.

6.3. Active and Reactive Power Dispatch

We use the POS obtained from stage 1 (SIM_i^{opt}, SIS_i^{opt}, tc_i^{opt}, and tp_i^{opt}) to solve a one-minute resolution power flow (D-OPF) problem for each SI. This determines the optimal reactive power injection/absorption (Q_i^G) and active power generation (P_i^G) for each SI. Figures 11 and 12 show the combined active and reactive power dispatch of all SIs on a cloudy day for D-OPF-1 and D-OPF-2, respectively. Negative values on the Q plot indicate reactive power absorption, while positive values show reactive power injection.

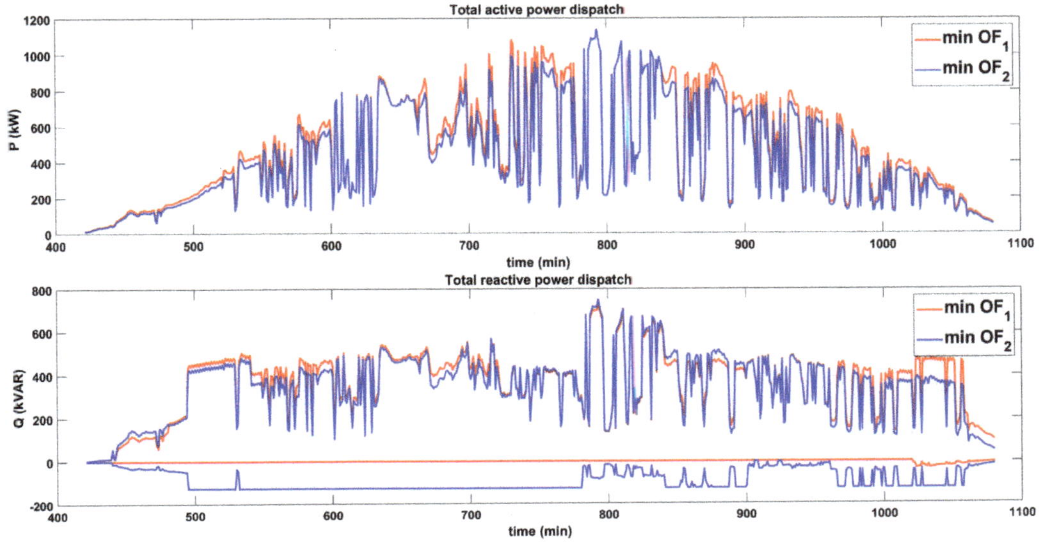

Figure 11. Active–reactive power dispatch for D-OPF-1.

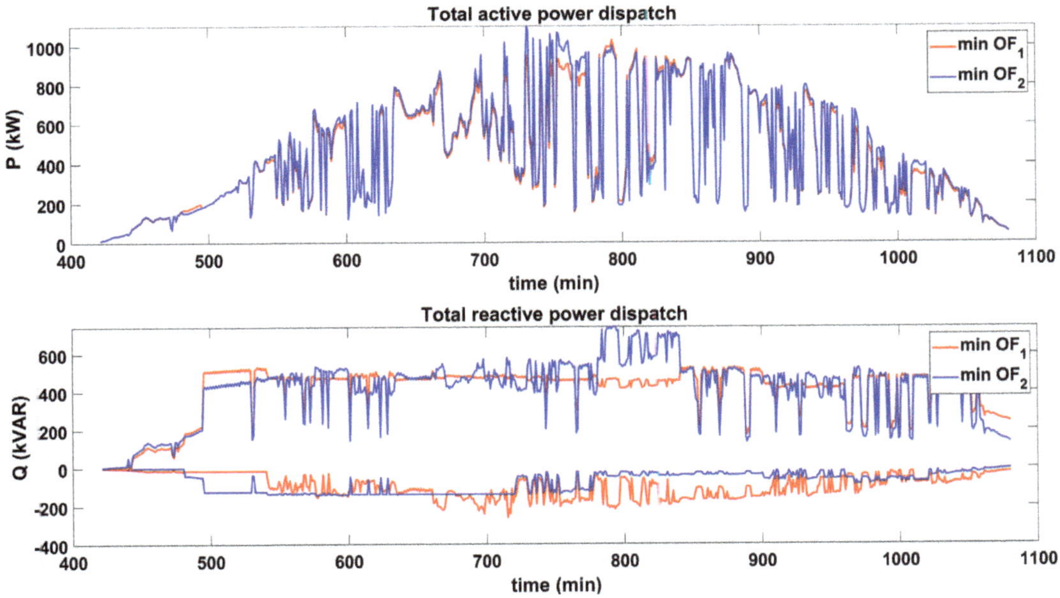

Figure 12. Active–reactive power dispatch for D-OPF-2.

Figures 13 and 14 also show the active power curtailment for both algorithms obtained from their POS for minimum values of OF_1 and OF_1.

Both D-OPF algorithms reduce active power generation (curtailment) in some intervals. They also inject or absorb reactive power to regulate voltage and minimize the active power losses effectively. Over a day, D-OPF-1 curtails a total of 26.96 kWh of energy to minimize OF_1 and 380.22 kWh of energy to minimize OF_2 from the POS obtained. For the case of

D-OPF-2, a total of 299.73 kWh of energy to minimize OF_1 and 150.43 kWh of energy to minimize OF_2 from the POS were obtained. The extent to which each algorithm limits active power is contingent upon the optimal settings and operational modes selected by the SI. Both algorithms use the VVC with Q-priority, enabling them to reduce/curtail active power generation as needed. As presented earlier in Tables 1 and 2, D-OPF-2 uses more VVC with Q-priority, causing it to curtail more active power in minimizing the voltage deviation OF_1.

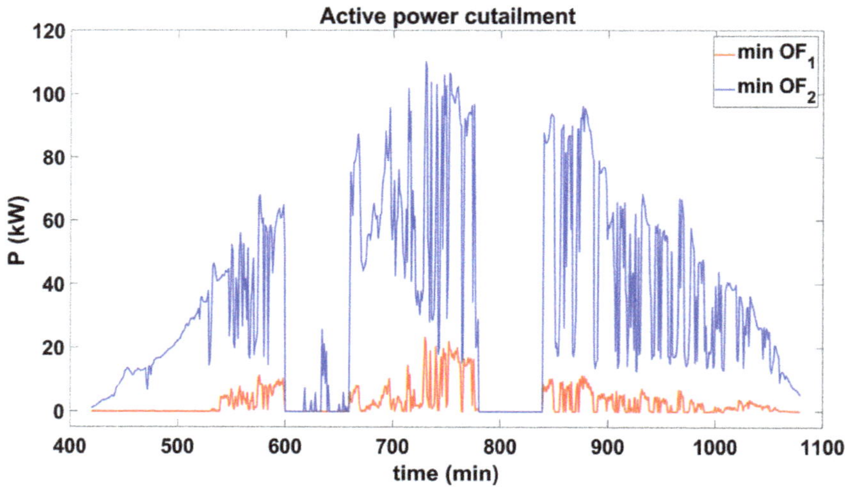

Figure 13. Active power curtailment for min OF_1 and min OF_2 using D-OPF-1.

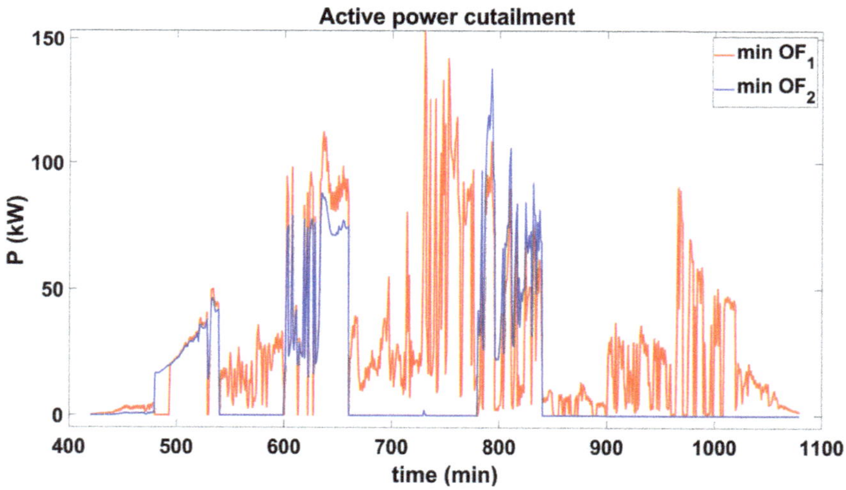

Figure 14. Active power curtailment for min OF_1 and min OF_2 using D-OPF-2.

6.4. Comparison with Existing Method

To further investigate the benefits of the proposed D-OPF-1 and D-OPF-2 models, these algorithms are compared with an existing conventional D-OPF approach, given in [28]. The optimal tap positions, SI modes, and SI settings obtained by using the approach in [28] are shown in Figures 15 and 16. The tap-changing and CAP operational summary by using this approach based on Figure 15 are presented in Table 7.

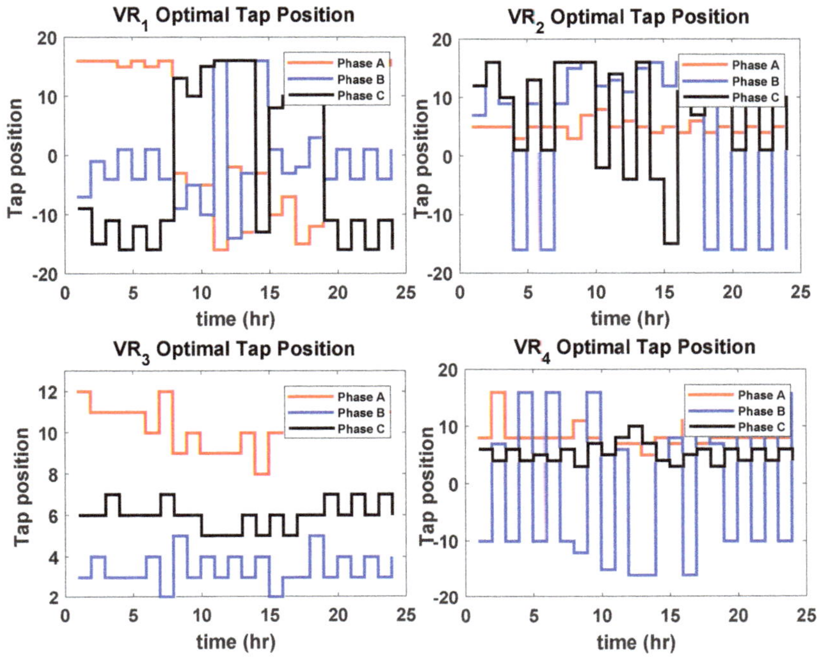

Figure 15. Optimal tap positions for VR_1, VR_2, VR_3, and VR_4 [28].

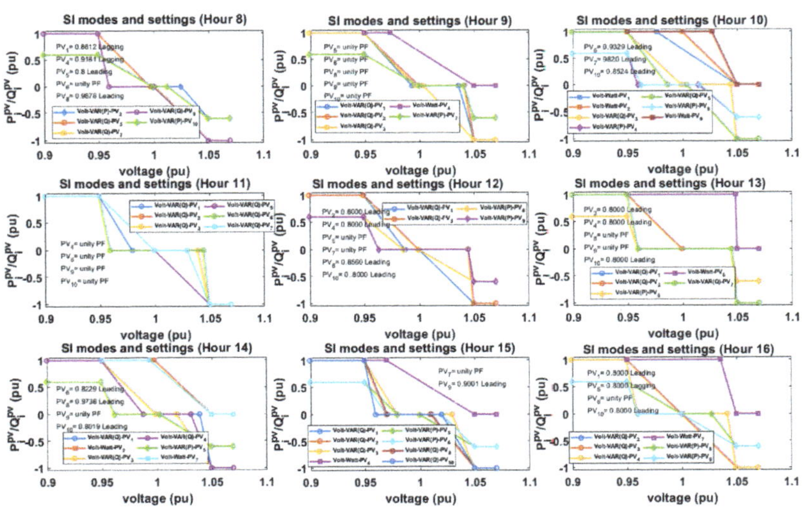

Figure 16. Optimal SI modes and settings [28].

Table 7. Tap-changing summary based on [28].

	Ph A	Ph B	Ph C
VR_1	20	23	21
VR_2	19	23	21
VR_3	10	20	16
VR_4	12	22	23
Total		230	

As seen in Table 7, the total tap changes for all four VRs over the time period considered is 230, which is higher than $\min|OF_1| = 217$ and $\min|OF_2| = 160$ for D-OPF-1 and $\min|OF_1| = 189$ and $\min|OF_2| = 116$ for D-OPF-2. The reduction in the VR tap operations by using the proposed D-OPF models compared with the one proposed in [28] further shows the effectiveness of the proposed multi-objective D-OPF models in this paper, increases the possibility of a longer lifespan of the VRs over a longer period, and indicates a more efficient use of the SIs, which can allow for better voltage regulation.

Also comparing the total number of CB ON/OFF status presented in Table 8 (obtained from [28]) with that of D-OPF-1 and D-OPF-2 (as presented in Tables 5 and 6), it can be seen that the number of CB ON statuses (76) obainted from [28] is more than that of D-OPF-2 ($\min|OF_1| = 9$ and $\min|OF_2| = 64$). This is due to the priotization of the use of SIs in D-OPF-2 in providing reactive power support in the network. The number of CB ON status ($\min|OF_1| = 96$ and $\min|OF_2| = 96$) in D-OPF-1 is more (than 76 obtained using the method proposed in [28]), since the use of SIs is not priotized for reative power control in the D-OPF-1 algorithm. This further highlight the benefits of priotizing the use of SIs in reactive power injection for voltage regulation.

Table 8. Total number of CAP ON/OFF statuses.

	$\sum_1^{24} tc_{i,T}^{opt}$
ON	76
OFF	20

7. Conclusions

With the increasing penetration of renewable-energy systems, particularly those that rely on power electronics, the role of smart inverters (SIs) in voltage control and grid management has grown significantly. These systems, driven by clean energy sources such as solar and wind, necessitate advanced strategies to maintain grid stability and efficiency. A critical aspect of this development is the coordination of voltage control efforts between modern SIs and legacy voltage control devices, such as capacitor banks and voltage regulators, which were traditionally used in grid operations.

This study addresses these challenges by proposing two innovative distribution optimal power flow (D-OPF) frameworks designed to optimize the configuration and operation of SIs in accordance with IEEE 1547 standards. The first framework, referred to as D-OPF-1, minimizes reliance on SIs for voltage regulation and active power loss minimization, ensuring that the system remains flexible and can accommodate a variety of operational scenarios. In contrast, the second framework prioritizes the use of SIs, leveraging their advanced capabilities to enhance voltage regulation and minimize power losses.

To validate the proposed frameworks, this study utilizes the IEEE 123-node test feeder, incorporating ten solar power generation units equipped with SIs. For each solar photovoltaic (PV) plant, four SI operational modes are modeled: VW, VV (P-priority), VV (Q-priority), and CPF. These modes enable dynamic switching, ensuring optimal SI operation under varying grid conditions. Furthermore, binary control of legacy capacitor banks and discrete voltage regulator settings are integrated as control variables, enabling a comprehensive and coordinated approach to grid management.

The results of this analysis highlight the effectiveness of the proposed D-OPF algorithms in optimizing SI droop settings and mode selection in conjunction with legacy grid control devices. Notably, prioritizing SIs for voltage control demonstrated superior voltage regulation capabilities and a significant reduction in the system's overall active power loss. These findings underscore the potential of advanced SI-based control frameworks to revolutionize voltage/var control strategies in modern power systems, paving the way for more efficient and sustainable grid operations in the era of renewable-energy dominance.

Author Contributions: Conceptualization, T.O.O. and O.O.; methodology, T.O.O. and O.O.; original draft preparation, T.O.O. and O.O.; writing—review and editing, T.O.O. and O.O. All authors have read and agreed to the published version of the manuscript.

Funding: This research received no external funding.

Data Availability Statement: The raw data supporting the conclusions of this article will be made available by the authors on request.

Conflicts of Interest: The authors declare no conflicts of interest.

Abbreviations

The following abbreviations are used in this manuscript:

t	minute time index $t \in [1, \ldots, 1440]$
i	index of buses
ik	index of lines
v_{st}	substation transformer voltage
$\mathcal{G}_{ik}, \mathcal{B}_{ik}$	elements of admittance matrix's real and imaginary parts ik
T	hourly time index $T \in [1, \ldots, 24]$
R_i^{eq}, X_i^{eq}	equivalent resistance and reactance at node i
P_i^{pv}	PV active power generation based on irradiance
Q_i^{pv}	available reactive power based on SI rating at node i
P_i^L, Q_i^L	load at node i active and reactive power
β_i, ψ_i	slope of volt-VAr and volt-Watt curve at node i
P_i^{curt}	curtailed PV generation at node i
v_i	instantaneous voltage at node i
$\mathcal{N}$	set of all network nodes i
SIS, SIS^{opt}	set of SI setting and optimal setting $[V_i^L, \ldots V_i^H; \phi_i]$
SIM_i	set of SI modes at node i
SIM_i^{opt}	set of SI optimal modes at node i
$\mathcal{L}$	set of branches (i, k)
P_{ki}, Q_{ki}	active and reactive power flow on branch (ki)
P_i^{min}	minimum active power dispatch from PV at node i
P_i^G, Q_i^G	instantaneous active and reactive power dispatch of PV at node i
$\Delta v_i^{tp}, \Delta v_i^c, \Delta v_i^{pv}$	voltage deviations due to OLTC/VR tap change, change in status of CAPs and PV active–reactive power injection at node i
$V_i^L \ldots V_i^H$	SI curve break points for SI at node i
m	SI modes
α, α^{max}	tap ratio and maximum tap ratio
P_0^{st}, Q_0^{st}	active and reactive power delivered from substation
Y_0, Z_0, I_0	admittance matrix, impedance matrix, and current injection before tap change
$\Delta Z(tp_i), \Delta Y(tp_i), \Delta I(tp_i)$	admittance matrix, impedance matrix and current injection change after change in tap tp on node i
δ_i	node i's voltage angle
v^{min}, v^{max}	minimum and maximum nodal voltages
q_i^c	capacitor bank at node i reactive power rating
Q_i^c	capacitor bank at node i reactive power injection
S_i^{SI}	SI kVA rating at node i
$\mathcal{N}_C$	set of nodes with CAPs
$\mathcal{N}_{tp}$	set of branches with OLTC/VR
$\mathcal{N}_{pv}$	set of nodes with PVs

ϕ_i	power factor of PV's SI at node i
ϕ_i^{min},ϕ_i^{max}	minimum and maximum power factor SI setting at node i
tp_i,tp_i^{opt},tp_i^{max}	tap and optimal tap position and maximum tap position for $\forall i \in \mathcal{N}_{tp}$
tc_i,tc_i^{opt}	CAP status and optimal CAP status $\forall i \in \mathcal{N}_C$

References

1. Li, P.; Ji, H.; Yu, H.; Wang, C. Chapter 2—Components and modeling of flexible distribution networks. In *Flexible Distribution Networks*; Li, P., Ji, H., Yu, H., Wang, C., Eds.; Elsevier: Amsterdam, The Netherlands, 2024; pp. 13–64. [CrossRef]
2. Cunha, V.C.; Andrade, J.; Ricciardi, T.R.; Trindade, F.C.L.; Freitas, W.; Rosolen, R.A.G.; Vilibor, H.O. Assessment of Capacitor Banks Control Practices in Distribution Systems with High PV Penetration. In Proceedings of the 2020 IEEE Power & Energy Society General Meeting (PESGM), Montreal, QC, Canada, 2–6 August 2020; pp. 1–5. [CrossRef]
3. Khan, H.A.; Zuhaib, M.; Rihan, M. Voltage fluctuation mitigation with coordinated OLTC and energy storage control in high PV penetrating distribution network. *Electr. Power Syst. Res.* **2022**, *208*, 107924. [CrossRef]
4. Li, C.; Disfani, V.R.; Pecenak, Z.K.; Mohajeryami, S.; Kleissl, J. Optimal OLTC voltage control scheme to enable high solar penetrations. *Electr. Power Syst. Res.* **2018**, *160*, 318–326. [CrossRef]
5. Fard, A.Y.; Shadmand, M.B. Multitimescale Three-Tiered Voltage Control Framework for Dispersed Smart Inverters at the Grid Edge. *IEEE Trans. Ind. Appl.* **2021**, *57*, 824–834. [CrossRef]
6. Lyu, X.; Du, W.; Mohiuddin, S.M.; Nandanoori, S.P.; Elizondo, M. Improved Transient Stability Analysis of Multi-Loop Droop-Controlled Grid Forming Inverters with Current Limiter. In Proceedings of the 2024 IEEE Power & Energy Society Innovative Smart Grid Technologies Conference (ISGT), Washington, DC, USA, 19–22 February 2024; pp. 1–5. [CrossRef]
7. *1547-2018*; IEEE Standard for Interconnection and Interoperability of Distributed Energy Resources with Associated Electric Power Systems Interfaces. IEEE: Piscataway, NJ, USA, 2018; pp. 1547–2018.
8. Babak, E.; Richard, B.; Michael, R.; Michael, H.; Julio, R.A.; Tom, K.; Frances, C.; Mark, S.; Ryan, Q.; Charlie, V.; et al. *Impact of IEEE 1547 Standard on Smart Inverters and the Applications in Power Systems*; IEEE PES Industry Technical Support Leadership Committee: Piscataway, NJ, USA, 2020.
9. Olowu, T.O.; Inaolaji, A.; Paudyal, S.; Sarwat, A. Two-Timescale Control of Smart Inverters and Legacy Devices in Unbalanced Distribution Feeders. In Proceedings of the 2023 IEEE Industry Applications Society Annual Meeting (IAS), Nashville, TN, USA, 29 October–2 November 2023; pp. 1–6. [CrossRef]
10. Dall'Anese, E.; Dhople, S.V.; Giannakis, G.B. Optimal Dispatch of Photovoltaic Inverters in Residential Distribution Systems. *IEEE Trans. Sustain. Energy* **2014**, *5*, 487–497. [CrossRef]
11. Jabr, R.A. Linear Decision Rules for Control of Reactive Power by Distributed Photovoltaic Generators. *IEEE Trans. Power Syst.* **2017**, *33*, 2165–2174. [CrossRef]
12. Su, X.; Masoum, M.A.; Wolfs, P.J. Optimal PV Inverter Reactive Power Control and Real Power Curtailment to Improve Performance of Unbalanced Four-wire LV Distribution Networks. *IEEE Trans. Sustain. Energy* **2014**, *5*, 967–977. [CrossRef]
13. Xiong, K.; Cao, D.; Zhang, G.; Chen, Z.; Hu, W. Coordinated volt/VAR control for photovoltaic inverters: A soft actor-critic enhanced droop control approach. *Int. J. Electr. Power Energy Syst.* **2023**, *149*, 109019. [CrossRef]
14. Takayama, S.; Ishigame, A. Volt–Var curve determination method of smart inverters by multi-agent deep reinforcement learning. *Int. J. Electr. Power Energy Syst.* **2024**, *157*, 109888. [CrossRef]
15. Jha, R.R.; Dubey, A.; Liu, C.; Schneider, K.P. Bi-Level Volt-VAR Optimization to Coordinate Smart Inverters with Voltage Control Devices. *IEEE Trans. Power Syst.* **2019**, *34*, 1801–1813. [CrossRef]
16. Farivar, M.; Clarke, C.R.; Low, S.H.; Chandy, K.M. Inverter VAR Control for Distribution Systems with Renewables. In Proceedings of the International Conference on Smart Grid Communications (SmartGridComm), Brussels, Belgium, 17–20 October 2011; pp. 457–462.
17. Ding, F.; Baggu, M. Coordinated Use of Smart Inverters with Legacy Voltage Regulating Devices in Distribution Systems with High Distributed PV Penetration—Increase CVR Energy Savings. *IEEE Trans. Smart Grid* **2023**, *14*, 1804–1813. [CrossRef]
18. Inaolaji, A.; Savasci, A.; Paudyal, S. Distribution Grid Optimal Power Flow in Unbalanced Multi-phase Networks with Volt-VAr and Volt-Watt Droop Settings of Smart Inverters. *IEEE Trans. Ind. Appl.* **2022**, *58*, 5832–5843. [CrossRef]
19. Savasci, A.; Inaolaji, A.; Paudyal, S. Two-Stage Volt-VAr Optimization of Distribution Grids with Smart Inverters and Legacy Devices. *IEEE Trans. Ind. Appl.* **2022**, *58*, 5711–5723. [CrossRef]
20. Lusis, P.; Andrew, L.L.; Liebman, A.; Tack, G. Interaction Between Coordinated and Droop Control PV Inverters. In Proceedings of the Eleventh ACM International Conference on Future Energy Systems, Virtual, 22–26 June 2020; pp. 314–324.
21. Gush, T.; Kim, C.H.; Admasie, S.; Kim, J.S.; Song, J.S. Optimal Smart Inverter Control for PV and BESS to Improve PV Hosting Capacity of Distribution Networks Using Slime Mould Algorithm. *IEEE Access* **2021**, *9*, 52164–52176. [CrossRef]

22. Olowu, T.O.; Inaolaji, A.; Sarwat, A.; Paudyal, S. Optimal Volt-VAR and Volt-Watt Droop Settings of Smart Inverters. In Proceedings of the 2021 IEEE Green Technologies Conference (GreenTech), Denver, CO, USA, 7–9 April 2021; pp. 89–96. [CrossRef]
23. Dao, V.T.; Ishii, H.; Hayashi, Y. Optimal parameters of volt-var functions for photovoltaic smart inverters in distribution networks. *IEEJ Trans. Electr. Electron. Eng.* **2019**, *14*, 75–84. [CrossRef]
24. Inaolaji, A.; Savasci, A.; Paudyal, S. Optimal Droop Settings of Smart Inverters. In Proceedings of the 48th Photovoltaic Specialists Conference (PVSC), Fort Lauderdale, FL, USA, 20–25 June 2021; pp. 2584–2589.
25. Murzakhanov, I.; Gupta, S.; Chatzivasileiadis, S.; Kekatos, V. Optimal Design of Volt/VAR Control Rules for Inverter-Interfaced Distributed Energy Resources. *arXiv* **2022**, arXiv:2210.12805. [CrossRef]
26. Rylander, M.; Reno, M.J.; Quiroz, J.E.; Ding, F.; Li, H.; Broderick, R.J.; Mather, B.; Smith, J. Methods to determine recommended feeder-wide advanced inverter settings for improving distribution system performance. In Proceedings of the IEEE 43rd Photovoltaic Specialists Conference (PVSC), Portland, OR, USA, 5–10 June 2016; pp. 1393–1398. [CrossRef]
27. Li, H.; Rylander, M.; Smith, J. *Analysis to Inform CA Grid Integration: Methods and Default Settings to Effectively Use Advanced Inverter Functions in the Distribution System*; EPRI: Palo Alto, CA, USA, 2015.
28. Olowu, T.O.; Inaolaji, A.; Paudyal, S.; Sarwat, A. Optimal Mode and Droop Setting of Smart Inverters. In Proceedings of the 2023 IEEE Power & Energy Society General Meeting (PESGM), Orlando, FL, USA, 16–20 July 2023; pp. 1–5. [CrossRef]
29. Giraldez, J.; Emmanuel, M.; Hoke, A.; Suryanarayanan, S. Impacts of Voltage-based Grid Support Functions on Energy Production of PV Customers. In Proceedings of the 2019 IEEE Power Energy Society General Meeting (PESGM), Atlanta, GA, USA, 4–8 August 2019; pp. 1–5. [CrossRef]
30. Yang, Z.; Zhong, H.; Bose, A.; Xia, Q.; Kang, C. Optimal Power Flow in AC–DC Grids with Discrete Control Devices. *IEEE Trans. Power Syst.* **2018**, *33*, 1461–1472. [CrossRef]
31. Jabr, R.A. Robust Volt/VAr Control with Photovoltaics. *IEEE Trans. Power Syst.* **2019**, *34*, 2401–2408. [CrossRef]
32. Seuss, J.; Reno, M.J.; Broderick, R.J.; Grijalva, S. *Analysis of PV Advanced Inverter Functions and Setpoints Under Time Series Simulation*; Sandia National Laboratories: Albuquerque, NM, USA, 2016.

electronics

MDPI

Article

Virtual Inertia Methods for Supporting Frequency Stabilisation in Autonomous AC/DC Microgrids

Faysal Hardan [1,*], Rosemary Norman [2] and Pietro Tricoli [1]

1 School of Engineering, The University of Birmingham, Edgbaston, Birmingham B15 2TT, UK; p.tricoli@bham.ac.uk
2 School of Engineering, Newcastle University, Newcastle Upon Tyne NE1 7RU, UK; rose.norman@newcastle.ac.uk
* Correspondence: f.hardan@bham.ac.uk

Abstract: Isolated microgrids have long been considered alternative power system entities that can integrate various types of distributed energy sources such as diesel and renewable power generators including energy storage. Renewable energy sources, such as wind and solar PV, introduce low inertia and high intermittency to the microgrid. For this reason, coordinated control and frequency stabilisation are crucial for maintaining higher service levels in the microgrid. This paper reports on the design and development of two proposed methods for virtual inertia provision, namely model-based and filter-based methods, which support the frequency stability of AC/DC microgrids. The inertial power produced by these methods was implemented through power-controlled voltage source converters, associated with a Li-ion battery energy storage system. To derive and develop the functions for the virtual inertia providers using these methods, a new electromechanical power-speed model was developed to represent the interaction between the microgrid AC/DC-sides and its generators. Small-signal analysis using the linearised form of this model was carried out, in addition to deriving the law for the model-based virtual inertia method. Detailed physical-system simulation and tests were performed, and performance analysis of the resulting generator speed-responses using the proposed methods illustrated their merits compared with other methods, namely the standard df/dt and frequency-event techniques.

Keywords: AC/DC microgrid; energy storage; renewable power generators; virtual inertia; voltage source converter

1. Introduction

Research into microgrids and their application is being carried out globally, with various projects around the world integrating different types of distributed energy resources (DERs), such as wind and solar PV, energy storage systems (ESS), and rotating generators [1–3]. These microgrids can provide better reliability, stability, and resilience to the connecting grid and within their service sites operating in both grid-connection and islanded modes [4–6]. Moreover, they can be designed with different hybrid topologies, including DC distribution networks, integrated battery ESSs, and advanced capabilities in terms of control, protection, and operational coordination [7–9]. Grid-forming capabilities with local microgrid generators and ESSs, mainly based on power-electronic (PE) systems such as the voltage source converter (VSC), are the minimum requirements for steady/uninterruptible operation of such microgrid systems when switching to islanded mode. In addition, adding a DC network to a microgrid increases its flexibility for integrating battery energy storage systems (BESS) and multi-level DC capability using DC-DC converters [10,11].

As most DERs with PE systems provide power to the grid without adding any rotating mechanical inertia to the large connecting power system, they have a considerable effect on the transient frequency response of such systems [12]. However, in response to extreme loading events or partial loss of power generation, this inertia is important for stabilising the system frequency, as it allows crucial time for dispatching other generators/sources to balance power generation with consumption. More importantly, when the microgrids are islanded, they inherently become low inertia systems, and their frequency stability is more affected. Therefore, attempts have been made to address and resolve this issue [13,14], suggesting and adopting control methods for synthetic inertia (SI) or virtual inertia (VI) emulation through PE inverters/converters [15–21], and even through connected electric vehicles and data centres, as reported in [22]. The research reported in [23] compared two SI methods, active and passive synthetic inertia, through simulating their behaviour in a small microgrid by modelling converter-based generators. It was proved that both types of SI realisation help in stabilising microgrid frequency. Reference [24] proposed a VI emulation strategy which is based on model reference control for diesel-wind systems, where a typical model representing the frequency response was used for realising the desired inertial response. The authors in [25] compared the VI control for frequency stability in both islanded microgrids and grid-connected power systems, and it was shown that, in a power system with island-optimised areas, dynamic behaviour can result in stability issues. In [26] a model predictive control-based VI system, integrated into a microgrid control loop, stabilised the microgrid frequency under high penetration of renewable energy sources (RESs). For microgrids with DC distribution networks, VI strategies for stabilising DC voltage, which adopt a similar philosophy to frequency or generator-speed stabilisation, have also been reported [27,28]. Whilst the inertia constant in the AC system is based on the total mechanical moment of inertia of the connected rotating generators, in the DC system it is based on the total capacitance of the DC sources connected to the DC network [27].

In this paper, we propose two virtual inertia methods—model-based and filter-based—-with design, analysis, and mathematical modelling, including tests, result analyses, and comparison with two other methods reported in the literature. An electromechanical power-speed model of a power system AC/DC microgrid, connecting diesel generators and interfacing VSCs, was also derived to support virtual inertia analysis, control, and design. The case-study microgrid is powered by two 0.5 MVA diesel engine generators (DEGs) with a relatively low inertia constant ($H = 0.247$ s), which is small in comparison to a typical high-power generator with $H \cong 3.5$ s [29]. Therefore, the DEG speed/frequency can easily deviate in response to load events, as the reaction of the diesel engine is comparatively slow. As such, electrical synchronous generators mechanically coupled to small diesel engines should be able to provide grid-forming and minimal stabilisation capabilities to the microgrid. For improved stabilisation performance, this paper addresses the challenge of developing the most appropriate virtual inertia methods to stabilise generator speed in an AC/DC microgrid, utilising a power-controlled VSC [30], with 0.5 MW Li-ion BESS. It also reports on the performances of the proposed VI methods for frequency support, in addition to underlying dynamic modelling of the microgrid and its components, including non-linear analysis and comparison with two standard VI methods reported in the literature [31, 32]. In [31], for the considered df/dt VI control method with ESS, the authors reported that a first-order lag with a large time constant of around 1 s is required for a medium stiff point of common coupling (PCC) of the local network. Reference [32] reports on utilising the df/dt VI for under-frequency support but using an activation frequency limit, so that VI is only applied when the frequency drops below this limit or at this frequency event. The authors in [33] proposed a method for continuous VI by extracting/storing kinetic energy associated with the rotor inertia of the wind turbine, even when the inertia constant, H, is continuously changing due to the variable speed operation of the wind turbine.

The paper is organised in the following manner. Section 2 describes the configuration of the microgrid considered as a case study in this paper. Sections 3 and 4 report on the derivation of the non-linear electromechanical model which combines the dynamics of both the VSC, including its AC-bus interface, and the DEGs that supply this bus within the microgrid. Model linearisation and system small-signal analysis based on this model around the nominal operating point, which covers power–voltage and power–speed interactions, were derived and are reported in Section 5. Section 6 deals with the development and design aspects of the proposed VI methods and their implementation. System simulation, results, performance analysis, and cross-comparison of the proposed VI methods with the other techniques reported in the literature are discussed in Section 7. Section 8 draws conclusions from the reported research and summarises the advantages of the proposed VI methods.

2. The Proposed Structure of the Microgrid

To develop an effective structure for an AC/DC microgrid that considers power quality, a detailed study was carried out with a focus on islanded applications. Accordingly, it was considered that the microgrid should be capable of supplying the maximum consumer load, which also should be covered by the Li-ion BESS for a certain period to minimise the use of the DEGs. The microgrid structure shown in Figure 1 was adopted, in which two 0.5 MVA diesel generators were connected to the AC-bus of the microgrid, whilst the BESS, scaled at 1 MWh, was connected to the DC-bus along with any other RESs. The latter can be energised by the DEG through four identical 125 kVA VSCs connected in parallel, by the BESS, or by both sources. Each VSC is based on three-level NPC topology [34] and designed using IGBTs as the main power-electronic switches operating at 2.25 kHz PWM switching frequency. Although modern power electronic converters at this power level can operate at PWM frequencies of 10 to 20 kHz, switching at lower frequencies is highly desirable to reduce the switching losses of the converter PE devices and to increase system efficiency for energy saving. The waveforms of the VSCs' switching frequencies are shifted by $2\pi/4$ rad for interleaved operation of the VSCs to reduce the ripple in their total current waveforms injected into the AC-bus. With this operation, the frequency of the ripple peak will be equivalent to four times the VSC switching frequency (four VSCs), which permits the use of a VSC switching-frequency filter tuned at four times the switching frequency, leading to reduction in the rating and size of the filter components [35]. As such, the filter provides very low-impedance path for the current harmonic component at this tuned frequency, preventing it from propagation into the AC loads and rotating generators. Details about calculating/designing the components of a network single-tuned shunt-filter, considering a tuning frequency of 4×2.25 kHz, are provided in the Appendix B.

Due to the bidirectional power capability of the VSCs, the AC-bus can be energised by the BESS through the VSCs in addition to the DEG, exploiting the bidirectional power-control capability of the VSCs with other supporting functions for VI provision or frequency support. AC and DC loads can then be supplied or connected flexibly to the microgrid even when the DEG is idling. It should be noted that, whilst the DC-bus of the microgrid is regulated to 1 kV by the VSCs when the AC-bus is energised by the DEGs, it can also be regulated to this voltage level through the bi-directional buck-boost DC-DC converter of the BESS [36] when the DEGs are shut down and disconnected from the AC-bus.

To test the performance of the proposed VI methods for DEG frequency/speed support, the microgrid in Figure 1 was simulated in continuous time using the physical modelling tools of the Matlab/Simulink® software (ver. R2024b) to represent the actual components of the microgrid and their connections. The converter PWM controllers and VI power algorithms/controllers were also modelled and simulated within their relevant units at sampling rate equivalent to the converter switching frequency.

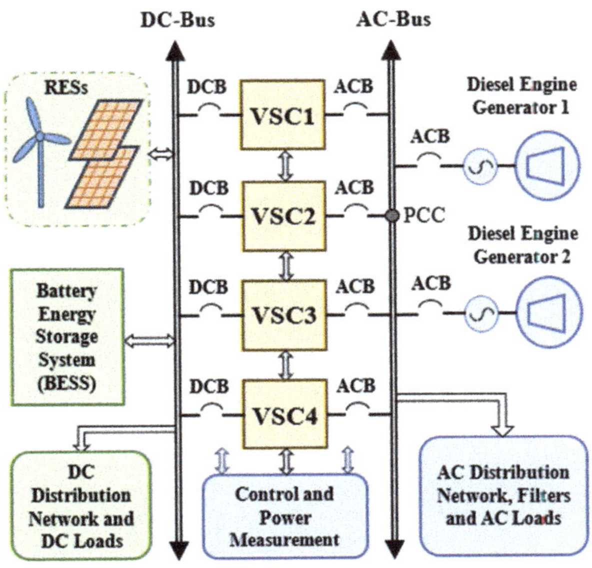

Figure 1. The proposed structure of the studied AC/DC microgrid.

3. VSC Network Connection Power Model

To gain insight into the power flow dynamics through the VSC AC connection, and to derive the proposed VI control functions, a dynamic power-flow model, when the power-electronic VSCs are interfaced to the AC- and DC-buses, was derived. The VSC connection to the PCC of the AC-bus can be simply accomplished through an equivalent inductor (for a transformer-less connection), which can be represented by two parameters, namely its inductance value L and its average internal resistance R, as illustrated by the connection circuit as shown in Figure 2a. The VSCs can be represented by four identical power sources connected in parallel, as illustrated in Figure 2b, which in turn results in an equivalent power-source circuit, shown in Figure 2c, as all VSCs have an identical fundamental voltage component of the PE converter output, $\vec{v}_{idq}$. Given that the voltage vector $\vec{v}_{dq}$ at the PCC is aligned with the d-axis of the synchronous d-q reference frame, a representation of the apparent dynamic power $\vec{S}$, in volt-ampere, which represents the total active and reactive power, P and Q, in a complex vector form, can be given as:

$$\vec{S} = P + jQ = \frac{3}{2}(v_d\, i_d + jv_d\, i_q), \text{ or} \tag{1}$$

$$\vec{S} = \frac{3}{2}\, V_n\, (i_d + j\, i_q);\ v_d = V_n,$$

where v_d is the d-axis voltage, V_n is the per-phase voltage magnitude at the PCC, and i_d, i_q are the d and q components of the inductance current vectors, respectively. The q voltage component v_q disappears (becomes zero) as its voltage vector forms a 90° angle with the q-axis of the reference frame. It should be noted that the above power representation in complex form is based on dynamic variables for dynamic modelling and should be differentiated from steady-state phasor expressions. The time derivative of the above equation is:

$$\vec{S}' = \frac{3}{2}\dot{V}_n\,(i_d + j\,i_q) + \frac{3}{2}V_n\,(\dot{i}_d + j\,\dot{i}_q), \tag{2}$$

where ′ denotes a variable's time-derivative. Based on the equivalent circuit of Figure 2c, the following voltage-current relationships at the PCC can be worked out:

$$\left.\begin{aligned} \dot{i}_d &= -\tfrac{R_{eq}}{L_{eq}}\,i_d + \omega_e i_q + \tfrac{1}{L_{eq}}\,Dv_d \\ \dot{i}_q &= -\tfrac{R_{eq}}{L_{eq}}\,i_q - \omega_e i_d + \tfrac{1}{L_{eq}}\,Dv_q \end{aligned}\right\}, \tag{3}$$

where $L_{eq} = L/4$ and $R_{eq} = R/4$ are the equivalent inductance and resistance of the VSCs' interfacing inductors, respectively, Dv_d and Dv_q are the d and q components of the applied voltage-difference vector $D\vec{v}_{dq}$, respectively, and ω_e is the angular frequency of the electrical network voltage or reference frame. Substituting $\dot{i}_d$ and $\dot{i}_q$ from (3) into (2), and then using (1) to substitute i_d and i_q with $2P/3V_n$ and $2Q/3V_n$, respectively, can result in the following power model:

$$\begin{bmatrix}\dot{P}\\ \dot{Q}\end{bmatrix} = \frac{\dot{V}_n}{V_n}\begin{bmatrix}P\\ Q\end{bmatrix} + \frac{3}{2}V_n\begin{bmatrix}-\frac{R_{eq}}{L_{eq}}\frac{2\,P}{3V_n} + \omega_e\frac{2\,Q}{3V_n} + \frac{1}{L_{eq}}\,Dv_d\\ -\frac{R_{eq}}{L_{eq}}\frac{2\,Q}{3V_n} - \omega_e\frac{2\,P}{3V_n} + \frac{1}{L_{eq}}\,Dv_q\end{bmatrix} \tag{4}$$

This model may be expressed using the canonical state-space form as:

$$\begin{bmatrix}\dot{P}\\ \dot{Q}\end{bmatrix} = \begin{bmatrix}\left(\frac{\dot{V}_n}{V_n} - \frac{1}{\tau_{eq}}\right) & \omega_e\\ -\omega_e & \left(\frac{\dot{V}_n}{V_n} - \frac{1}{\tau_{eq}}\right)\end{bmatrix}\begin{bmatrix}P\\ Q\end{bmatrix} + \begin{bmatrix}\frac{3V_n}{2L_{eq}} & 0\\ 0 & \frac{3V_n}{2L_{eq}}\end{bmatrix}\begin{bmatrix}Dv_d\\ Dv_q\end{bmatrix} \tag{5}$$

where $\tau_{eq} = L_{eq}/R_{eq}$ is the time constant of the connecting inductor. The active and reactive power, P and Q, respectively, represent the states and the outputs of the model and are interdependent or cross-coupled at the PCC, whilst Dv_d and Dv_q represent the model inputs.

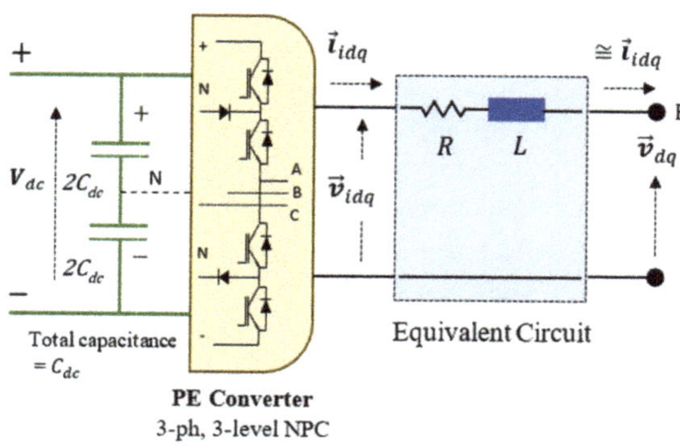

(a)

Figure 2. *Cont.*

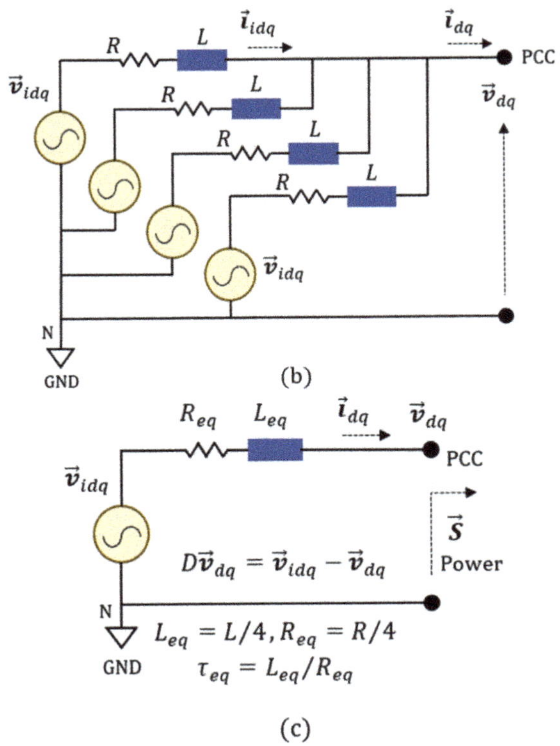

Figure 2. (**a**) Structure of the utilised VSC, including its network interface equivalent circuit. (**b**) Representation of the VSCs as paralleled voltage sources. (**c**) Equivalent voltage source of the VSCs.

4. The Electromechanical Power-Speed Model

For dynamic modelling, the two synchronised DEGs can be treated as single generator with a 1 MVA power rating. As such, the electromechanical dynamic model, which links this generator's electrical load, the VSCs' power P, and the engines' mechanical power to the generator speed, can be derived as the following power-speed form [30,37]:

$$\frac{d\omega_m}{dt} = K_H \omega_m \,(P + P_a - P_{ls}); \tag{6}$$

$$K_H = 1/(2\,H\,S_m),\; P_a = P_m - P_e; \tag{7}$$

where ω_m, H, and S_m are the generator's speed, inertia constant, and rated VA power, respectively, P_a, P_{ls}, and P_m are the generator's accelerating/decelerating power, total power losses, and mechanical power, respectively, and P_e is the load electrical power. Integrating Equation (6) into Equation (5) results in an expanded state-space model:

$$\begin{bmatrix} \dot{P} \\ \dot{Q} \\ \dot{\omega}_m \end{bmatrix} = \underbrace{\begin{bmatrix} (\frac{\dot{V}_n}{V_n} - \frac{1}{\tau_{eq}}) & \omega_m N_p & 0 \\ -\omega_m N_p & (\frac{\dot{V}_n}{V_n} - \frac{1}{\tau_{eq}}) & 0 \\ K_H \omega_m & 0 & -K_H P_{ls} \end{bmatrix}}_{A} \begin{bmatrix} P \\ Q \\ \omega_m \end{bmatrix} + \underbrace{\begin{bmatrix} \frac{3V_n}{2L_{eq}} & 0 & 0 \\ 0 & \frac{3V_n}{2L_{eq}} & 0 \\ 0 & 0 & K_H \omega_m \end{bmatrix}}_{B} \begin{bmatrix} Dv_d \\ Dv_q \\ P_a \end{bmatrix}; \tag{8}$$

where $\omega_m = \omega_e\, N_p$, and N_p is the generator number of pole pairs. In general, this system model is non-linear, as its transient and output matrices, **A** and **B**, depend on the state and exogenous variables, mainly the generator speed ω_m and the AC side voltage component

V_n (or v_d) that affects the input voltage components Dv_d and Dv_q. Therefore, the model matrices are not constant, and the system transient matrix can be affected by voltage magnitude variations at the PCC because of loading events. This model includes all state and output variables that interact with each other in transient and steady-state operations, where the control inputs can affect the generator speed directly or indirectly through the injected VSC power P.

5. Model Linearisation and System Analysis

Model linearisation of the above power system around an operating point (OP) (i.e., around operating speed, active and reactive power, and when considering the voltage deviation at the PCC) results in the following approximate perturbation model:

$$\begin{bmatrix} \Delta\dot{P} \\ \Delta\dot{Q} \\ \Delta\dot{\omega}_m \end{bmatrix} \cong \begin{bmatrix} \left(\frac{\dot{V}_n}{V_n} - \frac{1}{\tau_{eq}}\right) & \omega_m\, N_p & N_p\, Q \\ -\omega_m\, N_p & \left(\frac{\dot{V}_n}{V_n} - \frac{1}{\tau_{eq}}\right) & -N_p\, P \\ K_H\, \omega_m & 0 & K_H(P + P_a - P_{ls}) \end{bmatrix}_{OP} \begin{bmatrix} \Delta P \\ \Delta Q \\ \Delta\omega_m \end{bmatrix} + \begin{bmatrix} \frac{3(V_n - Dv_d)}{2L_{eq}} & 0 & 0 \\ -\frac{3Dv_q}{2L_{eq}} & \frac{3V_n}{2L_{eq}} & 0 \\ 0 & 0 & K_H\, \omega_m \end{bmatrix}_{OP} \begin{bmatrix} \Delta Dv_d \\ \Delta Dv_q \\ \Delta P_a \end{bmatrix} \tag{9}$$

It should be noted that, at some point of the derivation process, V_n was reconsidered as v_d and equal to $v_{id} - Dv_d$, according to Figure 2c. The time domain model in (9) can be converted to a model in the frequency or Laplace domain at any fixed operating point. For the analysis in this section, an operating point is considered when the VSC does not contribute any continuous power to the AC-bus, and therefore all its power outputs and the voltage difference components Dv_d and Dv_q are null. Also, at this point the DEG speed ω_m is considered nominal (which is normally the case) without an accelerating power P_a. Applying these conditions to the model in (9), which conforms to the dynamic VI provision, the converted model in the Laplace domain can be written as:

$$Y(s) = G(s)\, U(s); \tag{10}$$

$$Y(s) = \begin{bmatrix} \Delta P(s) & \Delta Q(s) & \Delta\omega_m(s) \end{bmatrix}^T, \tag{11}$$

$$U(s) = \begin{bmatrix} \Delta Dv_d(s) & \Delta Dv_q(s) & \Delta P_a(s) \end{bmatrix}^T; \tag{12}$$

where T indicates a matrix transpose, and:

$$G(s) = \begin{bmatrix} g_{11}(s) & g_{12}(s) & g_{13}(s) \\ g_{21}(s) & g_{22}(s) & g_{23}(s) \\ g_{31}(s) & g_{32}(s) & g_{33}(s) \end{bmatrix}; \tag{13}$$

$$g_{11}(s) = g_{22}(s) = \frac{\frac{3\, V_n}{2\, L_{eq}} \left(s + \left(\frac{1}{\tau_{eq}} - \frac{\dot{V}_n}{V_n}\right)\right)}{D_1(s)}, \tag{14}$$

$$g_{12}(s) = -g_{21}(s) = \frac{\frac{3\, V_n}{2\, L_{eq}} \left(\omega_m\, N_p\right)}{D_1(s)}, \tag{15}$$

$$g_{31}(s) = \frac{\frac{3\, V_n}{2\, L_{eq}}\, K_H\, \omega_m \left(s + \left(\frac{1}{\tau_{eq}} - \frac{\dot{V}_n}{V_n}\right)\right)}{D_2(s)}, \tag{16}$$

$$g_{32}(s) = \frac{\frac{3\,V_n}{2\,L_{eq}}\,K_H\,\omega_m^2\,N_p}{D_2(s)}, \tag{17}$$

$$g_{33}(s) = \frac{K_H\,\omega_m}{s+\,K_H\,P_{ls}}; \tag{18}$$

$g_{13}(s) = g_{23}(s) = 0$ and:

$$D_1(s) = s^2 + 2\left(\frac{1}{\tau_{eq}} - \frac{\dot{V}_n}{V_n}\right)s + \left(\left(\frac{1}{\tau_{eq}} - \frac{\dot{V}_n}{V_n}\right)^2 + \omega_m^2\,N_p^2\right), \tag{19}$$

$$D_2(s) = (s+\,K_H\,P_{ls})\,D_1(s). \tag{20}$$

As the parameters of the linearised model can be within a range of values, they can lead to a large set of Laplace or frequency-domain models which result in different system characteristics. Investigating all possible numerical models is beyond the scope of this paper. However, for system analysis and designing the proposed model-based VI method, the system parameters given in Appendix A were considered.

Using the characteristic polynomials $D_1(s)$ and $D_2(s)$ above, the open-loop system poles can be found as functions of the model parameters:

$$p_1 = \left(\frac{\dot{V}_n}{V_n} - \frac{1}{\tau_{eq}}\right) - j\,\omega_m\,N_p, \tag{21}$$

$$p_2 = \left(\frac{\dot{V}_n}{V_n} - \frac{1}{\tau_{eq}}\right) + j\,\omega_m\,N_p, \tag{22}$$

$$p_3 = -K_H\,P_{ls}. \tag{23}$$

As K_H is always positive, p_3 has a negative real value, and it is always in the left half of the complex plane. However, this is not the case with poles p_1 and p_2, as they can easily move to the right half of the plane when $\dot{V}_n/V_n$ becomes higher than $1/\tau_{eq}$. Therefore, to ensure VSC connection stability under extreme events, the condition $1/\tau_{eq} > \dot{V}_n/V_n$ must always be maintained. Since p_1 and p_2 are conjugate poles, the system is oscillatory at a frequency equal to the angular frequency of the AC network, $\omega_e = \omega_m N_p$. Using the values of the system parameters given in Appendix A and for a constant V_n, the numerical values for the system poles become $p_{1,2} = -20 \pm \mathrm{j}314.16$, $p_3 = -0.04$. These poles characterise the model with an oscillatory power response at 314.16 rad/s and a sluggish, non-oscillatory speed response, as demonstrated below.

Based on the system transfer functions in (13), the frequency and step responses, which give insights into the model characteristics and stability issues for system analysis, can be depicted as shown in Figures 3 and 4. The two graphs in the top of Figure 3 represent the frequency responses of the active power, where the traces show the oscillatory response at 314.16 rad/s for both g_{11} and g_{12}. The related step responses confirm their oscillatory nature, as shown in the lower graph, and it can be noticed that the active power is mainly affected by the deviation in the quadrature component of the control voltage-difference Dv_q of the VSC. Figure 4 shows the responses associated with g_{31}, g_{32}, and g_{33}, where the three graphs represent the frequency responses of the generator speed to VSC voltages ΔDv_d and ΔDv_q, and to the accelerating power ΔP_a, respectively. It is demonstrated that the responses g_{31} and g_{32} have very low oscillation around 314 rad/s due to their links to the VSC active and reactive power deviation, whereas the response g_{33} ($\Delta P_a \rightarrow \Delta\omega_m$) does not exhibit such an oscillation. The traces of the corresponding step responses for the generator speed, as given in the lower graph, illustrate that a much higher speed response or gain can be obtained when using the quadrature voltage-difference Dv_q in comparison with Dv_d and P_a effects. In other words, whilst speed can be changed considerably in response to a small change in Dv_q, its sensitivity to changes in Dv_d and P_a is very low, and therefore a very large change in P_a must be made to make any significant change in speed or frequency. For further insight into the system closed-loop characteristics, Table 1 summarises the model's gain and phase margins based on its transfer functions in (13).

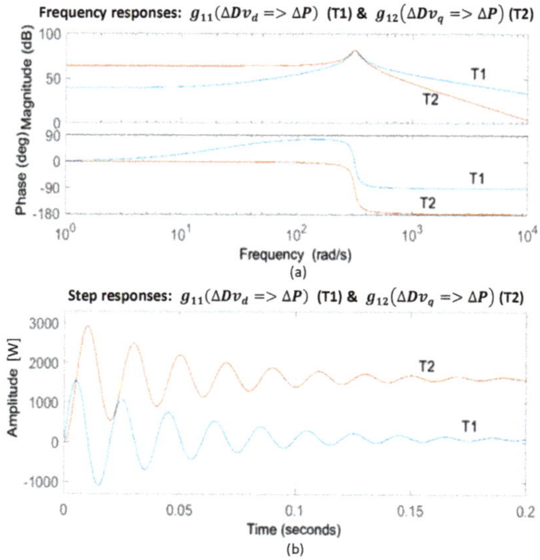

Figure 3. (**a**) Frequency responses of the VSC active power deviation ΔP using $g_{11}(s)$ and $g_{12}(s)$, (**b**) the corresponding step responses with unit steps of inputs ΔDv_d and ΔDv_q.

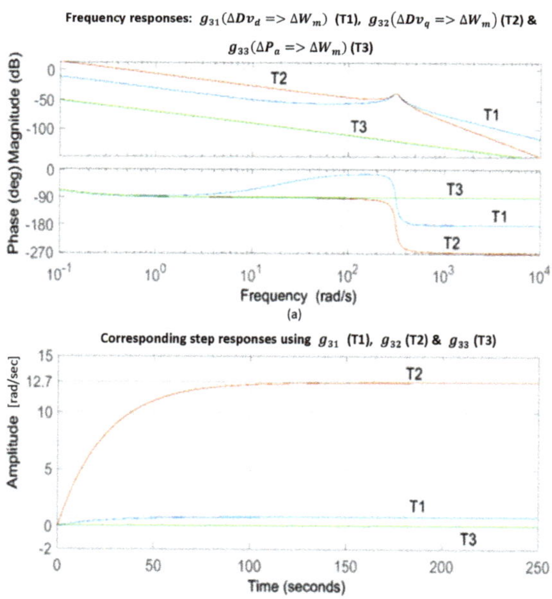

Figure 4. (**a**) Frequency responses of generator speed via $g_{31}(s)$, $g_{32}(s)$, and $g_{33}(s)$, (**b**) the corresponding step responses of the speed with unit steps of inputs ΔDv_d and ΔDv_q.

Table 1. Gain and phase margins of the system transfer functions.

Function	Gain Margin	Freq. [rad/s]	Phase Margin	Freq. [rad/s]
$g_{11}(s)g_{22}(s)$	inf. [1]	n/a [2]	90	500 K
$g_{12}(s)$	inf.	inf.	0.18	13 K
$g_{21}(s)$	0.0006	0	−180	13 K
$g_{31}(s)$	inf.	inf.	inf.	n/a
$g_{32}(s)$	78	314.8	94.5	0.51
$g_{33}(s)$	inf.	n/a	inf.	n/a

[1] infinite, [2] not applicable.

6. The Proposed Methods for Frequency Support

As the VSC can be controlled as a dynamic power-source/sink, it can be employed to act effectively as another synchronous generator (with similar or different power scale) connected in parallel with the DEG, adding an extra AC virtual inertia to the microgrid power system through its AC-bus. As such, it enhances frequency stabilisation for the existing power system, especially when the VSC is equipped with fast-response power-control capability, in addition to suitable VI functions. The methods proposed below are considered for designing and providing VI of similar order to that of the DEG through the static VSC.

6.1. The Model-Based Virtual Inertia Method

In this proposed method, the VI power can be defined as the change in the acceleration power ΔP_a, which can be derived using the last line of the matrix in (9) around the generator's speed and power operating point. Substituting $K_H(P + P_a - P_{ls})$ from (6) and replacing H with its expression $J\omega_m^2/(2\,S_m)$ within K_H into the third matrix line in (9), the law for calculating the required VI power can be derived as:

$$P_{VI}^{*} = \Delta P_a = J\,\omega_m\,\Delta\dot{\omega}_m - J\,\dot{\omega}_m\,\Delta\omega_m - \Delta P; \qquad (24)$$

where J is the generator's moment of inertia considered for calculating the VI power. Accordingly, a VI model-based provider based on Equation (24) can be realised within the VSC power controller by passing the calculated P_{VI}^{*} as part of the total VSCs' power demand or reference, as illustrated in Figure 5. A switching function SW was introduced to apply the VI power only when the speed is diverging away from its nominal value, ensuring that the VI power does not oppose the action of the DEG governor when the generator speed is approaching its set point. This SW condition can be derived as follows:

$$If\ \left[(\omega_m^{*} - \omega_m)\cdot P_{VI}^{*}\right] > 0\ then,$$
$$SW\ is\ closed,\ else$$
$$SW\ is\ open.$$

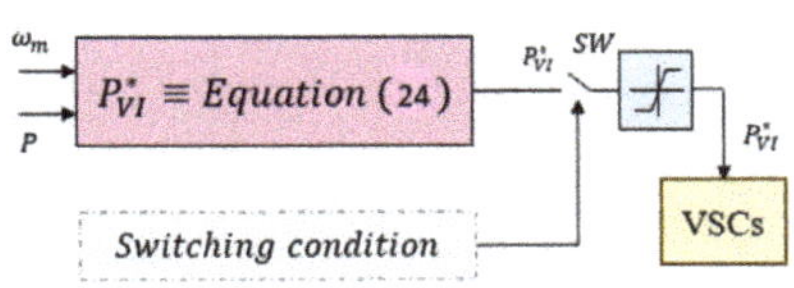

Figure 5. Simplified block diagram of the proposed model-based VI method and its switching mechanism.

6.2. The Filter-Based Virtual Inertia Method

A change in the microgrid transient AC load is the main cause of speed/frequency deviation of the DEGs, whilst the transient DC load can directly affect the DC-bus voltage.

The aspect of regulating the DC-bus voltage through the BESS and the VSC converters is beyond the scope of this paper. Accordingly, the idea for this method is to divert the impact of the transient AC load away from the rotating generator by absorbing the load variations utilising the VSCs and the integrated BESS. This mechanism was achieved by extracting the high-frequency components of the load perturbation ΔP_e, which can become a part of the accelerating/decelerating power as in Equation (6), utilising a low-pass filter (LPF). As such, the impact of the mechanism using the LFP is similar to adding extra inertia or VI to the DEGs or the AC power system. The proposed structure for this filter-based VI method, illustrated in Figure 6, demonstrates how the high-frequency components of the load VI power demand can be extracted. By demanding these power components (assigned as ΔP^*) through the VSC power controller, the required transient power can be generated to offset the changes in the AC load and to alleviate their effect on the generator speed. It is worthwhile to mention that frequency-division methods, using high-pass and low-pass filters, were presented in [38–40], for a different purpose: to produce a power reference for short-term energy storage systems. However, in [39,40], high-pass filtering was applied to the DC-voltage error, whilst in [38] it was applied to wind turbine power.

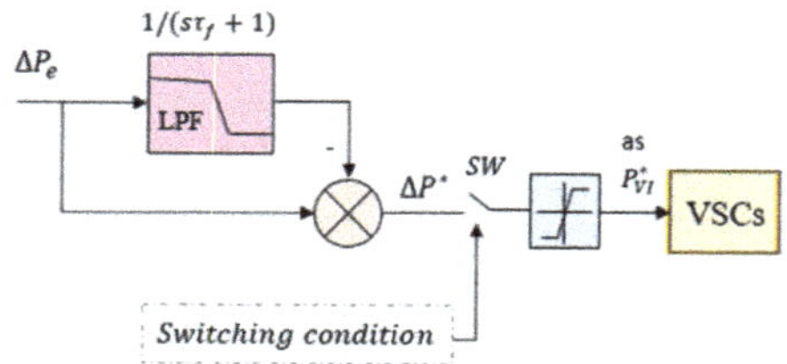

Figure 6. Proposed structure of the filter-based VI method for generating a VI power demand.

When utilising the proposed filter-based method, load uncertainty will not be an issue, as AC-load power can be accessed for direct measurement within the microgrid. The LPF used to derive the high-pass filtering function can be designed with different time constants T_f to satisfy the speed/frequency deviation limits, considering the response delay of the DEG speed-governor control loop and minimal usage of stored energy for VI purpose. In addition to covering these limits, the recommended approach to selecting the best value for the filter time constant can also be based on design and protection factors, such as the rate of change of frequency (RoCoF) limits and available peak power by the VSCs for VI provision.

The electromechanical model in (6) was used to evaluate the speed response at various values of the time constant T_f. When the change in AC load is ΔP_e, the VSC initially provides the same amount of power ΔP_e by injecting ΔP^*, as illustrated in Figure 6. Based on Equation (6), the power-speed dynamics (around the synchronous speed) at the instant of applying the load step can be found as:

$$\frac{d\omega_m}{dt} + K_H\, \omega_m\, P_{ls} \cong K_H\, \omega_{ms}\, (\Delta P^* - \Delta P_e); \tag{25}$$

where ω_{ms} is the synchronous speed of the rotating generator. In the s-domain, Equation (25) takes the following form:

$$\omega_m(s)\, [s + K_H\, P_{ls}] = K_H\, \omega_{ms}\, [\Delta P^*(s) - \Delta P_e(s)]. \tag{26}$$

Obtaining the relationship between ΔP_e and ΔP^* in the s-domain from Figure 6 and inserting it into (26), the following transfer function between the speed and the change in the electrical AC load can be derived:

$$\frac{\omega_m(s)}{\Delta P_e(s)} = \frac{-K_H\,\omega_{ms}}{(s + K_H\,P_{ls})\left(s\,T_f + 1\right)}. \tag{27}$$

Using this expression, an expression for calculating the time-domain response of the DEG speed as a function of AC-load change ΔP_e, filter time-constant T_f, and time can be found:

$$\Delta\omega_m\left(\Delta P_e,\, T_f,\, t\right) = \frac{\Delta P_e\,\omega_{ms}}{P_{ls}}\left\{\frac{\frac{1}{T_f}\,e^{-(K_H\,P_{ls})\,t} - K_H\,P_{ls}\,e^{-t/T_f}}{\left(\frac{1}{T_f} - K_H\,P_{ls}\right)} - 1\right\}. \tag{28}$$

For comparison with the case without using the filter, Equation (27) also gives the following time-domain expression:

$$\Delta\omega_m(\Delta P_e,\, t) = \frac{\Delta P_e\,\omega_{ms}}{P_{ls}}\left(e^{-(K_H\,P_{ls})\,t} - 1\right) \tag{29}$$

A numerical solution to (28) for the filter time constant can be found for maximum ΔP_e, as a worst-case scenario, which limits DEG speed deviation to the allowed range within a period Δt covering the delay of the DEG speed loop. From the expressions in (28) and (29), speed responses to a ± 0.1 MW AC-load step can be predicted at various values of the filter time constant T_f, as illustrated via the curves in Figure 7, which can help with (29) in selecting the best value of T_f according to specified design and optimisation criteria. It is evident that, the higher the value of the time constant, the lower the speed deviation that results at the period of transient AC-load.

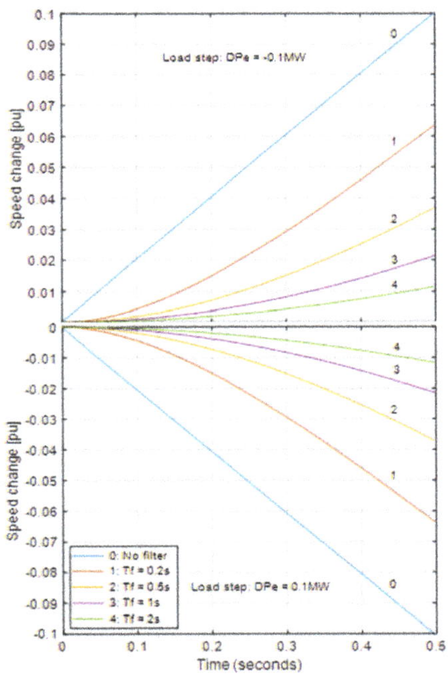

Figure 7. Predicted speed responses to $\Delta P_e = \pm 0.1$ MW AC-load step at various values of the filter time constant.

7. System Simulation and Test Results

To assess the performance of the proposed VI methods versus those associated with other methods reported in the literature, the AC/DC microgrid depicted in Figure 1 was simulated in continuous time to represent the actual or physical system. The DEG control system considered in the simulated microgrid is illustrated in the block diagram shown in Figure 8, which includes both the DEG speed governor and magnetic-excitation control loops [41,42]. The VI methods, including the proposed ones, were then embedded within the power control unit of the VSC, which can provide a power response-time between 1 ms and 2 ms. Two methods reported in the literature were considered for comparison, namely the standard df/dt VI method in [30,31], implemented in an external control loop, and the frequency-event VI method in which the VI power was generated using the standard method but applied only when the frequency/speed exceeds certain limits [31]. The block diagram in Figure 9 shows the main elements of these VI methods as embedded within the external control loop of the VSC power controller. The dead-band box shown within the diagram was only utilised for the frequency-event method to detect when frequency/speed hits the pre-determined lower or higher frequency limits, allowing VI provision only outside these limits.

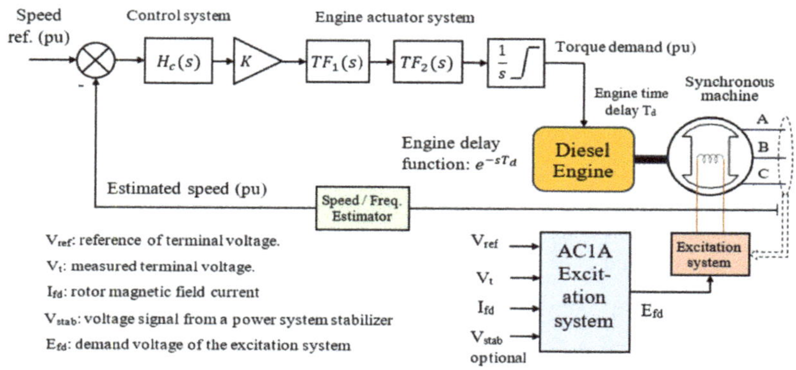

Figure 8. Block diagram of the control system of the diesel engine generator considered for testing the VI methods.

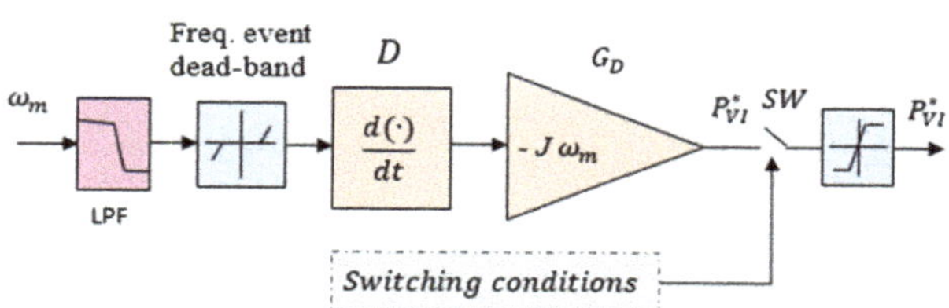

Figure 9. Block diagram of the standard external VI method, including frequency dead-band as an event detector, in addition to the switching function SW.

Before performing the tests to obtain the VI results, the functionality of the AC/DC microgrid and its components, including the VSCs and their power controllers, were checked. The structure and specifications given in Section 2 and in Appendix A were considered for obtaining the functionality test results, which are illustrated in Figure 10. The traces in graph (a) represent the microgrid three-phase voltages and their magnitude during VI provision using the standard df/dt method, whereas graph (b) illustrates the corresponding VSCs' currents in response to 0.1 pu change in the microgrid load. In graph (c), longer time records of the VSCs' currents are shown to demonstrate the decaying nature of the produced VI power by the VSCs for mitigating the deviation in the microgrid frequency or DEG speed.

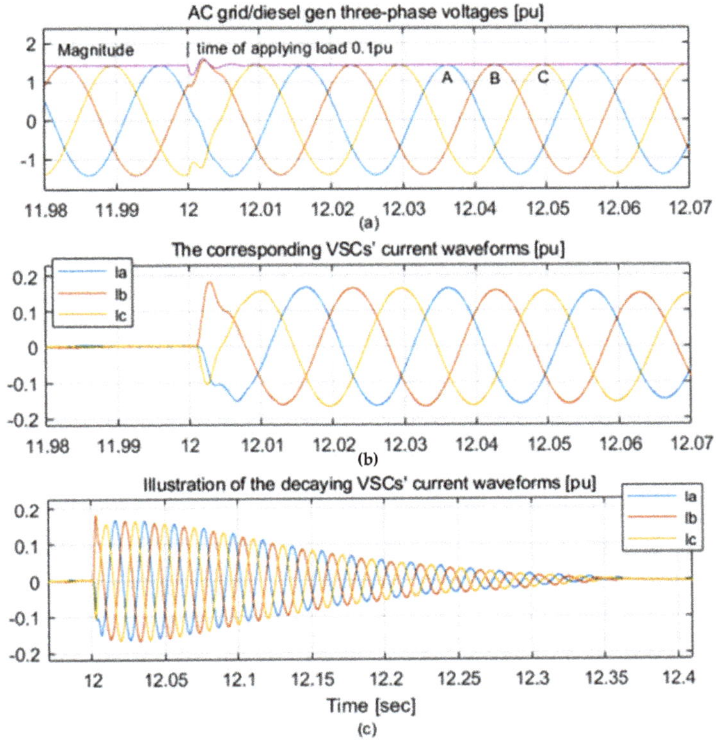

Figure 10. (**a**) Microgrid three-phase voltages and their magnitude during VI provision using the standard df/dt method, (**b**) corresponding VSCs' current waveforms, and (**c**) the same VSCs' current waveforms decaying over a longer timespan after a 0.1 pu change in the microgrid load.

As the model-based and filter-based VI methods rely also on AC-load power measurement, as illustrated in Figures 5 and 6, they were embedded directly within the power control loop of the VSC system, whilst the other VI methods from the literature were inserted in an external speed control loop, as they only rely on speed measurement. When testing all the VI methods, the AC side of the microgrid was subjected to a ±0.1 pu load step of the two DEGs' nominal power (i.e., ±0.1 MW), and this condition was applied with the same timing sequence for testing each method independently.

The graphs in Figure 11a show the speed response results and the corresponding VSC power over the timespan 12–16 s, obtained by running the system simulation for all the VI methods. As illustrated, all methods improved the generator's speed response compared with the response without VI provision, bringing its nadirs and peaks closer to the nominal speed line when the load was stepped up and down by ±0.1 MW (or 0.2 pu of the VSC nominal power) at 12 s and 14 s. The lower two graphs in Figure 11b are expanded sections of the previous graphs and clearly demonstrate that the filter-based method provided the best speed response performance at $T_f = 0.2s$. However, with this method, the associated VSC power is the highest, which requires more energy to be sourced or stored through the DC-bus. The differences between the proposed model-based method and the other two methods are subtle. The trace representing the speed response without VI shows that the generator speed dropped to a level close to 98% of the nominal speed in response to the 0.1 pu load step. The model-based and standard external VI methods resulted in almost identical responses: the VSC power peak generated by the model-based VI method was limited to approximately ±0.05 pu, whilst the peak power produced by the external VI method reached a level of ±0.063 pu. Therefore, the model-based VI can save part of the VSC VA-power for other necessary demands. The event-based VI method does not compare favourably in this respect. As the VI in this method is activated at speed levels below 0.995 pu and above 1.005 pu, it generated a comparatively high VI power, exceeding 0.1 pu to recover the speed to its set point. Furthermore, this method did not contribute to any reduction in the RoCoF within the speed band from 0.995 to 1.005 pu.

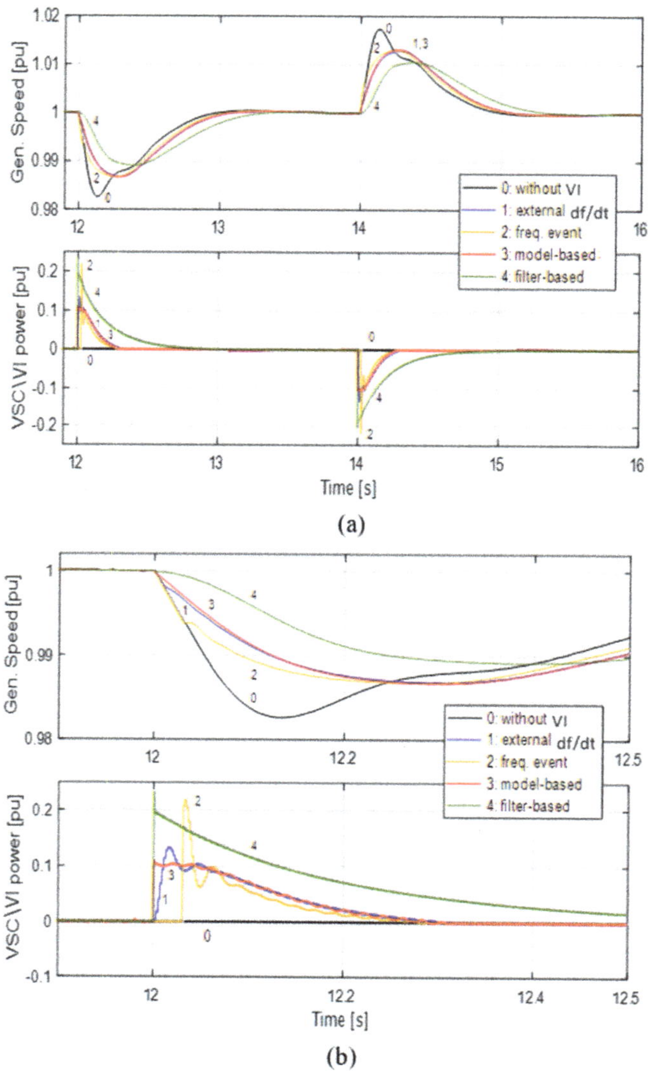

Figure 11. (**a**) Speed responses and their VSC power obtained for all VI methods (filter method at $T_f = 0.2s$). (**b**) Expanded sections of graphs in (**a**).

The filter-based VI method (at $T_f = 0.2s$) had a superior speed/frequency response compared to the other techniques, and even better results were obtained at higher time constants. This is clearly demonstrated through the simulation results obtained using different values of T_f, as shown in Figure 12a. However, VI provision using this method required more drawn energy from the BESS for better speed-response performance at higher T_f, which may be considered a disadvantage of the method unless a filter with a relatively low time constant is utilised. As such, the performance of the filter-based VI will be compromised in comparison with the model-based VI method. Figure 12b shows an expanded section of the graphs of Figure 12a. Table 2 details the energy provided through the VSC during the period from 12 s to 14 s when implementing each of the VI methods for comparison. In terms of performance with minimal VSC power and energy, the proposed model-based VI method seems to be a balanced choice for supporting the frequency of the microgrid.

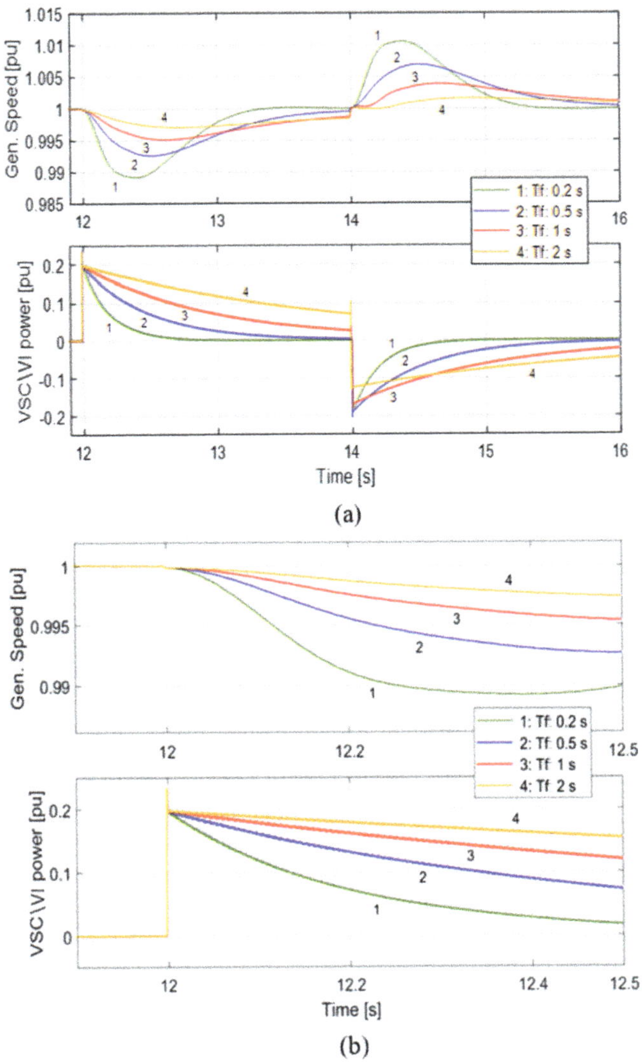

Figure 12. (**a**) Speed responses and their VSC power when using the filter-based VI method at various values of time-constant T_f. (**b**) Expanded sections of graphs in (**a**).

Table 2. The energies in [MJ] provided by the VSC during the period from 12 s to 14 s for all the virtual VI methods.

T_f [s]	Filter-Based	Model-Based	External *df/dt*	Frequency Event
0.2	0.01953			
0.5	0.04813	0.007	0.007	0.005
1.0	0.08508			
2.0	0.12477			

8. Conclusions

This paper has proposed two virtual inertia (VI) methods to support frequency stabilisation for power systems through an AC/DC microgrid utilising a VSC and a battery energy storage system. Firstly, a new non-linear electromechanical model for connecting the VSC to the microgrid was derived and developed to its linearised form for analysing and designing a model-based VI method. A switching function to apply the generated VI power using the VSC, only when the generator's speed is diverging away from its nominal speed, was introduced and utilised throughout the testing procedures reported in this paper. This function was adopted to prevent the VI power from competing with the action of the DEG governors, which are the main speed/frequency regulator.

An extended simulation of the microgrid and its power system was carried out, where all the proposed VI methods and functions were embedded within the VSC control system for testing and assessment. The performance of each proposed method was compared against other existing VI techniques, namely the standard df/dt and the frequency-event methods. It was demonstrated through the design and the obtained results that, whilst the proposed filter-based method is easier to realise, with a wider degree of freedom to select the filter time constant for higher performance, the proposed model-based VI method can provide a sufficient performance for speed response, with minimal VI power required by the VSC.

Author Contributions: Conceptualisation, F.H.; methodology, F.H.; software, F.H.; validation, F.H.; R.N.; P.T.; formal analysis, F.H.; investigation, F.H.; resources, F.H. and R.N.; data curation, F.H.; writing—original draft preparation, F.H.; writing—review and editing, F.H. and R.N.; visualisation, F.H.; funding acquisition, F.H. All authors have read and agreed to the published version of the manuscript.

Funding: The first author was sponsored by Newcastle University when he was performing the research and studies reported in this article.

Data Availability Statement: No new data were created or analyzed in this study. Data sharing is not applicable to this article.

Acknowledgments: The research reported in this paper was supported in part by Newcastle University and the University of Birmingham in the UK.

Conflicts of Interest: The authors declare no conflicts of interest.

Nomenclature

BESS	Battery energy storage system
DEG	Diesel engine generator
DER	Distributed energy resource
ESS	Energy storage system
IGBT	Integrated gate bipolar transistor
LPF	Low pass filter
NPC	Neutral point clamping

OP	Operating point
PCC	Point of common coupling
PE	Power electronic
PV	Photovoltaic
PWM	Pulse width modulation
RES	Renewable energy source
SI	Synthetic inertia
VI	Virtual inertia
VSC	Voltage source converter
H	Generator inertia constant [s]
J	DEG moment of inertia [kg.m^2]
L	Connection circuit inductance [Henry]
L_{eq}	Equivalent VSC connection inductance [Henry]
N_p	DEG pole pair number
P	VSC active power [W]
P_a	Generator accelerating power [W]
P_e	Load electrical power [W]
P_{ls}	Power losses of the generators [W]
P_m	Mechanical power of the generator engines [W]
Q	VSC reactive power [VAr]
R	Connection circuit resistance [Ohm]
R_{eq}	Equivalent VSC connection resistance [Ohm]
$\vec{S}$	Apparent total power vector of the VSCs [VA]
S_m	DEG rated power [VA]
T_f	Filter time constant [s]
V_n	Per-phase voltage magnitude at the PCC [V]
v_d, v_q	d- and q-axis voltage components at the PCC [V]
τ_{eq}	Equivalent time constant of the connected VSCs [s]
ω_e	Microgrid/system angular frequency [rad/s]
ω_m	Speed of the DEG [rad/s]

Appendix A

The per-unit (PU) base system

Base diesel genset power: 1 MVA
Base VSC power: 500 kVA
Base DC-link voltage V_{dc}: 1 kV
Base network phase voltage: 240 V
Base network phase current: 1380 A
Base genset speed: 50·π rad/s

Voltage Source Converter (VSC)

Type: three-phase, three-level, neutral point clamping
DC voltage bus: 1000 V
Nominal line voltage: 415 V
Total nominal power: 0.5 MVA
Switching frequency F_c: 2250 Hz
Equivalent connection inductance L_{eq} : $0.001H$
Equivalent connection resistance R_{eq} : 0.005Ω

Diesel genset unit

Nominal line voltage: 415 V
Speed: 1500 rpm
Nominal frequency: 50 Hz

Nominal power: 0.5 MVA
Generator Inertia Constant H: 0.247 s
Mechanical moment of inertia J: 10 kg/m^2
Considered average power losses P_{ls}: 10 kW.
Generator constant K_H : 4.05×10^{-6}
Speed governor control functions and parameters:
$H_c(s) = \frac{1+0.2s}{1+0.01s+0.0002s^2}$,
$K = 40$,
$TF_1(s) = \frac{1+0.25s}{1+0.009s}$,
$TF_2(s) = \frac{1}{1+0.0384s}$,
$T_d = 0.024$.

Appendix B

Calculation for RLC components of network single-tuned shunt-filter

Tuning frequency: $f_n = \frac{1}{2\pi}\sqrt{\frac{1}{LC}}$,
Filter quality factor: $Q = 2\pi\, f_n \frac{L}{R}$,
Filter frequency bandwidth: $B = \frac{f_n}{Q}$,
Generated reactive power at fundamental (grid) frequency f_1: $Q_c = 2\pi\, f_1 C\, V^2 \frac{n^2}{(n^2-1)}$ where $n = f_n/f_1$ and V is the nominal applied voltage to the filter terminals.
Filter designed at f_n equivalent to four times of the converter switching frequency F_c:
Given per-phase $V = 240$ V (Y connected three-phase filter), $f_1 = 50$ Hz, $n = 180$, $Q = 50$, and considering $C = 1$ μF, gives: $L = 0.3127$ mH, $R = 0.354$ Ω; reduced impedance at a frequency equivalent to four times the converter switching frequency due to parallel connection of four interleaved converters.

References

1. Hatziargyriou, N.; Asano, H.; Iravani, R.; Marnay, C. Microgrids. *IEEE Power Energy Mag.* **2007**, *5*, 78–94. [CrossRef]
2. Hirsch, A.; Parag, Y.; Guerrero, J. Microgrids: A review of technologies, key drivers, and outstanding issues. *Renew. Sustain. Energy Rev.* **2018**, *90*, 402–411. [CrossRef]
3. AbdelHady, R. Modeling and simulation of a micro grid-connected solar PV system. *Water Sci.* **2017**, *31*, 1–10. [CrossRef]
4. Yukita, K. AC/DC microgrids. In *Integration of Distributed Energy Resources in Power Systems: Implementation, Operation and Control*; Funabashi, T., Ed.; Academic Press: Cambridge, MA, USA, 2016; pp. 236–260.
5. Jiang, Q.; Xue, M.; Geng, G. Energy Management of Microgrid in Grid-Connected and Stand-Alone Modes. *IEEE Trans Power Syst.* **2013**, *28*, 3380–3389. [CrossRef]
6. Ortiz, L.; Orizondo, R.; Águila, A.; González, J.W.; López, G.J.; Isaac, I. Hybrid AC/DC microgrid test system simulation: Grid-connected mode. *Heliyon* **2019**, *5*, e02862. [CrossRef] [PubMed]
7. Navarro-Rodríguez, Á.; García, P.; Gómez-Aleixandre, C.; Blanco, C. Cooperative Primary Control of a Hybrid AC/DC Microgrid Based on AC/DC Virtual Generators. *IEEE Trans. Energy Convers.* **2022**, *37*, 2837–2850. [CrossRef]
8. Hu, J.; Shan, Y.; Xu, Y.; Guerrero, J.M. A coordinated control of hybrid ac/dc microgrids with PV-wind-Battery under variable generation and load conditions. *Int. J. Electr. Power Energy Syst.* **2019**, *104*, 583–592. [CrossRef]
9. Mahmoud, M. (Ed.) *MICROGRID—Advanced Control Methods and Renewable Energy System Integration*, 1st ed.; Butterworth-Heinemann: Oxford, UK, 2016; ISBN 9780081017531.
10. Chen, Y.; Zhao, S.; Li, Z.; Wei, X.; Kang, Y. Modeling and Control of the Isolated DC–DC Modular Multilevel Converter for Electric Ship Medium Voltage Direct Current Power System. *IEEE J. Emerg. Sel. Top. Power Electron.* **2017**, *5*, 124–139. [CrossRef]
11. Eajal, A.A.; Shaaban, M.F.; Ponnambalam, K.; El-Saadany, E.F. Stochastic Centralized Dispatch Scheme for AC/DC Hybrid Smart Distribution Systems. *IEEE Trans. Sustain. Energy* **2016**, *7*, 1046–1059. [CrossRef]
12. Khatibi, M.; Ahmed, S. Impact of Distributed Energy Resources on Frequency Regulation of the Bulk Power System. In Proceedings of the IEEE Conference on Power Electronics and Renewable Energy (CPERE), Aswan, Egypt, 23–25 October 2019. [CrossRef]
13. Bose, U.; Chattopadhyay, S.K.; Chakraborty, C.; Pal, B. A Novel Method of Frequency Regulation in Microgrid. *IEEE Trans. Ind. Appl.* **2019**, *55*, 111–121. [CrossRef]

14. Zhang, Y.; Sun, Q.; Zhou, J.; Li, L.; Wang, P.; Guerrero, J.M. Coordinated Control of Networked AC/DC Microgrids With Adaptive Virtual Inertia and Governor-Gain for Stability Enhancement. *IEEE Trans. Energy Convers.* **2021**, *36*, 95–110. [CrossRef]
15. Rapizza, M.R.; Canevese, S.M. Fast frequency regulation and synthetic inertia in a power system with high penetration of renewable energy sources: Optimal design of the required quantities. *Sustain. Energy Grids Netw.* **2020**, *24*, 100407. [CrossRef]
16. Skiparev, V.; Machlev, R.; Chowdhury, N.R.; Levron, Y.; Petlenkov, E.; Belikov, J. Virtual Inertia Control Methods in Islanded Microgrids. *Energies* **2021**, *14*, 1562. [CrossRef]
17. Mandal, R.; Chatterjee, K. Virtual inertia emulation and RoCoF control of a microgrid with high renewable power penetration. *Electr. Power Syst. Res.* **2021**, *194*, 107093. [CrossRef]
18. Kerdphol, T.; Rahman, F.S.; Watanabe, M.; Mitani, Y. Robust Virtual Inertia Control of a Low Inertia Microgrid Considering Frequency Measurement Effects. *IEEE Access* **2019**, *7*, 57550–57560. [CrossRef]
19. Ali, H.; Magdy, G.; Li, B.; Shabib, G.; Elbaset, A.A.; Xu, D.; Mitani, Y. A New Frequency Control Strategy in an Islanded Microgrid Using Virtual Inertia Control-Based Coefficient Diagram Method. *IEEE Access* **2019**, *7*, 16979–16990. [CrossRef]
20. Fini, M.H.; Hamedani Golshan, M.E. Determining optimal virtual inertia and frequency control parameters to preserve the frequency stability in islanded microgrids with high penetration of renewables. *Electr. Power Syst. Res.* **2018**, *154*, 13–22. [CrossRef]
21. Farmer, W.J.; Rix, A.J. Optimising Power System Frequency Stability Using Virtual Inertia from Inverter-based Renewable Energy Generation. *IET Renew. Power Gener.* **2020**, *14*, 2820–2829. [CrossRef]
22. Michael, N.E.; Hasan, S.; Mishra, S. Virtual inertia provision through data centre and electric vehicle for ancillary services support in microgrid. *IET Renew. Power Gener.* **2020**, *14*, 3792–3801. [CrossRef]
23. Marchgraber, J.; Alács, C.; Guo, Y.; Gawlik, W.; Anta, A.; Stimmer, A.; Lenz, M.; Froschauer, M.; Leonhardt, M. Comparison of Control Strategies to Realize Synthetic Inertia in Converters. *Energies* **2020**, *13*, 3491. [CrossRef]
24. Zhang, Y.; Melin, A.M.; Djouadi, S.M.; Olama, M.M.; Tomsovic, K. Provision for Guaranteed Inertial Response in Diesel-Wind Systems via Model Reference Control. *IEEE Trans. Power Syst.* **2018**, *33*, 6557–6568. [CrossRef]
25. Lerch, K.S.; Kerdphol, T.; Mitani, Y.; Turschner, D. Frequency Stability Assessment on Virtual Inertia Control Strategy in Connected and Islanded Multi-Area Power Systems. In Proceedings of the IEEE International Conference on Environment and Electrical Engineering and IEEE Industrial and Commercial Power Systems Europe (EEEIC/I&CPS Europe), Madrid, Spain, 9–12 June 2020. [CrossRef]
26. Kerdphol, T.; Rahman, F.; Mitani, Y.; Hongesombut, K.; Küfeoğlu, S. Virtual Inertia Control-Based Model Predictive Control for Microgrid Frequency Stabilization Considering High Renewable Energy Integration. *Sustainability* **2017**, *9*, 773. [CrossRef]
27. Wang, Y.; Wang, C.; Xu, L.; Meng, J.; Hei, Y. Adjustable Inertial Response from the Converter With Adaptive Droop Control in DC Grids. *IEEE Trans. Smart Grid* **2019**, *10*, 3198–3209. [CrossRef]
28. Samanta, S.; Prakash Mishra, J.; Krishna Roy, B. Implementation of a virtual inertia control for inertia enhancement of a DC microgrid under both grid connected and isolated operation. *Comput. Electr. Eng.* **2019**, *76*, 283–298. [CrossRef]
29. Brisebois, J.; Aubut, N. Wind Farm Inertia Emulation to Fulfil Hydro-Quebec's Specific Need. In Proceedings of the 2011 IEEE Power and Energy Society General Meeting, Detroit, MI, USA, 24–28 July 2011. [CrossRef]
30. Hardan, F.; Norman, R. Balancing loads of rotating generators utilizing VSC direct power controllers in a ship AC/DC smartgrid. *Electr. Power Syst. Res.* **2020**, *182*, 106200. [CrossRef]
31. Duckwitz, D.; Fischer, B. Modeling and Design of df/dt-Based Inertia Control for Power Converters. *IEEE J. Emerg. Sel. Top. Power Electron.* **2017**, *5*, 1553–1564. [CrossRef]
32. Liu, H.; Chen, Z. Contribution of VSC-HVDC to frequency regulation of power systems with offshore wind generation. *IEEE Trans. Energy Convers.* **2015**, *30*, 918–926. [CrossRef]
33. Gloe, A.; Jauch, C.; Cracium, B.; Winkelmann, J. Continuous provision of synthetic inertia with wind turbines: Implications for the wind turbine and for the grid. *IET Renew. Power Gener.* **2019**, *13*, 668–675. [CrossRef]
34. Rodriguez, J.; Lai, J.S.; Peng, F.Z. Multilevel inverters: A survey of topologies, controls, and applications. *IEEE Trans. Ind. Electron.* **2002**, *49*, 724–738. [CrossRef]
35. Bloemink, J.M.; Green, T.C. Reducing passive filter sizes with tuned traps for distribution level power electronics. In Proceedings of the 14th European Conference on Power Electronics and Applications, Birmingham, UK, 30 August–1 September 2011.
36. Rashid, M.H. *Power Electronics: Devices, Circuits, and Applications, International Edition*, 4th ed.; Pearson Education Limited: Essex, UK, 2014.
37. Glover, J.D.; Sarma, M.S.; Overbye, T.J. Transient Stability. In *Power System Analysis and Design*, 5th ed.; Stamford, C.T., Ed.; Cengage Learning: Boston, MA, USA, 2012.
38. Hardan, F.; Bleijs, J.A.M.; Jones, R.; Bromley, P. Bi-directional power control for flywheel energy storage system with vector-controlled induction machine drive. In Proceedings of the Seventh International Conference on Power Electronics and Variable Speed Drives (IEE Conf. Publ. No. 456), London, UK, 21–23 September 1998. [CrossRef]
39. Jin, Z.; Meng, L.; Guerrero, J.M.; Han, R. Hierarchical Control Design for a Shipboard Power System with DC Distribution and Energy Storage Aboard Future More-Electric Ships. *IEEE Trans. Ind. Inform.* **2018**, *14*, 703–714. [CrossRef]

40. Xiao, Z.; Li, H.; Fang, H.; Guan, Y.; Liu, T.; Hou, L.; Guerrero, J.M. Operation Control for Improving Energy Efficiency of Shipboard Microgrid Including Bow Thrusters and Hybrid Energy Storages. *IEEE Trans. Transp. Electrif.* **2020**, *6*, 856–868. [CrossRef]
41. Song, Q.; Grigoriadis, K.M. Diesel engine speed regulation using linear parameter varying control. In Proceedings of the American Control Conference, Denver, CO, USA, 4–6 June 2003. [CrossRef]
42. *421.5-2016;* IEEE Recommended Practice for Excitation System Models for Power System Stability Studies. (Revision of IEEE 521.5-1992). IEEE Standard: Piscataway, NJ, USA, 2005; Volume 421.

electronics

MDPI

Article

Adaptive Virtual Inertial Control and Virtual Droop Control Coordinated Control Strategy for Hybrid Energy Storage Taking into Account State of Charge Optimization

Chao Xing [1], Jiajie Xiao [2,*], Peiqiang Li [2], Xinze Xi [1], Yunhe Chen [2] and Qi Guo [2]

1 Electric Power Research Institute of Yunnan Power Grid Co., Ltd., Kunming 650217, China; xingchao@yn.csg.cn (C.X.)
2 College of Electrical and Information Engineering, Hunan University, Changsha 410082, China
* Correspondence: xiaojiajie33@hnu.edu.cn

Abstract: For energy-storage-assisting conventional units to participate in the primary frequency regulation of a power system, firstly, based on the frequency regulation mechanism of virtual inertial control (VIC) and virtual droop control (VDC) of energy storage, we analyze the effect of the action timing of energy storage on the frequency deviation of the grid under two control methods and put forward a reasonable combination of the two control methods; on this basis, we also put forward hybrid energy storage adaptive VIC and VDC based on the demand of VIC and VDC on the power and capacity of energy storage. On this basis, based on the demand of VIC and VDC on the power and capacity of energy storage, a hybrid energy storage adaptive VIC and VDC coordinated control strategy based on supercapacitor–lithium batteries is proposed, whereby a high-power storage supercapacitor responds to inertial control signals to rapidly suppress a drop in frequency, and the high-capacity lithium battery responds to droop control signals to perform long-time droop control. The high-capacity lithium battery responds to the sagging control signal and is used to perform a long-time sagging power response; finally, in order to avoid the state of charge (SOC) of energy storage falling into a low/high working condition and losing the subsequent frequency regulation ability, an adaptive power control strategy of energy storage based on the improved logistic function is proposed. The simulation results show that under typical load disturbance, the SOC level of the proposed strategy increases by 19.17% and 30.16%, respectively, compared with that of the single-lithium strategy and no energy storage, and the SOC level of the supercapacitor and lithium battery increases by 29.4% and 2.1%, respectively, compared with that of logistic optimization.

Keywords: hybrid energy storage; SOC optimization; primary frequency regulation; coordinated control

1. Introduction

The growing demand for energy has aggravated environmental pollution caused by the burning of fossil fuel resources, and the development of renewable and low-carbon energy sources such as wind and hydrogen has become an inevitable trend, and our country has also put forward the "3060 goal" and incorporated it into the "14th Five-Year Plan". However, the mismatch between source and load caused by the continuous increase in the proportion of wind and solar installations has also brought about serious challenges to the frequency adjustment of power systems. At the same time, in recent years, the rapid development of an energy storage system as a new type of power device that can be charged and discharged, with fast power response characteristics, has become an indispensable part of the security and stability of today's power systems [1,2].

Energy storage can be widely used in various aspects of a power system due to its power flexibility [3,4], which can participate in power system peak and frequency regulation but also smooth new energy power fluctuations [5,6] to ensure the security and stability of large-scale wind and other new energy grid-connected power systems. Among them, lithium-battery-based electrochemical energy storage, due to its mature technology and high degree of scale, has been widely adopted by most experts and scholars [7]. In the field of power system frequency regulation, due to the power system frequency regulation time scale, the response time of energy storage equipment is more in line with the use of energy storage to assist the conventional unit involved in power system frequency regulation, a topic which has become the focus of many scholars. The authors of [8] constructed a mechanism model for energy storage to participate in wind power leveling, laying a theoretical foundation for energy-storage-assisted power system frequency regulation. The authors of [9] used energy storage to assist wind turbines to jointly participate in grid frequency regulation, so that the wind turbine has a similar traditional unit of primary frequency regulation ability, and they constructed a storage economy to reach the goal of establishing an optimization model so as to find the optimal capacity configuration of energy storage. The authors of [10] analyzed the characteristics of SOC that may fall into poor working conditions after energy storage is involved in one-time FR with VDC, and they proposed an energy storage SOC reconfiguration control strategy after one-time FR which can significantly improve the service life of energy storage. The authors of [11] analyzed the characteristics of wind power output and proposed an optimized VDC control strategy for energy storage participation in wind-power-containing grids, which can effectively suppress wind power fluctuations. The authors of [12], with VIC, simulated the inertial response characteristics of conventional units and energy storage because of its fast response capability to obtain a better frequency regulation effect. The authors of [13] produced a mechanism model between VIC and grid frequency for energy storage using VIC when parameter tuning and control strategy selection are used to lay a theoretical foundation. In order to overcome the problem of the frequent penetration of distributed energy in a microgrid environment, the authors of [14] proposed the concept of a virtual power plant and proposed an adaptive VDC control strategy with a remarkable frequency modulation effect.

Most of the existing strategies only consider a virtual control strategy or simply the superposition of VIC and VDC [15–17], which ignores the intrinsic frequency regulation mechanism of the two. The authors of [18], based on the energy storage and frequency regulation outputs of VIC and VDC, conclude that the simultaneous use of VIC and VDC before frequency deviation reaches an extreme value is beneficial to inhibit the rapid fall of frequency, while after frequency deviation reaches an extreme value, the energy storage outputs of VIC and VDC present mutual counteracting characteristics to weaken the effectiveness of part of the frequency regulation, meaning that adopting VIC and VDC before the frequency deviation reaches the extreme value is proposed, considering only the VDC energy storage output strategy after it reaches an extreme value.

However, virtual inertia control focuses on the support of system inertia, which is proportional to the rate of change in the frequency deviation; virtual sag control focuses on the frequency deviation itself, which is proportional to the size of the value of the frequency deviation, and the use of a type of energy storage to participate in FR often does not fully match the FR mechanism of the two controls. Therefore, on the basis of clarifying the action timing of energy storage, it is necessary to choose the appropriate type of energy storage to achieve the best frequency regulation effect based on the frequency regulation demand for energy storage of the two.

Therefore, this paper proposes a hybrid energy storage adaptive VIC and VDC coordinated control strategy taking into account SOC optimization for energy-storage-assisting conventional units to participate in the primary frequency regulation of the power system. Firstly, based on the energy storage output mechanism of VIC and VDC, we analyze the influence of the action timing of the two control methods on the grid frequency deviation and put forward a control method that reasonably combines the two; secondly, based on the virtual inertia and the sag control demand for energy storage power and capacity, we put forward a hybrid energy storage adaptive VIC and VDC coordination control strategy based on supercapacitor–lithium batteries and put forward a hybrid energy storage adaptive VIC and VDC coordination control strategy based on super capacitors and lithium batteries, in which the power-type power storage supercapacitor, which has high power, responds to the inertia control signal and quickly suppresses the energy storage capacity. The energy storage supercapacitor responds to the inertia control signal to quickly suppress the frequency drop, while the high-capacity lithium battery responds to the sag control signal and performs a long-duration sag power response; finally, in order to avoid the energy storage SOC from falling into the lower/higher severe operating conditions and losing the subsequent frequency regulation capability, the energy storage adaptive power control strategy based on the improved logistic function is proposed. The simulation verifies the effectiveness of the proposed strategy.

2. This Paper's Hybrid Energy Storage FR Framework

Figure 1 shows the framework of energy storage frequency regulation under the control strategy of this paper. First, when the system generates frequency deviation Δf due to power shortage, calculate its frequency deviation change rate $\mathrm{d}(\Delta f)/\mathrm{d}t$ and input the two into the energy storage artificial setting dead zone module; then, when Δf or $\mathrm{d}(\Delta f)/\mathrm{d}t$ exceeds the deadband range, judge the positivity and negativity of the product of Δf and $\mathrm{d}(\Delta f)/\mathrm{d}t$, if $\Delta f \cdot \mathrm{d}(\Delta f)/\mathrm{d}t > 0$, then adopt the control strategy of superimposing the VIC and the VDC at the same time; if $\Delta f \cdot \mathrm{d}(\Delta f)/\mathrm{d}t < 0$, then adopt the control strategy of the VDC only. In this regard, the supercapacitor with higher power density is used to assume the FR power command of the VIC, while the lithium battery with higher energy density is used to assume the FR power command of the VDC. The details will be described and analyzed in detail in Sections 2 and 3 of this paper.

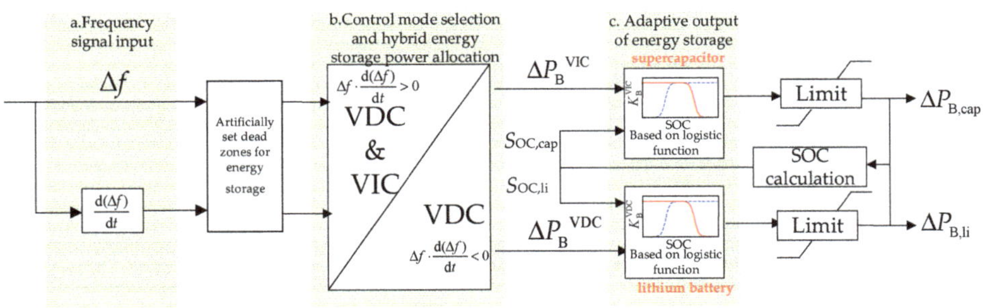

Figure 1. The current energy storage frequency regulation framework.

3. Primary FR model of a Regional Power System with Energy Storage

3.1. Primary FR Model for Conventional Thermal Power Units

In this paper, a thermal power unit with larger capacity is taken as a conventional FR unit, the governor and reheat turbine are mainly considered in the FR process, and the corresponding FR transfer function can be expressed as G_{gen} [19].

$$G_{\mathrm{gen}}(s) = \frac{1 + sF_{\mathrm{HP}}T_{\mathrm{RH}}}{(1 + sT_{\mathrm{CH}}) \cdot (1 + sT_{\mathrm{RH}})} \cdot \frac{1}{1 + sT_{\mathrm{g}}} \tag{1}$$

3.2. Generalized Control Model for Primary FR of Energy Storage

Compared with conventional FR units, in order to reflect the fast response characteristics of energy storage, the first-order inertia link is generally used to equate its output characteristics. It has been pointed out in [20] that appropriate simplification of the energy storage model can increase the simulation speed while ensuring the simulation accuracy and its transfer function $B(s)$ can be expressed as

$$B(s) = \frac{1}{1 + sT_{\mathrm{B},i}} \tag{2}$$

In addition, in order to fully characterize the remaining FR capacity of the energy storage to reflect the advantages and disadvantages of its own state, this paper further introduces the state of charge (SOC) of the energy storage to evaluate it as

$$S_{\mathrm{OC}}(t) = S_{\mathrm{OC}}(t - \Delta t) + \frac{\int_{t-\Delta t}^{t} \Delta P_{\mathrm{B}}(t)\mathrm{d}t}{E_{\mathrm{BN}}} \tag{3}$$

In summary, this paper combines Formulas (1) and (2) and takes into account the power and SOC constraints to construct a generalized control model for energy storage participating in primary FR as shown in Figure 2, where $E_0(s)$ is the primary FR signal accepted by the energy storage and S_{OC0} is the initial SOC of the energy storage.

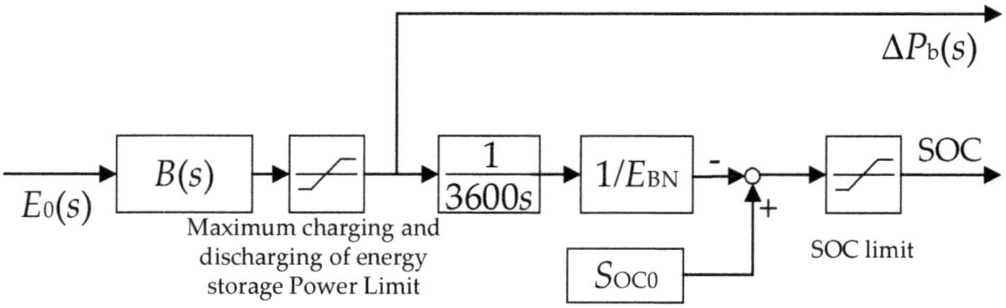

Figure 2. General control model of energy storage and primary frequency regulation.

3.3. *Dynamic Modeling of Primary Frequency Regulation of Regional Power Systems with Energy Storage*

Currently, the common control methods for energy storage to participate in primary FR include VIC and VDC. When the system has a load disturbance $\Delta P_{\mathrm{L}}(s)$, the system frequency will change abruptly and, at this time, the dynamic model of the primary FR of the regional power system containing energy storage can be expressed as [21]

$$\Delta P_{\mathrm{G}}(s) = -\Delta f(s) \cdot K_{\mathrm{gen}} \cdot G_{\mathrm{gen}}(s) \tag{4}$$

$$\Delta P_{\mathrm{G}}(s) + \Delta P_{\mathrm{B}}(s) - \Delta P_{\mathrm{L}}(s) = (2H \cdot s + D) \cdot \Delta f(s) \tag{5}$$

the output power and the corresponding frequency deviation are as shown in Formulas (6) and (7), respectively, when the storage battery adopts VIC and VDC.

$$\begin{cases} \Delta P_{\mathrm{B}}{}^{\mathrm{VIC}}(s) = -sK_{\mathrm{B}}^{\mathrm{VIC}} B(s)\Delta f(s) \\ \Delta f^{\mathrm{VIR}}(s) = \frac{-\Delta P_{\mathrm{L}}(s)}{2Hs+D+K_{\mathrm{gen}}G_{\mathrm{gen}}(s)+K_{\mathrm{B}}^{\mathrm{VIV}}B(s)} \end{cases} \tag{6}$$

$$\begin{cases} \Delta P_{\mathrm{B}}{}^{\mathrm{VDC}}(s) = -K_{\mathrm{B}}^{\mathrm{VDC}} B(s)\Delta f(s) \\ \Delta f^{\mathrm{VDC}}(s) = \frac{-\Delta P_{\mathrm{L}}(s)}{2Hs+D+K_{\mathrm{gen}}G_{\mathrm{gen}}(s)+K_{\mathrm{B}}^{\mathrm{VDC}}B(s)} \end{cases} \tag{7}$$

Combining the frequency deviation in Formulas (6) and (7), the literature [18] has analyzed the initial frequency rate of change $\Delta f_{\mathrm{ov}}^{\mathrm{VIC}}$ and the steady state frequency deviation of the system $\Delta f_{\mathrm{ss}}^{\mathrm{VIC}}$ under the two control modes as follows:

$$\begin{cases} \Delta f_{\mathrm{ov}}^{\mathrm{VIC}} = \lim\limits_{s\to\infty} s \cdot [s \cdot \Delta f^{\mathrm{VIC}}(s)] = \frac{-\Delta P_{\mathrm{L}}}{2H+K_{\mathrm{B}}^{\mathrm{VIC}}} \\ \Delta f_{\mathrm{ss}}^{\mathrm{VIC}} = \lim\limits_{s\to 0}[s \cdot \Delta f^{\mathrm{VIC}}(s)] = \frac{-\Delta P_{\mathrm{L}}}{D+K_{\mathrm{gen}}} \end{cases} \tag{8}$$

$$\begin{cases} \Delta f_{\mathrm{ov}}^{\mathrm{VDC}} = \lim\limits_{s\to\infty} s \cdot [s \cdot \Delta f^{\mathrm{VDC}}(s)] = \frac{-\Delta P_{\mathrm{L}}}{2H} \\ \Delta f_{\mathrm{ss}}^{\mathrm{VDC}} = \lim\limits_{s\to 0}[s \cdot \Delta f^{\mathrm{VDC}}(s)] = \frac{-\Delta P_{\mathrm{L}}}{D+K_{\mathrm{gen}}+K_{\mathrm{B}}^{\mathrm{VDC}}} \end{cases} \tag{9}$$

From Formulas (8) and (9), it can be seen that when VIC is used, the initial rate of change of frequency depends on the inertia coefficient H of the thermal unit itself and the virtual inertia coefficient K_B^{VIC} of the energy storage under the Δf_{ov}^{VIC} system load perturbation; the steady state frequency deviation is irrelevant with K_B^{VIC}. It can be seen that VIC can quickly suppress the initial rate of change of frequency but cannot improve the system steady-state frequency deviation. When VDC is used, Δf_{ov}^{VDC} depends only on the thermal unit's own H; Δf_{ss}^{VDC} depends on both K_{gen} and K_B^{VDC}. It can be seen that VDC can improve the steady-state frequency deviation but does not work on the initial frequency deviation rate of change.

In addition, the literature [22] also utilizes the sensitivity analysis to further conclude that the use of VDC for energy storage can effectively improve the FR effect throughout the FR cycle, while the use of VIC can only play a role in recovering the frequency deviation before the frequency reaches the lowest point but will inhibit the recovery of the frequency after the frequency deviation reaches the maximum value. Therefore, it is necessary to combine the FR characteristics of different types of energy storage and select suitable FR control methods for them to realize the control effects of Δf_{ov} and Δf_{ss} at the same time.

4. Primary FR Control Strategy for Supercapacitor–Lithium Battery Counting and SOC Optimization

4.1. Adaptive Coordinated Control of VIC and VDC Based on Supercapacitor–Lithium Battery

From the analysis in Section 3.3, it can be seen that the energy storage involved in VIC can increase the system inertia, the energy storage output is proportional to the rate of change of frequency when a sudden change in the system frequency needs to respond quickly to inhibit the change of frequency, and at the same time there is the disadvantage of inhibiting the recovery of the frequency after the frequency deviation reaches the extreme value. Energy storage involved in the VDC when the output and the frequency deviation is proportional to the frequency in the entire FR cycle can be effective in inhibiting the deterioration of the frequency, however, VDC requires energy storage to produce power for a long period of time, which requires energy storage to be configured with a higher FR capacity. It can be seen that VIC and VDC should be used before the frequency reaches the extreme value, and only VDC should be used after the extreme value point to ensure the effect of frequency regulation. However, the literature [18–21] suggests that a single energy storage to implement the above strategies at the same time cannot reflect the characteristics of the energy storage involved in the VIC and VDC at the same time, which ignores that the characteristics of the VIC and VDC for the energy storage power and capacity requirements are different, therefore, in order to give full play to the advantages of the power-type and energy-type energy storage in the two modes of FR control, this paper further puts forward a supercapacitor–lithium-battery-based adaptive VIC and VDC control. An adaptive coordinated control strategy of VIC and VDC is further proposed in this paper.

At present, the most commonly used and the most mature energy storage is based on lithium batteries as the main energy-based energy storage, which is characterized by high energy density and low power density. Supercapacitor-based power-type energy storage is less used in the field of FR, which is characterized by high power density and small energy density, which is due to the high cost and small capacity of supercapacitors, which are not suitable for today's drastic load disturbances in systems. Table 1 demonstrates the performance comparison between the typical power-type and energy-type energy storage [14]. It can be seen that power-based energy storage is suitable for VIC due to its high power density and fast response to frequency-changing signals, while energy-based energy storage is suitable for VDC due to its high energy density, which can leave a sufficient FR margin to perform FR tasks throughout the FR cycle.

Table 1. Comparison of power-type and energy-type energy storage performance.

Type of Energy Storage	Charging Time	Discharging Time	Power Density (W/kg)	Energy Density (Wh/kg)
Power Energy Storage-Supercapacitor	1~30 s	1~30 s	1000~2000	1~10
Energy Storage-Lithium Battery	1~5 h	0.3~3 h	50~200	20~100

In summary, in order to solve the limitations of a single control method on the frequency control effect and make full use of the complementary advantages of VIC and VDC, this paper firstly proposes an adaptive coordinated control strategy of VIC and VDC. When the system frequency changes abruptly (i.e., energy storage is needed for inertial support), since VDC can improve the frequency control effect throughout the frequency control cycle, the control strategy of superposition of VIC and VDC should be adopted at this time; when the frequency deviation reaches the maximum value, since VIC will inhibit the frequency recovery instead, only the VDC control strategy should be adopted at this time. When the frequency deviation reaches the maximum value, the VIC will inhibit the frequency recovery, and only the VDC control strategy should be adopted. Then, considering that power-based energy storage is applicable to VIC and energy-based energy storage is applicable to VDC, the output power of hybrid energy storage and its auxiliary thermal power units to participate in the primary FR of the regional grid are obtained in the dynamic model as shown in Formula (10) and Figure 3.

$$\begin{cases} \Delta P_{\mathrm{B,cap}}(s) = -sK_{\mathrm{B,cap}}^{\mathrm{VIC}} B_{\mathrm{cap}}(s)\Delta f(s)\ , \Delta f \cdot \frac{\mathrm{d}(\Delta f)}{\mathrm{d}t} > 0 \\ \Delta P_{\mathrm{B,li}}(s) = -K_{\mathrm{B,li}}^{\mathrm{VDC}} B_{\mathrm{li}}(s)\Delta f(s),\ \text{all time} \end{cases} \tag{10}$$

where $B_{\mathrm{cap}}(s)$ and $B_{\mathrm{li}}(s)$ are the FR transfer functions of the supercapacitor and lithium battery, respectively.

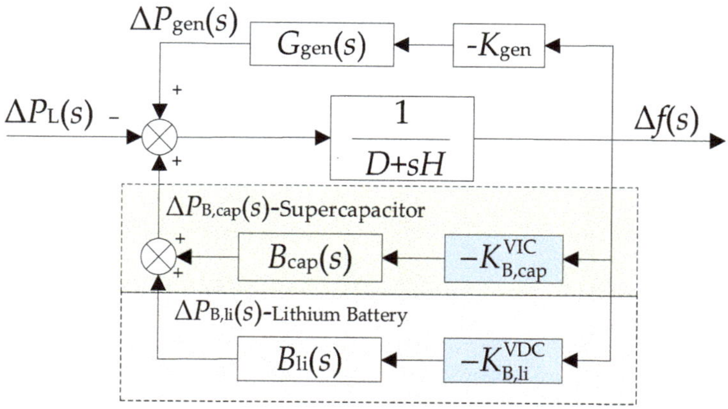

Figure 3. Hybrid energy storage auxiliary thermal power units participate in the primary frequency regulation model of the regional power grid.

From Formula (10) and Figure 3, the frequency deviation under the control strategy of this paper is given as

$$\Delta f(s) = -\Delta P_{\mathrm{L}}(s)/(2H \cdot s + D + K_{\mathrm{gen}} \cdot G_{\mathrm{gen}}(s) + s \cdot B_{\mathrm{cap}}(s) \cdot K_{\mathrm{B,cap}}^{\mathrm{VIC}} + B_{\mathrm{li}}(s) \cdot K_{\mathrm{B,li}}^{\mathrm{VDC}}) \tag{11}$$

Further, the initial frequency rate of change and the steady-state frequency deviation can be calculated as

$$\begin{cases} \Delta f_{ov} = \lim_{s \to \infty} s \cdot [s \cdot \Delta f^{VIC}(s)] = \frac{-\Delta P_L}{2H + K_{B,cap}^{VIC}} \\ \Delta f_{ss} = \lim_{s \to 0} [s \cdot \Delta f^{VDC}(s)] = \frac{-\Delta P_L}{D + K_{gen} + K_{B,li}^{VDC}} \end{cases} \quad (12)$$

4.2. Improved Generalized Logistic Function Design Method Based on Energy Storage Output Constraints

When the energy storage battery responds to the demand of system frequency regulation, if it has been operating with a given inertia/sag coefficient, it is very easy to saturate or exhaust the energy storage capacity due to the excessive frequency deviation in some time periods. In previous studies, many experts and scholars have improved the fixed inertia/sag coefficient control [22,23], and the more typical ones are the variable coefficient control [24], however, the previous control strategies focus on establishing a simple linear constraint between the storage inertia/sag coefficients K_B^{VIC} and K_B^{VDC} and the SOC. This alleviates the problem of overcharge and overdischarge of the energy storage battery to a certain extent, but at the same time, there are two main problems: one is weakening of the fast response characteristics of the energy storage battery, and the other is a secondary perturbation of the system frequency when its own SOC reaches the critical point.

Therefore, when introducing the variable coefficient control, it is necessary to consider that when the storage SOC is in a relatively good interval, it should be a priority to ensure the FR demand and participate in the FR control with a high inertia/sag coefficient; and when the SOC is in a relatively poor interval, the storage should be less charged and less discharged or even not charged and not discharged in order to ensure that the SOC is maintained in a reasonable range and to leave the FR margin for the subsequent cycle and reduce the storage life loss. This paper is based on a generalized logistic curve, which is often used as an S-shaped curve that mimics similar population growth, as shown in Formula (13). The curve initially grows slowly, with the fastest growth rate at the center point, and then gradually slows down, and the overall growth rate is characterized by slow–fast–slow. In this paper, the generalized logistic function is improved by combining the relationship between the energy storage output and SOC, so that the discharge/charge of the energy storage can be limited as much as possible when its SOC is small/large, i.e., its corresponding inertia/sag control coefficients are lowered, and the output of the energy storage can be unrestricted as much as possible when the SOC is in a better state.

$$P(t) = \frac{KP_0 e^{nt}}{K + P_0(e^{nt} - 1)} \quad (13)$$

where K is the maximum value; P_0 is the initial value, affecting the logistic function slope at the maximum (hereinafter referred to as the center point) of the location, P0. The smaller the function of the center of the curve from the S_{OC} = 0.5, the closer the curve, so this paper takes the P_0 to 0.01; n represents the curve change as fast or slow (i.e., the slope of the center point); the appropriate transformation of Formula (13) can be obtained with the curve shown in Figure 4, and its expression is as follows:

$$\begin{cases} K_{B,dis}^{VIC/VDC} = \frac{K_{max} \cdot P_0 \cdot e^{n(S_{OC} - 0.1)/0.4}}{K_{max} + P_0(e^{n(S_{OC} - 0.1)/0.4} - 1)} \\ K_{B,ch}^{VIC/VDC} = \frac{K_{max} \cdot P_0 \cdot e^{n(0.9 - S_{OC})/0.4}}{K_{max} + P_0(e^{n(0.9 - S_{OC})/0.4} - 1)} \end{cases} \quad (14)$$

In summary, the flowchart of the hybrid energy storage adaptive VIC and VDC coordinated control strategy proposed in this paper taking into account SOC optimization is shown in Figure 5.

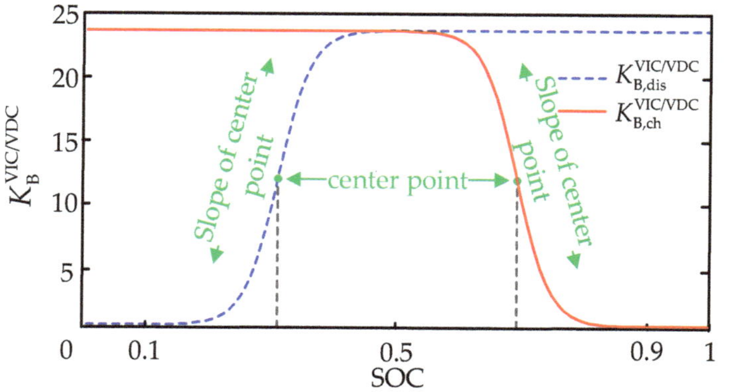

Figure 4. Relationship between the virtual inertia/droop control coefficient of energy storage and the SOC.

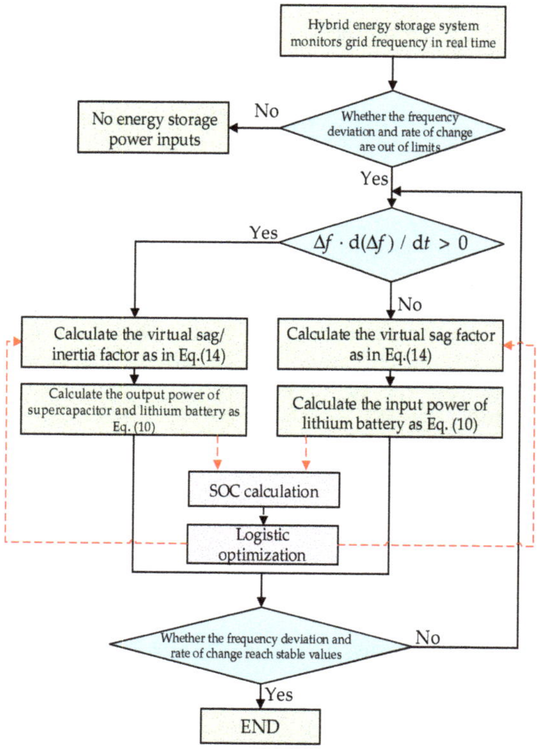

Figure 5. Strategy flowchart of this article.

In order to quantitatively analyze the effectiveness of the strategies proposed in this paper, this paper defines the evaluation indexes of maximum frequency deviation Δf_m, frequency decline rate V_m, and steady-state frequency deviation Δf_s to evaluate the effectiveness of FR and the SOC of energy storage under different strategies, as follows.

(1) Primary FR Effectiveness Indicators

In the typical step load disturbance, so that the initial moment is t_0, the initial frequency deviation is Δf_0. When the frequency deviation reaches the maximum value of the moment for the frequency deviation that at this time is set to t_m, it can be defined as the frequency rate of descent for the Δf_m; the frequency deviation begins to recover until it reaches the steady-state moment V_m when the steady-state frequency deviation is set to $V_m = (\Delta f_0 - \Delta f_m)/(t_m - t_0)$. Δf_0, t_0 are taken to be 0. Obviously, the smaller the value of Δf_m, V_m, and Δf_s, the more pronounced the effect of frequency regulation.

Under continuous load disturbance, all frequency deviation values in the sampling points can be measured. So that there are n sampling points, regarding Δf_i for the i_{th} sampling point of the frequency deviation value, the root mean square value of $\Delta f_{RMS} = \sqrt{\frac{1}{n}\sum_{i=1}^{n}(\Delta f_i - \Delta f_0)^2}$ is calculated and, obviously, a smaller value of the frequency regulation effect is better.

(2) SOC Maintenance Effectiveness Indicators

In the same way as for the frequency deviation metric under continuous perturbation, the root mean square value of the SOC deviation (i.e., the degree of deviation from the optimal $S_{OC0} = 0.5$) is taken to be $S_{RMS} = \sqrt{\frac{1}{n}\sum_{i=1}^{n}(S_{OCi} - S_{OC0})^2}$, and it is clear that the smaller the value, the better the SOC maintenance effect.

5. Simulation Verification

A hybrid energy-storage-assisted thermal power unit primary FR simulation model for a regional power grid is built by combining Figures 1 and 2.

Among them, in order to highlight the characteristics of power-type energy storage with high power but small capacity and energy-type energy storage with low power but large capacity, after fully considering the capacity configuration range of the two types of energy storage, the economic costs of the two types of energy storage are further considered, as shown in Table 2. It can be seen from Table 2 that compared with lithium batteries, the power cost of supercapacitors is lower and the capacity cost is higher. Therefore, this paper comprehensively considers the feasibility and economy of the strategy and sets the power and capacity configuration of supercapacitors in the hybrid energy storage system as 2.5 MW and 0.5 MWh and that of lithium batteries as 1 MW and 5 MWh.

Table 2. Comparison of the economic costs of energy storage.

Type of Energy Storage	Investment Cost of Energy Storage per Unit of Power	Investment Cost per Unit Capacity of Energy Storage
Power Energy Storage—Supercapacitor	CNY 1000~2000/kW	CNY 40 × 10⁴~100 × 10⁴/kWh
Energy Storage—Lithium Battery	CNY 8400~28,000/kW	CNY 2500~3000/kWh

In addition, in order to make the energy storage have a relatively ideal charge and discharge capacity at the initial moment, the initial energy storage SOC is generally set to 0.5, the power in the system is normalized to 100 MW, and the rest of the simulation parameters are shown in Table 3.

Table 3. Simulation parameters.

Parameters	Numerical Value	Parameters	Numerical Value
H	5	D	1
K_{gen}	20	K_B^{VIC}, K_{B1}^{VDC}	12, 24
$T_{B,cap}$, $T_{B,li}$	0.08, 0.2	T_G	0.1
F_{HP}	0.5	T_{CH}, T_{RH}	0.3, 10
F_{HP}	0.5	T_{CH}, T_{RH}	0.3, 10

5.1. Analysis of Load Step Disturbance FR Effect

In order to verify the effectiveness of this paper's strategy, a step load perturbation of 0.05 p.u. is set at 2 s. The system frequency deviation, the thermal unit output, the storage output, the SOC change curves of this paper's strategy and the single-battery control strategy proposed in the literature [10] (a common lithium battery is used in this paper, hereinafter referred to as the single-lithium strategy), and the strategy of no energy storage are shown in Figures 6–9, and the evaluation indexes are shown in Table 4.

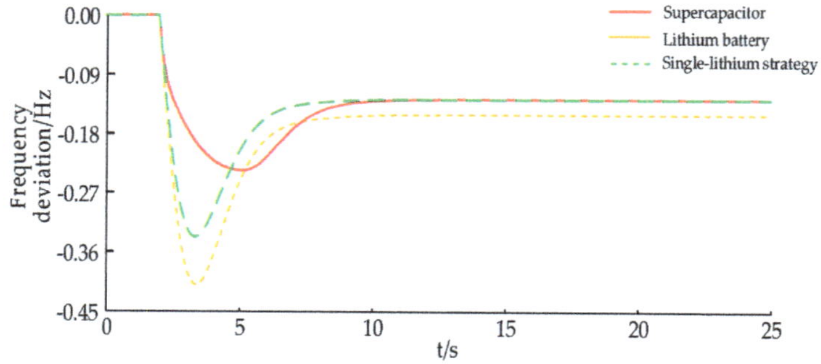

Figure 6. System frequency deviation under step perturbation.

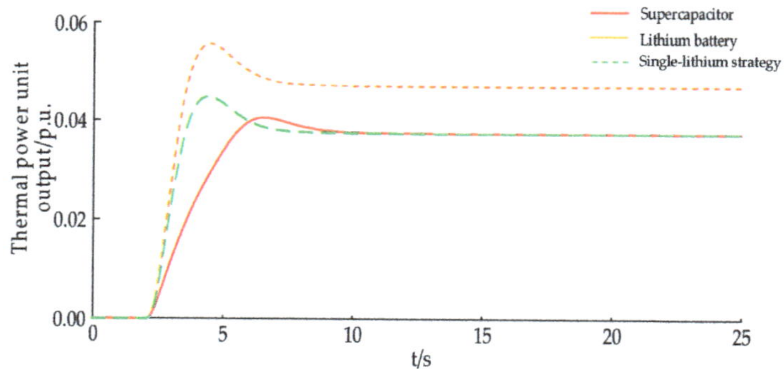

Figure 7. Output of thermal power unit under step jump disturbance.

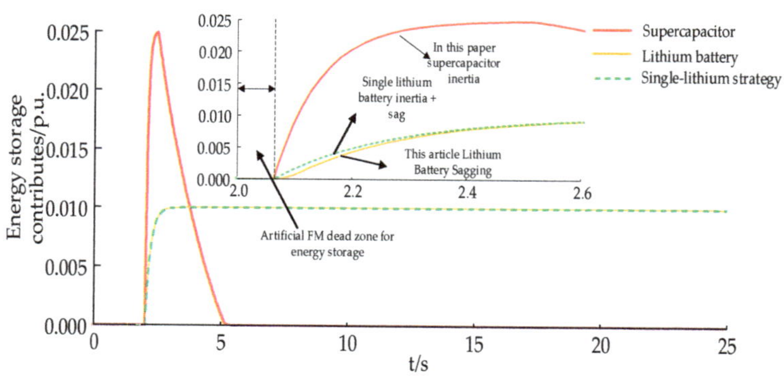

Figure 8. Energy storage output under step-jump disturbance.

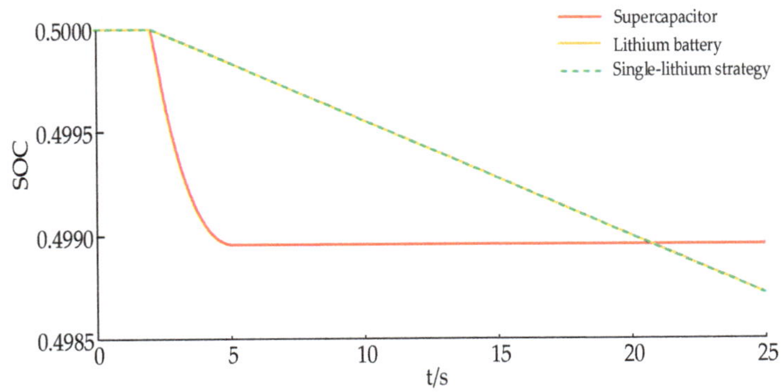

Figure 9. Energy storage SOC under step perturbation.

Table 4. Frequency regulation index under different strategies under step perturbation.

Methodologies	Δf_m/Hz	Δf_s/Hz	V_m/(Hz·s^{-1})	S_{RMS}(cap, li)	
Paper strategy	−0.2355	−0.1267	0.2842	8.27×10^{-4}	4.23×10^{-4}
Single-lithium strategy	−0.3366	−0.1267	0.2430	4.24×10^{-4}	
No energy storage	−0.3366	−0.1267	0.2430	/	

From Figures 6 and 7, it can be seen that comparing the system frequency deviation and thermal unit output curve, the absolute value of the maximum frequency deviation under this paper's strategy is the smallest, followed by the single-lithium strategy, and it is the worst without energy storage; for the steady-state frequency deviation, the current strategy is consistent with the single-lithium strategy, and the no energy storage is the worst. The reason for this is that in the frequency fall when the energy storage displays its fast response characteristics, in the system VIC and VDC quickly make up for the active deficit of the system and then inhibit the frequency fall, so compared with the no storage frequency regulation effect it is better and, at the same time, the improvement of the frequency regulation effect will also directly reduce the thermal power unit's primary frequency regulation output. From Figures 8 and 9, it can be seen that under the single-lithium strategy, due to the use of only lithium batteries in the VIC and VDC at the same time, it cannot effectively respond to the superimposed power command signal because of the limitation of its power amplitude. In contrast, the strategy in this paper considers the use of supercapacitor high-power short-time response to the VIC signal and the use of lithium battery high-energy long-time response to the VDC signal. When the load disturbance occurs in the 2 s system, after the energy storage set up in this paper and after the frequency regulation dead zone, the high-power supercapacitor responds quickly to the VIC signal with the rated power, the Li-ion battery only assumes the VDC signal, and the supercapacitor quits operation as the frequency deviation reaches the lowest point, which effectively avoids the supercapacitor SOC from falling into the bad state and at the same time also facilitates the restoration of the system frequency, as shown in Figure 6. The steady-state frequency deviation is consistent with the single-lithium strategy.

Analyzing from the perspective of data, in this paper, the improvement under strategy Δf_m is 30.04% and 42.49% compared to the single-lithium strategy and no storage, respectively; Δf_s is 15.82% compared to no storage, which is consistent with the single-lithium strategy; and V_m is 14.50% and 73.72% compared to the single-lithium strategy and no storage, respectively.

5.2. Analysis of the Effect of Continuous Load Perturbation FR

The 2 h continuous load disturbance data of a regional power grid are used as the research object, the sampling interval is 1 min, and the corresponding load disturbance curve is shown in Figure 10. In order to verify the superiority of the logistic-based variable coefficient control proposed in this paper, the strategy of this paper is compared with the no energy storage, single-lithium strategy, and the strategy of this paper with no logistic optimization (hereinafter referred to as without logistic). The system frequency deviation, storage output, and SOC variation curves under different strategies are shown in Figures 11–13, and the evaluation indexes are shown in Table 5.

As can be seen from Figure 11, the strategy in this paper maintains the best FR effect in the continuous disturbance scenario compared with no energy storage and the single-lithium strategy. From Figures 12 and 13, it can be seen that the supercapacitor under the strategy of this paper can exit the inertial response in time after the frequency reaches the frequency extreme value, while the lithium battery participates in the sagging response of the system in the whole FR cycle; compared with the strategy presented in this paper and without logistic, the SOC state of the supercapacitor and lithium battery under the strategy presented in this paper is significantly better than that of the lithium battery without logistic. If the without logistic control strategy is adopted, the SOC of the supercapacitor will fall into a bad condition of 0.1/0.9 when the load disturbance is large (e.g., 92~100 min in Figure 13) and lose the ability of the subsequent FR, which can be seen in Figure 11, and the FR effect is the worst in this time period.

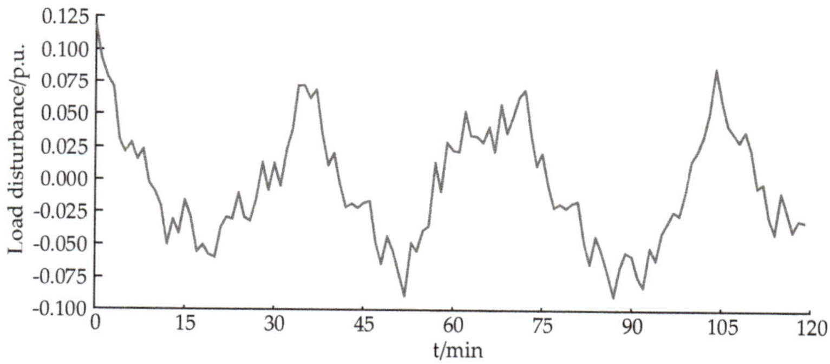

Figure 10. A 2 h continuous load perturbation curve.

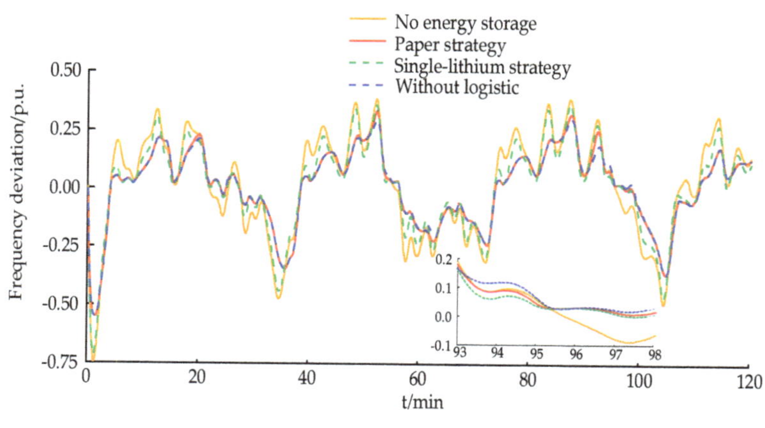

Figure 11. System frequency deviation under continuous load disturbance.

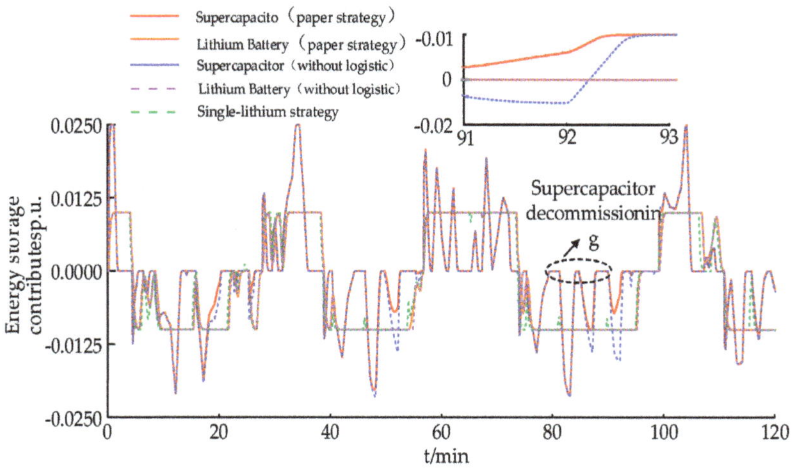

Figure 12. Energy storage output under continuous load disturbance.

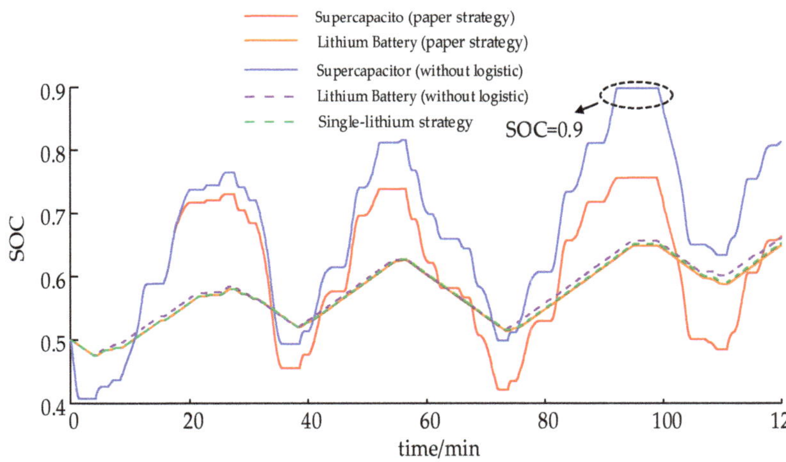

Figure 13. Energy storage SOC under continuous load disturbance.

Table 5. Frequency regulation index under different strategies under continuous load disturbance.

Methodologies	Δf_{RMS}/Hz	S_{RMS}(cap, li)	
Paper strategy	0.1547	0.1454	0.0803
Without logistic	0.1524	0.2057	0.0820
Single-lithium strategy	0.1914	0.0856	
No energy storage	0.2215	/	

Analyzing from the data point of view, compared with the single-lithium strategy and no energy storage, with the strategy in this paper the Δf_{RMS} improved 19.17% and 30.16%, respectively, slightly worse than the without logistic strategy, which is due to the limitation by the logistic curve in some periods of time to reduce the coefficients of storage involved in the inertia and sagging response but effectively maintaining the SOC of the energy storage, and with the supercapacitor and lithium batteries, the SOC level improved 29.4% and 2.1%, respectively.

6. Conclusions

The influence of the action timing of the two control modes on the frequency deviation of the grid based on the output mechanism of VIC and VDC is analyzed, putting forward a control method that reasonably combines the two and elaborating on the different requirements of the two control modes regarding the energy storage output characteristics.

According to the demand of VIC and VDC for energy storage output power and capacity, a hybrid energy storage adaptive VIC and VDC coordination control strategy is proposed, which will have a high-power energy storage supercapacitor responding to inertia control signal and a high-capacity lithium battery responding to sag control signal. It makes the output characteristics of VIC and VDC fit with different types of energy storage, effectively reducing frequency differences and preventing frequency deterioration. Compared with the single-lithium strategy and no energy storage, Δf_{m} increases by 30.04% and 42.49%, respectively, under step disturbance, and Δf_{RMS} increases by 19.17% and 30.16%, respectively, under continuous disturbance.

In order to avoid the energy storage SOC from falling into lower/higher severe operating conditions and losing the subsequent FR capability, an adaptive output control strategy for energy storage based on an improved logistic function is proposed, which improves the SOC levels of the supercapacitor and Li-ion batteries by 29.4% and 2.1%, respectively, under continuous perturbation as compared to that of the supercapacitor and Li-ion batteries without logistic.

At present, the price of energy storage equipment is high, but with the rapid development of related technologies, its economy will be continuously improved, and the application of energy storage will be more large-scale. The price used in this paper may somewhat deviate from the current energy storage price, but its capacity allocation method and cost–benefit model are the same, which have certain reference value.

Author Contributions: Conceptualization, C.X.; methodology, J.X.; software, J.X.; validation, X.X.; formal analysis, C.X.; investigation, X.X.; resources, C.X.; data curation, X.X.; writing—original draft preparation, C.X.; writing—review and editing, X.X.; visualization, Y.C.; supervision, P.L.; project administration, Q.G.; funding acquisition, P.L. All authors have read and agreed to the published version of the manuscript.

Funding: The paper has been funded by National Key Research and Development Program (NKRDP) (2019YFE0118000); National Key Program for Basic Research and Development (973 program) (2021YFB2601504).

Data Availability Statement: All data underlying the results are available as part of the article.

Conflicts of Interest: The authors Chao Xing and Xinze Xi were employed by the company Electric Power Research Institute of Yunnan Power Grid Co., Ltd. The author declares that the research was conducted in the absence of any commercial or financial relationships that could be construed as a potential conflict of interest.

Nomenclature

Variables	
Δf	frequency deviation
t	time index
$S_{\mathrm{OC}}(t)$	energy storage SOC at moment t
Δt	one FR cycle
$\Delta P_{\mathrm{B}}(t)$	output power of the energy storage at moment t
$\Delta P_{\mathrm{G}}(s)$	output power of the thermal power unit
$\Delta P_{\mathrm{B}}(s)$	output power of the energy storage
$\Delta P_{\mathrm{B,cap}}(s)$	output power of the supercapacitor
$\Delta P_{\mathrm{B,li}}(s)$	output power of the lithium battery
Parameters	
T_{G}	thermal unit governor time constant
T_{CH}	turbine time constant

T_{RH}	heater time constant
F_{HP}	heater gain
$T_{B,i}$	energy storage time constant
$T_{B,cap}$	time constants of supercapacitor
$T_{B,li}$	time constants of lithium battery
D	system damping coefficient
H	inertia coefficient of the thermal power unit
K_B^{VIC}	inertia coefficient when VIC is used for energy storage
K_B^{VDC}	sag coefficient when VDC is used for energy storage
E_{BN}	total capacity of the energy storage system
K_{gen}	primary unit regulation power of the thermal power unit
Abbreviations	
VIC	virtual inertial control
VDC	virtual droop control
SOC	state of charge
FR	frequency regulation

References

1. Cirio, D.; Conte, F.; Gabriele, B.; Gandolfi, C.; Rapizza, S.M.M.; Silvestro, F. Fast Frequency Regulation from a Wind Farm-BESS Unit by Model Predictive Control: Method and Hardware-in-The-Loop Validation. *IEEE Trans. Sustain. Energy* **2023**, *14*, 2049–2061. [CrossRef]
2. Li, C.; Li, J.; Li, J.; Zhang, X.; Hou, T. Optimization strategy of secondary Frequency regulation based on dynamic loss model of the energy storage unit. *J. Energy Storage* **2022**, *51*, 104425. [CrossRef]
3. Xiao, J.; Li, P.; Mao, Z. Battery Energy Storage Participating in Secondary Frequency Regulation Based on Fuzzy Allocation Factor. *Power Syst. Technol.* **2023**, *47*, 3568–3577. [CrossRef]
4. Li, B.; Li, M.; Yan, S.; Zhang, Y.; Shi, B.; Ye, J. An optimal energy storage system sizing determination for improving the utilization and forecasting accuracy of photovoltaic (PV) power stations. *Front. Energy Res.* **2023**, *10*, 1074916. [CrossRef]
5. Liu, Y.; Wang, Y.; Xu, M.; Xiao, Z. Wind Power Fluctuation Smooth Strategy Based on Digital Twin Hybrid Energy Storage. *Power Syst. Technol.* **2021**, *45*, 2503–2514. [CrossRef]
6. Barra, P.H.A.; de Carvalho, W.C.; Menezes, T.S.; Fernandes, R.A.S.; Coury, D.V. A review on wind power smoothing using high-power energy storage systems. *Renew. Sustain. Energy Rev.* **2021**, *137*, 110455. [CrossRef]
7. Wang, L.; Tan, L.; Xu, Z.; Tan, X.; Wu, C.; Ye, H.; Li, A. Lithium battery energy storage power station primary Frequency regulation designoptimization and verification. *Energy Storage Sci. Technol.* **2022**, *11*, 3862–3871. [CrossRef]
8. Xiong, L.; Xiu, L.; Wang, H.; Xu, Z. Mechanism of Energy Storage System to Suppress Grid Power Oscillations. *Trans. China Electrotech. Soc.* **2019**, *34*, 4373–4380. [CrossRef]
9. Zhao, J.; Xu, C.; Lu, X.; Xu, C. Optimization of Micro-grid Primary Frequency Regulation Reserve Capacity and Energy Storage System. *Proc. CSEE* **2017**, *37*, 4324–4332. [CrossRef]
10. Liu, S.; Zhao, L.; Huang, S.; You, H.; Li, J.; Yang, L. Research on the mixed control strategy of the battery energy storage considering frequency modulation, peak regulation, and SOC. *Energy Sci. Eng.* **2022**, *10*, 3459–3470. [CrossRef]
11. Xing, W.; Wang, H.; Lu, L.; Han, X.; Sun, K.; Ouyang, M. An adaptive virtual inertia control strategy for distributed battery energy storage system in microgrids. *Energy* **2021**, *233*, 121155. [CrossRef]
12. Hasan, N.S.; Rosmin, N.; Nordin, N.M.; Abd Bakar, S.; Aman, A.H.M. Dynamic response of hybrid energy storage based virtual inertial support in wind application. *J. Energy Storage* **2022**, *53*, 105181. [CrossRef]
13. Xing, P.; Shi, Q.; Wang, G.; Fu, L.; Wang, Y.; Wu, Y. Response Characteristics and Mechanism Analysis About Virtual lnertia Control of Wind Generators. *High Volt. Eng.* **2018**, *44*, 1302–1310. [CrossRef]
14. Gupta, S.; Mukhopadhyay, S.; Banerji, A.; Sanki, P.; Biswas, S.K. Adaptive Droop Control for Voltage and Frequency Regulation in Virtual Power Plant under Power System Contingencies. *Jordan J. Electr. Eng.* **2023**, *9*, 618–649. [CrossRef]
15. Obaid, Z.A.; Cipcigan, L.M.; Muhssin, M.T.; Sami, S.S. Control of a population of battery energy storage systems for frequency response. *Int. J. Electr. Power Energy Syst.* **2020**, *115*, 105463. [CrossRef]
16. Xiao, J.; Jia, Y.; Jia, B.; Li, Z.; Pan, Y.; Wang, Y. An inertial droop control based on comparisons between virtual synchronous generator and droop control in inverter-based distributed generators. *Energy Rep.* **2020**, *6*, 104–112. [CrossRef]
17. Hernández, J.C.; Sanchez-Sutil, F.; Vidal, P.G.; Rus-Casas, C. Primary frequency control and dynamic grid support for vehicle-to-grid in transmission systems. *Int. J. Electr. Power Energy Syst.* **2018**, *100*, 152–166. [CrossRef]
18. Tan, Z.; Li, X.; He, L.; Li, Y.; Huang, J. Primary frequency control with BESS considering adaptive SoC recovery. *Int. J. Electr. Power Energy Syst.* **2020**, *117*, 105588. [CrossRef]
19. Chen, C.; Li, X.; Zhang, B.; Yang, T. Energy storage peak and Frequency regulation cooperative control strategy based on multi-time-scale. *Dianli Xitong Baohu Yu Kongzhi/Power Syst. Prot. Control* **2022**, *50*, 94–105. [CrossRef]
20. Cao, Y.; Wu, Q.; Zhang, H.; Li, C. Optimal sizing of hybrid energy storage system considering power smoothing and transient frequency regulation. *Int. J. Electr. Power Energy Syst.* **2022**, *142*, 108227. [CrossRef]

21. Gao, S.; Li, J.; Song, H.; Ding, L.; Zhang, C. An integrated thermal power-energy storage system for improving primary frequency regulation performance of thermal power units. *Power Syst. Prot. Control* **2023**, *51*, 116–125. [CrossRef]
22. Zhang, X.; Qin, S.; Zhang, Y.; Hao, S.; Wu, Q.; Zhang, J. Wind Turbine and Storage Joint Frequency regulation Control Strategy Considering Energy Storage State of Charge. *High Volt. Eng.* **2023**, *49*, 4120–4130. [CrossRef]
23. Zhu, X.; Han, X. The adaptive control strategy of energy storage battery cooperating with conventional generating units to participate in primary frequency regulation. *Electr. Meas. Instrum.* **2023**, *60*, 34–42. [CrossRef]
24. Zhang, Y.; Zhao, C.; Tang, W.; Low, S. Profit-Maximizing Planning and Control of Battery Energy Storage Systems for Primary Frequency Control. *IEEE Trans. Smart Grid* **2018**, *9*, 712–723. [CrossRef]

electronics

MDPI

Article

Optimizing Microgrid Performance: Integrating Unscented Transformation and Enhanced Cheetah Optimization for Renewable Energy Management

Ali S. Alghamdi

Department of Electrical Engineering, College of Engineering, Majmaah University, Majmaah 11952, Saudi Arabia; aalghamdi@mu.edu.sa

Abstract: The increased integration of renewable energy sources (RESs), such as photovoltaic and wind turbine systems, in microgrids poses significant challenges due to fluctuating weather conditions and load demands. To address these challenges, this study introduces an innovative approach that combines Unscented Transformation (UT) with the Enhanced Cheetah Optimization Algorithm (ECOA) for optimal microgrid management. UT, a robust statistical technique, models nonlinear uncertainties effectively by leveraging sigma points, facilitating accurate decision-making despite variable renewable generation and load conditions. The ECOA, inspired by the adaptive hunting behaviors of cheetahs, is enhanced with stochastic leaps, adaptive chase mechanisms, and cooperative strategies to prevent premature convergence, enabling improved exploration and optimization for unbalanced three-phase distribution networks. This integrated UT-ECOA approach enables simultaneous optimization of continuous and discrete decision variables in the microgrid, efficiently handling uncertainty within RESs and load demands. Results demonstrate that the proposed model significantly improves microgrid performance, achieving a 10% reduction in voltage deviation, a 10.63% decrease in power losses, and an 83.32% reduction in operational costs, especially when demand response (DR) is implemented. These findings validate the model's efficacy in enhancing microgrid reliability and efficiency, positioning it as a viable solution for optimized performance under uncertain renewable inputs.

Keywords: microgrid management; renewable energy integration; operational efficiency improvement; unscented transformation; uncertainty propagation; enhanced cheetah optimization algorithm

1. Introduction

The increasing integration of renewable energy sources (RESs), particularly photovoltaic (PV) systems and wind turbines (WTs), is transforming the power sector, though it also introduces significant challenges due to their inherent variability. These sources, while clean, are intermittent and heavily influenced by weather conditions, which complicates the management of microgrids (MGs) and necessitates robust optimization approaches. Solar and wind energy availability fluctuates based on factors such as sunlight exposure and wind speed, while additional uncertainties stem from varying consumer loads, demand response (DR) programs, and electric vehicle (EV) charging behaviors [1–4]. The need to handle these uncertainties has led to advanced optimization techniques, with hybrid algorithms demonstrating promise in terms of enhancing MG operations. For instance, a combined PSO and Bat Algorithm (BAPSO) has been applied to optimize solar PV configurations in microgrids, effectively reducing transmission losses and enhancing efficiency [5]. Other approaches integrate various storage and backup systems, employing hybrid algorithms such as zebra optimization and artificial gorilla troops optimizers for maximum power point tracking, achieving substantial improvements in resource utilization and response times [6]. Likewise, demand-side management strategies incorporating hybrid techniques like SHO-MDACGAN have shown potential in minimizing operating costs and improving power stability in PV-based microgrids [7]. Additionally, a novel Gravitational Particle Swarm Optimization Algorithm (GPSOA) that combines the strengths of PSO and the gravitational search algorithm (GSA) has proven effective in addressing combined economic and emission dispatch (CEED) challenges in wind–thermal systems by balancing fuel costs and emission reduction objectives [8]. Another sequential hybrid algorithm, the PSO-GSA (HPSO-GSA), with dependent random coefficients has demonstrated enhanced exploration abilities, making it a robust approach for complex optimization scenarios where premature stagnation is a concern [9].

To address the operational uncertainty, accurate modeling techniques such as Monte Carlo Simulation (MCS) and the Point Estimate Method (PEM) are frequently used [10,11]. Integrating these uncertainty quantification methods with hybrid optimization algorithms can effectively optimize both operational costs and reliability in MGs, enhancing energy management while balancing efficiency and adaptability [12]. Among the widely used techniques for handling uncertainty in renewable energy sources and load demand is MCS. The technique involves the simulation of different scenarios generated from probability distributions of input variables like solar radiation, wind speed, and load profiles. Accurate results are indeed obtained with MCS, although with very high computational complexity, especially in large systems. For instance, Ahn et al. have used the sequential MCS method for scenario estimations, which involves electric, cooling, and heating demand combined with solar irradiance [13]. Other works have also discussed the effectiveness of MCS in regard to scenario generation and energy management in MGs [4,14,15].

In contrast, the PEM provides a computationally efficient alternative by approximating the statistical behavior of random variables using specific points. In fact, the PEM has been applied to several energy management systems of MGs when computation time with desired accuracy is of interest. Alavi et al. in [16] utilized the PEM for modeling power exchanges and optimizing the operations of MGs within uncertain conditions. Further studies [17,18] emphasize the potential of the application of the PEM to enhance performance while reducing computational burdens in various microgrid approaches. According to [19], the PEM cannot handle correlated variables and its error can increase with the standard deviation of a random input variable.

UT is a nonlinear analytical method that propagates the mean and covariance of input variables through a nonlinear function [20]. It relies on a reduced number of deterministically calculated samples (sigma points) that capture the statistical distribution of uncertain variables. The UT is simple to apply and requires low computational time to estimate the mean and covariance of desired variables [19]. In [21], a comparison of the UT method, MCS and the PEM is presented in detail. The previous works have presented the application of the UT method to solve the probabilistic power flow issues in an unbalanced three-phase microgrid [22], a balanced islanded microgrid with renewable sources [20], and to provide a power flow solution for a transmission and distribution system [23].

These uncertainties could be dispersed only by using advanced optimization techniques. The optimization methods available in the literature for microgrid energy management problems may be broadly classified into two groups: classical techniques and metaheuristics optimization algorithms. Classical techniques mainly comprise mixed-integer linear programming (MILP) and mixed-integer quadratic programming (MIQP), which are quite useful in modeling the on/off states of generators and loads [24]. Yet, most of them require some form of linearization in order to be able to handle the complexity; hence, there is the development of mixed-integer quadratically constrained programming (MIQCP), although, in many optimization scenarios, this may not be suitable [25].

Traditional methods include linear programming, quadratic programming, and MILP, which provide accuracy but are not capable of handling dynamic variables such as market prices and weather variability with ease [26–28]. The applications of gradient descent, Quasi-Newton, and Powell's method fall short because most are stuck to a local optimum point [29–31].

Compared to the limitations mentioned above, metaheuristic and heuristic algorithms can further allow for flexibility and efficiency in terms of handling complex nonlinear problems. Algorithms of particular note are genetic algorithms (GAs) [32], particle swarm optimization (PSO) [33], and evolutionary optimization techniques. Indeed, as an example of this potential, one hybrid evolutionary optimization algorithm was utilized with stochastic multi-objective optimization in microgrids within an approach where wind power forecasts were integrated to lessen the computational load and reduce time consumption [34]. GAs, indeed, have proven to be useful in applications, as they are able to slice optimization by time up to 23% [35].

Metaheuristic algorithms are exceptionally good at dealing with the intrinsic complexity of microgrid energy management. By their very nature, these algorithms are bio-inspired and do not make any assumptions about the problem being solved. They make use of probabilistic search, local investigation, and global optimization techniques. The technique known as PSO has been used to optimize microgrid performance by solving complex optimization problems regarding control and fault detection [36–38]. Ant colony optimization (ACO) has also been applied to energy management with the intent of minimizing operational costs and enhancing renewable energy utilization [39]. Other highlighted MHOAs are the salp swarm optimization algorithm (SSA) [40], GAs [41], the whales optimization algorithm (WOA) [42], and gray wolf optimization (GWO) [43], since each will come forth with varied benefits in regard to tackling microgrid optimization challenges.

They ensure convergence to the optimum solution; however, due to the generally nonconvex problems in microgrid energy management, traditional methods of mathematical programming cannot handle them [44]. The metaheuristic algorithms discussed above could be a flexible and computationally efficient alternative solutions for possibly dealing with nonlinearities–uncertainties possessed by microgrid systems. Hence, such approaches are currently desired for MGs because of their simplicity and ability towards reducing computational efforts.

Despite the emergence of various approaches to modeling uncertainties and optimization techniques in microgrids, most of the gaps still remain. MCS is able to model renewable energy sources and load demands accurately with high precision but also introduces high computational complexity, especially for large-scale systems. On the contrary, the PEM offers computational efficiency but limits handling correlated variables and increases error when higher standard deviations are experienced. Traditional methods in optimization, like MILP and quadratic programming, are based on linearization techniques that may not fit well with the complex nature of microgrid problems. On the other hand, a number of promises from metaheuristic algorithms like the GA and PSO can solve complex optimization issues. However, notwithstanding their promising potential, most of these conventional methods inherently suffer from a number of drawbacks, such as premature convergence, nonlinearities handling difficulty, and balancing exploration and exploitation processes. These result in mediocre near-global optima, especially for complex systems like microgrids with renewable energy sources that have a high variability and uncertainty. While both of these methods assure reasonable global search capabilities, solution fine-tuning in the latter stages of optimization is relatively poor, which leads to reduced system performance for adaptability and precision-intensive scenarios.

The original conventional cheetah optimization (CO) [45], inspired by the hunting behavior of cheetahs, overcame several challenges by increasing the balance between exploration and exploitation. However, like many other conventional algorithms, it still suffers from a number of limitations regarding global exploration, especially while dealing with complex multi-objective problems involving high uncertainty. In application to unbalanced three-phase distribution networks, appropriate optimization of both continuous and discrete decision variables might not be achieved by the CO due to its tendency towards premature convergence in view of fluctuating renewable energy and load demands. Advanced methods are thus of utmost necessity in handling their inherent uncertainties and improving, in general, the operational efficiency of microgrids. Combining these methods with advanced uncertainty modeling approaches for unbalanced three-phase distributions is still a challenge. In fact, there is a real need for better methods that are capable of integrating all the strengths of uncertainty modeling and different optimization techniques in order to enhance efficiency and reliability in microgrids.

This work integrates UT and the ECOA for optimal microgrid planning and operation. The system under investigation is an imbalanced three-phase distribution system and therefore bridges the highlighted research gaps. This paper extends earlier research by:

- Applying UT to model uncertainty propagation in nonlinear functions for the more accurate transmission of uncertainty in terms of renewable energy supply and load demand. This technique avoids a couple of major limitations associated with MCS and the PEM in the imbalanced three-phase distribution system, given the lower computational workload involved in modeling the uncertainties.
- The advanced metaheuristic optimization algorithm inspired by cheetahs' hunting nature proposes stochastic leaps and adaptive chase mechanism cooperation approaches that enhance the global exploration capability by removing it from premature convergence in order to converge into a robust solution for complicated optimization problems in unbalanced three-phase systems.
- UT and ECOA inclusions create a comprehensive framework within our approach to continuous and discrete optimization variables of microgrid operations. A combination of UT will improve resilience in managing uncertainties in renewable energy sources and load demand for efficient unbalanced three-phase distribution systems.
- The proposed framework can reduce network losses by 10.63%, operational costs by 83.32%, and voltage profile variation by 10% when simulation results are considered. In addition, the inclusions of DR programs further enhance the efficiency of the system and the highly effective and dependable paradigm for microgrid planning and operation in unbalanced three-phase distribution networks.
- The paper contributes by applying and comparing the performances of the GA, PSO, conventional CO, and two hybrid algorithms, such as GPSOA and HPSO-GSA, in the optimal management of microgrids. It further indicates the weaknesses of both the GA and PSO in dealing with uncertainties and falling into the local optima. The CO drains more, though it still shows its weakness in highly complex scenarios. These comparisons highlight the importance of the ECOA, which is able to yield better results in terms of minimizing power losses, voltage deviation, and overall cost of operation.

In a nutshell, this research further develops the state of the art in terms of microgrid optimization through a methodology that overcomes the main limitations affecting existing approaches to the problem thanks to advanced uncertainty modeling combined with sophisticated optimization techniques, conceived for unbalanced three-phase distribution systems.

The paper is organized as follows: Section 2 elaborates on problem formulation, including a description of the objectives and constraints of the microgrid optimization problem. Section 3 describes Unscented Transformation in uncertainty modeling while examining the use of UT in dealing with the intermittency of renewable energy sources. Section 4 elaborates on the proposed algorithm, investigating the integration of UT into the ECOA. Section 5 gives the simulation results, showing the performance comparison between the proposed method and conventional approaches. Conclusions are drawn in Section 6, which also summarizes the main findings and proposes future research directions.

2. Problem Formulation

The objective of the research is to find, out of several generation options, the most efficient energy management approach in a three-phase unbalanced MG. This means the minimization of operational critical metrics (i.e., the total cost, including costs caused by emission, operation, and maintenance) and assurance of voltage stability with a minimum of active power losses. In this regard, the problem is a multi-objective optimization problem subjected to microgrid operational and stability requirements.

2.1. Decision Variables

The problem includes the nature of the optimization procedure involved in a microgrid operation that possesses a mix of characteristics, including continuous and discrete decision variables. These are subsequently categorized below:

The continuous variables can be defined for this problem as the power outputs and the flows at every time-step t:

- $P_{PV}(t)$: Active power output from the PV system.
- $P_{WT}(t)$: Active power output from the WT.
- $P_{DR}(t)$: Power reduction achieved through DR.
- $P_{Grid}(t)$: Power purchased from the main grid.
- $P_{BESS}(t)$: Power associated with charging or discharging the battery energy storage system (BESS).
- $P_{DG}(t)$: Active power generated by the diesel generator (DG).
- $P_{MT}(t)$: Active power generated by the microturbine (MT).

These variables are collectively expressed as:

$$X_{cont} = [P_{PV}(t), P_{WT}(t), P_{DR}(t), P_{Grid}(t), P_{BESS}(t), P_{DG}(t), P_{MT}(t)] \quad \forall t \in \mathcal{T} \tag{1}$$

The discrete decision variables determine the bus and phase allocations for distributed energy resources (DERs), which are crucial for managing three-phase imbalances in the microgrid.

- Bus Allocation Variables (B_{bus}):

$$B_{bus} = [B_{PV}(t,b), B_{WT}(t,b), B_{DR}(t,b), B_{Grid}(t,b), B_{BESS}(t,b), B_{DG}(t,b), B_{MT}(t,b)] \quad \forall b \in \mathcal{B}, \forall t \in \mathcal{T} \tag{2}$$

where $B_{DER}(t,b)$ indicates the bus location b for each DER at time t.

- Phase Assignment Variables (B_{phase}):

$$B_{phase} = [\phi_{PV}(t,p), \phi_{WT}(t,p), \phi_{DR}(t,p), \phi_{Grid}(t,p), \phi_{BESS}(t,p), \phi_{DG}(t,p), \phi_{MT}(t,p)] \quad \forall p \in \mathcal{P}, \forall t \in \mathcal{T} \tag{3}$$

where $\phi_{DER}(t,p)$ denotes the phase p (A, B, or C) assigned to each DER at time t.

The full set of decision variables for the optimization problem is:

$$X = [X_{cont}, X_{disc}] \quad \forall t \in \mathcal{T} \tag{4}$$

2.2. Objective Functions

The optimization problem is formulated to minimize the following three objective functions:

2.2.1. Total Power Losses (J_1)

$$J_1 = \sum_{l=1}^{L_{lines}} \left(r_l^a \cdot I_l^a(t)^2 + r_l^b \cdot I_l^b(t)^2 + r_l^c \cdot I_l^c(t)^2\right) \quad \forall t \in \mathcal{T} \tag{5}$$

where r_l^a, r_l^b, and r_l^c are the resistances of line l in phases A, B, and C, respectively, and $I_l^a(t)$, $I_l^b(t)$, and $I_l^c(t)$ are the corresponding line currents.

2.2.2. Voltage Deviation (J_2)

This objective ensures voltage levels across all buses and phases remain within acceptable limits.

$$J_2 = \frac{1}{3} \times \left(\sqrt{\sum_{n=1}^{N_{bus}} (1-|V_n^a(t)|)^2} + \sqrt{\sum_{n=1}^{N_{bus}} (1-|V_n^b(t)|)^2} + \sqrt{\sum_{n=1}^{N_{bus}} (1-|V_n^c(t)|)^2}\right) \tag{6}$$

2.2.3. Total Cost (J_3)

$$J_3 = C_{PV}(t) + C_{WT}(t) + C_{BESS}(t) + C_{DG}(t) + C_{MT}(t) + C_{Grid}(t) + C_{DR}(t) \quad (7)$$

- Cost of PV and WT Systems.

  The total cost of PV and WT systems can be expressed as:

$$C_{RES,\zeta}(t) = C_{FIX,\zeta} + C_{VAR,\zeta} \times P_\zeta(t) \quad \forall t \in \mathcal{T}, \forall \zeta \in \{PV, WT\} \quad (8)$$

In this equation, $C_{RES,\zeta}(t)$ is the total cost of the renewable energy source ζ (which could be a PV system or WT) at time t. The term $C_{FIX,\zeta}$ represents the fixed cost, $C_{VAR,\zeta}$ is the variable cost of the renewable energy source ζ, while $P_\zeta(t)$ is the power output of source ζ at time t.

The fixed cost is calculated as:

$$C_{FIX,\zeta} = \frac{C_{INV,\zeta} \times r}{(1-(1+r)^{-T_{Life}}) \times 8760 \times CF_\zeta} \quad (9)$$

Here, $C_{INV,\zeta}$ represents the initial investment cost of source ζ, r is the discount rate, T_{Life} is the expected lifetime of the system in years, and CF_ζ is the capacity factor of source ζ. The number 8760 represents the total hours in a year.

The variable cost is given by:

$$C_{VAR,\zeta} = C_{O\&M,\zeta} \quad (10)$$

In this case, $C_{O\&M,\zeta}$ is the operation and maintenance cost for the renewable energy source ζ.

- Cost of the BESS

  The cost of the BESS (battery energy storage system) operation is expressed as:

$$C_{BESS}(t) = C_{FIX,BESS} + \left(C_{CHG} \times P^{ch}_{BESS}(t) + C_{DSG} \times P^{ds}_{BESS}(t)\right) + \gamma_{TOU}(t) \times P_{BESS}(t) \quad (11)$$

This equation then gives the overall cost of operating the BESS at time t. Note that $C_{BESS}(t)$ is the total cost of the BESS operation at time t, $C_{FIX,BESS}$ is the fixed cost of the BESS, and the terms C_{CHG} and C_{DSG} then represent the variable charging and discharging costs, respectively, while $P^{ch}_{BESS}(t)$ and $P^{ds}_{BESS}(t)$ are the power charged to and discharged from the BESS at time t. $\gamma_{TOU}(t)$ is the time-of-use (TOU) electricity price at time t; $P_{BESS}(t)$ is the total BESS power (both charging and discharging) at time t.

The fixed cost of the BESS is calculated as:

$$C_{FIX,BESS} = \frac{C_{INV,BESS} \times r}{(1-(1+r)^{-T_{Life,BESS}}) \times 8760 \times CF_{BESS}} \quad (12)$$

Here, $C_{INV,BESS}$ is the investment cost of the BESS, r is the discount rate, $T_{Life,BESS}$ is the expected lifetime of the BESS in years, and CF_{BESS} is the capacity factor of the BESS. The number 8760 represents the total hours in a year.

The variable costs of charging and discharging are expressed as:

$$C_{CHG} = C_{O\&M,CHG}\,, \quad C_{DSG} = C_{O\&M,DSG} \quad (13)$$

where $C_{O\&M,CHG}$ and $C_{O\&M,DSG}$ are the operation and maintenance costs for charging and discharging the BESS, respectively.

- Cost of DG, MT, and Grid Power

  The cost of operating DG, MT, and grid power systems is modeled as:

$$C_{GEN,\xi}(t) = C_{FUEL,\xi}(t) + C_{EMI,\xi}(t) + C_{O\&M,\xi}(t) \quad \forall t \in \mathcal{T}, \forall \xi \in \{DG, MT, Grid\} \quad (14)$$

where

$$C_{\text{FUEL},\xi}(t) = \alpha_\xi \times P_\xi(t) + \beta_\xi \times P_\xi^2(t) + \gamma_\xi \quad (15)$$

$$C_{\text{EMI},\xi}(t) = \left(\mu_{\text{CO}_2} + \mu_{\text{SO}_2} + \mu_{\text{NO}_x}\right) \times P_\xi(t) \quad (16)$$

$$C_{\text{O\&M},\xi}(t) = \kappa_\xi \times P_\xi(t) \quad (17)$$

Specific equations are as follows, in which $C_{\text{GEN},\xi}(t)$ is the total generation cost of technology ξ, at time t, which includes fuel cost $C_{\text{FUEL},\xi}(t)$, emission cost $C_{\text{EMI},\xi}(t)$, and operation and maintenance cost $C_{\text{O\&M},\xi}(t)$, while $P_\xi(t)$ is the power output of technology ξ at time t. The form of the fuel cost function contains a linear term with coefficient α_ξ and a quadratic term with coefficient β_ξ. The emission cost depends on the unit costs for the emissions of CO_2, SO_2, and NOx, labeled by μ_{CO_2}, μ_{SO_2}, and μ_{NO_x}, respectively. Operation and maintenance costs depend on coefficient κ_ξ. ξ is in use for different generation technologies, such as DG, MT, and the grid, while $\mathcal{T}$ represents the set of time periods under consideration.

Finally, $C_{DR}(t)$ represents the costs associated with DR at time t:

$$C_{DR}(t) = \beta_{\text{I/C}} \times P_{DR}(t) + \lambda \times \mathbb{E}[\text{VSI}(t)] \quad (18)$$

In this equation, $\beta_{\text{I/C}}$ represents the cost coefficient for interruptible or curtailable loads, while $P_{DR}(t)$ denotes the voluntary load reduction at time t. The parameter λ serves as a weighting factor for balancing cost and grid stability, and $\mathbb{E}[\text{VSI}(t)]$ stands for the expected voltage stability.

For unifying the scales of these objectives, each objective function is normalized:

$$J'_i = \frac{J_i - J_i^{\min}}{J_i^{\max} - J_i^{\min}} \quad \forall i \in \{1,2,3\} \quad (19)$$

The aggregate objective function to be minimized is:

$$J_{\text{total}} = J'_1 + J'_2 + J'_3 \quad (20)$$

2.3. Dynamic Pricing-Based DR

This DR program encourages a load reduction in response to real-time electricity prices. The reduction in power due to DR is modeled as:

$$P_{DR}(t) = \gamma \times \left(\frac{\pi(t)}{\pi_{\text{base}}}\right) \times \sum_{i=1}^{N} P_{\text{Load}}(t,i) \quad (21)$$

where $P_{DR}(t)$ is the power reduction at time t due to DR, γ is a proportionality constant reflecting the sensitivity of the load to price changes, $\pi(t)$ is the real-time electricity price at time t, and π_{base} is the base-line price. $\sum_{i=1}^{N} P_{\text{Load}}(t,i)$ represents the total load, which is the sum of the individual consumer loads at time t, where N reflects the total number of consumers.

This formulation would be subject to the following constraints:

$$-\alpha(t) \times \sum_{i=1}^{N} P_{\text{Load}}(t,i) \le P_{DR}(t) \le \alpha(t) \times \sum_{i=1}^{N} P_{\text{Load}}(t,i) \quad (22)$$

where $\alpha(t)$ is the utmost allowable load reduction or increase, ensuring that the DR does not exceed certain limits.

2.4. Adaptive Load Shedding Algorithm

The adaptive load shedding algorithm dynamically optimizes load reduction to minimize total operational costs while maintaining grid stability:

$$\min_{P_{DR}(t)} \left(C_{\text{total}}(t) + \lambda \times \mathbb{E}[\text{VSI}(t)]\right) \quad (23)$$

In this equation, $C_{\text{total}}(t)$ represents the total operational cost at time t, λ is a weighting factor that balances the cost and grid stability, and $\mathbb{E}[\text{VSI}(t)]$ is the expected value of the voltage stability index (VSI) at time t, which measures the grid's stability under the current load conditions.

2.5. Constraints

The optimization is subject to several constraints, ensuring the microgrid operates within its physical and operational limits:

2.5.1. Power Flow Constraints

- Active Power Balance:

$$P_{\text{PV}}(t) + P_{\text{WT}}(t) + P_{\text{DG}}(t) + P_{\text{MT}}(t) + P_{\text{Grid}}(t) + P_{\text{DR}}(t) \pm P_{\text{BESS}}(t) = P_{\text{Load}}(t) + P_{\text{loss}}(t) \tag{24}$$

This equation ensures that the total power supply equals the demand plus losses at all times. Here, $P_{\text{PV}}(t)$ is the power generated by the PV system at time t, $P_{\text{WT}}(t)$ is the WT power output, $P_{\text{DG}}(t)$ is the power from DG, and $P_{\text{MT}}(t)$ is the power from MT. The power imported from the main grid is denoted by $P_{\text{Grid}}(t)$, while the power reduction resulting from DR is denoted by $P_{\text{DR}}(t)$. The power of the BESS is denoted by the term $P_{\text{BESS}}(t)$, which can be either positive (discharging) or negative (charging). $P_{\text{Load}}(t)$ represents the entire load demand on the demand side, while $P_{\text{loss}}(t)$ represents the power losses in the system.

- Reactive Power Flow:

$$Q_{\text{flow}}(n,t,m,p) = \sum_{m\in\mathcal{B}} \left(|V_n(t,p)|\cdot|V_m(t,p)|\cdot|Y_{nm}|\cdot\sin(\theta_n(t,p) - \theta_m(t,p) - \phi_{nm}(t,p))\right) \tag{25}$$

This equation represents the reactive power flow between nodes in the power system. Here, $Q_{\text{flow}}(n,t,m,p)$ denotes the reactive power flowing from node n to node m at time t with phase p. The term $|V_n(t,p)|$ is the magnitude of the voltage at node n and $|V_m(t,p)|$ is the magnitude of the voltage at node m. $|Y_{nm}|$ represents the magnitude of the admittance between nodes n and m. The angle difference $\theta_n(t,p) - \theta_m(t,p)$ is adjusted by the phase angle $\phi_{nm}(t,p)$, which accounts for the phase shift introduced by the admittance. The sine function captures the reactive power component of the flow.

2.5.2. Voltage and Current Limits

- Line Current Limits:

$$|I_l(t,p)| \leq I_{l,\max} \quad \forall l \in L_{\text{lines}}, \forall t \in \mathcal{T}, \forall p \in \mathcal{P} \tag{26}$$

This constraint ensures that the current flowing through each transmission line l does not exceed its maximum allowable value $I_{l,\max}$ at any time t and for any phase p.

- Voltage Magnitude Limits:

$$V_{\min} \leq |V_n(t,p)| \leq V_{\max} \quad \forall n \in \mathcal{B}, \forall t \in \mathcal{T}, \forall p \in \mathcal{P} \tag{27}$$

This constraint ensures that the voltage magnitude at each node n remains within the specified minimum $V_{\min}$ and maximum $V_{\max}$ limits at all times t and under all conditions p.

2.5.3. DER Operational Limits

$$0 \leq P_\zeta(t) \leq P_{\zeta,\max} \quad \forall \zeta \in \{\text{PV}, \text{WT}, \text{BESS}, \text{DG}, \text{MT}, \text{Grid}\}, \forall t \in \mathcal{T} \tag{28}$$

This constraint ensures that the power output $P_\zeta(t)$ from each type of distributed energy resource (DER) ζ is within the range from 0 to its maximum allowable output $P_{\zeta,\max}$ at all times t.

2.5.4. Integer Constraints

$$1 \leq B_{\text{bus}}, B_{\text{phase}} \leq N_{\text{bus}}, N_{\text{phase}} \tag{29}$$

This constraint ensures that the integer variables B_{bus} and B_{phase} are within the allowable range, where N_{bus} and N_{phase} represent the total number of buses and phases, respectively.

3. UT for Uncertainty Modeling

In modern microgrids, the integration of renewable energy sources, like PV systems and WTs, introduces considerable uncertainty in terms of their operation due to the variability of weather conditions. Consequently, there is intrinsic uncertainty in load demand, which makes energy management even more complicated [34]. We employ UT as a robust way of propagating uncertainty through nonlinear functions describing the operation of the microgrid.

UT is a statistical methodology that is used to calculate the mean and covariance of a random variable that undergoes a nonlinear transformation. Unlike traditional methods, such as MCS, UT relies on a deterministic sampling technique, called sigma points, to accurately capture the mean and variance of the transformed variable using minimal computational resources [19,46].

In the context of the modeling of the microgrid, the most important uncertainties are considered to be produced by a case of renewable energy generation and one of load demand. The generation of renewable energy, such as from PVsystems and WTs, varies due to changes in irradiance and wind speed, respectively. Similarly, load demand follows a temporal pattern that changes with the time of day, climatic and meteorological conditions, as well as with consumers' patterns of behavior.

3.1. UT Procedure

Let $\mathbf{Z}$ be the vector of uncertain variables comprising renewable generation and load demand. UT is applied to propagate this uncertainty through nonlinear power flow equations as well as the overall optimization problem.

For an n-dimensional random variable, $\mathbf{Z}$ with mean, $\overline{\mathbf{Z}}$ and a covariance matrix, $\mathbf{P}_Z$ UT determines $2n + 1$ sigma points $\{\mathbf{Z}_i\}_{i=0}^{2n}$ according to the following formulae:

$$\mathbf{Z}_0 = \overline{\mathbf{Z}} \tag{30}$$

$$\mathbf{Z}_i = \overline{\mathbf{Z}} + \left(\sqrt{(n+\lambda)\mathbf{P}_Z}\right)_i, \quad i = 1,2,\ldots,n \tag{31}$$

$$\mathbf{Z}_{i+n} = \overline{\mathbf{Z}} - \left(\sqrt{(n+\lambda)\mathbf{P}_Z}\right)_i, \quad i = 1,2,\ldots,n \tag{32}$$

where λ is a scaling parameter given by $\lambda = \alpha^2(n+\kappa) - n$, where α is a small positive constant and κ is a secondary scaling parameter.

Each of the sigma points, $\mathbf{Z}_i$, is propagated through the nonlinear functions modeling the power flow equations and operational constraints of the microgrid. For a nonlinear function $\mathbf{f}(\cdot)$, the transformed sigma points are computed as:

$$\mathbf{Y}_i = \mathbf{f}(\mathbf{Z}_i), \quad i = 0,1,\ldots,2n \tag{33}$$

The mean and covariance of the transformed variable $\mathbf{Y}$ are estimated as:

$$\overline{\mathbf{Y}} = \sum_{i=0}^{2n} W_i^{(m)} \mathbf{Y}_i \tag{34}$$

$$\mathbf{P}_Y = \sum_{i=0}^{2n} W_i^{(c)} (\mathbf{Y}_i - \overline{\mathbf{Y}})(\mathbf{Y}_i - \overline{\mathbf{Y}})^T \tag{35}$$

where $W_i^{(m)}$ and $W_i^{(c)}$ are weights for the mean and covariance, respectively:

$$W_0^{(m)} = \frac{\lambda}{n+\lambda}, \quad W_0^{(c)} = \frac{\lambda}{n+\lambda} + (1-\alpha^2+\beta) \tag{36}$$

$$W_i^{(m)} = W_i^{(c)} = \frac{1}{2(n+\lambda)}, \quad i = 1,\ldots,2n \tag{37}$$

Here, β is a parameter that incorporates prior knowledge of the distribution of $\mathbf{Z}$, such as $\beta = 2$ for Gaussian distributions.

3.2. Incorporating UT into Microgrid Optimization

This is where, in the microgrid optimization problem, the UT algorithm enhances the problem capability for handling uncertainties involved in renewable generation and load demand. The incorporation of UT will hence make the optimization framework capable of handling the mentioned uncertainty in the objective functions and constraints effectively. This can be written as considering the robustness in the final solution against the inherent variability in the system for UT-generated sigma points in the objective functions, which include total cost, voltage deviation, and power losses.

Putting this into a modified objective function, including the uncertainty provided by the following:

$$J_{\text{total}}^{\text{UT}} = \sum_{i=0}^{2n} W_i^{(m)} J_{\text{total}}(\mathbf{X}, \mathbf{Z}_i) \tag{38}$$

where $J_{\text{total}}(\mathbf{X}, \mathbf{Z}_i)$ is the total objective function evaluated at the i-th sigma point $\mathbf{Z}_i$ and $\mathbf{X}$ is the vector of decision variables.

Furthermore, the constraints are checked at all the sigma points to ensure that the constraints are valid along the range of the uncertain variables. This may be expressed as:

$$\text{Subjectto} : g_k(\mathbf{X}, \mathbf{Z}_i) \leq 0, \forall k, \forall i = 0, 1, \ldots, 2n \tag{39}$$

This concept keeps the microgrid's operational states within safe limits of operations, given the uncertainties in renewable energy generation and load demand. The incorporation of UT into the optimization problem gives robustness to the problem for the more reliable and efficient operation of microgrids.

4. The Proposed Algorithm

4.1. Overview of the Cheetah Optimizer (CO)

The CO is an optimization algorithm inspired by the hunting strategy of cheetahs [45]. It is based on the different strategies that a cheetah deploys in locating its prey, such as scanning, stalking, and rapid acceleration, that enable it to efficiently locate and capture the hunted object. The CO could demonstrate promising performance in regard to various optimization problems due to its appropriate balance between exploration and exploitation through different intelligent search strategies. However, while the CO is good at finding near-optimal solutions, there are some issues in terms of evading the local optima and keeping diversity in the search space. These might engender premature convergence and poor performance in complex high-dimensional optimization problems.

We try to address these challenges by proposing the Enhanced Cheetah Optimization Algorithm, which tries to improve upon the original CO by introducing dynamic and adaptive mechanisms that amplify global exploration and finesse in the exploitation phase. The ECOA avoids such risk from the local optima, further increasing the convergence speed and enhancing the overall robustness of the algorithm in solving such complex optimization tasks using introduced stochastic processes and cooperative strategies.

4.2. *Enhanced Cheetah Optimization Algorithm (ECOA)*

The ECOA is an enhanced version of the original Cheetah Optimizer. In the enhanced variant of the Cheetah Optimization Algorithm, the mechanism of exploration and exploitation has become much more sophisticated, involving several major adaptive mechanisms and stochastic processes that enhance performance and avoid the local optima. Below, the key elements of the ECOA are outlined.

4.2.1. Exploration by Stochastic Jump Strategy

The ECOA's exploration is based on a stochastic jump technique inspired by Lévy flights. This method allows for big, dynamic jumps in the search space, which contributes to improved global exploration. During iteration t + 1, the position of the i-th cheetah in the j-th dimension is updated:

$$X_{i,j}^{(t+1)} = X_{\text{leader},j}^{t} + \epsilon \cdot \text{Lévy}(j) \cdot \gamma_{i,j}^{t}, \tag{40}$$

where

- $X_{\text{leader},j}^{t}$ is the leader position in the j-th dimension at iteration t.
- ϵ is a small positive constant controlling the jump intensity.
- $\text{Lévy}(j)$ is a random vector derived from the Lévy distribution.
- $\gamma_{i,j}^{t}$ is the step size at iteration t, dynamically adjusted based on the cheetahs' relative position.

The step size of $\gamma_{i,j}^{t}$ in Equation (40) changes dynamically in the ECOA, using the cheetah's role and position in the population, i.e., it naturally balances the exploration and utilization through this adaptative mechanism:

$$\gamma_{i,j}^{t} = \kappa \cdot (X_{\text{prey},j}^{t} - X_{i,j}^{t}), \tag{41}$$

where κ is a scaling factor that depends on the amount of distance that one is from optimum.

The second-best solution guides the search in Equation (40). In every iteration, $X_{\text{leader},j}^{t}$ drives the search. In such a way, the ECOA refines the search in regions around the best-known solutions, improving convergence:

4.2.2. Stationary Ambush Strategy

If the ECOA finds promising regions, then a stationary ambush strategy is employed which saves energy by reducing all unnecessary movements. Early convergence is prevented, with the position update rule being defined as:

$$X_{i,j}^{(t+1)} = X_{i,j}^{t} \tag{42}$$

This would imply that the cheetah does not move from its position to maintain stability.

4.2.3. Adaptive Chase Mechanism

The ECOA introduces an adaptive chase mechanism during the exploitation phase, wherein cheetahs adjust their positions based on the optimal solution (prey). The position update is defined as:

$$X_{i,j}^{(t+1)} = X_{\text{prrey},j}^{t} + \eta_{i,j} \cdot \zeta_{i,j}^{t}, \tag{43}$$

where $\eta_{i,j}$ is the randomness factor, simulating unpredictable movements during the chase. $\zeta_{i,j}^{t}$ represents the cooperative interaction factor.

The ECOA enhances the performance by incorporating a mechanism of cooperative interaction among cheetahs. This mechanism helps to increase cooperation inside the population, significantly enhancing convergence rates and robustness against local optima:

$$\zeta_{i,j}^{t} = X_{i,j}^{t} - X_{k,j}^{t}, \tag{44}$$

Here, $X^t_{k,j}$ is upper neighborhoods (other cheetah position at the same dimension developing the cooperative search).

The psedue-code of ECOA is summarized in Table 1.

Table 1. Pseudo-code for the ECOA.

Step	Description	Equation/Details
1	Define parameters	D: dimensionality, n: population size, MaxIt: maximum iterations
2	Generate initial population	Generate $\mathbf{X}_i^0$ for $i = 1, 2, \ldots, n$ and evaluate fitness
3	Initialize iteration counter	$t \leftarrow 1$
4	Main loop	While $t \leq$ MaxIt do
5	Sort and select	Sort population, select prey $\mathbf{X}_{\text{prey}}$ and leader $\mathbf{X}_{\text{leader}}$
6	Iterate over each cheetah	For each cheetah i do
7	Select neighbor	Select neighbor cheetah k
8	Randomly select dimensions	$\mathcal{J} \subset \{1, 2, \ldots, D\}$, size $d = \left\lceil \frac{D}{\text{rand}_i \lceil \frac{D}{3} \rceil} \right\rceil$
9	Stationary ambush	$X^t_{i,j} \leftarrow X^{(t-1)}_{i,j}$ for all $j \in \mathcal{J}$
10	Calculate parameters	Update H_i^t
11	Ambush condition	If $H_i^t > 0.25$ then
12	Exploration strategy	$X^{(t+1)}_{i,j} \leftarrow X^t_{\text{Prey},j} + \epsilon \cdot \text{Lévy}(j) \cdot \gamma^t_{i,j}$
13	Otherwise, use exploitation	Else
14	Exploitation strategy	$X^{(t+1)}_{i,j} = X^t_{\text{prrey},j} + \eta_{i,j} \cdot \zeta^t_{i,j}$
15	End condition check	End If
16	Update position	Update $X^{(t+1)}_{i,j}$ if new position is better
17	End cheetah iteration	End For
18	Update best solutions	Update prey $\mathbf{X}_{\text{prey}}$ and leader $\mathbf{X}_{\text{leader}}$
19	Increment iteration	$t \leftarrow t + 1$
20	End main loop	End While
21	Return best solution	Return $\mathbf{X}_{\text{prey}}$ as the best solution

The ECOA achieves automatic optimal parameter acquisition through several adaptive strategies that dynamically adjust as the algorithm progresses. First, the stochastic jump strategy enhances global exploration by using Lévy flights to allow for large, adaptive jumps in the search space. The step size $\gamma^t_{(i,j)}$ adjusts automatically based on each cheetah's position and role, facilitating a smooth transition from exploration in early stages to exploitation as optimal solutions are approached.

In the adaptive chase mechanism, cheetahs modulate their positions relative to the optimal (prey) solution, using both a randomness factor $\eta_{(i,j)}$ and a cooperative interaction factor $\zeta^t_{(i,j)}$. This dual adjustment creates an inherent balance between the exploration and exploitation phases without the need for manual tuning, as it adapts in real-time to the search landscape.

Finally, cooperative interaction among cheetahs contributes to efficient convergence by fostering collaboration between individuals in the population, helping to prevent premature convergence to local optima. This adaptive cooperation among neighboring solutions automatically tunes parameters to enhance both convergence speed and robustness. Together, these strategies enable the ECOA to perform optimally across complex, unbalanced three-phase distribution networks, facilitating both global and local searches without manual parameter settings.

4.3. Implementation of the Proposed Methodology

The steps for solving the problems involved in operation and planning in microgrids using the ECOA and UT for handling uncertainties are shown in Figure 1 and briefly explained as follows:

- Step 1: Define input data.
    - Initialization of the microgrid data includes the prices for the electric energy, the load data, and the characteristics of the DERs.
    - All the parameters of the algorithm have to be determined: the size of the population, the number of iterations, and the parameters of UT, i.e., the number of uncertain variables and their respective probability density functions.
- Step 2: Reduce the problem to an unrestricted form.

  Convert the constrained optimization problem into an unconstrained one, penalizing all constraint violations by using a penalty function so that all the solutions satisfy the operational and safety limits of the microgrid.
- Step 3: Generation of initial population.

  Generate an initial population of solutions for the ECOA. Each solution is a feasible microgrid operation configuration that are generated within prescribed limits; in cases when constraint violations occur, the solutions are repaired to make them feasible.
- Step 4: Evaluate the objective function using UT.

  For each population solution, use UT to propagate the uncertainty in renewable energy generation and load demand through the power flow equations. Determine the sigma points for each parameter that is unknown. Transform the sigma points using the nonlinear microgrid functions and then calculate the goal function 2n + 1 times, where m represents the number of unknown parameters. Calculate the expected value of the objective function based on the results of UT.
- Step 5: Choose the best solution.

  Keep the best solution, which has the lowest predicted cost or objective value among the analyzed population.
- Step 6: Apply the ECOA Improvisation Stage.

  Update the population using ECOA improvisation mechanisms; explore and exploit the solution space through interactions between the prey, leader, and other cheetahs.
- Step 7: Improvements to the CO.

  Improve the quality of the solutions and increase the convergence and variety of the population by applying the ECOA's specific variation operators to the newly updated population.
- Step 8: Conclusion.

  The algorithm terminates if any of the termination criteria (the utmost number of iterations or the stated objective function value) are satisfied. If they are not, proceed to Step 4 to continue refining and iterating the solutions.

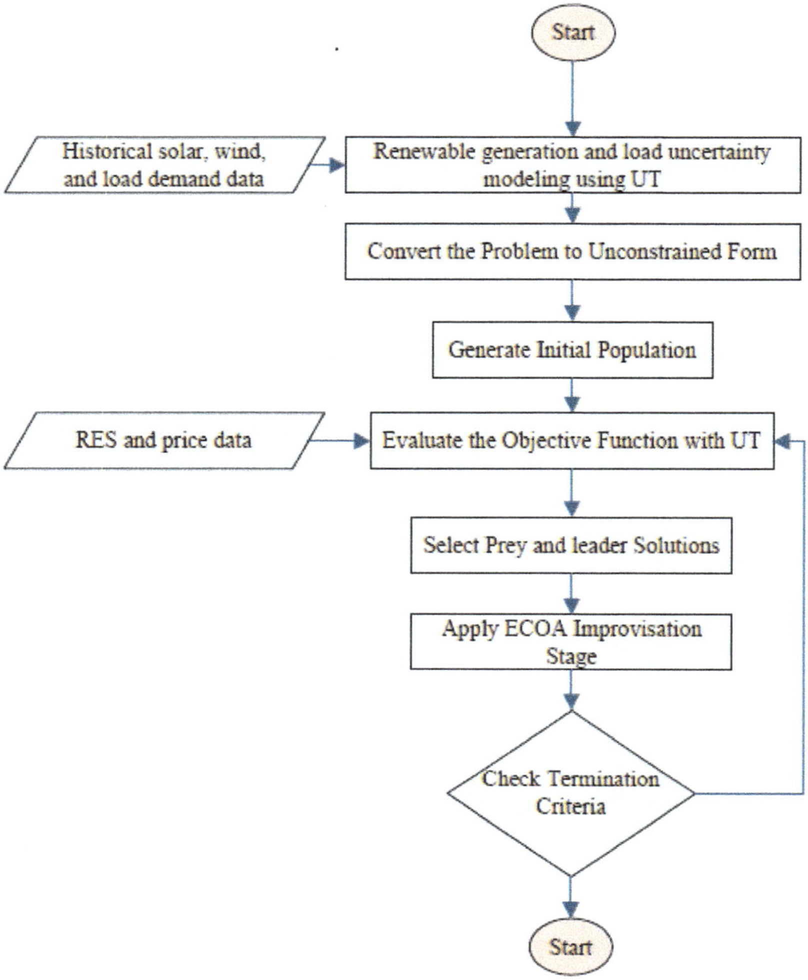

Figure 1. Flowchart of the proposed UT-based ECOA for optimal solving of EM problems.

5. Simulation Results

5.1. Overview of the Simulated Microgrid System

The difficulties of integrating large amounts of RESs, including PV systems and WTs, into an imbalanced three-phase distribution network are handled by the simulated microgrid system. The model records the dynamic interactions between changing renewable energy and changing load needs using the IEEE 13-bus test system [47]. Essential technical details of the generation units can be found in Table 2 [48–50]. In addition, the hourly active power production of the WT and PV and the hourly load demand forecast are shown in Figure 2. Based on current laws, particularly those set by the Ontario energy board, the model's DR behavior should be affected by its use of a TOU pricing system. In this scenario, grid electricity accounts for 50% of the base load, MTs provide 10%, and the remaining demand is met by PV systems, WTs, and the BESS. Stochastic weather patterns and fluctuating demand are used to simulate the real-time fluctuation of RES output and load situations. The methodology will find the best power-generating mix—PV systems, WTs, DGs, MTs, and the BESS—by integrating DR programs and striking a balance between supply and demand. The optimization was performed using the ECOA in conjunction with UT for uncertainty management.

Table 2. Parameters of the utilized resources in the model.

Generation Source	Parameter (Unit) Value					
MT	μ_{CO_2} (kg/kWh)	μ_{SO_2} (kg/kWh)	μ_{NO_x} (kg/kWh)	β_ξ (USD/kWh)	α_ξ (USD/kWh)	γ_ξ (USD/h)
	0.72	0.002	0.091	0.000018	0.03872	1.356
DG	μ_{CO_2} (kg/kWh)	μ_{SO_2} (kg/kWh)	μ_{NO_x} (kg/kWh)	β_ξ (USD/kWh)	α_ξ (USD/kWh)	γ_ξ (USD/h)
	0.65	0.093	4.483	0.0002	0.03632	1.65
PV	$C_{INV,\zeta}$ (USD)	$C_{O\&M,\zeta}$ (USD/kW)	CF_ζ	T_{Life} (years)		
	6675 × 400	0.012	0.3	25		
WT	$C_{INV,\zeta}$ (USD)	Oper.cost (USD/kW)	CF	T_{Life} (years)		
	1500 × 400	0.00952	0.2	20		
BESS	$C_{INV,BESS}$ (USD)	$C_{O\&M,CHG}$ /$C_{O\&M,DSG}$ (USD/kW)	CF_{BESS}	$T_{Life,BESS}$ (years)		
	1775 × 200	0.05	0.25	25		
Grid	μ_{CO_2} (kg/kWh)	μ_{SO_2} (kg/kWh)	μ_{NO_x} (kg/kWh)			
	0.85	2.14	9.723			
DR	$\beta_{I/C}$ (USD/kWh)					
	0.1					

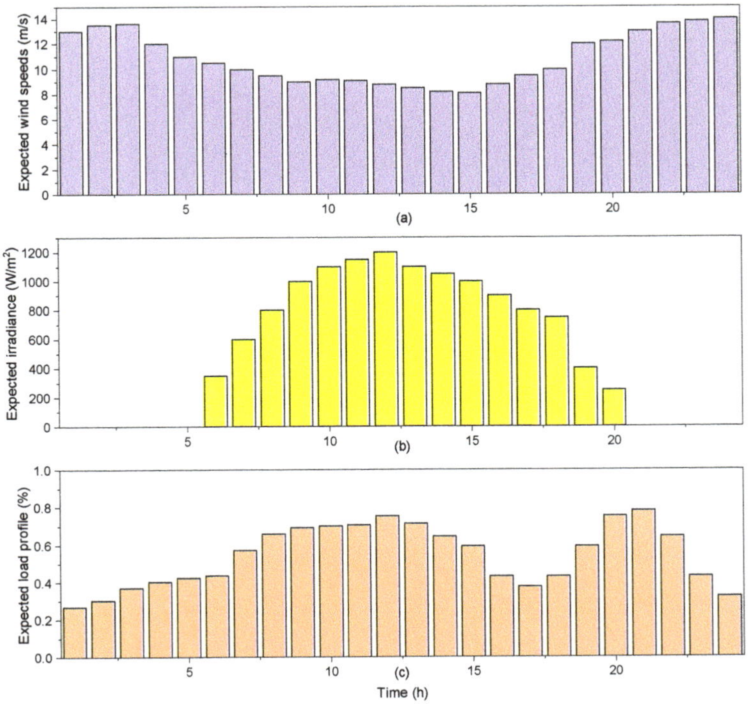

Figure 2. The mean values of (**a**) wind speed, (**b**) solar irradiance, and (**c**) load demand.

The simulation parameters are chosen to ensure that the results are robust: 25 runs with 100 iterations and a population size of 50 for each procedure. The conventional CO, PSO, and GA, as well as two hybrid algorithms (GPSOA and HPSO-GSA), were compared, and testing revealed that the ECOA performed better in terms of cost minimization and energy supply optimization under uncertainty. All simulations were run in MATLAB R2021b on a PC with an Intel i5 processor, 4 GB RAM, and a CPU speed of 2.5 GHz. This can assure steady performance in computationally intensive optimization processes. The established approach can provide practical and scalable insights into how a microgrid operates, allowing for a full assessment of system efficiency and dependability.

5.2. Optimal Power Generation and Resource Contribution (Continuse Variables)

The proposed methodology is used to find the optimal power generation performances inside a microgrid provided with distributed sources like PV systems, WTs, DGs, MTs, and the BESS. It also implements various DR programs. The proposed model captures the temporal variability of the power output of these sources to simulate the performance under variant conditions, as depicted in Figure 3.

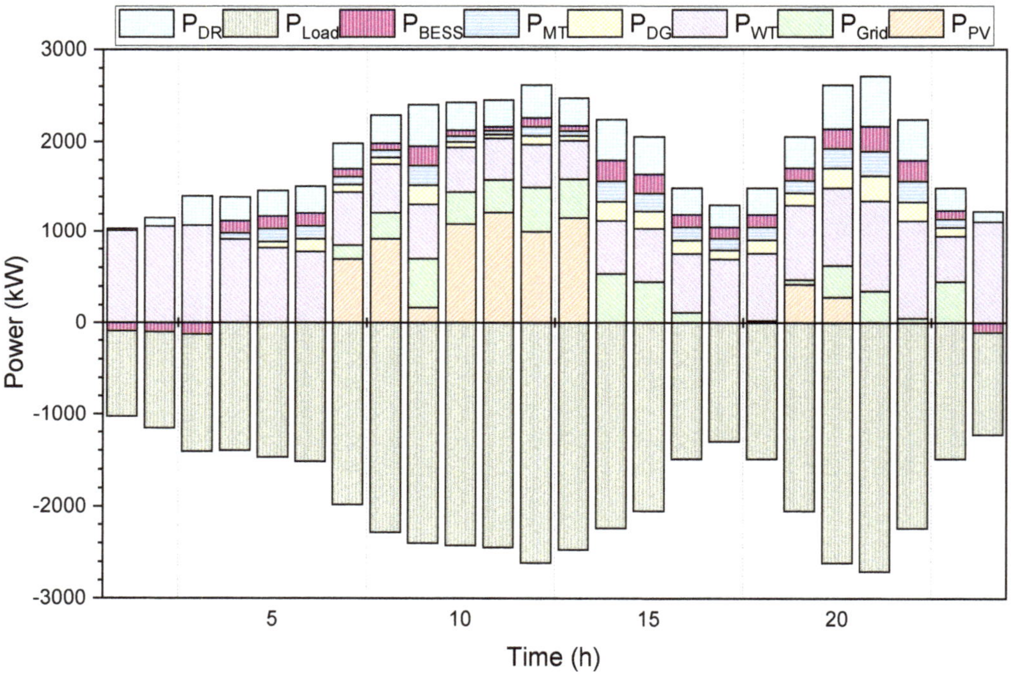

Figure 3. Microgrid's Optimal generation scheduling.

The generated PV generation is very negligible in most of the night-time hours (for example, from hour 1 to hour 5). However, from hour 6, it increases further to midday, reaching a maximum output of 991.2684 kW at hour #12. The WTs produce a decent amount relatively stably throughout the day; their peak generation is at hour 19, which is 837.6866 kW, showing that wind energy is one of the most reliable renewable resources inside the microgrid. Utility grid power usage varies by the availability of the local generation. In this case, during low output of the renewable energy sources (from hour 1 to 5 h), utility grid power contributes a minimum, peaking at hour 14 with 531.3536 kW, where demand and the amount drawn from the microgrid are higher. Dispatchable sources such as DGs and MTs are of paramount importance for backup power, especially during periods when the generation from renewable sources is at an all-time low. The peak output for diesel generators falls at hour 9, with 211.9006 kW, while peak output for microturbines falls at hour 5, with 146.9435 kW, contributing quite effectively during these high-demand periods.

The BESS has a significant role that involves smoothing the microgrid energy fluctuations, absorbing the surplus power during low demand, and releasing the stored energy at peak demand. From this perspective, negative values in the BESS output signify charging states, while positive values denote discharging states. For instance, a BESS charged during hours 1–5, when renewable generation is limited, gains extra energy by absorbing the negative values of −93.24 kW, −104.79 kW, and −128.02 kW to enhance its storage capacity. On the other hand, this extra stored energy is utilized efficiently during peak demand periods to meet the high demand while easing the dependence on the grid, as in hour 10, where the BESS discharges 62.78 kW. This interaction between the BESS and other sources, such as WT and PV systems, provides an optimum balance between energy supply and demand during the day, thus improving the general efficiency and reliability of the microgrid.

In fact, DR programs have proven to be efficient in reshaping the load curve, especially during peak demand hours. The modified load, after the application of DR, is shown in Figure 4 and indicates a critical reduction in demand from hours 7 to 21. For example, the original load of hour 9 was 1949.73 kW, which throttles down to 1242.08 kW after the application of DR. This provided a smoothening effect on the load profile and will reduce the strain that would otherwise have been placed on the generation units, thus adding to the system efficiency.

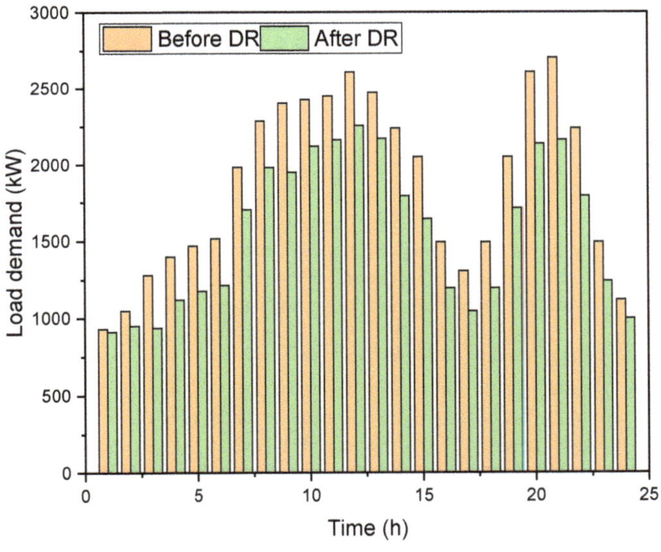

Figure 4. DR's effect on the hourly load curve.

5.3. Impact of TOU Pricing on Microgrid Operations

TOU pricing, therefore, grants influence over the optimization of microgrid operations through segmenting electricity costs into three distinct periods: off-peak at 8.7 ¢/kWh, mid-peak at 12.2 ¢/kWh, and on-peak at 18.2 ¢/kWh. This form of pricing encourages the shifting of power generation and consumption for both economic efficiency and operational reliability.

- Off-Peak Period (Hours 1–6 and 20–24)

From Figure 3, when the off-peak hour is considered to have the lowest price of electricity, the microgrid prioritizes charging the BESS. The negative values of the BESS in those periods show the state of the BESS while charging, absorbing excess generation. Still, at hour 1, −93.24 kW is utilized from the BESS due to the optimization of energy costs.

The wind generation also starts to kick in, albeit at a lower level; therefore, WTs can supply base energy. This will pretty effectively minimize reliance on the more expensive grid electricity as the ESS builds its charge for later use.

- Mid-Peak Period: 7–10 h; 17–19 h

During the mid-peak hours of the TOU pricing, WTs and the BESS begin to become more important in terms of maintaining generation and load balance (see Figure 3). A WT will start giving its contribution fully with an approximate 630.3 kW output during such mid-peak periods to help reduce reliance on expensive grid electricity. Meanwhile, the BESS moves into a discharge state, providing power to reduce further costs. For instance, in hour 8, it supplies 76.22 kW, which shows that it utilizes the stored energy efficiently to meet the demand in the period of high pricing.

- On-Peak Period [Hours 11–16]

It is at this point that the microgrid operates when the electricity prices are highest; it concentrates its strategy on maximizing the deployment of local renewable energy sources, such as PV systems. As such, during hour 10, for example, the output of PV reaches its peak of 1083.23 kW and thus reduces dependence on highly expensive grid power. The WT still provides a considerable amount of energy, with the output reaching 452.73 kW at hour 11; hence, it is important for the minimization of the operational costs because of peak pricing. Combining these sources will enable the microgrid system to function during periods of high demand without incurring many utility costs.

DR programs also emerge as one of the most essential strategies in all periods. For instance, as shown in Figure 4, at hour 12, DR reduces the load from 2255.29 kW to 1127.64 kW and, hence, works effectively while shifting consumption at a time when the prices are high. Throughout the process, as shown, the cumulative effect of DR enhances the microgrid to match the consumption based on the pricing structure. This saves stress on the grid and also encourages economic efficiency. The efficient TOU pricing methodology, the optimal generation of WTs, charging and discharging of the energy storage system, and DR programs together effectively enhance the operational efficiency of the microgrid. An overall cost reduction in operational costs amounting to 80% from scenarios without such optimization measures is a testament to the success of this integrated approach in fostering economic viability while ensuring a reliable and sustainable energy supply.

5.4. *Optimal Location-Specific Performance Analysis (Discrete Variables)*

The analysis of the optimal generation results points to the performance of not only each generation option but also specific locations and phases that are important from the point of view of the distribution network. In that respect, efficiency in the delivery of power can be guaranteed together with system reliability at different load conditions during the course of the day.

According to the results indicated in Figure 5, regarding PV generation, the optimal performance is in Bus 2, Phase A, where solar energy generation is at its peak in the afternoon. This area receives ideal sunlight, which makes it a perfect site to mount a PV system. Other phases have relatively smaller outputs of PV, especially Bus 1 and Bus 4, which further illustrates the need for proper site selection to tap into the full potential of solar resources effectively.

Figure 6 also identifies Bus 2 as one of the highly contributing buses across all phases, especially in Phase A and Phase B, which can contribute significant robustness to utility supply in the stabilization process during periods of smaller renewable outputs.

Figure 7 shows the results of the dispersed WT generation; it can be seen that the highest contributions are obtained at Bus 2 in Phase A, particularly during the early hours of the morning. This sort of pattern characterizes the good wind conditions of this point; it means that wind installations must consider the geographical aspect of siting. The other buses, like Bus 1 and Bus 4, show low wind output; hence, siting becomes an important aspect to be considered in regard to WTs.

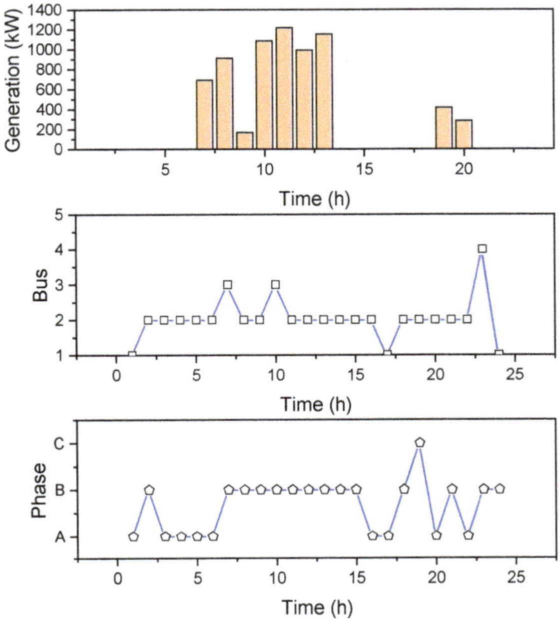

Figure 5. Optimal results of the PV's power generation, bus, and phase locations.

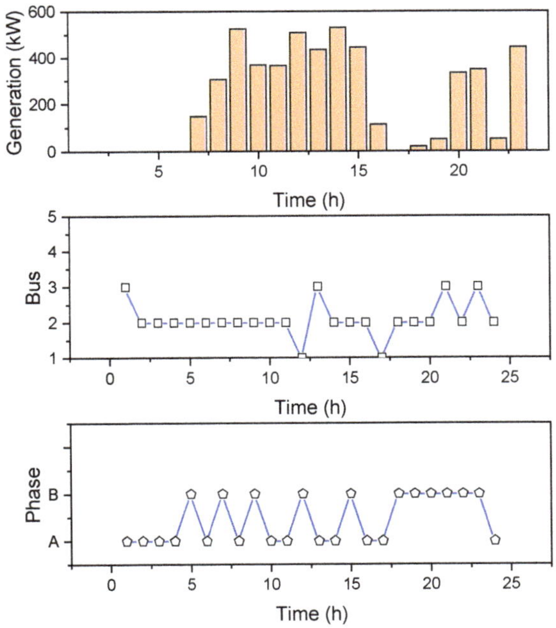

Figure 6. Optimal results of the grid's power generation, bus, and phase locations.

Figure 8 further presents the DG at Bus 2, Phase A, which features continued optimal output, resonating with its capability of delivering steady power throughout the day. The installations of DGs within the subject phase are appropriately positioned to support fluctuating renewable generation and hence deliver a steady supply during peak demand periods.

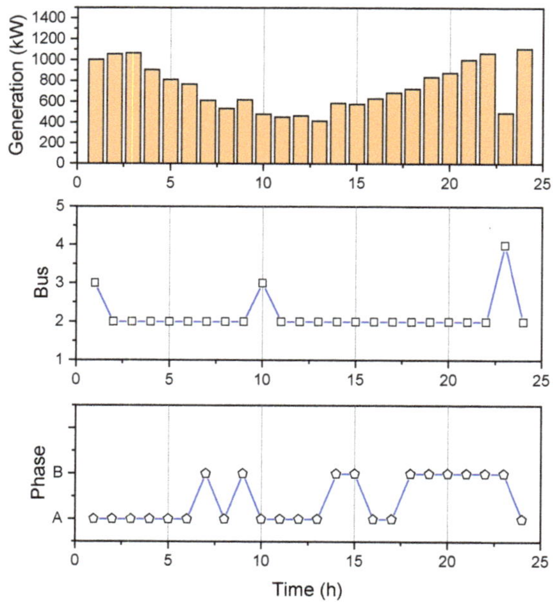

Figure 7. Optimal results of the WT's power generation, bus, and phase locations.

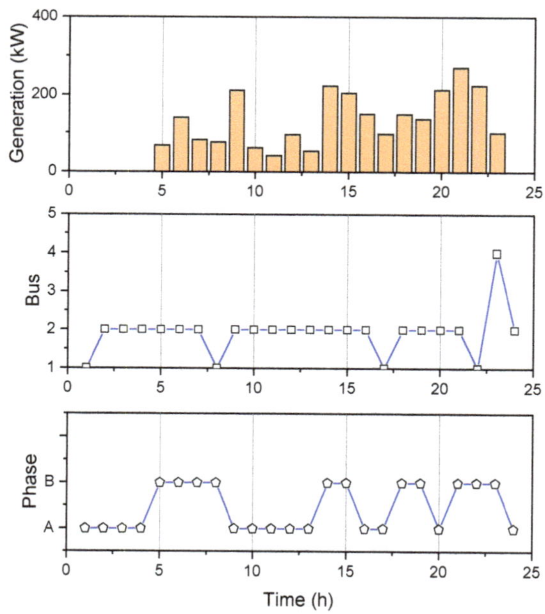

Figure 8. Optimal results of the DG's power generation, bus, and phase locations.

It can be observed from Figure 9 that the MT outputs are maximized at Bus 2, and Phase B in particular has been found to be very effective during peak hours. The fact that MT generation coincides with demand peaks points to its potential contribution to enhancing the resilience of the system. Future plans of expansion should be biased toward areas with higher MT outputs so as to maintain a balanced energy supply.

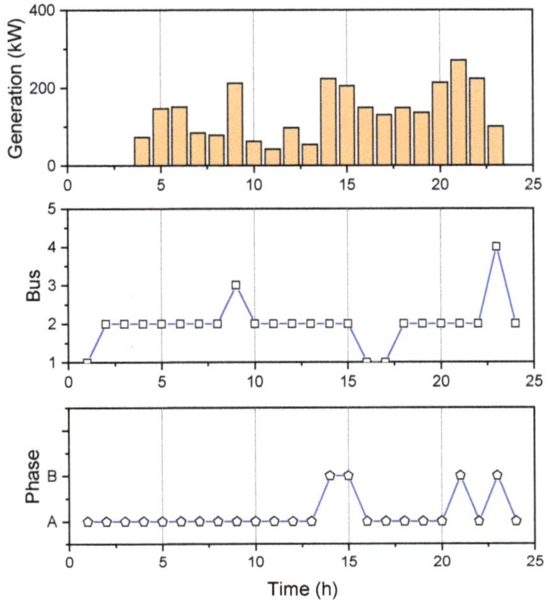

Figure 9. Optimal results of the MT's power generation, bus, and phase locations.

The BESS is, according to Figure 10, most effective at Bus 2. It is capable of providing power during evening peak loads and replenishes its capacity by charging during the day. Placement of the ESS in this location serves to enable superior management of the energy flows to ensure that any excess generation from the PV and WT is valorized effectively.

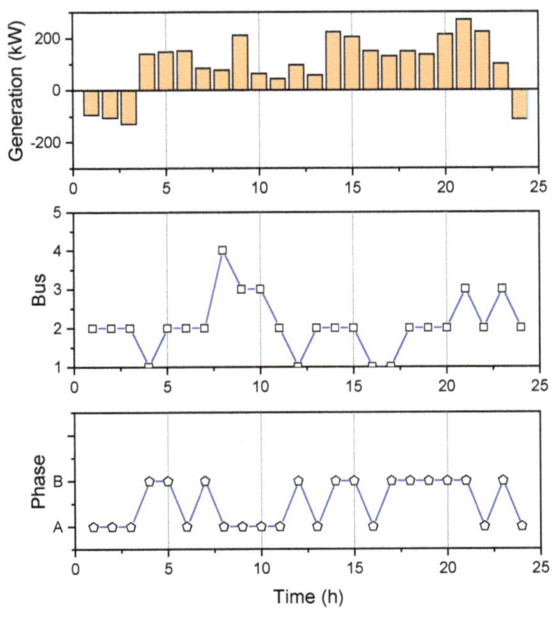

Figure 10. Optimal results of the BESS's power generation, bus, and phase locations.

On the whole, the results depict Bus 2 as the best place of location for most of the generation options, especially in Phase A. This means that there is a need to have a proper site assessment during the actual planning of renewable energy projects and also at the deployment of the DGs. The final result of the outcome is used in showing how critical location and phase selection are in enhancing the efficiency and reliability of the energy system and therefore in ensuring a more robust and sustainable power network.

5.5. Impact of the Proposed Model on Voltage Deviation, Energy Losses, and Costs

The proposed model has huge consequences for, but not limited to, key performance indicators like voltage deviation, energy losses, and operational costs. A deeper look into such impacts is presented which proves the effectiveness of the model in enhancing the overall stability, efficiency, and, therefore, economic viability of the network.

The significant voltage stability improvements were realized after the execution of the proposed model within the studied periods. Figure 11 clearly shows that the values of VD (voltage deviation) were always about 1.489 to 1.490, representing a significant deviation with significant questions about the reliability of the distribution network. After implementation, the VD was proven to reduce significantly during peak operational hours. For example, in the first hour, the VD reduced from 1.48966502 p.u to 1.340699 p.u, and similar patterns continued throughout the entire day, as depicted with the 24th hour reducing from 1.48970899 p.u to 1.340902 p.u. The foregoing results totally represent how well the model works in controlling voltage deviations, thus enhancing power delivery stability and reliability within microgrids.

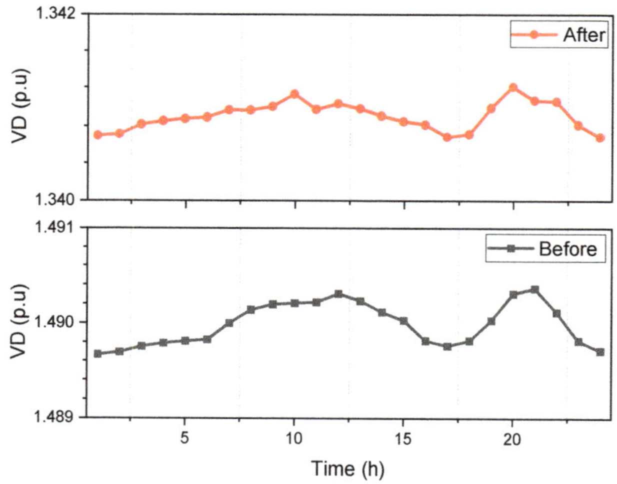

Figure 11. Voltage deviations before and after the proposed optimization EM model.

Among other elements, the model has provided voltage stability. There was also a significant reduction in the loss of energy. Figure 12 shows the reduction in energy loss to be constant both before and after the implementation of the model. Before optimization, the losses were at their peak of 37.10 kW during hour 21. However, with the use of the proposed model, energy losses went down considerably: in hour 1, energy losses went from 9.61 kW to 8.66 kW, while, by hour 24, the extra losses were from 11.44 kW to 9.95 kW, reflecting improved system efficiency. Total energy losses, which were 499.76 kW before implementation, went down to 446.65 kW after implementation, underlining the capability of the model for optimization in energy distribution and indicating that less energy loss occurs.

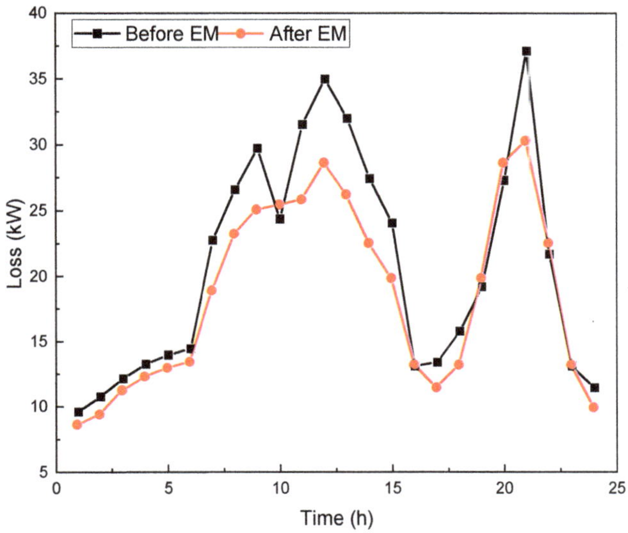

Figure 12. Microgrid losses before and after the proposed optimization EM model.

In the end, the model contributed to a significant impact on the total costs related to different sources in the microgrid generations.

The total microgrid operation cost with RESs is USD 220,921.39, as represented in Table 3. Compared to a non-RES scenario, which was a total of USD 1,324,587, this is a huge cost reduction. In detail, the breakdown includes the following: PV at USD 8667.05, grid power at USD 20,756.11, WTs at USD 40,142.04, and distributed generation at USD 38,145.56. Added to this are microturbines at USD 8715.11 and ESS at USD 61,655.70, while DR accounted for USD 42,839.82.

Table 3. Daily costs of the test system using proposed method.

	PV	Grid	WT	DG	MT	BESS	DR	Total
Cost (USD)	8667.055	20,756.11	40,142.04	38,145.56	8715.105	61,655.7	42,839.82	220,921.3863

The huge cost difference underlines the financial advantages of integrating RES into the microgrid, thus proving it feasible both from an environmental and economic point of view. These findings should be able to validate the proposed model with voltage stability improvement, energy loss reductions, and huge cost savings. Thus, a robust solution for the optimization of MGs can be derived.

5.6. Results Comparison with Other Algorithms and Stochastic Methods

This section compares the performance of the proposed ECOA with other techniques that traditionally have been used in microgrid management. For this purpose, the main performance indicator under study is the fitness function values $J_{\text{total}}^{\text{UT}}$.

In optimization algorithms, various parameters play a crucial role in determining their performance and effectiveness. For instance, in the HPSO-GSA, the parameters are set as follows: $w_{\max} = 0.9$ and $w_{\min} = 0.4$ (inertia weights), and $c_{1\min} = 0.5$, $c_{1\max} = 2.5$, $c_{2\min} = 0.5$, and $c_{2\max} = 2.5$ (cognitive and social coefficients) [9]. The parameter $Tf = 20$ represents the iteration interval cycle and $S_{\min} = 4$ refers to the minimum size of the swarm. Similarly, the GPSOA employs parameters such as $c_3 = c_4 = 0.5$ (acceleration factors) along with $\omega_{\max} = 0.9$, $\omega_{\max} = 0.4$, and $c_1 = c_2 = 2$ (cognitive and social factors) [8]. The PSO algorithm typically uses $\omega = 0.7$ (inertia weight) with $c_1 = c_2 = 2$ (cognitive and social coefficients). The GA is characterized by a crossover rate of 0.8 and a mutation rate of 0.1. In contrast, our proposed ECOA does not require parameter settings, similar to the traditional CO algorithm [45], which streamlines its implementation and enhances its applicability in diverse optimization scenarios.

The simulation results show that the ECOA outperforms the applied traditional and hybrid optimization techniques. As reflected in Table 4, the best minimum fitness value gained by the ECOA was 2.090865, with its maximum value being 2.101378, featuring a mean of 2.089554 and a small standard deviation of 0.01126, representing the high stability and reliability of the ECOA. For its part, the GA has an average fitness of 2.412126; PSO and the conventional CO algorithm reached means of 2.33533 and 2.225719, respectively.

Table 4. Comparative results of the fitness function values (J_{total}^{UT}) obtained by the applied Optimization Algorithms.

Algorithm	Min	Max	Mean	SD	CPU Time (s)
GA	2.346278	2.46748	2.412126	0.031894	67.65
PSO	2.283724	2.393773	2.33533	0.02973	42.15
CO	2.187906	2.286634	2.225719	0.02943	55.02
GPSOA	2.120345	2.210457	2.145893	0.025870	51.47
HPSO-GSA	2.105437	2.202345	2.135672	0.020110	57.85
ECOA	2.090865	2.101378	2.089554	0.01126	46.26

Furthermore, the GPSOA and HPSO-GSA hybrid algorithms also demonstrated competitive performance, with mean fitness values of 2.145893 and 2.135672, respectively. These results underline the effectiveness of hybrid algorithms in enhancing optimization outcomes in microgrid management. In addition, as can be seen from Table 4, the ECOA's CPU time of 46.26 s indicates that this algorithm has good competitiveness in terms of effectively throwing a balance between its efficiency and performance. In contrast, the GA took 67.65 s, meaning that, besides producing worse optimal results, it was even slower compared to the ECOA.

Moreover, the GPSOA and HPSO-GSA also exhibited competitive results, with minimum fitness values of 2.120345 and 2.105437, respectively, and CPU times of 51.47 s and 57.85 s. Although both hybrid algorithms showed improved performance over the GA, the ECOA remains the superior choice regarding both optimal results and computational efficiency.

Looking closer, the convergence characteristic is somewhat different, as shown in Figure 13. The highest value of 3.5 was initiated by the GA, which showed relatively slow convergence with noticeable fluctuations during the first iterations that then stabilized after the thirty-fifth iteration, with a value of around 2.346278. Thus, the final value was approached rather gently, and the convergence behavior followed a step-like pattern. Meanwhile, PSO started from a relatively lower point, 3.0, and decreased much more smoothly before converging to 2.283724 until iteration 60. This one turned out to be an immensely smoother algorithm where the fitness curve did not suddenly increase erroneously often, which indicates the strength of this algorithm in spreading its coverage over the solution space. Meanwhile, the behavior of the conventional CO algorithm was similar, starting at approximately 3.1, and, after 40 iterations, it converged to 2.187906. Yet, its convergence was marked by slower fitness reductions compared to the ECOA, and hence it was less efficient in its approach to the optimal solution.

While all other algorithms were showing convergence attitudes, the ECOA, however, showed the fastest convergence rate among the compared algorithms. This algorithm, with an initial value starting at around 3.0, has successfully fallen below its fitness to 2.090865 within just 30 iterations and has kept that value constant, with a maximum of limited fluctuation thereafter. Thus, the ECOA's convergence curve is comparatively smooth, indicating that this algorithm executes accordingly in minimizing an objective function in uncertain renewable generation conditions.

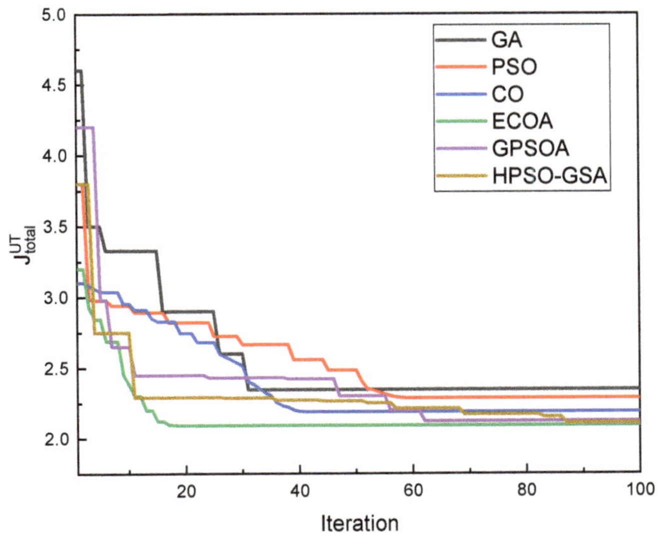

Figure 13. Convergence curves of the comparative algorithms in solving the problem.

The hybrid algorithms, GPSOA and HPSO-GSA, also exhibit competitive convergence characteristics. However, the ECOA's overall speed and stability remain advantageous: the GPSOA starts from a higher initial value of 4.2 and shows a gradual reduction, reaching a fitness of around 2.12035 by iteration 70. Additionally, while the GPSOA maintains a smooth convergence curve, its rate of descent is slower than the ECOA's. This indicates that the GPSOA, though efficient, is less aggressive in its convergence approach compared to the ECOA's rapid optimization. The HPSO-GSA, which begins at an initial value of 3.8, demonstrates a slightly faster convergence than the GPSOA, reaching a final fitness of 2.10544 by iteration 60. The HPSO-GSA's stable curve and minimal fluctuations indicate an effective performance; however, the ECOA still outpaces it, reaching an optimal value in fewer iterations.

These findings show that the ECOA is indeed a powerful and effective technique for the management of microgrid systems, especially when their operation is under uncertain renewable generations.

Finally, to better understand the performance of the proposed UT approach, simulation results are further analyzed together with other stochastic frameworks such as MCS and the PEM. In Table 5, the proposed UT has a cost function value of $J^{UT}_{total} = 2.090865$, which is competitive in accuracy against the MCS at a higher value of 2.091234.

Table 5. Comparison of Cost Function Value in Different Stochastic Frameworks.

Method	J^{UT}_{total}	CPU Time (s)
MCS [10]	2.091234	137.00
PEM [18]	2.090950	37.10
Proposed UT	2.090865	46.26

While the MCS indeed gave a comprehensive representation of the uncertainties, it required significantly more computational time (137 s) compared with both the PEM at 37.10 s and the proposed UT method at 46.26 s. The results show that the UT approach does strike a good balance in terms of performance for both computational efficiency and accuracy, hence being robust in performing microgrid management under uncertainty.

5.7. Discussion

The results from this work identified the ECOA in terms of optimizing microgrid management. The results clearly depict the ECOA as outperforming all traditional and hybrid optimization techniques, such as the GA, PSO, the conventional CO, the GPSOA, and the HPAO-GSA, by most of the key performance indicators. The statistical analysis of the fitness function provides supportive evidence for the ECOA in terms of performance, where the average value is 2.090865 with a low standard deviation of 0.01126. Therefore, it proves that the algorithm is capable of dealing with various operating conditions in a consistent and reliable mandate, which plays an important role in microgrid applications due to the high variability and uncertainty associated with the integration of renewable sources.

Table 6 presents a comparison of optimization algorithms applied to microgrid management, focusing on key performance attributes such as convergence speed, final fitness value, stability, computational efficiency, and exploration–exploitation balance. The ECOA stands out with the fastest convergence rate, the lowest final fitness value, and a consistently smooth stability profile, indicating its effectiveness in achieving optimal solutions under uncertain conditions. The GPSOA and HPSO-GSA exhibit a balanced exploration–exploitation approach and moderate stability, though they converge at a slower rate than the ECOA. This comparison highlights the ECOA's robust performance in minimizing objective functions, which is especially suited for renewable energy scenarios that require rapid, stable convergence.

Table 6. Comparison of optimization algorithms based on key performance attributes.

Algorithm	Convergence Speed	Final Fitness Value	Stability	Computational Efficiency	Exploration–Exploitation Balance
GA	Moderate	Moderate	Moderate	Moderate	Low
PSO	Moderate	High	High	Moderate	Moderate
CO	Moderate	Moderate	Moderate	High	Low
ECOA	High	Very Low	Very High	High	Balanced
GPSOA	Moderate	Low	Moderate	Moderate	Balanced
HPSO-GSA	Moderate	Low	High	Moderate	Balanced

The reduction in voltage deviation, as realized across the run, demonstrates how effective the proposed model is in enhancing voltage stability across the microgrid. The VD values in the network before optimization were relatively high and consistent, which could tend to pose potential instability and reliability concerns in practical scenarios. With the application of the ECOA in conjunction with UT, it can be observed that the model minimized these VD values substantially by up to 10%. These improved voltage profiles in all time periods indeed confirm the relevance of using advanced optimization techniques in order to provide flat voltage profiles, especially in unbalanced distribution networks with high RES penetration. This will be very important when trying to handle intermittent systems like PV systems and WTs, which may create disturbances in generation and further deteriorate voltage imbalances.

The model applied in this contribution has reduced energy losses significantly. Before optimization, high energy losses were witnessed, especially during peak demand periods. In this context, the presented model performs an efficient dispatch of resources of generation, including PV systems, WTs, DG, MTs, and the BESS, which enhances overall efficiencies within the microgrid system. This is the capability of the model in streamlining the process of energy distribution to cut down on wastages to improve efficiency. The reduction from 499.77 kW to 446.65 kW is proof of the reduction in losses, which is very important in the sense that reduced losses translate into more efficient energy usage, reduced environmental impacts, and lower operating costs for microgrid operators in the long term.

Perhaps the most striking result is the drastic reduction in total operational costs when renewable energy sources were integrated into the microgrid. This put the total cost of operation with RES integration at USD 220,921.39 against USD 1,324,587 without RES. The large difference in cost indicates the economic viability of integrating renewable energy into microgrids. Furthermore, the breakdown of the costs shows a relative decrease in overall operational costs with the inclusion of RESs like PV systems and WTs. Additionally, the BESS and DR programs will serve to optimize energy consumption patterns and further reduce dependence on these relatively highly valued external sources of energy. Results also show that further investment in RESs and storage technologies could lead to larger long-term cost savings, hence supporting the transition toward more sustainable and cost-effective energy management frameworks.

A number of limitations exist in the present model. The price modeling for the optimization process is simplified and does not fully capture the complexities inherent in real-world electricity pricing structures, which are influenced by market fluctuations, time-of-use (TOU) rates, and demand elasticity. This simplification may reduce the model's ability to accurately reflect the economic signals that drive consumer behavior and energy consumption patterns.

While the model has demonstrated significant improvements across various performance metrics, concerns regarding the computational efficiency of the Enhanced Cheetah Optimization Algorithm (ECOA) persist, particularly when applied to larger-scale microgrid systems. The algorithm's performance may vary under different environmental and demand conditions, especially when confronted with extreme fluctuations in renewable energy outputs or sudden changes in load patterns. This variability could affect both the convergence speed of the algorithm and the quality of the optimal solution.

Furthermore, the issues of improving computational efficiency and scalability will be paramount in ensuring the practical applications of the model. Future research could explore adaptive mechanisms within the ECOA to enhance its responsiveness to changing conditions as well as strategies for managing scalability. Hierarchical optimization approaches and parallel processing capabilities could extend the applicability of the proposed method across diverse microgrid scenarios.

In general, the ECOA has proven to be a robust optimization tool for microgrid management, effectively addressing the challenges posed by renewable energy integration and load variability. The notable improvements in voltage stability, energy efficiency, and cost reduction exhibited by the proposed model contribute to the development of more resilient and sustainable power systems. Addressing the identified limitations will be crucial for enhancing the model's performance and applicability in real-world scenarios. Overall, the results offer a promising pathway for sustainable energy management practices and the integration of higher levels of renewable energy into the grid, underscoring the need for continuous research and innovation in this field.

6. Conclusions

This paper presents a comprehensive framework for optimizing microgrid management by integrating the Enhanced Cheetah Optimization Algorithm (ECOA) with Unscented Transformation (UT). The proposed model effectively addresses the complexities associated with variable renewable energy sources and fluctuating load demands, resulting in significant improvements in key performance metrics. Specifically, the optimization yielded an overall cost of operation of USD 220,921.39, representing a reduction of over 83% reduction, a decrease in overall power losses to 446.65 W (more than 10%), and an excellent voltage stability deviation of 10%. The integration of UT enhances decision-making under uncertainty, allowing for the precise propagation of variability through nonlinear functions, which is critical for effective energy management in environments with fluctuating renewable generation. While the proposed optimization framework demonstrates significant effectiveness, the performance of the Enhanced Cheetah Optimization Algorithm (ECOA) may fluctuate under varying environmental and demand conditions, especially during extreme shifts in renewable energy output or sudden changes in load patterns. Such variability can influence the algorithm's convergence speed and the quality of the optimal solution. Future research could investigate the development of adaptive mechanisms within the ECOA to improve its responsiveness to these changing conditions. Additionally, although the current framework shows effectiveness in terms of the specified microgrid configurations, its scalability to larger systems or more complex microgrid architectures requires further exploration. Strategies for managing scalability—such as hierarchical optimization approaches or parallel processing capabilities—could be developed to broaden the applicability of the proposed method across diverse microgrid scenarios. To enhance the model's performance in dynamic energy environments, future research should also consider integrating adaptive forecasting techniques and investigating emerging technologies.

Funding: This research was funded by Deanship of Postgraduate Studies and Scientific Research at Majmaah University, project number: ER-2024-1406.

Data Availability Statement: The data presented in this study are available in this article.

Acknowledgments: The author extends the appreciation to the Deanship of Postgraduate Studies and Scientific Research at Majmaah University for funding this research work through the project number (ER-2024-1406).

Conflicts of Interest: The authors declare no conflicts of interest.

References

1. Zia, M.F.; Elbouchikhi, E.; Benbouzid, M. Microgrids Energy Management Systems: A Critical Review on Methods, Solutions, and Prospects. *Appl. Energy* **2018**, *222*, 1033–1055. [CrossRef]
2. Kumari, B.A.; Vaisakh, K. Integration of Solar and Flexible Resources into Expected Security Cost with Dynamic Optimal Power Flow Problem Using a Novel DE Algorithm. *Renew. Energy Focus* **2022**, *42*, 48–69. [CrossRef]
3. Fathi, R.; Tousi, B.; Galvani, S. A New Approach for Optimal Allocation of Photovoltaic and Wind Clean Energy Resources in Distribution Networks with Reconfiguration Considering Uncertainty Based on Info-Gap Decision Theory with Risk Aversion Strategy. *J. Clean. Prod.* **2021**, *295*, 125984. [CrossRef]
4. Ghadikolaei, E.R.; Ghafouri, A.; Sedighi, M. Probabilistic Energy Management of DGs and Electric Vehicle Parking Lots in a Smart Grid Considering Demand Response. *Int. J. Energy Res.* **2024**, *2024*, 5543500. [CrossRef]
5. Almadhor, A.; Rauf, H.T.; Khan, M.A.; Kadry, S.; Nam, Y. A Hybrid Algorithm (BAPSO) for Capacity Configuration Optimization in a Distributed Solar PV Based Microgrid. *Energy Rep.* **2021**, *7*, 7906–7912. [CrossRef]
6. Elymany, M.M.; Enany, M.A.; Elsonbaty, N.A. Hybrid Optimized-ANFIS Based MPPT for Hybrid Microgrid Using Zebra Optimization Algorithm and Artificial Gorilla Troops Optimizer. *Energy Convers. Manag.* **2024**, *299*, 117809. [CrossRef]
7. Kumar, S.S.; Iruthayarajan, M.W.; Saravanan, R. Hybrid Technique for Optimizing Charging-Discharging Behaviour of EVs and Demand Response for Cost-Effective PV Microgrid System. *J. Energy Storage* **2024**, *96*, 112667. [CrossRef]
8. Jiang, S.; Zhang, C.; Wu, W.; Chen, S. Combined Economic and Emission Dispatch Problem of Wind-thermal Power System Using Gravitational Particle Swarm Optimization Algorithm. *Math. Probl. Eng.* **2019**, *2019*, 5679361. [CrossRef]
9. Jiang, S.; Zhang, C.; Chen, S. Sequential Hybrid Particle Swarm Optimization and Gravitational Search Algorithm with Dependent Random Coefficients. *Math. Probl. Eng.* **2020**, *2020*, 1957812. [CrossRef]
10. Hashish, M.S.; Hasanien, H.M.; Ji, H.; Alkuhayli, A.; Alharbi, M.; Akmaral, T.; Turky, R.A.; Jurado, F.; Badr, A.O. Monte Carlo Simulation and a Clustering Technique for Solving the Probabilistic Optimal Power Flow Problem for Hybrid Renewable Energy Systems. *Sustainability* **2023**, *15*, 783. [CrossRef]
11. Saha, A.; Bhattacharya, A.; Das, P.; Chakraborty, A.K. A Novel Approach towards Uncertainty Modeling in Multiobjective Optimal Power Flow with Renewable Integration. *Int. Trans. Electr. Energy Syst.* **2019**, *29*, e12136. [CrossRef]
12. Ullah, Z.; Qazi, H.S.; Alferidi, A.; Alsolami, M.; Lami, B.; Hasanien, H.M. Optimal Energy Trading in Cooperative Microgrids Considering Hybrid Renewable Energy Systems. *Alex. Eng. J.* **2024**, *86*, 23–33. [CrossRef]
13. Ahn, H.; Rim, D.; Pavlak, G.S.; Freihaut, J.D. Uncertainty Analysis of Energy and Economic Performances of Hybrid Solar Photovoltaic and Combined Cooling, Heating, and Power (CCHP + PV) Systems Using a Monte-Carlo Method. *Appl. Energy* **2019**, *255*, 113753. [CrossRef]
14. Shahbazi, A.; Moradi CheshmehBeigi, H.; Abdi, H.; Shahbazitabar, M. Probabilistic Optimal Allocation of Electric Vehicle Charging Stations Considering the Uncertain Loads by Using the Monte Carlo Simulation Method. *J. Oper. Autom. Power Eng.* **2023**, *11*, 277–284.
15. Ahmed, D.; Ebeed, M.; Kamel, S.; Nasrat, L.; Ali, A.; Shaaban, M.F.; Hussien, A.G. An Enhanced Jellyfish Search Optimizer for Stochastic Energy Management of Multi-Microgrids with Wind Turbines, Biomass and PV Generation Systems Considering Uncertainty. *Sci. Rep.* **2024**, *14*, 15558. [CrossRef] [PubMed]

16. Alavi, S.A.; Ahmadian, A.; Aliakbar-Golkar, M. Optimal Probabilistic Energy Management in a Typical Micro-Grid Based-on Robust Optimization and Point Estimate Method. *Energy Convers. Manag.* **2015**, *95*, 314–325. [CrossRef]
17. Xiao, F.; Ai, Q. New Modeling Framework Considering Economy, Uncertainty, and Security for Estimating the Dynamic Interchange Capability of Multi-Microgrids. *Electr. Power Syst. Res.* **2017**, *152*, 237–248. [CrossRef]
18. Habibi, S.; Effatnejad, R.; Hedayati, M.; Hajihosseini, P. Stochastic Energy Management of a Microgrid Incorporating Two-Point Estimation Method, Mobile Storage, and Fuzzy Multi-Objective Enhanced Grey Wolf Optimizer. *Sci. Rep.* **2024**, *14*, 1667. [CrossRef]
19. Aien, M.; Fotuhi-Firuzabad, M.; Aminifar, F. Probabilistic Load Flow in Correlated Uncertain Environment Using Unscented Transformation. *IEEE Trans. Power Syst.* **2012**, *27*, 2233–2241. [CrossRef]
20. Lin, X.; Shu, T.; Tang, J.; Yang, Y.; Liu, F.; Zheng, J.; Peng, S. An Unscented Transformation Based Probabilistic Power Flow for Autonomous Hybrid AC/DC Microgrid with Correlated Uncertainty Sources. In Proceedings of the 2018 2nd IEEE Conference on Energy Internet and Energy System Integration (EI2), Beijing, China, 20–22 October 2018; pp. 1–6.
21. Kavousi-Fard, A.; Niknam, T.; Akbari-Zadeh, M.-R.; Dehghan, B. Stochastic Framework for Reliability Enhancement Using Optimal Feeder Reconfiguration. *J. Syst. Eng. Electron.* **2014**, *25*, 901–910. [CrossRef]
22. Oke, O.; Ozgonenel, O.; Thomas, D.W.P.; Ataseven, M.S. Probabilistic Load Flow of Unbalanced Distribution Systems with Wind Farm. *Teh. Vjesn.* **2019**, *26*, 1260–1266.
23. Kurt, U.; Ozgonenel, O.; Ayvaz, B.B. Probabilistic Power Flow Analysis Using Matlab Graphical User Interface (GUI). *J. Electr. Eng. Technol.* **2022**, *17*, 929–943. [CrossRef]
24. Zhang, Y.; Gatsis, N.; Giannakis, G.B. Robust Energy Management for Microgrids With High-Penetration Renewables. *IEEE Trans. Sustain. Energy* **2013**, *4*, 944–953. [CrossRef]
25. Carli, R.; Cavone, G.; Pippia, T.; Schutter, B.D.; Dotoli, M. Robust Optimal Control for Demand Side Management of Multi-Carrier Microgrids. *IEEE Trans. Autom. Sci. Eng.* **2022**, *19*, 1338–1351. [CrossRef]
26. Balderrama, S.; Lombardi, F.; Riva, F.; Canedo, W.; Colombo, E.; Quoilin, S. A Two-Stage Linear Programming Optimization Framework for Isolated Hybrid Microgrids in a Rural Context: The Case Study of the "El Espino" Community. *Energy* **2019**, *188*, 116073. [CrossRef]
27. Yoon, C.; Park, Y.; Sim, M.K.; Lee, Y.I. A Quadratic Programming-Based Power Dispatch Method for a DC-Microgrid. *IEEE Access* **2020**, *8*, 211924–211936. [CrossRef]
28. Nemati, M.; Braun, M.; Tenbohlen, S. Optimization of Unit Commitment and Economic Dispatch in Microgrids Based on Genetic Algorithm and Mixed Integer Linear Programming. *Appl. Energy* **2018**, *210*, 944–963. [CrossRef]
29. Ruder, S. An Overview of Gradient Descent Optimization Algorithms. *arXiv* **2016**, arXiv:1609.04747.
30. Dennis John, E.J.; Moré, J.J. Quasi-Newton Methods, Motivation and Theory. *SIAM Rev.* **1977**, *19*, 46–89. [CrossRef]
31. Riley, J.; Butler, R.; Ogilvie, D.; Finniear, R.; Jenner, D.; Powell, S.; Anand, R.; Smith, J.C.; Markham, A.F. A Novel, Rapid Method for the Isolation of Terminal Sequences from Yeast Artificial Chromosome (YAC) Clones. *Nucleic Acids Res.* **1990**, *18*, 2887–2890. [CrossRef]
32. Goldberg, D.E. *Genetic Algorithms*; Pearson Education India: Bangalore, India, 2013; ISBN 817758829X.
33. Kennedy, J.; Eberhart, R. Particle Swarm Optimization. In Proceedings of the ICNN'95-International Conference on Neural Networks, Perth, WA, Australia, 27 November–1 December 1995; Volume 4, pp. 1942–1948.
34. Karimi, H.; Jadid, S. Optimal Energy Management for Multi-Microgrid Considering Demand Response Programs: A Stochastic Multi-Objective Framework. *Energy* **2020**, *195*, 116992. [CrossRef]
35. Li, Z.; Zang, C.; Zeng, P.; Yu, H.; Li, H. Two-Stage Stochastic Programming Based Model Predictive Control Strategy for Microgrid Energy Management under Uncertainties. In Proceedings of the 2016 International Conference on Probabilistic Methods Applied to Power Systems (PMAPS), Beijing, China, 16–20 October 2016; pp. 1–6.
36. Yu, D.; Zhu, H.; Han, W.; Holburn, D. Dynamic Multi Agent-Based Management and Load Frequency Control of PV/Fuel Cell/Wind Turbine/CHP in Autonomous Microgrid System. *Energy* **2019**, *173*, 554–568. [CrossRef]
37. Anand, H.; Ramasubbu, R. A Real Time Pricing Strategy for Remote Micro-Grid with Economic Emission Dispatch and Stochastic Renewable Energy Sources. *Renew. Energy* **2018**, *127*, 779–789. [CrossRef]
38. Gholami, K.; Dehnavi, E. A Modified Particle Swarm Optimization Algorithm for Scheduling Renewable Generation in a Micro-Grid under Load Uncertainty. *Appl. Soft Comput.* **2019**, *78*, 496–514. [CrossRef]
39. Kreishan, M.Z.; Zobaa, A.F. Mixed-Integer Distributed Ant Colony Optimization of Dump Load Allocation with Improved Islanded Microgrid Load Flow. *Energies* **2023**, *16*, 4257. [CrossRef]
40. Jumani, T.A.; Mustafa, M.W.; Rasid, M.M.; Memon, Z.A. Dynamic Response Enhancement of Grid-Tied Ac Microgrid Using Salp Swarm Optimization Algorithm. *Int. Trans. Electr. Energy Syst.* **2020**, *30*, e12321. [CrossRef]
41. Sen, S.; Gupta, K.D.; Poudyal, S.; Ahsan, M.M. A Genetic Algorithm Approach to Optimize Dispatching for a Microgrid Energy System with Renewable Energy Sources. In Proceedings of the CS & IT Conference Proceedings, Dubai, United Arab Emirates, 23–24 February 2019; Volume 9.
42. Qazi, S.H.; Mustafa, M.W.; Sultana, U.; Mirjat, N.H.; Soomro, S.A.; Rasheed, N. Regulation of Voltage and Frequency in Solid Oxide Fuel Cell-Based Autonomous Microgrids Using the Whales Optimisation Algorithm. *Energies* **2018**, *11*, 1318. [CrossRef]
43. Tayab, U.B.; Lu, J.; Taghizadeh, S.; Metwally, A.S.M.; Kashif, M. Microgrid Energy Management System for Residential Microgrid Using an Ensemble Forecasting Strategy and Grey Wolf Optimization. *Energies* **2021**, *14*, 8489. [CrossRef]

44. Bukar, A.L.; Tan, C.W.; Lau, K.Y. Optimal Sizing of an Autonomous Photovoltaic/Wind/Battery/Diesel Generator Microgrid Using Grasshopper Optimization Algorithm. *Sol. Energy* **2019**, *188*, 685–696. [CrossRef]
45. Akbari, M.A.; Zare, M.; Azizipanah-Abarghooee, R.; Mirjalili, S.; Deriche, M. The Cheetah Optimizer: A Nature-Inspired Metaheuristic Algorithm for Large-Scale Optimization Problems. *Sci. Rep.* **2022**, *12*, 1–20. [CrossRef]
46. Nowdeh, S.A.; Naderipour, A.; Davoudkhani, I.F.; Guerrero, J.M. Stochastic Optimization–Based Economic Design for a Hybrid Sustainable System of Wind Turbine, Combined Heat, and Power Generation, and Electric and Thermal Storages Considering Uncertainty: A Case Study of Espoo, Finland. *Renew. Sustain. Energy Rev.* **2023**, *183*, 113440. [CrossRef]
47. Kersting, W.H. Radial Distribution Test Feeders. In Proceedings of the 2001 IEEE Power Engineering Society Winter Meeting. Conference Proceedings (Cat. No.01CH37194), Columbus, OH, USA, 28 January–1 February 2001; Volume 2, pp. 908–912.
48. Alilou, M.; Nazarpour, D.; Shayeghi, H. Multi-Objective Optimization of Demand Side Management and Multi Dg in the Distribution System with Demand Response. *J. Oper. Autom. Power Eng.* **2018**, *6*, 230–242.
49. Sedighizadeh, M.; Fazlhashemi, S.S.; Javadi, H.; Taghvaei, M. Multi-Objective Day-Ahead Energy Management of a Microgrid Considering Responsive Loads and Uncertainty of the Electric Vehicles. *J. Clean. Prod.* **2020**, *267*, 121562. [CrossRef]
50. Abou El-Ela, A.A.; El-Sehiemy, R.A.; Kinawy, A.-M.; Ali, E.S. Optimal Placement and Sizing of Distributed Generation Units Using Different Cat Swarm Optimization Algorithms. In Proceedings of the 2016 Eighteenth International Middle East Power Systems Conference (MEPCON), Cairo, Egypt, 27–29 December 2016; pp. 975–981.

electronics

MDPI

Article

The Robust Optimization of Low-Carbon Economic Dispatching for Regional Integrated Energy Systems Considering Wind and Solar Uncertainty

Mingguang Zhang, Bo Wang * and Juan Wei

College of Electrical and Information Engineering, Lanzhou University of Technology, Lanzhou 730050, China; zhangmg@lut.edu.cn (M.Z.); 222085801008@lut.edu.cn (J.W.)

* Correspondence: 222080802004@lut.edu.cn

Abstract: In this paper, a two-stage robust optimization approach is employed to address the variability in renewable energy output by accounting for the uncertainties associated with wind and solar energy. The model aims to achieve a balanced system that is both low-carbon and economically efficient while also being resilient to uncertainties. Initially, a regional integrated energy system model is developed, integrating electricity, gas, and heat. The variability of wind and photovoltaic power outputs is represented using a modifiable uncertainty set. A resilient optimal scheduling model is formulated in two stages, with the objective of minimizing costs under worst-case scenarios. This model is solved iteratively through a column and constraint generation approach. Additionally, the scheduling model incorporates horizontal time shifts and vertical complementary substitutions for carbon trading costs and demand-side loads to avoid excessive conservatism and to manage carbon emissions and energy trading in the regional integrated energy system (RIES). Results show that the two-stage robust optimization approach significantly enhances the system's resilience to risks and minimizes economic losses. The inclusion of carbon trading mechanisms and the demand response prevents the system from becoming overly robust, which could impede economic growth, while also reducing carbon emissions. The proposed method effectively achieves balanced optimal scheduling for a robust, economical, and low-carbon system.

Keywords: wind and solar uncertainty; carbon trading mechanism; demand response; robust optimization

1. Introduction

The escalating severity of global warming and the depletion of fossil resources highlight the urgent need for advanced solutions. Addressing the energy crisis and environmental pollution requires comprehensive advancements in renewable energy, which represents the primary direction for future energy development [1]. However, renewable sources like photovoltaics and wind power are notably unpredictable and irregular, leading to potential imbalances between the energy supply and demand. Additionally, the integration of wind and solar energy complicates the operation and dispatch of traditional power systems [2]. Consequently, developing a robust, efficient, and environmentally conscious energy infrastructure is crucial.

By integrating a diverse array of energy sources, the Regional Integrated Energy System (RIES) overcomes barriers between different energy types, facilitates the coordinated operation of various energy subsystems, achieves efficient energy conversion, and significantly enhances the utilization of renewable energy. However, the dispatching of resources, including renewable energy, within the RIES is marked by considerable uncertainty, posing significant challenges to the overall safety and operational efficiency of the system [3]. Therefore, it is essential to explore the impact of wind and solar energy uncertainty on RIES dispatch [4].

Currently, the predominant methods for addressing uncertainties can be classified as stochastic optimization or robust optimization [5]. Robust optimization provides a reliable and effective solution by constructing an indeterminate set based on the characteristics of uncertain variables without the need for precise modeling of their distribution functions [6]. This approach ensures that if the values of the variables fall within the defined set, the robust optimization model will deliver a feasible solution, thereby significantly simplifying the problem-solving process [7]. For power dispatching between the main grid and a microgrid incorporating solar panels and batteries, a robust algorithm has been proposed that ensures resilience to uncertainty while managing the variability in solar power generation and avoiding an excessively conservative dispatch through time-varying soft constraints [8]. Additionally, a two-stage robust multi-period economic dispatch problem, as discussed in [9], is addressed using a column and constraint generation method. This method employs robust optimization to handle uncertainties related to renewable energy generation and load consumption in active distribution networks.

To maximize the economic benefits of renewable energy systems, it is essential to consider demand. The demand response (DR) effectively harnesses the user-side potential, ensuring operational safety [10], enhancing the integration of renewable energy sources [11], and improving the economics and flexibility of the RIES operation [12]. Reference [13] proposed a two-stage stochastic p-robust optimal energy transaction management strategy that accounts for uncertainties in microgrid operations and integrates a hybrid demand response to improve robustness and to minimize operating costs. Reference [14] introduced a two-stage robust scheduling model for optimizing the operation of solar–wind–hydro–thermal pump (SWHTP) systems in a competitive environment, which enhances system flexibility by implementing demand response programs. This approach leverages adjustable and curtailable elastic load characteristics to improve the continuous energy supply reliability and to reduce overall energy costs in a deregulated market.

Nevertheless, the aforementioned research primarily focuses on model resilience and cost-effectiveness, often neglecting the goal of achieving low-carbon operation for the system. To ensure robust scheduling, it is necessary to anticipate and prepare for increased unit production and energy acquisition to enhance the resilience against risks [15], which, in turn, may lead to higher carbon emissions. Carbon trading allows for the exchange of predetermined CO_2 emissions allowances [16], facilitating a low-carbon operation by incorporating the cost of carbon trading into the overall cost, thereby leveraging economic incentives [17]. Reference [18] introduced a two-stage adjustment robust optimization model for multi-energy virtual power plants, accounting for various uncertainties and carbon trading. This model optimizes costs in the worst-case scenario, dynamically balancing economic and robust dispatching strategies while guiding energy conservation and emission reduction efforts. Reference [19] proposed a method for planning economic and environmental power distribution network assets, incorporating carbon emissions trading and the demand response. This approach mitigates carbon emissions at the distribution level and addresses the uncertainties of distributed generation and power demand through a two-stage robust optimization model.

In summary, existing studies often fall short in addressing the equilibrium between reduced carbon emissions, economic considerations, and system resilience. Moreover, there is a lack of research evaluating the balanced optimization of these three factors. Integrating carbon trading mechanisms and the demand response into the robust optimization model for Regional Integrated Energy Systems (RIESs) is crucial for achieving a well-balanced system that is low-carbon, cost-effective, and resilient [20]. This paper employs a robust optimization approach to tackle the uncertainties associated with wind power and photovoltaics. Additionally, it incorporates a carbon trading system and demand response to control carbon emissions, enhance energy flexibility, and optimize the balance for low-carbon, cost-effective, and resilient operations [21]. The proposed system model is validated with a concrete example, demonstrating its strong resilience and cost-effectiveness while effectively maintaining a low-carbon operation.

2. The RIES Operational Architecture

The RIES is a complex structural system that integrates multiple energy sources. It optimizes the interplay between electricity, heat, and gas to improve energy efficiency and meet diverse customer needs while ensuring a reliable and consistent energy supply [22]. This study constructs an RIES framework that incorporates the demand response for electricity, gas, and heat. As shown in Figure 1, the framework is based on a carbon trading mechanism.

The energy supply side of the RIES includes an advanced power grid and an advanced gas grid, as well as wind power and solar panels [23]. There are four main types of energy conversion equipment: power-to-gas (P2G) systems, gas boilers (GBs), gas turbines (GTs), and waste heat boilers (WHBs) [24]. The system operates in a combined heat and power (CHP) coupling mode, allowing it to adapt to various system conditions. Additionally, there are three types of energy storage: electric, gas, and thermal. The energy demand-side-response resources within the system consist of the electricity load, gas load, and heat load [25].

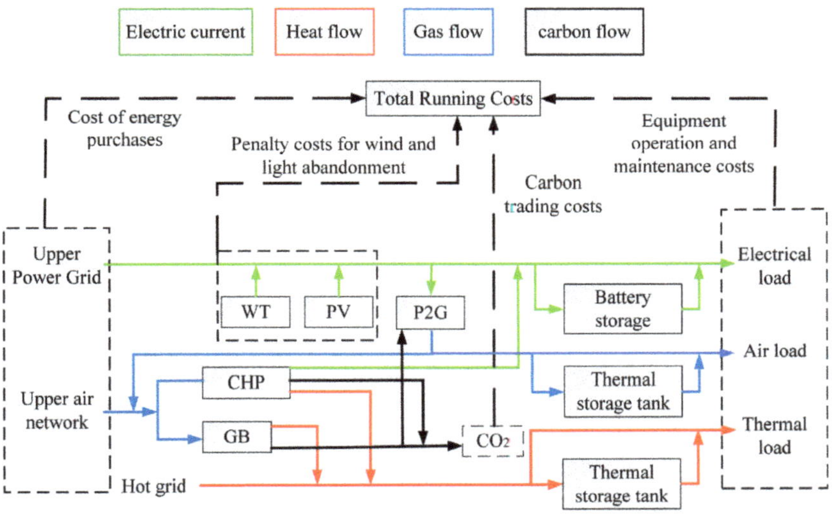

Figure 1. The RIES architecture coupled with electricity, gas, and heat multi-energy sources.

2.1. *Ladder Carbon Trading Mechanism*

This article evaluates solar power, wind turbines, and natural gas as energy sources through a life cycle assessment perspective. It converts carbon emissions generated throughout the entire life cycle of energy and energy storage equipment—encompassing production, transportation, operation, and disposal—into carbon emission costs. These costs are integrated into optimization objectives to examine their impact on system optimization outcomes when considering full life-cycle carbon emission costs. For detailed measurement procedures of each component, please refer to reference [26].

The government or relevant agencies allocate carbon emission allowances to each carbon emission source within an RIES [27]. If a manufacturer's actual CO_2 emissions are below the allocated quota, they can sell the surplus allowances for cash. Conversely, if their emissions exceed the quota, manufacturers must purchase additional allowances to cover the excess. The carbon trading mechanism model mainly consists of three components:

2.1.1. The Initial Quota Model

The primary sources of carbon emissions in the RIES include superior power purchases, gas turbines (GTs), gas boilers (GBs), and demand-side gas loads. The initial quota model is as follows:

$$\begin{cases} E_c = E_{c,buy} + E_{c,GT} + E_{c,GB} + E_{g,load} \\ E_{c,buy} = \beta_e^* \sum\limits_{t=1}^{T} P_{e,buy}(t)\Delta t \\ E_{c,GT} = \beta_h^* \left(\sum\limits_{t=1}^{T} P_{GT,h}(t)\Delta t + \varphi_{e,h} \sum\limits_{t=1}^{T} P_{GT,e}(t)\Delta t \right) \\ E_{c,GB} = \beta_h^* \sum\limits_{t=1}^{T} P_{GB,h}(t)\Delta t \\ E_{g,load} = \varphi_{g,load} \sum\limits_{t=1}^{T} P_{g,load}(t)\Delta t \end{cases} \tag{1}$$

where E_c, $E_{c,buy}$, $E_{c,GT}$, $E_{c,GB}$, and $E_{g,load}$ are the free carbon emission rights quotas of the RIES and the system purchase of electricity from superiors, gas turbines, gas boilers, and demand-side gas loads, respectively; β_e^* and β_h^* represent the allocated quotas for carbon emissions that are gained by generating electricity and heat, respectively; $\varphi_{g,load}$ represents the allocated amount of carbon emissions that is permitted for each unit of gas load used; and $\varphi_{e,h}$ is the conversion parameter of gas turbine power generation to a calorific value. During the *t* period, $P_{e,buy}(t)$ represents the power that the system purchases from the superior, $P_{GT,h}(t)$ represents the amount of electric power that the gas turbine supplies, $P_{GT,e}(t)$ is the amount of thermal power that the gas turbine supplies, $P_{GB,h}(t)$ is the amount of thermal power that the gas boiler supplies, $P_{g,load}(t)$ is the amount of gas load that is consumed during the *t* period, and T is the period that the system is scheduled to run.

2.1.2. The Actual Model

During the power-to-gas conversion process, some CO_2 will be utilized by the system, and significant CO_2 emissions will be generated by the demand-side gas load. Given these factors, the original model requires adjustments. The revised model can be summarized as follows:

$$\begin{cases} E_{c,a} = E_{e,buy,a} + E_{GBGT,a} + E_{g,load,a} - E_{P2G,a} \\ E_{e,buy,a} = \sum\limits_{t=1}^{T} [a + bP_{e,buy}(t) + cP_{e,buy}^2(t)]\Delta t \\ E_{GTGB,a} = \sum\limits_{t=1}^{T} [d + eP_{GTGB}(t) + fP_{GTGB}^2(t)]\Delta t \\ P_{GTGB}(t) = P_{GT,h}(t) + P_{GT,e}(t) + P_{GB,h}(t) \\ E_{g,load,a} = \varphi \sum\limits_{t=1}^{T} P_{g,load}(t)\Delta t \\ E_{P2G,a} = \theta \sum\limits_{t=1}^{T} P_{P2G,g}(t)\Delta t \end{cases} \tag{2}$$

where the actual carbon emissions from the RIES are denoted by $E_{c,a}$, while those from the power the system bought from the superior are denoted by $E_{e,buy,a}$, and the real carbon emissions from gas turbines and boilers are denoted by $E_{GBGT,a}$. In the *t* period, $P_{GTGB}(t)$ stands for the equivalence of the gas turbine and gas boiler output power; in the case of the demand-side gas load, $E_{g,load,a}$ represents the real carbon emission, $E_{P2G,a}$ stands for the CO_2 absorbed by the P2G equipment during the electricity-to-gas conversion process, $P_{P2G,g}(t)$ represents the power output from natural gas produced by the P2G equipment in the given time period, and φ is the equivalent carbon emission parameter per unit gas load. For specific values of the θ parameters for absorbing CO_2 in this process, please refer to reference [27]. The carbon emission parameters for coal power units are represented by the variables a, b, and c, while the carbon emission characteristics for units consuming natural gas are represented by the variables d, e, and f.

2.1.3. The Calculation Model

The carbon emissions trading share in the carbon trading market can be determined by considering the following two components:

$$E_{RIES,c} = E_{c,a} - E_c \tag{3}$$

where $E_{RIES,c}$ represents the system's share in the carbon trading market for carbon emissions.

The carbon trading cost calculation model uses a hierarchical approach to classify CO_2 emissions into various ranges. As CO_2 emissions rise, both transaction prices and costs increase. A mathematical model representing carbon trading costs is formulated as a piecewise linear function with multiple intervals for different emission levels. Surplus carbon can be sold at a specified price at the corresponding trading center, generating a subsidy. The carbon trading cost can be expressed using a piecewise function as follows:

$$F_{co_2} = \begin{cases} cE_{RIES,c}, & E_{RIES,c} < d \\ c(1+\lambda)(E_{RIES,c} - d) + cd, & d \le E_{RIES,c} < d \\ c(1+2\lambda)(E_{RIES,c} - 2d) + c(2+\lambda)d, & 2d \le E_{RIES,c} < 3d \\ c(1+3\lambda)(E_{RIES,c} - 3d) + c(3+3\lambda)d, & 3d \le E_{RIES,c} < 4d \\ c(1+4\lambda)(E_{RIES,c} - 4d) + c(4+6\lambda)d, & 4d \le E_{RIES,c} < 5d \\ c(1+5\lambda)(E_{RIES,c} - 5d) + c(5+10\lambda)d, & 5d \le E_{RIES,c} \end{cases} \tag{4}$$

where F_{co_2} represents the stepped carbon trading cost, λ indicates the price increase rate, c is the carbon trading benchmark price, and d is the interval length for carbon emissions.

2.2. IDR

This paper introduces a demand response approach utilizing a horizontal time shift and vertical complementary alternative strategy for managing electricity, gas, and heat [28]. Operators can develop various scheduling plans in response to energy price fluctuations to accommodate different energy demands and to ensure the economical, flexible, and efficient operation of the RIES. In line with traditional power demand categorization, the demand response loads in the RIES—electricity, gas, and heat—are classified into three groups based on their interactive response characteristics: fixed load, transferable load, and replaceable load.

The transferable load versus alternative load models are shown below:

$$\begin{cases} P^{n*}_{k,load} = P^{n}_{k,load}(t) + \Delta P^{n}_{k,load}(t) \\ \Delta P^{n}_{k,load}(t) = v^{n}_{k,in} P^{n}_{k,in}(t) - v^{n}_{k,out} P^{n}_{k,out}(t) \\ v^{n}_{k,in} + v^{n}_{k,out} = 1 \\ \sum\limits_{t=1}^{T} \Delta P^{n}_{k,load}(t) = 0 \\ P^{n,\min}_{k} \le \Delta P^{n}_{k,load}(t) \le P^{n,\max}_{k} \end{cases} \tag{5}$$

where n represents the different types of loads ($n = p$, c); when $n = p$, $P^{p*}_{k,load}$ denotes the magnitude of the transferable load's power during the k-th load t time following its involvement in the demand response; when $n = c$, $P^{c*}_{k,load}$ represents the power after the substitutable load in the k-th load t period participates in the demand response. $v^{n}_{k,in}$ and $v^{n}_{k,out}$ are the binary parameters that reflect the transfer-in and transfer-out for the k-th load in the t period, respectively; $P^{n}_{k,in}(t)$ and $P^{n}_{k,out}(t)$ denote, correspondingly, that power is transported to and from the k-th load in t periods. For the k-th load taking part in the demand response, $P^{n,\min}_{k}$ and $P^{n,\max}_{k}$ represent the minimum and maximum values, respectively.

In summary, the DR model can adjust the energy demand through various types of demand responses, as detailed below:

$$\begin{cases} P^{*}_{k,load} = P_{k,load}(t) + \Delta P_{k,load}(t) \\ = P_{k,load}(t) + \Delta P^{p}_{k,load}(t) + \Delta P^{c}_{k,load}(t) \end{cases} \tag{6}$$

where $P^{*}_{k,load}$ denotes the power level following the involvement of the k-th load in the DR during the time t, while $\Delta P_{k,load}(t)$ illustrates the strength of the k-th load taking part in the DR within the given time period t.

3. Robust Optimization

This research addresses the unpredictability in wind power and solar output using a two-stage robust optimization approach. The model's min–max–min structure aims to minimize the worst-case day-ahead dispatching cost. The first phase involves an outer minimization problem, optimizing variables such as energy storage equipment and the energy purchase state to ensure that the system can handle any scenario within the uncertain set [29]. The second phase involves an internal maximum–minimum problem, where, after determining the first-stage optimization variables, the goal is to identify a set of worst-case scenarios that maximize the day-ahead scheduling cost. This stage includes finding the worst-case scenario and deriving a deterministic optimization problem from it [30]. The objective is to maximize the output of each unit under the worst-case scenario while minimizing the potential day-ahead scheduling cost. Figure 2 illustrates the detailed framework of the optimization model.

The costs of RIES scheduling and the benefits of low-carbon emissions are predominantly influenced by the worst-case scenario simulated in the second stage of the robust optimization. When the actual wind power and photovoltaic outputs deviate from expected levels, the RIES employs robust optimization strategies to address the risks associated with uncertainty, especially in worst-case scenarios [31]. Generally, using robust optimization results in a more cautious approach, leading to higher anticipated unit outputs and increased energy purchases. This approach raises dispatching costs and reduces the benefits of low-carbon energy [32].

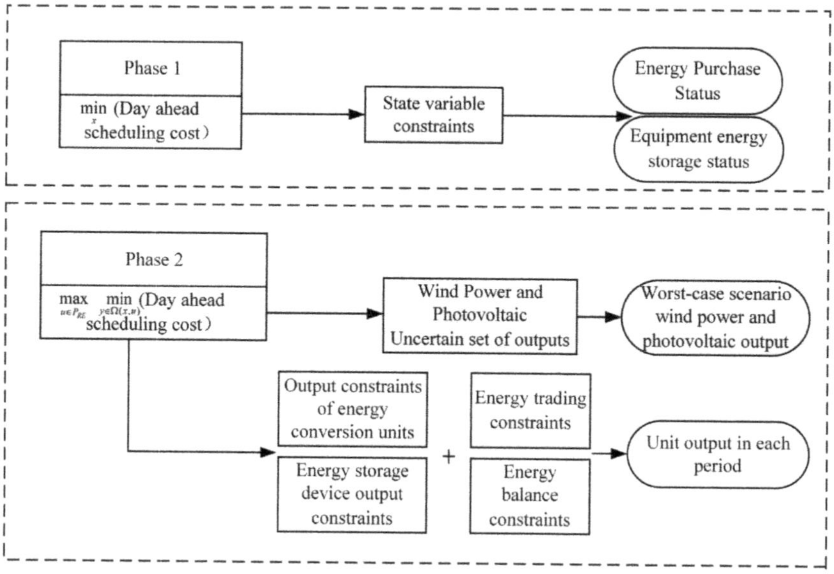

Figure 2. Robust optimization model framework.

3.1. Uncertainty Ensemble Modeling

Due to factors such as illumination, wind speed, and geographical location, renewable energy production is highly variable. Therefore, this study employs a robust optimization strategy to model the uncertainty in wind and solar outputs, typically using an uncertainty set to characterize the nature of this uncertainty. The optimization outcomes are significantly influenced by the choice of uncertainty sets. Accurate modeling of the uncertainty set ensures that the optimization results closely reflect the actual scenarios, although it can make the model more complex and difficult to solve. Conversely, if the uncertainty set is too loose, the optimization approach becomes overly conservative. To effectively manage the level of conservatism in the scheduling plan, a flexible uncertainty set is used to describe the variability in energy output, as indicated in Equation (7).

$$P^{RE}=\begin{cases} P_{ele,t}^{WT}=\overline{P_{ele,t}^{WT}}-u_{\mathrm{wt,t}}d_{wt,t} \\ \sum\limits_{t} u_{\mathrm{wt,t}} \leq \Gamma^{WT} \\ P_{ele,t}^{PV}=\overline{P_{ele,t}^{PV}}-u_{\mathrm{pv,t}}d_{pv,t} \\ \sum\limits_{t} u_{\mathrm{pv,t}} \leq \Gamma^{PV} \\ \forall t \in T \end{cases} \tag{7}$$

where $\overline{P_{ele,t}^{WT}}$ and $\overline{P_{ele,t}^{PV}}$ are the predicted values of wind and solar outputs in period t, and the wind power and solar output fluctuations have upper and lower limitations, ΔP_{+}^{RE} and ΔP_{-}^{RE}, respectively. In the t period, $d_{wt,t}$ and $d_{pv,t}$ stand for the wind power and photovoltaic fluctuation values, respectively, and $u_{\mathrm{wt,t}}$ and $u_{\mathrm{pv,t}}$, which are 0–1 variables, indicate the wind power and photovoltaic fluctuation states, respectively. The worst-scenario output is denoted by $P_{ele,t}^{WT}$ and $P_{ele,t}^{PV}$; Γ^{WT} and Γ^{PV} are the unknowns, which stand for the total amount of time periods when the cycle's wind and solar output changes. The larger the uncertainty value, the more periods during which the wind and solar output fluctuates, and the more conservative the scheduling scheme is.

3.2. Objective Function

The overall operational cost F is taken into account as the objective function of the RIES in this study and includes the energy purchase cost F_{buy}, carbon trading cost F_{CO_2}, demand response compensation cost F_{DR}, and equipment operation and maintenance cost F_{MC}, which are specifically expressed as follows:

$$F_{\min}=F_{buy}+F_{CO_2}+F_{MC}+F_{DR} \tag{8}$$

1. Energy purchase cost F_{buy}.

$$F_{buy}=\sum_{t=1}^{T}\alpha_t P_{e,buy}(t)+\sum_{t=1}^{T}\beta_t P_{g,buy}(t) \tag{9}$$

   where $P_{g,buy}(t)$ represents the gas purchase volume in period t, and α_{t} and β_t reflect the prices of electricity and gas, respectively, during the time period t.
2. F_{CO_2} is shown in Equation (4).
3. Demand response compensation cost F_{DR}.

$$F_{DR}=\sum_{k=1}^{3}\sum_{t=1}^{T}\left(\mu_p\left|\Delta P_{k,load}^{p}(t)\right|+\mu_c\left|\Delta P_{k,load}^{c}(t)\right|\right) \tag{10}$$

   where μ_p and μ_c, respectively, represent the unit compensation coefficients of transferable loads and replaceable loads participating in the demand response.

4. Equipment operation and maintenance cost F_{MC}.

$$F_{MC}=\sum_{t=1}^{T}(\sum_{i=1}^{n}P_i(t)\cdot C_i+\sum_{j=1}^{m}P_j(t)\cdot C_j+P_{p2g}(t)\cdot C_{p2g}+|P_{es}(t)|\cdot C_{es}) \tag{11}$$

where C_i stands for the cost of operating and maintaining the i-th coupling equipment per unit of power, and C_j is the cost of operating and maintaining new energy equipment per unit of power. The operational power of the i-th coupling equipment is represented by $P_i(t)$, whereas the operating power of the new energy equipment is represented by $P_j(t)$; C_{p2g} shows how much the power-to-gas technology costs to run and maintain per unit of power; $P_{es}(t)$ represents the energy storage device's operational power for a given time period; and C_{es} shows how much money it costs the energy storage systems to run and maintain each unit.

3.3. System Operation Constraints

1. Constraints on electricity generation from wind and solar systems.

$$\begin{cases} 0\le P_{WT}(t)\le P_{WT}^{\max} \\ 0\le P_{PV}(t)\le P_{PV}^{\max} \end{cases} \tag{12}$$

where $P_{WT}^{\max}$ is the wind power's maximum output power, and $P_{PV}^{\max}$ is the highest power that can be produced by photovoltaics.

2. CHP operation constraints.

$$\begin{cases} P_{CHP,e}(t)=\varepsilon_{CHP}^{e}P_{g,CHP}(t) \\ P_{CHP,h}(t)=\varepsilon_{CHP}^{h}P_{g,CHP}(t) \\ P_{g,CHP}^{\min}\le P_{g,CHP}(t)\le P_{g,CHP}^{\max} \\ \Delta P_{g,CHP}^{\min}\le P_{g,CHP}(t+1)-P_{g,CHP}(t)\le \Delta P_{g,CHP}^{\max} \\ \omega_{CHP}^{\min}\le P_{CHP,h}(t)/P_{CHP,e}(t)\le \omega_{CHP}^{\max} \end{cases} \tag{13}$$

where $P_{g,CHP}(t)$ represents the quantity of natural gas consumed by the CHP over the given time period t. $P_{CHP,e}(t)$ and $P_{CHP,h}(t)$ are the thermal and electrical power generated by the CHP in the time period t, while ε_{CHP}^{e} and ε_{CHP}^{h} demonstrate the relative effectiveness of transforming the CHP into thermal energy and electricity; the two extremes of the power input to CHP from natural gas are denoted as $P_{g,CHP}^{\min}$ and $P_{g,CHP}^{\max}$, respectively. As for the CHP, its lower ascending limit is $\Delta P_{g,CHP}^{\min}$, and its maximum climbing limit is $\Delta P_{g,CHP}^{\max}$; the lowest and highest values for the thermoelectric ratio of the CHP are given by $\omega_{CHP}^{\min}$ and $\omega_{CHP}^{\max}$, respectively.

3. GB running constraints.

$$\begin{cases} P_{GB,h}(t)=\varphi_{GB}P_{GB,g}(t) \\ P_{GB,g}^{\min}\le P_{GB,g}(t)\le P_{GB,g}^{\max} \\ \Delta P_{GB,g}^{\min}\le P_{GB,g}(t+1)-P_{GB,g}(t)\le \Delta P_{GB,g}^{\max} \end{cases} \tag{14}$$

where φ_{GB} indicates how well the GB converts energy, and $P_{GB,g}(t)$ shows the GB input power from natural gas in the time period t; the two extremes of the power input to GB from natural gas are denoted as $P_{GB,g}^{\min}$ and $P_{g,CHP}^{\max}$, respectively. As for the GB, its lower ascending limit is $\Delta P_{GB,g}^{\min}$, and its maximum climbing limit is $\Delta P_{g,CHP}^{\max}$.

4. Electric power balance constraints.

$$\begin{cases} P_{e.\mathrm{buy}}(t)=P_{e.load}(t)+P_{e.P2G}(t)+P_{ES,cha}^{e}(t) \\ -P_{PV}(t)-P_{ES,dis}^{e}(t)-P_{DG}(t)-P_{GT,e}(t) \\ 0\le P_{e.buy}(t)\le P_{e,buy}^{\max} \end{cases} \tag{15}$$

where $P_{e.load}(t)$ stands for the electric load in period t, $P_{e.P2G}(t)$ stands for the electric power of input P2G in period t, $P^{e}_{ES,cha}(t)$ stands for the power input to the electric energy storage system in period t, $P^{e}_{ES,dis}(t)$ stands for the power output from the electric energy storage system in period t, and $P^{\max}_{e,buy}$ stands for the maximum purchased power from the superior power grid.

5. Gas power balance constraints.

$$\begin{cases} P_{g.\text{buy}}(t) = P_{g.load}(t) + P_{GT,g}(t) + P^{g}_{ES,cha}(t) \\ \qquad + P_{GB,g}(t) - P_{e.P2G}(t) - P^{g}_{ES,dis}(t) \\ 0 \leq P_{g.buy}(t) \leq P^{\max}_{g,buy} \end{cases} \tag{16}$$

where in the given time period, $P^{g}_{ES,cha}(t)$ represents the power input to the natural gas energy storage system, and $P^{g}_{ES,dis}(t)$ denotes the power output from the same system. $P^{\max}_{g,buy}$ is the highest possible power of the gas purchase from the superior natural gas network, and $P_{GT,g}(t)$ is the gas power rate of the input gas turbine in time t.

6. Heat power balance constraints.

$$P_{CHP,h}(t) + P_{GB,h}(t) = P_{h.load}(t) + P^{h}_{ES,cha}(t) - P^{h}_{ES,dis}(t) \tag{17}$$

where in each given time period, $P_{h.load}(t)$ stands for the heat load, $P^{h}_{ES,cha}(t)$ stands for the power input to the thermal energy storage system, and $P^{h}_{ES,dis}(t)$ stands for the power output from the system.

7. Energy storage operation constraints.

$$\begin{cases} P^{\min}_{es,cha}(t) \leq P_{es,cha}(t) \leq P^{\max}_{es,cha}(t) \\ P^{\min}_{es,dis}(t) \leq P_{es,dis}(t) \leq P^{\max}_{es,dis}(t) \\ S_{es,\min}(t) \leq S_{es}(t) \leq S_{es,\max}(t) \\ S_{es}(0) = S_{es}(24) \end{cases} \tag{18}$$

where $P^{\max}_{es,cha}(t)$ and $P^{\min}_{es,cha}(t)$ stand for the maximum and minimum values of the power storage capacity for electricity, heat, and gas during the time period t; $P^{\max}_{es,dis}(t)$ and $P^{\min}_{es,dis}(t)$ represent the maximum and minimum values of the power supply for these energy storage mediums during the time period t; $S_{es,\max}(t)$ and $S_{es,\min}(t)$ stand for the maximum and minimum percentages of the energy condition of gas, heat, and electricity storage throughout the specified time period; and $S_{es}(0)$ and $S_{es}(24)$ stand for the percentages representing the energy storage device's initial and final states during the scheduling cycle.

3.4. Model Solution

The RIES two-stage resilient optimal scheduling model is formulated using constraints and objective functions. For ease of solution, the model is presented in a streamlined format, as shown in Equation (19).

$$\begin{cases} \min\limits_{x}(\max\limits_{u \in P_{RE}} \min\limits_{y \in \Omega(x,u)} c^{T}y) \\ s.t. \\ Dy \geq d \\ Ey = 0 \\ Fx + Gy \geq h \\ I_{u}y = u \end{cases} \tag{19}$$

$$\begin{cases} x = [u_{ess,t}, u_{tss,t}, v_{ele,t}, v_{ther,t}] \\ y = [P^{w}_{e,t}, P^{i}_{dis,t}, P^{i}_{ch,t}, P^{buy}_{e,t}, P^{sell}_{e,t}, P^{l}_{e,t}] \\ u = [P^{PV}_{ele,t}, P^{WT}_{ele,t}] \end{cases} \tag{20}$$

where c is the symbol for the objective function's corresponding coefficient matrix; x and y represent the vectors of system output variables and state variables, respectively; the coefficient matrices D, E, F, and G correspond to the restrictions; and the system operational parameters d and h are shown.

The initial problem (Equation (19)) is decomposed into the primary problem (outer minimization problem) as shown in Equation (21), and the secondary problem (inner max–min problem) is depicted in Equation (22). The column and constraint generation (C&CG) algorithm is used to solve it iteratively. This approach accelerates convergence by systematically integrating variables and constraints related to the sub-problems while solving the main problem. Initially, a set of numerical values is designated as the uncertainty parameters, which are then input into the sub-problem. The solution of the sub-problem identifies the factors that lead to the worst-case scenario. These parameters are iteratively reintroduced into the main problem for further resolution. The sub-issue is reformulated as a single-layer maximization problem to determine the worst-case scenario's uncertainty parameters. The procedure is detailed in Figure A1 (Appendix B), while the equipment specifications and energy prices are listed in Tables A1 and A2. The anticipated values of wind and photovoltaic power, representing the initial uncertainty parameters, are shown in Figure A3 in Appendix B.

$$MP\begin{cases} \min\limits_{x}\theta \\ s.t. \\ \theta \geq c^T y_k \\ Dy_k \geq d \\ Fx + Gy_k \geq h \\ I_u y_k = u_k^* \end{cases} \tag{21}$$

where the sub-problem solution after the k-th iteration is denoted as y_k, and u_k^* is the value of the wind power and solar outputs in the most unfavorable scenario computed after the k-th iteration.

$$SP\begin{cases} \max\limits_{u\in P_{RE},\gamma,\lambda,\nu,\pi} \quad d^T y + (h - Fx)^T \nu + u^T \pi \\ s.t. \\ D^T y + K^T \lambda + G^T \nu + I_u^T \pi \leq c \\ \gamma \geq 0, \nu \geq 0, \pi \geq 0 \end{cases} \tag{22}$$

Because the maximization problem obtained by dual transformation of the max–min problem of the sub-problem has the bilinear term $u^T\pi$, it is linearized using the large M approach. In Equation (23), we can see the sub-problem in its ultimate linearized form.

$$\begin{cases} \max\limits_{u\in P_{RE},\gamma,\lambda,\nu,\pi} \quad d^T y + (h - Fx)^T \nu + \overline{u^T \pi} + \Delta u^T B' \\ s.t. \\ D^T y + K^T \lambda + G^T \nu + I_u^T \pi \leq c \\ 0 \leq B' \leq M \cdot B \\ \pi - M \cdot (1 - B) \leq B' \leq \pi \\ \gamma \geq 0, \nu \geq 0, \pi \geq 0 \end{cases} \tag{23}$$

where the dual variable has the upper bound of a big-enough positive real number M, and the auxiliary variables $\Delta u = [d^{pv,t}, d^{wt,t}]$ and $B' = [B'_{pv,t}, B'_{wt,t}]$ are also included.

4. Example Analysis

This article presents a case study utilizing the MATLAB environment and the CPLEX solver from the YALMIP toolkit to address the issue. The study develops the unit output, energy storage charging and discharging, and energy acquisition strategies for each time period using a two-stage robust optimization approach. The objective is to ensure that the RIES achieves the lowest cost even under adverse wind and solar output conditions. The model's capacity to balance robustness, economics, and low-carbon benefits is validated by the operation scheduling results. The example data are configured as follows: The characteristics of the energy conversion and energy storage units are selected from the referenced source [33]; specifically, Table A1 in Appendix A. Real-time energy prices from the energy trading market are listed in Table A2 in Appendix A. Load prediction values, along with photovoltaic and wind power output figures, are depicted in Figures A2 and A3 in Appendix B. References for actual carbon emission model parameters are given in [34]; the carbon emission right quota β_e^* per unit of electricity generated is 0.798 kg/(kW · h), and the carbon emission right quota β_h^* per unit of heat generated is 0.385 kg/(kW · h); the carbon emission quota $\varphi_{g,load}$ = 0.180 kg/(kW · h) per unit gas load consumption; the interval length of the carbon emission d = 2000 kg, the price increase range λ = 0.25, and the base price of carbon trading c = 0.251 CNY/kg. The transferable load on the demand side accounts for 10% of the total load, and the substitutable load accounts for 5% of the total load. The robust optimization model considers the extreme values of wind and solar production as the most unfavorable scenarios. However, the actual output may vary, falling within or outside this range. Consequently, during the robust optimization process, there can be a discrepancy between the wind and solar production predicted by the uncertainty set and the actual output of wind and solar sources.

4.1. Comparison and Analysis of Scene Optimization Results

To demonstrate the efficacy of the RIES robust optimization model in conjunction with the carbon trading mechanism and the demand response, this study constructs four optimization scenarios:

Scenario 1: This scenario uses deterministic optimization based on a stepped carbon trading mechanism without incorporating the demand response and considering the costs of carbon trading.

Scenario 2: This scenario focuses on carbon trading costs within the stepped carbon trading system using deterministic optimization while excluding the demand response.

Scenario 3: Deterministic optimization is employed in this scenario, which accounts for demand-side loads and carbon emission costs related to horizontal time shifts and vertical complementary substitutions.

Scenario 4: Robust optimization is applied in this scenario with the stepped carbon trading mechanism, considering both horizontal time shifts and vertical complementary substitutions of carbon emission costs and demand-side loads.

This section provides a comparison and analysis of the most efficient scheduling outcomes across the four scenarios. Power balance diagrams for all the scenarios are detailed in Appendix C. Scenario 1, which uses deterministic optimization and excludes the demand response and carbon trading prices, results in the highest carbon trading costs and actual carbon emissions. Scenario 2, by focusing solely on carbon trading costs and using deterministic optimization, achieves a 12.93% reduction in actual carbon emissions and a 29.37% decrease in carbon trading costs compared with Scenario 1. Additionally, reduced energy purchases in Scenario 2 further enhance the low-carbon benefits.

Scenario 3 incorporates both carbon emission costs and the demand response, resulting in a 12.49% reduction in energy procurement expenses and a 7.01% decrease in carbon trading costs compared with Scenario 2. This scenario also achieves a 6.17% reduction in system carbon emissions. The use of multiple energy sources in a complementary manner enhances the system's economic and low-carbon performance.

In Scenario 4, robust optimization is applied to Scenario 3, leading to a 6.11% increase in energy purchase costs and a 3.42% rise in real carbon emissions. However, this robust model improves the system's resilience by increasing energy purchases to mitigate the risks associated with fluctuations in wind and solar outputs. It effectively balances resilience and low-carbon objectives while maintaining a robust system operation.

4.2. Analyzing the Effects of the Base Price and Range Length in Carbon Trading

Since carbon trading is a key component of the objective function, fluctuations in its cost will influence the overall operating cost of the system. Figure 3 explores the relationships among variables such as the carbon trading base price (c) and interval length (d) and the system's actual carbon emissions and trading costs. By analyzing these relationships, one can assess how changes in the carbon trading base price and interval length impact both the real carbon emissions and the costs associated with carbon trading.

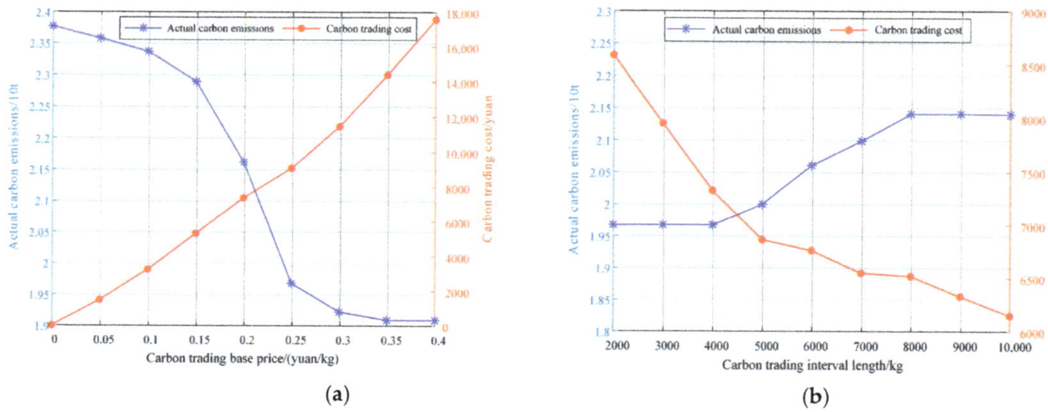

Figure 3. (**a**) Impact of the carbon trading base price on the RIES; (**b**) Impact of the carbon trading interval length on the RIES.

The proportion of carbon trading costs increases directly with the value of c. As the carbon price rises, the system quickly adjusts, with the aim of minimizing operational expenses by significantly reducing carbon emissions. However, when the price exceeds the actual cost, carbon trading costs will continue to rise steadily with the increase in c. When c reaches or exceeds 0.35 CNY/kg, the operation of system components stabilizes, and carbon emissions drop to their lowest and remain steady. Beyond this point, further increases in c do not affect the system's behavior but lead to higher overall running costs due to increased carbon trading expenses. In the two-stage robust optimization, the treatment of carbon trading costs needs to take into account the uncertainty of wind and solar resources. In the first stage, policymakers may need to set aside some flexibility to cope with possible future fluctuations in carbon trading costs. In the second stage, when faced with the actual wind and solar resource situation, the operation strategy needs to be adjusted to minimize the overall cost, including the carbon trading cost. The dispatch of wind and solar power generation should be optimized while meeting the needs and stability requirements of the system. Through these integrations, the model is able to optimize the overall performance of the system in the face of uncertainty, while considering the impact of carbon trading costs, thereby achieving the dual goals of economic and environmental benefits.

When the weight of d falls between 2000 kg and 4000 kg, the system needs to purchase additional carbon emission quotas at higher pricing tiers, leading to a spike in carbon trading costs. Despite the system's prompt adjustment to minimize carbon emissions and reduce operational expenses, this higher pricing impacts overall costs. For weights between 4000 kg and 8000 kg, the system acquires extra quotas at lower prices, resulting in increased carbon emissions but significantly reduced trading costs. When the weight exceeds 8000 kg, the system continues to purchase quotas at the lower end of the carbon price scale, with minimal price escalation and stable carbon emission levels.

Properly adjusting the interval duration and carbon trading benchmark pricing can optimize economic benefits while simultaneously reducing carbon emissions.

4.3. Impact Analysis of the Demand Response Strategy Based on Horizontal Time Shifts and Vertical Complementary Substitutions

The data presented in Table 1 indicate that Scenario 3 results in a reduction of 1278 kg in carbon emissions compared with Scenario 2, along with a further savings of 1331 CNY in the system's overall running cost. The demand response outcomes for Scenario 3 are illustrated in Figure 4. For detailed balance diagrams of electric, heat, and gas power in both scenarios, refer to Figures A5 and A6 in Appendix C.

Table 1. Optimized scheduling outcomes for each scenario.

Scenario	Total Cost/CNY	F_{buy}/CNY	F_{CO_2}/CNY	F_{MC}/CNY	F_{DR}/CNY	Actual Carbon Emissions/kg
1	19,418	5396	12,904	1118	0	23,807
2	15,658	5429	9114	1116	0	20,728
3	14,327	4751	8475	1019	82	19,450
4	15,051	5060	8819	1055	117	20,138

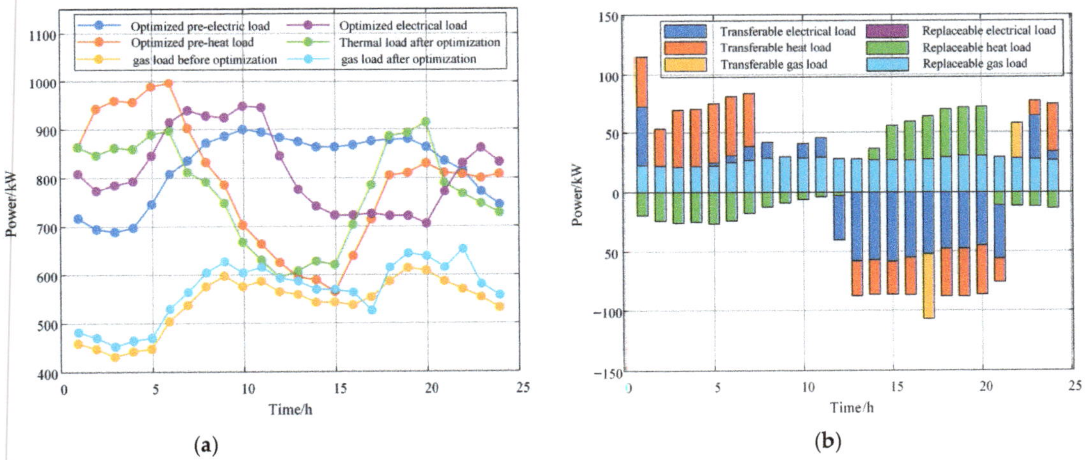

Figure 4. (**a**) Line chart of each load before and after the demand response; (**b**) Histogram of each load in the demand response process.

To reduce wind and solar curtailment, demand-side measures such as horizontal time shifts and vertical complementary alternatives are employed. Since electricity is more expensive during the day compared with gas, gas loads can replace some of the electricity and heat loads. Conversely, at night, electric loads can substitute for part of the gas and heat loads. Horizontal time shift demand responses achieve "peak shaving and valley filling" by redistributing load power from peak to lower demand periods. Vertical complementary substitutions allow for flexible consumption options across multiple energy loads. This demand response strategy adjusts the load curve, eases energy supply pressures, and fully leverages the optimization potential in the RIES operation.

4.4. Research on How New Energy Output Uncertainty Adjustment Parameters Affect Scheduling Results

Figure 5 illustrates the deviations between the photovoltaic power generation, wind power generation, and predicted values ($\Gamma^{PV} = \Gamma^{WT} = 0$) at Γ^{PV} values of 6 and 12 and Γ^{WT} values of 12 and 24 under the scenario of 150% photovoltaic power.

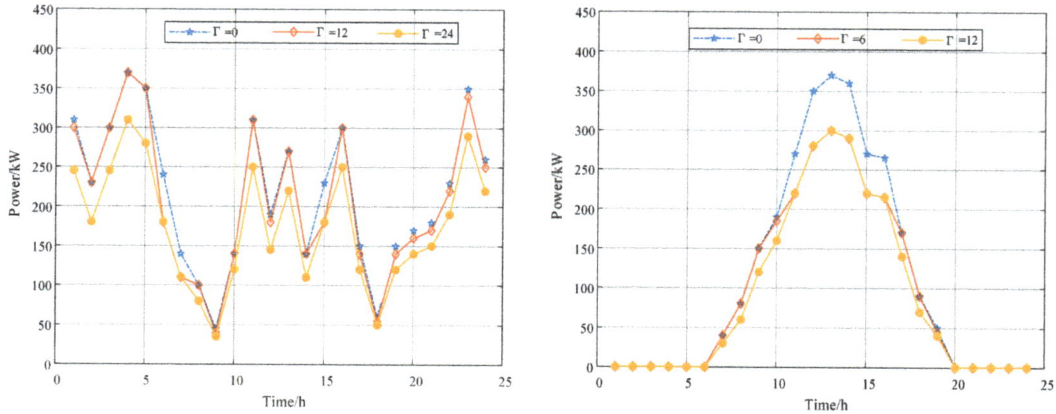

(**a**) (**b**)

Figure 5. (**a**) Deviation between the wind output and the predicted value; (**b**) Deviation between the solar output and the predicted value.

Figure 5 illustrates that the most unfavorable scenario occurs at the minimum expected scenario output. When wind and solar outputs hit their lowest values within the specified interval, the RIES shows its highest operational cost, reflecting the "worst" scenario described in the study.

For comparison, three sets of new energy output uncertainty adjustment parameters have been selected. The values of these adjustment parameters and their solution results are detailed in Table 2.

Table 2. The influence of different uncertainty adjustment parameters on various costs.

Optimization Method		Day-Ahead Cost/CNY	Intraday Cost/CNY	Total Cost/CNY
Deterministic optimization		14,327	4504	18,831
Two-stage robust optimization	$\Gamma^{WT} = 0,\ \Gamma^{PV} = 0$	15,051	3439	18,490
	$\Gamma^{WT} = 12,\ \Gamma^{PV} = 6$	15,937	2221	18,158
	$\Gamma^{WT} = 24,\ \Gamma^{PV} = 12$	16,819	1168	17,987

Table 2 shows that as the uncertainty adjustment parameters increase, the system scheduling scheme becomes more robust and conservative, which is reflected in the rising operating costs and economic losses. The model generates scheduling schemes based on the risk preference of the schedulers to ensure the system is sufficiently robust to handle various worst-case scenarios. The day-ahead scheduling cost for the resilient optimization approach with the highest level of uncertainty is 14.82% higher compared with the deterministic optimization scheme. However, during the intraday operation, the system's consideration of the worst-case scenario results in significant reductions in operational expenses. Consequently, the resilient optimization approach has a lower overall cost compared with the deterministic optimization approach.

The provided image demonstrates that the proposed model effectively reduces the overall system cost and improves economic performance by incorporating real wind and solar output scenarios and selecting appropriate uncertainty measures. This approach ensures the system's robustness while optimizing cost efficiency.

5. Conclusions

Taking carbon trading and the demand response into account, this work develops a scheduling model aimed at optimizing a low-carbon economy within the context of a RIES. To address the unpredictability of wind and solar production, a two-stage robust optimization strategy is employed. Analysis of the scheduling results reveals the following:

1. To enhance economic cooperation with reduced carbon emissions, a stepped carbon trading mechanism is utilized. This involves establishing a rational base price and range for carbon trading. A small trade-off in system resilience is made to improve low-carbon performance, thereby optimizing the balance between system resilience and low-carbon benefits.
2. Incorporating the demand response into the stepped carbon trading mechanism improves the RIES's efficiency in achieving energy savings and emission reduction goals. This approach also alleviates strain on the energy supply to some extent and effectively aligns the economic and low-carbon aspects of the system's operation.
3. By selecting an appropriate level of uncertainty for dispatching, considering the unique characteristics of wind and solar output, the system's resilience to uncertainty-related risks is enhanced. Dynamically adjusting the level of caution in the scheduling scheme allows for a rational balance between cost-effectiveness and resilience.

This study concentrates exclusively on the uncertainties related to new energy sources and does not address the uncertainties associated with energy demand. Future research will aim to enhance the joint optimization of both the supply and demand sides, facilitating more efficient and stable system operations while promoting low-carbon and cost-effective practices.

Author Contributions: Conceptualization, M.Z.; methodology, M.Z.; validation, M.Z. and B.W.; investigation, M.Z.; writing—original draft preparation, M.Z. and B.W.; writing—review and editing, M.Z. and B.W.; supervision, J.W.; project administration, J.W.; funding acquisition, M.Z. All authors have read and agreed to the published version of the manuscript.

Funding: This work was supported by the National Natural Science Foundation of China (Project Number: 71963024).

Data Availability Statement: Data supporting this study are included within the article.

Acknowledgments: All the authors are thanked for their contributions to this article.

Conflicts of Interest: All authors are employed by the School of Electrical and Information Engineering, Lanzhou University of Technology. They declare that the research was conducted in the absence of any commercial or financial relationships that could be construed as a potential conflict of interest.

Appendix A

Table A1. Main parameters of the RIES equipment.

Equipment	Total Capacity/kW	Conversion Efficiency/%	Climbing Constraints/%
P2G	500	60	20
GT	1000	22 (e→g), 72 (g→h)	20
WHB	600	80	20
GB	800	82	20
Electrical storage	450	— —	20
Thermal storage	500	— —	20
Gas storage	300	— —	20

Table A2. Time-sharing price of the grid network power.

Peak and Valley Periods	Peak Period	Ordinary Period	Valley Period
Time period	08:00–11:00 18:00–21:00	06:00–08:00婉11:00–18:00 21:00–22:00	22:00–06:00
Electricity purchase price [CNY(kW·h)$^{-1}$]	1.167	0.718	0.338

Appendix B

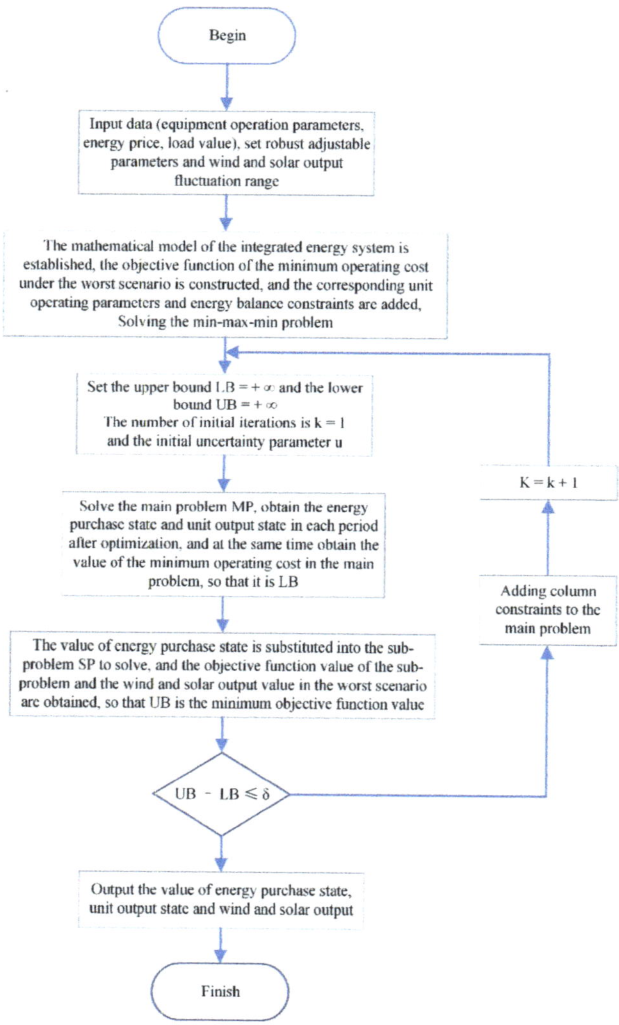

Figure A1. Flow of the C&CG algorithm.

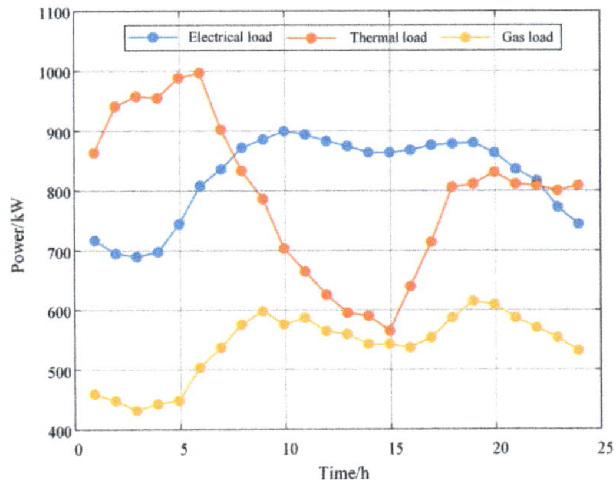

Figure A2. Load forecast value.

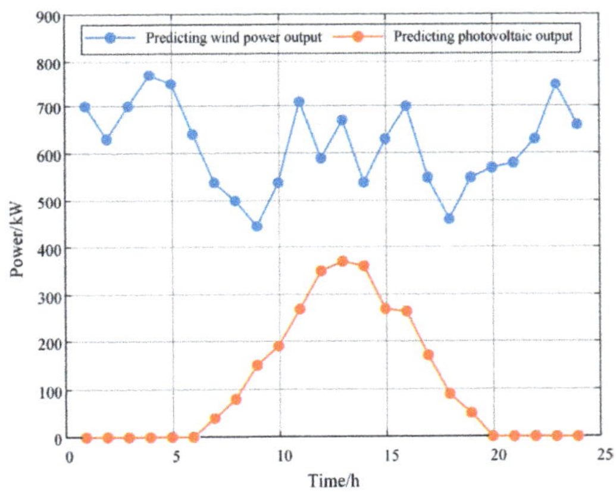

Figure A3. Wind power and photovoltaic output forecast values.

Appendix C

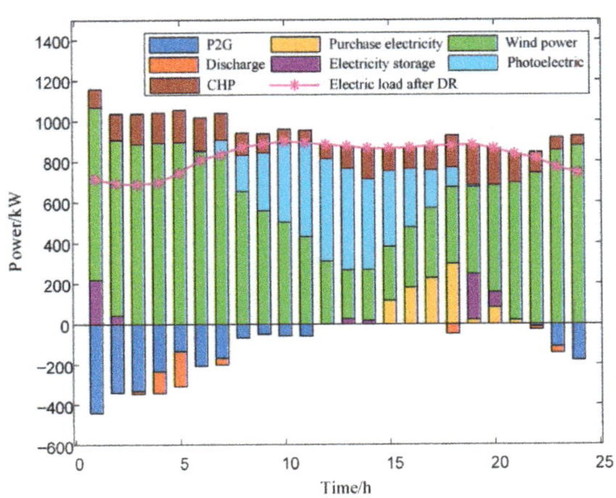

Figure A4. *Cont.*

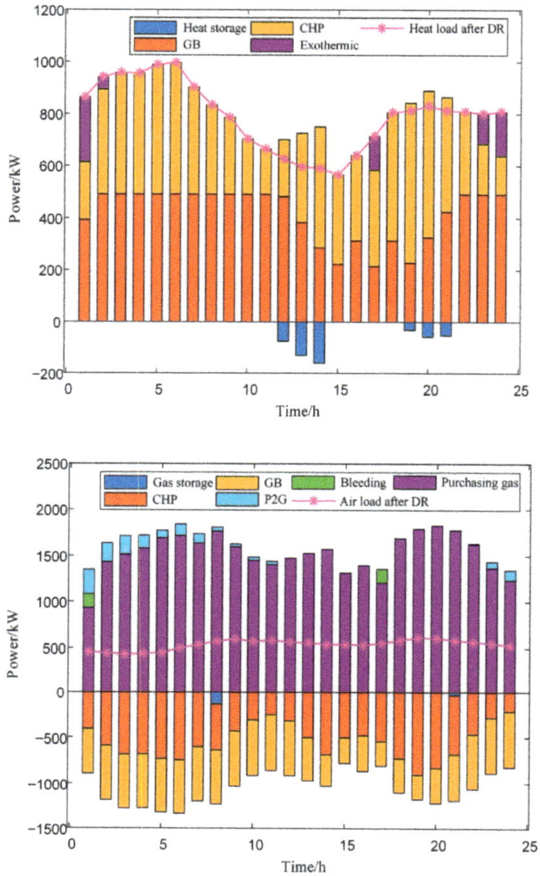

Figure A4. Scenario 1—Balance diagram of the electric, heat, and gas power.

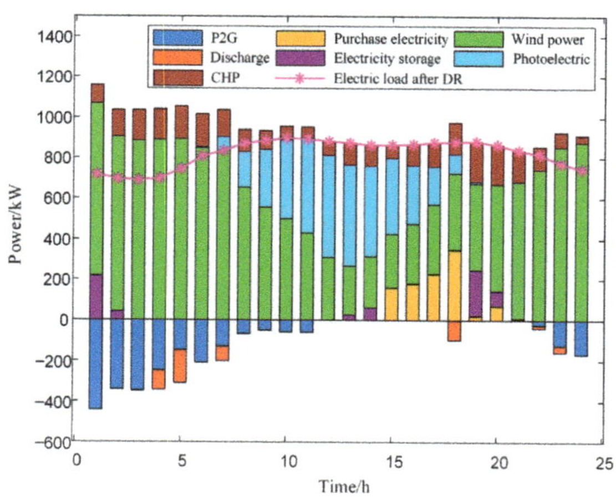

Figure A5. *Cont.*

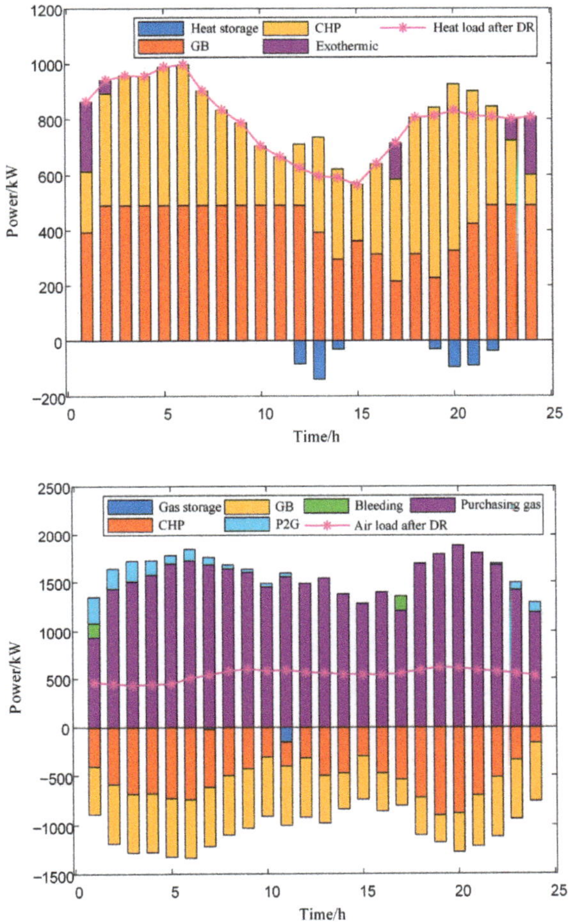

Figure A5. Scenario 2—Balance diagram of the electric, heat, and gas power.

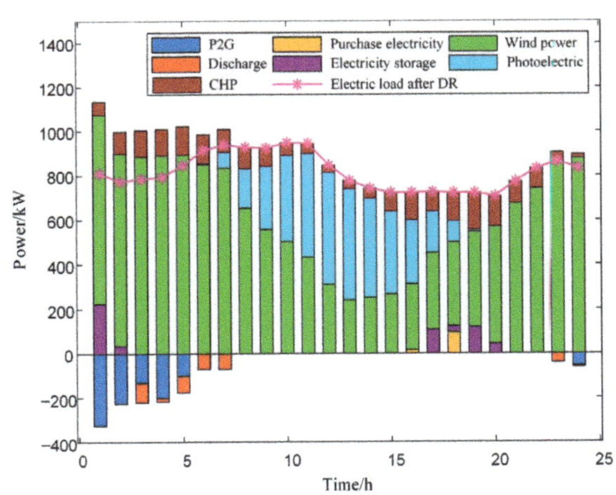

Figure A6. *Cont.*

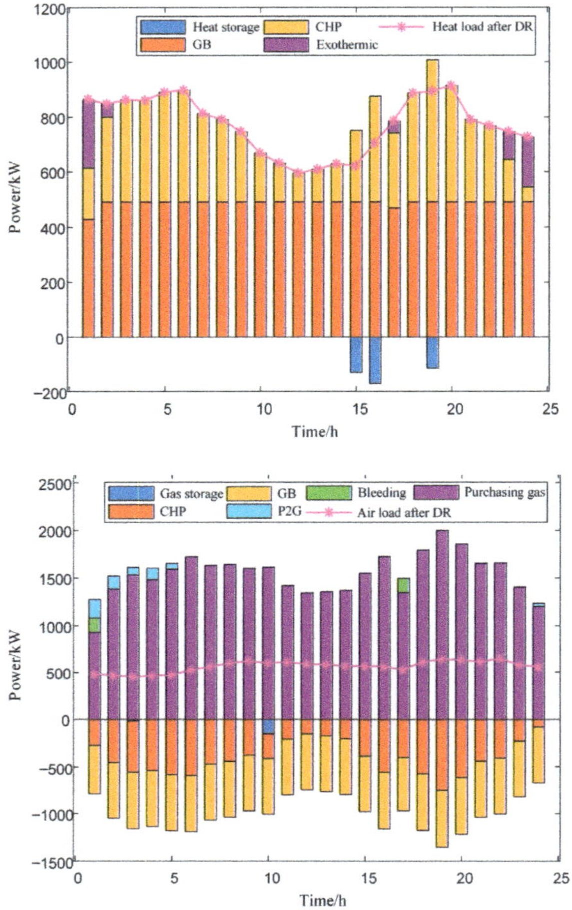

Figure A6. Scenario 3—Balance diagram of the electric, heat, and gas power.

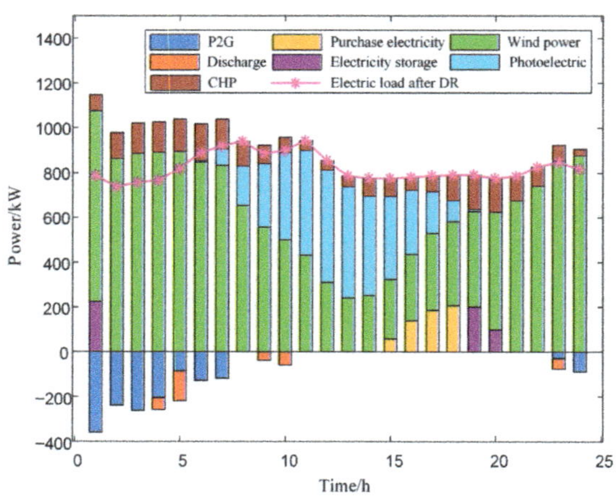

Figure A7. *Cont.*

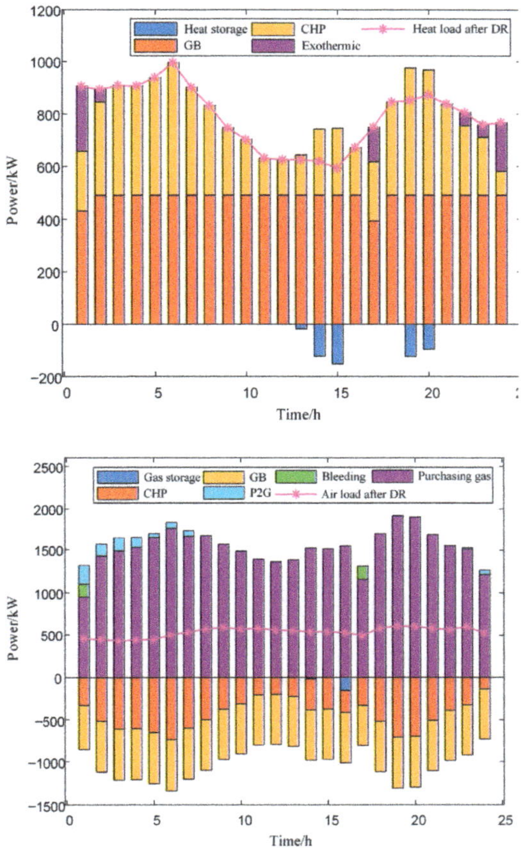

Figure A7. Scenario 4—Balance diagram of the electric, heat, and gas power.

References

1. Woon, K.S.; Phuang, Z.X.; Taler, J.; Varbanov, P.S.; Chong, C.T.; Klemeš, J.J.; Lee, C.T. Recent advances in urban green energy development towards carbon emissions neutrality. *Energy* **2023**, *267*, 126502. [CrossRef]
2. Dai, R.; Zhang, X.; Zou, H. Two-Stage Distributed Robust Optimal Allocation of Integrated Energy Systems under Carbon Trading Mechanism. *Processes* **2024**, *12*, 1044. [CrossRef]
3. Lu, S.; Gu, W.; Zhou, S.; Yao, S.; Pan, G. Adaptive robust dispatch of integrated energy system considering uncertainties of electricity and outdoor temperature. *IEEE Trans. Ind. Inform.* **2020**, *16*, 4691–4702. [CrossRef]
4. Qu, B.; Fu, L.; Xing, Z. Research on Collaborative Optimal Dispatching of Electric Heating Integrated Energy based on Wind Power Prediction Accuracy. *IEEE Access* **2023**, *11*, 145167–145184. [CrossRef]
5. Li, P.; Wang, Z.; Wang, N.; Yang, W.; Li, M.; Zhou, X.; Yin, Y.; Wang, J.; Guo, T. Stochastic robust optimal operation of community integrated energy system based on integrated demand response. *Int. J. Electr. Power Energy Syst.* **2021**, *128*, 106735. [CrossRef]
6. Zhang, Y.; Lan, T.; Hu, W. A Two-Stage robust optimization microgrid model considering carbon trading and demand response. *Sustainability* **2023**, *15*, 14592. [CrossRef]
7. Kong, X.; Xiao, J.; Liu, D.; Wu, J.; Wang, C.; Shen, Y. Robust stochastic optimal dispatching method of multi-energy virtual power plant considering multiple uncertainties. *Appl. Energy* **2020**, *279*, 115707. [CrossRef]
8. Valibeygi, A.; Habib, A.H.; De Callafon, R.A. Robust Power Scheduling for Microgrids with Uncertainty in Renewable Energy Generation. In Proceedings of the 2019 IEEE Power & Energy Society Innovative Smart Grid Technologies Conference (ISGT), Washington, DC, USA, 18–21 February 2019. [CrossRef]
9. Ghaemi, S.; Salehi, J.; Moeini-Aghtaie, M. Estimating abilities of distributed energy resources in providing flexible ramp products for active distribution networks. *Sustain. Cities Soc.* **2021**, *65*, 102593. [CrossRef]
10. Zheng, S.; Sun, Y.; Qi, B.; Li, B. Incentive-Based Integrated Demand Response Considering S&C Effect in Demand Side With Incomplete Information. *IEEE Trans. Smart Grid* **2022**, *13*, 4465–4482. [CrossRef]

11. Guo, W.; Xu, X. Comprehensive Energy Demand Response Optimization Dispatch Method based on carbon trading. *Energies* **2022**, *15*, 3128. [CrossRef]
12. Yi, T.; Ren, W. Low carbon economy scheduling of integrated energy system considering the mutual response of supply and demand. *Sustain. Energy Grids Netw.* **2024**, *38*, 101279. [CrossRef]
13. Kim, H.; Kim, M.; Lee, J. A two-stage stochastic p-robust optimal energy trading management in microgrid operation considering uncertainty with hybrid demand response. *Int. J. Electr. Power Energy Syst.* **2021**, *124*, 106422. [CrossRef]
14. Daneshvar, M.; Mohammadi-Ivatloo, B.; Zare, K. Two-stage optimal robust scheduling of hybrid energy system considering the demand response programs. *J. Clean. Prod.* **2020**, *248*, 119267. [CrossRef]
15. He, K.; Zeng, L.; Yang, J.; Gong, Y.; Zhang, Z.; Chen, K. Optimization Strategy for Low-Carbon Economy of Integrated Energy System considering Carbon Capture-Two Stage Power-to-Gas Hydrogen Coupling. *Energies* **2024**, *17*, 3205. [CrossRef]
16. Zhou, W.; Sun, Y.; Zong, X.; Zhou, H.; Zou, S. Low-carbon economic dispatch of integrated energy system considering carbon trading mechanism and LAES-ORC-CHP system. *Front. Energy Res.* **2023**, *11*, 1134221. [CrossRef]
17. Yang, P.; Jiang, H.; Liu, C.; Kang, L.; Wang, C. Coordinated optimization scheduling operation of integrated energy system considering demand response and carbon trading mechanism. *Int. J. Electr. Power Energy Syst.* **2023**, *147*, 108902. [CrossRef]
18. Yan, Q.; Zhang, M.; Lin, H.; Li, W. Two-stage adjustable robust optimal dispatching model for multi-energy virtual power plant considering multiple uncertainties and carbon trading. *J. Clean. Prod.* **2022**, *336*, 130400. [CrossRef]
19. Melgar-Dominguez, O.D.; Pourakbari-Kasmaei, M.; Lehtonen, M.; Mantovani, J.R.S. An economic-environmental asset planning in electric distribution networks considering carbon emission trading and demand response. *Electr. Power Syst. Res.* **2020**, *181*, 106202. [CrossRef]
20. Zhang, X.; Zhao, X.; Zhong, J.; Ma, N. Low carbon multi-objective scheduling of integrated energy system based on ladder light robust optimization. *Int. Trans. Electr. Energy Syst.* **2020**, *30*, e12498. [CrossRef]
21. Wang, Y.; Qin, Y.; Du, B.; Liu, Y.; Xu, S.; Li, D.; Wang, D. Operational optimization of an integrated electricity-thermal energy system considering demand response under a stepped carbon trading mechanism. *J. Phys. Conf. Ser.* **2022**, *2396*, 012056. [CrossRef]
22. Gao, J.; Meng, Q.; Liu, J.; Wang, Z. Thermoelectric optimization of integrated energy system considering wind-photovoltaic uncertainty, two-stage power-to-gas and ladder-type carbon trading. *Renew. Energy* **2024**, *221*, 119806. [CrossRef]
23. Long, X.; Liu, H.; Wu, T.; Ma, T. Optimal scheduling of Source–Load synergy in rural integrated energy systems considering complementary Biogas–Wind–Solar utilization. *Energies* **2024**, *17*, 3066. [CrossRef]
24. Xue, Y.; Zheng, Y.; Gómez-Expósito, A. (Eds.) *Proceedings of the 7th PURPLE MOUNTAIN FORUM on Smart Grid Protection and Control (PMF2022)*; Springer: Singapore, 2022. [CrossRef]
25. Shi, L.; Liang, C.; Zhou, J.; Li, Y.; Liu, J.; Wu, F. Optimal scheduling of integrated energy systems with a ladder-type carbon trading mechanism and demand response. *Front. Energy Res.* **2024**, *12*, 1363285. [CrossRef]
26. Sun, H.; Sun, X.; Kou, L.; Zhang, B.; Zhu, X. Optimal scheduling of park-level integrated energy system considering ladder-type carbon trading mechanism and flexible load. *Energy Rep.* **2023**, *9*, 3417–3430. [CrossRef]
27. Guo, R.; Ye, H.; Zhao, Y. Low carbon dispatch of electricity-gas-thermal-storage integrated energy system based on stepped carbon trading. *Energy Rep.* **2022**, *8*, 449–455. [CrossRef]
28. Ye, X.; Ji, Z.; Xu, J.; Liu, X. Optimal dispatch of integrated energy systems considering integrated demand response and stepped carbon trading. *Front. Electron.* **2023**, *4*, 1110039. [CrossRef]
29. Li, F.; Wang, D.; Guo, H.; Zhang, J. Distributionally Robust Optimization for integrated energy system accounting for refinement utilization of hydrogen and ladder-type carbon trading mechanism. *Appl. Energy* **2024**, *367*, 123391. [CrossRef]
30. Tan, H.; Yan, W.; Ren, Z.; Wang, Q.; Mohamed, M.A. A robust dispatch model for integrated electricity and heat networks considering price-based integrated demand response. *Energy* **2022**, *239*, 121875. [CrossRef]
31. Lu, J.; Huang, D.; Ren, H. Data-driven source-load robust optimal scheduling of integrated energy production unit including hydrogen energy coupling. *Glob. Energy Interconnect.* **2023**, *6*, 375–388. [CrossRef]
32. Yang, M.; Li, J.; Sun, J.; Xu, J.; Li, J. Robust optimal scheduling of EHG-IES based on uncertainty of wind power and PV output. *Int. Trans. Electr. Energy Syst.* **2022**, *2022*, 6587478. [CrossRef]
33. Gao, L.; Fei, F.; Jia, Y.; Wen, P.; Zhao, X.; Shao, H.; Feng, T.; Huo, L. Optimal dispatching of integrated agricultural energy system considering ladder-type carbon trading mechanism and demand response. *Int. J. Electr. Power Energy Syst.* **2024**, *156*, 109693. [CrossRef]
34. Wang, R.; Wen, X.; Wang, X.; Fu, Y.; Zhang, Y. Low carbon optimal operation of integrated energy system based on carbon capture technology, LCA carbon emissions and ladder-type carbon trading. *Appl. Energy* **2022**, *311*, 118664. [CrossRef]

electronics

MDPI

Article

Enhanced Sequence-to-Sequence Deep Transfer Learning for Day-Ahead Electricity Load Forecasting

Vasileios Laitsos [1], Georgios Vontzos [1], Apostolos Tsiovoulos [1], Dimitrios Bargiotas [1,*] and Lefteri H. Tsoukalas [2]

1 Department of Electrical and Computer Engineering, University of Thessaly, 383 34 Volos, Greece; vlaitsos@uth.gr (V.L); gvontzos@uth.gr (G.V.); atsiov@uth.gr (A.T.)

2 Center for Intelligent Energy Systems (CiENS), School of Nuclear Engineering, Purdue University, West Lafayette, IN 47906, USA; tsoukala@purdue.edu

* Correspondence: bargiotas@uth.gr

Abstract: Electricity load forecasting is a crucial undertaking within all the deregulated markets globally. Among the research challenges on a global scale, the investigation of deep transfer learning (DTL) in the field of electricity load forecasting represents a fundamental effort that can inform artificial intelligence applications in general. In this paper, a comprehensive study is reported regarding day-ahead electricity load forecasting. For this purpose, three sequence-to-sequence (Seq2seq) deep learning (DL) models are used, namely the multilayer perceptron (MLP), the convolutional neural network (CNN) and the ensemble learning model (ELM), which consists of the weighted combination of the outputs of MLP and CNN models. Also, the study focuses on the development of different forecasting strategies based on DTL, emphasizing the way the datasets are trained and fine-tuned for higher forecasting accuracy. In order to implement the forecasting strategies using deep learning models, load datasets from three Greek islands, Rhodes, Lesvos, and Chios, are used. The main purpose is to apply DTL for day-ahead predictions (1–24 h) for each month of the year for the Chios dataset after training and fine-tuning the models using the datasets of the three islands in various combinations. Four DTL strategies are illustrated. In the first strategy (DTL Case 1), each of the three DL models is trained using only the Lesvos dataset, while fine-tuning is performed on the dataset of Chios island, in order to create day-ahead predictions for the Chios load. In the second strategy (DTL Case 2), data from both Lesvos and Rhodes concurrently are used for the DL model training period, and fine-tuning is performed on the data from Chios. The third DTL strategy (DTL Case 3) involves the training of the DL models using the Lesvos dataset, and the testing period is performed directly on the Chios dataset without fine-tuning. The fourth strategy is a multi-task deep learning (MTDL) approach, which has been extensively studied in recent years. In MTDL, the three DL models are trained simultaneously on all three datasets and the final predictions are made on the unknown part of the dataset of Chios. The results obtained demonstrate that DTL can be applied with high efficiency for day-ahead load forecasting. Specifically, DTL Case 1 and 2 outperformed MTDL in terms of load prediction accuracy. Regarding the DL models, all three exhibit very high prediction accuracy, especially in the two cases with fine-tuning. The ELM excels compared to the single models. More specifically, for conducting day-ahead predictions, it is concluded that the MLP model presents the best monthly forecasts with MAPE values of 6.24% and 6.01% for the first two cases, the CNN model presents the best monthly forecasts with MAPE values of 5.57% and 5.60%, respectively, and the ELM model achieves the best monthly forecasts with MAPE values of 5.29% and 5.31%, respectively, indicating the very high accuracy it can achieve.

Keywords: deep transfer learning; electricity load forecasting; multilayer perceptron; convolutional neural network; ensemble deep learning; multi-task deep learning; exploratory data analysis

1. Introduction

Research interest in the electricity sector has been growing for several key reasons. First, there is a global shift toward using sustainable and renewable energy sources like solar and wind. This shift has led researchers to find ways to better incorporate these technologies into existing power grids. Second, the increasing need for electricity due to population growth and industrialization requires new and efficient solutions for the transmission, distribution, and consumption of electricity. Third, the development of smart grids and improvements in energy storage technologies have opened up opportunities to improve the resilience, reliability, and responsiveness of power grids. Additionally, concerns about the environment and climate change have motivated researchers to explore cleaner and more eco-friendly energy options. The digitization of the electricity sector, thanks to advances in data analytics, machine learning, and Internet of Things (IoT) technologies, has also spurred research to create smarter and more efficient energy systems. In general, these factors have created a dynamic landscape, encouraging researchers to explore new approaches and technologies to tackle the changing challenges and opportunities in the electricity sector.

Deep transfer learning (DTL) refers to the use of deep neural networks (DNNs) in the domain of transfer learning. DTL utilizes the knowledge learned from one task to improve the performance of another related task. Usually, in DTL a pre-trained DNN model is fine-tuned for a different task. This domain improves data efficiency, reduces training time, and allows models to generalize well, capturing the underlying patterns in electricity consumption. The ability to adapt to dynamic conditions and the improved accuracy stemming from pre-trained models make transfer learning a valuable tool for addressing the challenges of forecasting in the energy sector. With limited and heterogeneous datasets, transfer learning enables models trained in one domain to be adapted to another, addressing the challenges of varying temporal scales and spatial characteristics. This approach proves essential for optimizing electricity forecasting models, particularly in the face of emerging technologies, changing infrastructures, and the need for resource-efficient solutions. By reusing pre-trained models and enhancing adaptability, transfer learning contributes significantly to robust and accurate predictions, ultimately supporting more effective energy management in the dynamically evolving landscape of the electricity sector.

More generally, in recent years, transfer learning has attracted increasing scientific interest. In this regard, an extensive literature review follows, aiming to highlight the most relevant papers that have addressed this specific field. Meng et al. in [1] propose a transfer learning-based method for abnormal electricity consumption detection, where a pre-trained model is fine-tuned using a small amount of data from the target domain. Antoniadis et al. in [2] discuss the use of transfer learning techniques for electricity load forecasting, specifically in the context of leveraging information from finer scales to improve forecasts at wider scales. Yang et al. in [3] discuss the implementation of a transfer learning strategy to address the multi-parameter coupling problem in the design of water-flow-enabled generators. Dong et al. in [4] propose a transfer learning model based on the Xception neural network for electrical load prediction, which is trained using pre-trained models and fine-tuned in the training process. Li et al. in [5] explore a transfer learning scheme for non-intrusive load monitoring (NILM) in smart buildings, which involves transferring a well-trained model to estimate power consumption in another dataset for all appliances. Peirelinck et al. in [6] discuss the use of transfer learning techniques in the context of demand response in the electricity sector. The study shows that transfer learning can improve performance by over 30% in various tasks. Laitsos et al. in [7] use a transfer learning technique with several deep learning models to predict the energy consumption of Greek Islands and the pre-trained model demonstrates outstanding flexibility when adapting to a new and unknown dataset. Wu et al. in [8] propose an attentive transfer framework for efficient residential electric load forecasting using transfer learning and graph neural networks. Kamalov et al. in [9] introduce an NBEATS model in order to test its effectiveness in medium-term electricity forecasting for zero-shot transfer learning. Syed et al. in [10] propose a reliable inductive transfer learning (ITL) method for load forecasting in electrical networks, which uses knowledge from existing deep learning models to develop accurate ITL models at other distribution nodes. Laitsos et al. in [11] propose an automated deep learning application for electricity load forecasting. Santos et al. in [12] propose a novel methodology that combines transfer learning and deep learning techniques to enhance short-term load forecasting for buildings with limited electricity data. Arvanitidis et al. in [13] propose clustering MLP models for short-term load forecasting. Luo et al. in [14] discuss the use of transfer learning techniques for load, solar, and wind power predictions, but their study does not specifically refer to the application of transfer learning to load prediction. Li et al. in [15] discuss a short-term load forecasting framework that adopts transfer learning. Transfer learning is used to train learnable parameters based on trend components and then to transfer them to the load forecasting model. Chan et al. in [16] introduce a hybridized modeling approach, using a convolutional neural network (CNN) and a support vector machine (SVM), for short-term load forecasting. Gontijo et al. in [17] examine the hourly power generation data in Brazil from 2018 to 2020, categorized based on the different electrical subsystems and their corresponding energy sources. The aim was to assess the precision of key methods for combining and splitting forecasts generated by the autoregressive integrated moving average (ARIMA) and the error, trend, seasonal (ETS) models.

Jung et al. in [18] propose a monthly electricity load forecasting framework for smart cities, using transfer learning techniques. Collecting data from multiple districts, they selected similar data based on correlation coefficients, and fine-tuned the model using target data. Al-Hajj et al. in [19] report a survey of transfer learning in renewable energy systems, specifically in the prediction of solar and wind power, the prediction of load, and the diagnosis of faults. Nivarthi et al. in [20] discuss the use of transfer learning in renewable energy systems, specifically in power forecasting and anomaly detection. The authors propose a transfer learning framework and a feature embedding approach to handle missing sensor data. Miraftabzadeh et al. in [21] present a framework based on transfer learning and deep neural networks for the prediction of day-ahead photovoltaic power. Dakovic et al. in [22] report an extensive review of machine learning applications aimed at addressing energy-related issues through the examination of various energy types and opportunities for energy reduction. Vontzos et al. in [23] propose a data-driven short-term forecasting method for electricity consumption in airports. Yang et al. in [24] propose an innovative monthly DNN approach for load forecasting in urban and regional areas. In order to draw more secure conclusions, an extended comparison with other machine learning models was performed. Li et al. in [25] propose a building electricity load forecasting method based on the maximum mean discrepancy (MMD) and an improved TrAdaBoost algorithm (iTrAdaBoost). Gontijo et al. in [26] introduce a dynamic time scan forecasting (DTSF) technique as a novel approach to predict hourly energy prices in Brazil. By identifying similarity patterns in time series data, DTSF demonstrated competitive advantages over traditional forecasting methods presented in prior research.

The relentless growth of electricity demand, coupled with the dynamic and often unpredictable nature of energy consumption patterns, requires advanced forecasting methods for effective grid management. In contrast to the aforementioned studies, which investigate transfer learning only within a few hours or so, the forecasting strategies developed in this paper target day-ahead prediction, thereby enhancing the effectiveness of the models. Additionally, none of the aforementioned studies explore ensemble deep learning, as is done in this study. This deep learning domain includes models that have attracted significant scientific interest due to their high performance, as well as the potential they offer for further investigations and improvements. So, in this context, transfer learning emerges as a promising paradigm to address the challenges associated with limited and disparate data sources. This research paper delves into the application of deep transfer learning techniques in the domain of electricity forecasting, with the aim of exploiting the knowledge gained from a source domain to improve predictive accuracy in a target domain. By leveraging preexisting models trained on related datasets or domains, transfer learning seeks to enhance the adaptability and robustness of forecasting models, ultimately contributing to more accurate and reliable predictions in the complex and ever-evolving landscape of electricity demand. This paper explores the theoretical foundations, methodologies and practical implications of transfer learning in the specific context of electricity forecasting, shedding light on its potential to revolutionize the field and pave the way for more resilient and efficient energy management systems.

With respect to the contributions of this paper, the following points are emphasized:

- For the first time, a high accuracy results implementation of sequence-to-sequence (Seq2seq) ensemble deep transfer learning for day-ahead (1–24 h) forecasting is conducted on three distinct datasets from islands using the Greek power system. Although the Rhodes training dataset exhibits somewhat different behavior compared to the other two datasets, the proposed algorithms provide very satisfactory results. This fact further enhances the performance of the proposed strategies and models. The characteristics of the Rhodes dataset lead to a more robust and comprehensive evaluation of the models, as it introduces variability and challenges that may not be present in the other datasets. This diversity in behavior across datasets provides a more realistic and thorough assessment of the models' capabilities.
- The results obtained indicate that deep transfer learning (DTL) could be of particular value to both transmission system operators (TSO) and distribution system operators (DSO) within various regions of the Greek system.
- The application of the models is performed on actual load data with minimal data preprocessing, a fact that leads to optimistic conclusions regarding their applicability under real-time conditions.

This paper is organized as follows: First, in Section 2, exploratory dataset analysis and feature creation is used. Then, in Section 3, the forecasting strategies are analyzed. In Section 4, the emergent results for each algorithm are presented, along with a discussion about their performance. Finally, in Section 5, the main conclusions are drawn and future study proposals are presented.

2. Materials and Methods

2.1. Dataset Analysis

In this section, all the features, behaviors, and correlations of the three datasets used are investigated and analyzed. Initially, the three time series under study are presented, and then their monthly and daily average fluctuations are considered.

Figure 1 illustrates the three power time series fluctuations for each of the three datasets at an hourly resolution. What is noteworthy is that Rhodes, especially during the summer months, experiences a substantial increase in demand. The other two islands exhibit several similarities between them, with both the average and extreme values behaving relatively similarly. This consistency in behavior across the two datasets suggests common characteristics or patterns in the energy-related dynamics of these islands. The shared trends in both average and extreme values contribute to a more coherent and comparable analysis between the two datasets, helping to develop and evaluate models for these specific island environments.

Figure 2 presents the monthly boxplots for each of the three datasets in hourly resolution.

In addition, Figure 3 visualizes the average daily electricity consumption (1–24 h) for each island.

This figure clearly illustrates that the average hourly values for Rhodes are significantly higher than those of the other two islands. However, it should be noted that the three patterns exhibit strong similarities among them, a fact demonstrated by the common peak and off-peak demand hours among the three islands.

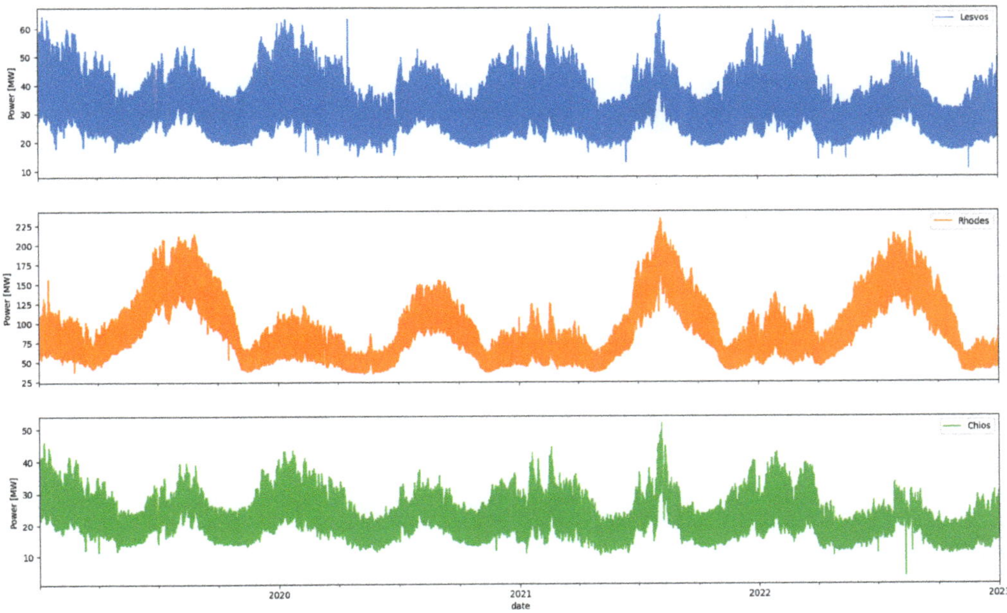

Figure 1. Hourly consumption for Lesvos, Rhodes and Chios.

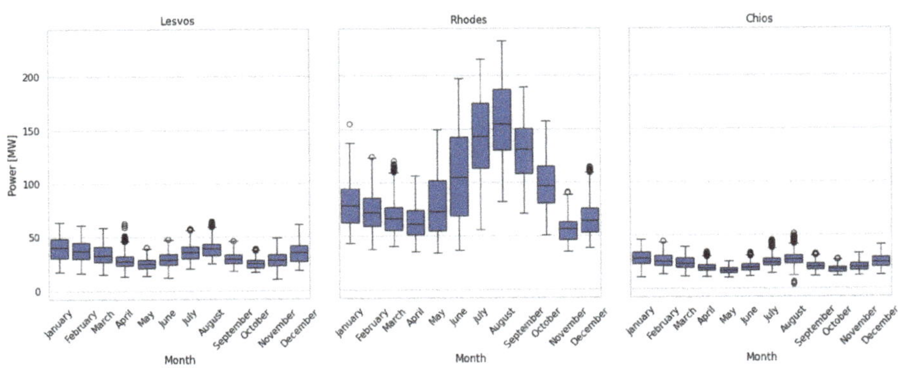

Figure 2. Monthly average consumption for every island.

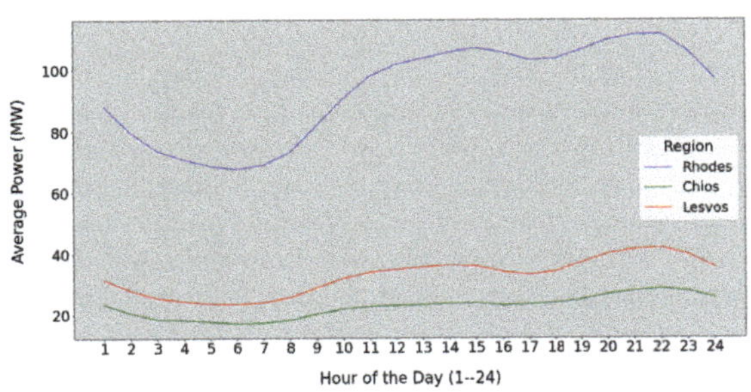

Figure 3. Hourly average consumption per day for each island.

2.2. Data Preprocessing

In order to shape the raw data into a suitable format capable of being used as input in deep learning models, the preprocessing involved the following four stages:

1. Anomaly Detection: Anomaly detection in time series involves establishing a baseline of normal behavior through statistical methods or machine learning algorithms, extracting relevant features, and training a model on labeled data to distinguish normal patterns from anomalies. Due to instances of zero consumption during specific hours, probably caused by network faults, these particular values were set equal to the corresponding values from one week prior. This adjustment was made to address the challenge of unexpected situations in the data, as algorithms may struggle to account for such anomalies. The goal is to ensure optimal training for each model by handling these irregularities in the dataset.
2. Filling Missing Values: Filling missing values in time series data is a crucial preprocessing step for anomaly detection. Since anomalies are often identified based on patterns and trends in the data, it is essential to address gaps caused by missing values.
3. Min-Max Scaling: The preprocessing method applied to all datasets in this paper involves min-max scaling, which normalizes the data points to a range between 0 and 1. To achieve this, two distinct scalers were employed, one for the input and another for the output dataset. The primary rationale behind utilizing min-max scaling is its ability to enhance the efficiency of training deep learning models during the training phase, facilitating faster convergence to the optimal solution of the loss function.
4. One-Hot Encoding: With this process, numerical data are transformed to cyclical data through trigonometric equations. In this study, day of the week, hour of the day, and month of the year were converted to sin and cosine formulations. Figure 4 represents the day of the week transformed in sin and cosine format. The periodicity that appears with this method helps the models to better understand the patterns presented by each time series studied.

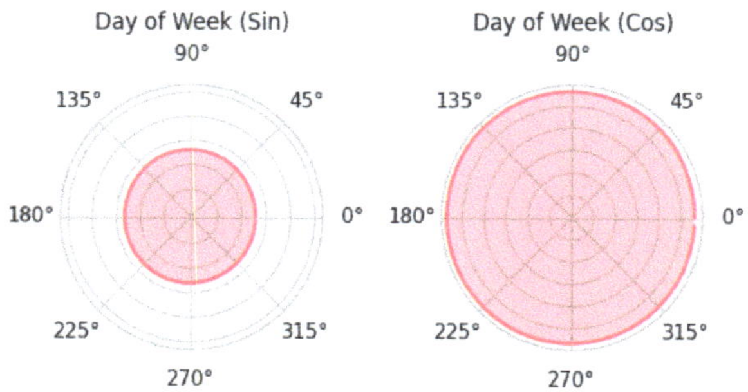

Figure 4. Day of the week in sine and cosine formulation.

2.3. Feature Creation

In this subsection, the generation of the parameters used as inputs for the models is described. For this purpose, several input characteristics were studied and evaluated in order to understand the most significant for predicting electricity demand. After several trials, we determined the best combination of features which would provide us with the most accurate prediction results. So, eight input characteristics were selected and investigated to forecast the day-ahead electricity demand. More specifically, variables 2–8 involve the methodology of one-hot encoding, which was introduced earlier and helps the DL models to better understand and adapt to time series exhibiting seasonal patterns, such as those studied in the paper. The input variables used for all the deep learning models remain consistent and are presented in detail below.

1. Power in hourly resolution: The sequence of 168 h of load values for 7 days/one week.
2. Cos of Day of Week: The sequence of 168 values of Day of the Week (0–6) converted by one-hot encoding to cosine type.
3. Sin of Day of Week: The sequence of 168 values of Day of the Week (0–6) converted by one-hot encoding to sin type.
4. Cos of Hour of Day: The sequence of 168 values of Hour of Day (0–23), converted by one-hot encoding to cosine type.
5. Sin of Hour of Day: The sequence of 168 values of Hour of Day (0–23), converted by one-hot encoding to sin type.
6. Cos of Month of Year: The sequence of 168 values of Month of the Year (1–12) converted by one-hot encoding to cosine type.
7. Sin of Month of Year: The sequence of 168 values of Month of the Year (1–12) converted by one-hot encoding to sin type.
8. IsWeekend: The sequence of 168 values of a dummy variable named "Is Weekend", with value equal to 0 for working days and 1 for weekends and holidays.

Our target was to utilize a historical sequence of 168 h from the aforementioned eight features, and to create day-ahead predictions for the load, i.e., a sequence of 24 values of the power. Figure 5 visualizes the Seq2seq prediction technique.

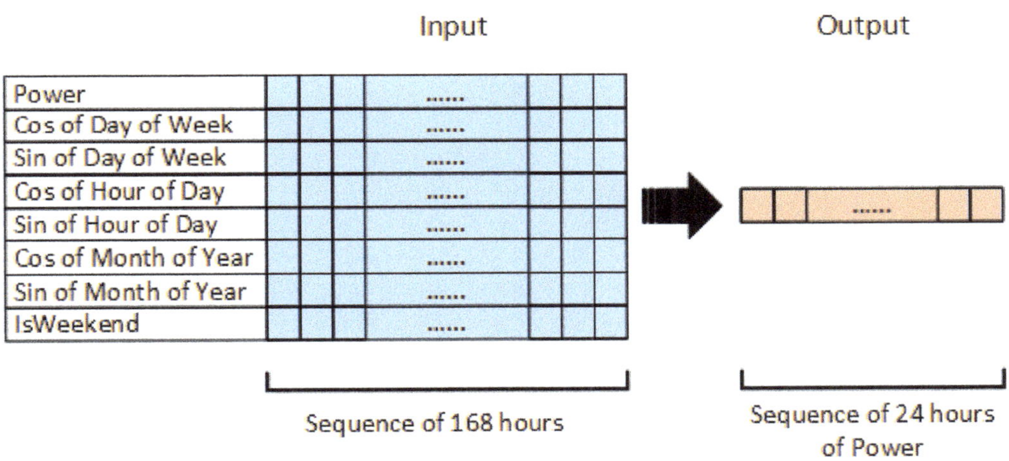

Figure 5. Sequence-to-sequence forecasting technique.

Finally, a correlation heatmap is presented in Figure 6 in order to highlight the relationships between the power time series for each of the three islands.

What is noteworthy is that the target time series of Chios shows a correlation of 0.37 with that of Rhodes and 0.94 with that of Lesvos. This particular characteristic further enhances the reliability of the implemented applications, demonstrating generality and robustness through the results that will be presented below. The correlation between the target time series of different islands suggests some level of interdependence or shared patterns, which can contribute to the generalizability and effectiveness of the models developed for these islands.

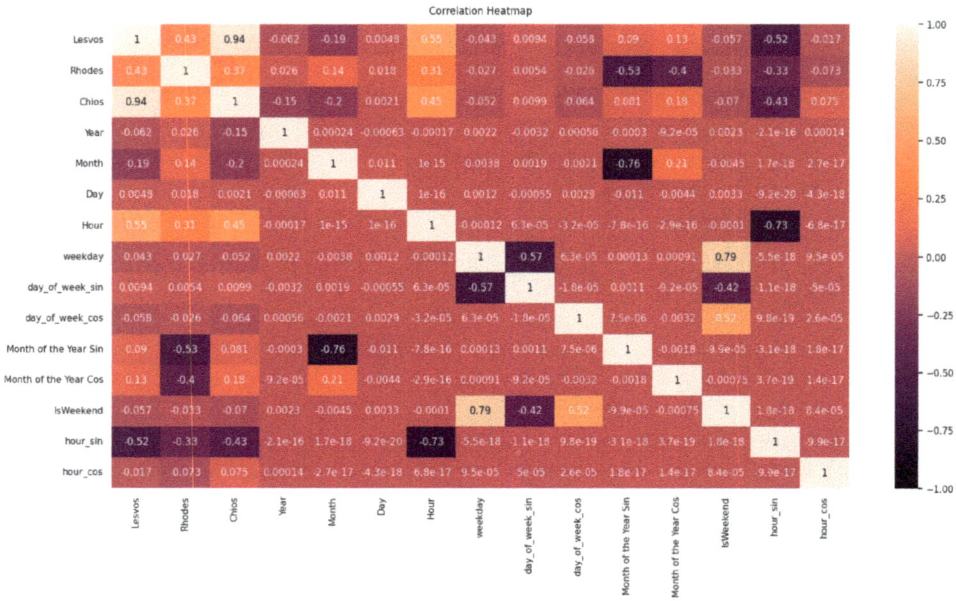

Figure 6. Correlation heatmap.

3. Methodology

In this section, the fundamental methodology followed in the paper is presented. Initially, an introduction to the three DL models is presented. Then, the forecasting strategies are analyzed in depth, emphasizing all the details of each, highlighting the way datasets are utilized for each strategy. Subsequently, reference is made to the optimization framework which was used for enhancing the performance of the DL models. Additionally, the evaluation metrics which were used to evaluate and compare the performance of the algorithms of each strategy are presented. Finally, the software tools that were used are analyzed.

3.1. Deep Learning Models

In this subsection, the functionality and architectures of the three DL models used in the paper, MLP, CNN, and ELM, are analyzed. The main reason why the specific models were chosen is both because they are very powerful and advanced architectures of the deep learning domain in the field of time series forecasting, and because they are of high scientific interest in comparison to simpler machine learning algorithms.

3.1.1. Multilayer Perceptron

A multilayer perceptron (MLP), as presented in Figure 7, is a versatile artificial neural network architecture employed for learning and modeling complex relationships within data. Comprising an input layer (X), one or more hidden layers (H_i), and an output layer (Y), an MLP is characterized by its capacity to capture intricate non-linear patterns. During forward propagation, the input data are transformed through weighted connections and activation functions (σ) in the hidden layers, generating progressively more abstract representations. The hidden layer outputs (H_i) can be mathematically expressed as:

$$H_i = \sigma(W_i H_{i-1} + b_i), \quad (1)$$

where W_i denotes the weight matrix connecting layer $i-1$ to i, H_{i-1} is the output of the previous layer, and b_i represents the bias term for layer i. The activation function introduces non-linearity, enabling the network to capture complex mappings.

The final output (Y) is computed through similar operations in the output layer:

$$Y = \sigma(W_{\text{out}} H_{\text{last}} + b_{\text{out}}) \tag{2}$$

During training, the network adjusts its weights to minimize a defined loss function (L), which quantifies the disparity between the predicted output and the actual target values. The weights are updated using an optimization algorithm, typically gradient descent, with the weight update rule expressed as:

$$W_{\text{new}} = W_{\text{old}} - \eta \frac{\partial L}{\partial W}, \tag{3}$$

where η is the learning rate. This iterative process, known as back propagation, refines the model's weights to improve its predictive accuracy.

In the domain of time series forecasting, MLPs exhibit efficiency owing to their inherent ability to capture temporal dependencies and non-linear patterns. The adaptability of the model enables it to discern and model various temporal structures, including trends and seasonality. The hidden layers serve as dynamic feature extractors, automatically learning relevant temporal features from the time series data. This feature learning capability, coupled with the tunability of the model parameters, positions MLPs as robust and effective tools for a wide array of time series forecasting tasks.

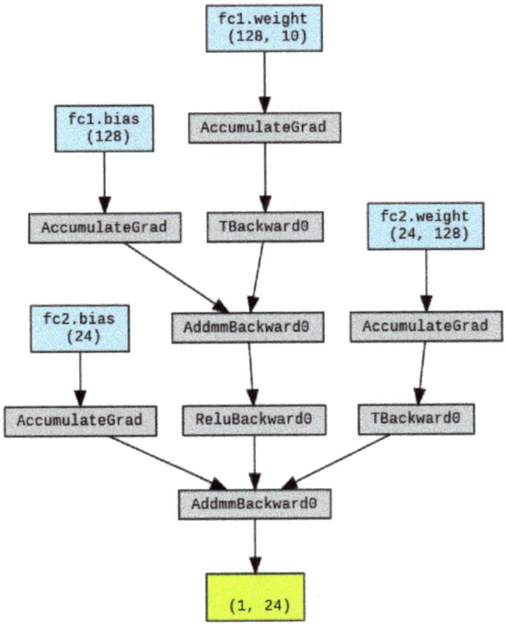

Figure 7. Multilayer perceptron architecture.

3.1.2. Convolutional Neural Network

Convolutional neural networks (CNNs), the architecture of which is presented in Figure 8, constitute a class of sophisticated deep learning architectures specifically designed for the analysis and processing of visual data. The principal structure of CNNs encompasses multiple layers, notably including convolutional layers, pooling layers, and fully connected layers. Convolutional layers assume a pivotal role in feature extraction from input images through the application of convolutional operations utilizing trainable filters. These filters adeptly identify patterns and features at various spatial scales, enabling the network to discern intricate details within the data. Accompanying pooling layers serve to diminish the spatial dimensions of feature maps, thereby diminishing computational complexity and augmenting the model's capacity for generalization. The culmination of these operations transpires in fully connected layers positioned at the conclusion of the network, where the amalgamated features facilitate conclusive predictions. The applicability of CNNs extends across diverse computer vision domains, manifesting notable success in tasks such as image classification, object detection, and image segmentation.

In time series forecasting, CNNs adapt to sequential data using 1D convolutional layers. These layers analyze temporal patterns, aided by pooling layers for downsizing. CNNs efficiently capture short- and long-term dependencies, making them valuable for tasks such as stock price prediction and weather forecasting, showcasing their versatility across diverse data types.

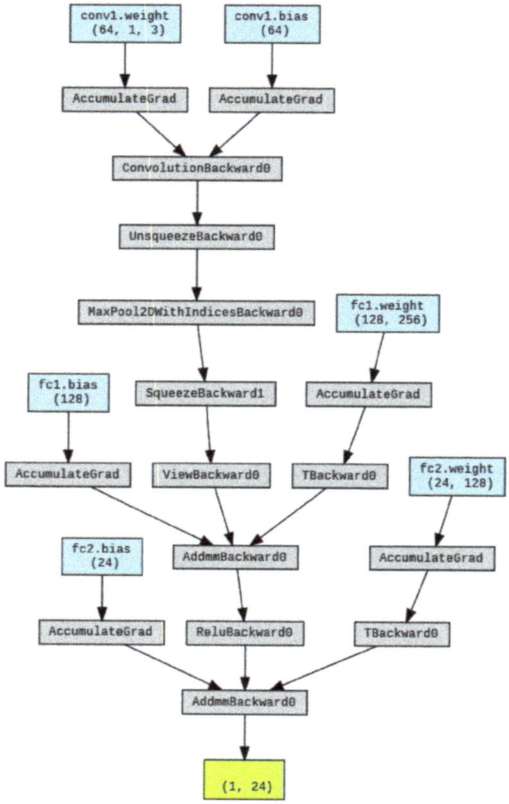

Figure 8. CNN model architecture.

3.1.3. Ensemble Learning Model

The ensemble learning model (ELM), comprising a multilayer perceptron (MLP) and a convolutional neural network (CNN), operates by independently training both models on a given dataset and then combining their predictions through weighted averaging. The MLP focuses on learning non-linear relationships, while the CNN excels at extracting hierarchical features. The weights assigned to each model in the ensemble are determined based on their performance, enhancing the contribution of the better-performing model. The final prediction is generated by summing the weighted predictions, aiming to capitalize on the complementary strengths of the MLP and CNN for improved the predictive accuracy and robustness across diverse data patterns. Figure 9 visualizes the main body of ELM created in this paper.

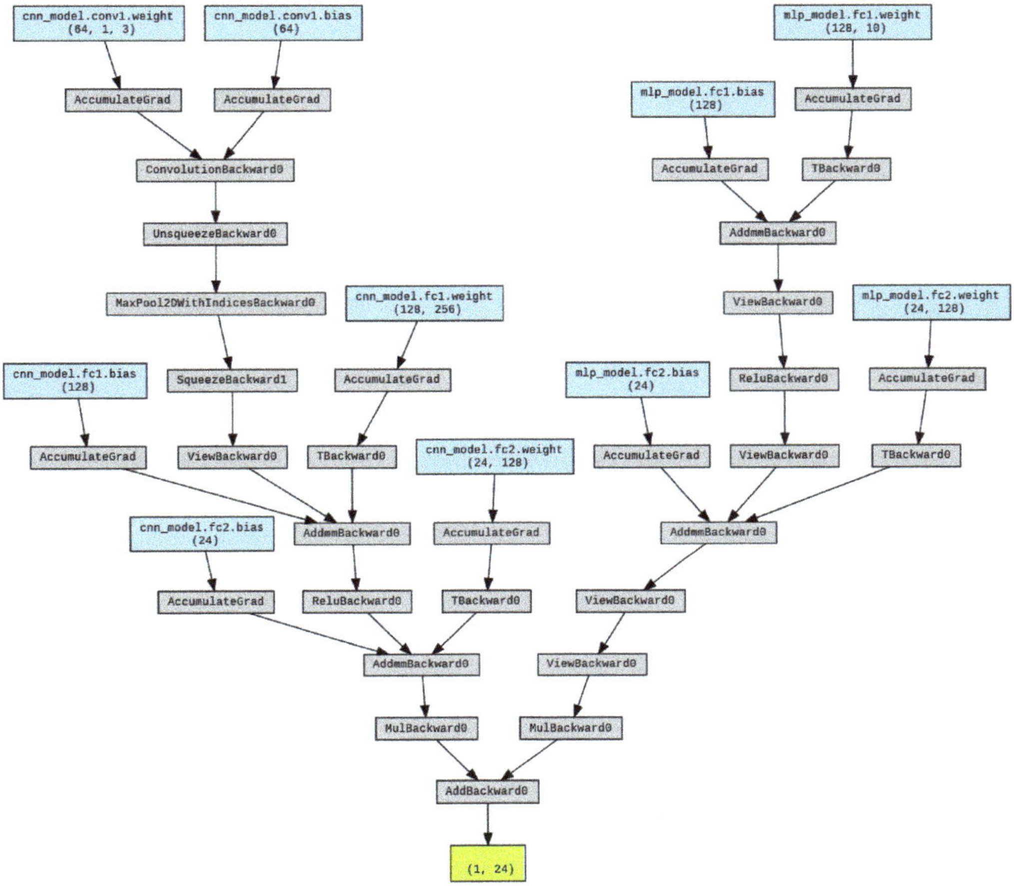

Figure 9. Ensemble deep learning model architecture.

3.2. Deep Transfer Learning Forecasting Strategies

The basic idea on which the four implemented strategies are based is the extended application and exploration of deep transfer learning in three different datasets, finding the best method, in combination with the creation of powerful DTL forecasting strategies and tools with robust generalization capabilities. The first forecasting strategy, named Deep Transfer Learning Case 1 (DTL Case 1), involves training each of the three DL models exclusively on the Lesvos dataset, with fine-tuning carried out using the Chios dataset. In the second strategy, Deep Transfer Learning Case 2 (DTL Case 2), both the Lesvos and Rhodes datasets are used concurrently during the DL model training phase, followed by fine-tuning using the Chios dataset. The third strategy, Deep Transfer Learning Case 3 (DTL Case 3), entails training the DL models solely on the Lesvos dataset, with the testing phase conducted directly on the Chios dataset, without any fine-tuning. Lastly, in the Multi-task Deep Learning (MTDL) application strategy, each of the three DL models are trained simultaneously on all three datasets, with final predictions made on the unused portion of the Chios dataset.

The way in which the available data were used in order to implement the DTL strategies and the MTDL strategy is described below:

- For DTL Case 1, only the dataset of Lesvos was used for the first training phase of the models, specifically for the time period from 2019-01-01 01:00:00 to 2021-12-31 23:00:00.

Then, for the second phase, i.e., fine-tuning, the time period from 2019-01-01 01:00:00 to 2021-12-31 23:00:00 of the Chios dataset was utilized.

- For DTL Case 2, the first training phase of the models was based on the datasets from Lesvos and Rhodes, and more specifically, for the time period from 2019-01-01 01:00:00 to 2021-12-31 23:00:00 for each dataset. Then, for the second phase, i.e., fine-tuning, the same time period as in DTL Case 1 was used, from 2019-01-01 01:00:00 to 2021-12-31 23:00:00 of the Chios dataset.
- For DTL Case 3, the training dataset that was used covers the time period from 2019-01-01 01:00:00 to 2022-12-31 23:00:00 from the Lesvos dataset without fine tuning. The choice of Lesvos alone was made because it exhibited a higher correlation and similarity with the corresponding dataset of the Chios time series compared to the Rhodes island.
- For MTDL, the datasets from all three islands were used simultaneously for training. Specifically, the datasets of Rhodes and Lesvos for the time period 2019-01-01 01:00:00 to 2022-12-31 23:00:00, and for Chios dataset, for the time period 2019-01-01 01:00:00 to 2021-12-31 23:00:00, were used.

Deep transfer learning is a field of transfer learning that entails utilizing knowledge gained from solving one task to make predictions on another related task, employing deep neural networks. This field often consists of pre-training a neural network on a source task with abundant labeled data and subsequently applying the acquired knowledge to a target task characterized by a foreign dataset. Two prevalent scenarios in transfer learning include domain adaptation, where the source and target tasks have the same input space but differ in output spaces, and feature extraction, where the source and target tasks share similar input and output spaces.

In the realm of time series approaches, deep transfer learning is valuable for several reasons.Time series data often exhibit complex patterns, trends, and seasonality, and acquiring labeled data for training deep models can be challenging due to limited availability. Deep transfer learning allows a model pre-trained on a source time series task to capture generic temporal features and representations that can be beneficial for a target task.The learned features can serve as a useful initialization for the target task, reducing the need for extensive training data and potentially enhancing the model's ability to generalize to new patterns. Additionally, transfer learning is particularly advantageous when the source and target tasks share similar temporal characteristics, enabling the model to transfer knowledge effectively and improve its performance on the target task. This approach is especially relevant in situations where collecting large amounts of labeled data for every specific task is impractical or costly.

In this paper, DTL strategies with and without fine-tuning are developed. Initially, the two cases of DTL with fine-tuning are examined (DTL Case 1 and 2) and are analyzed below. Subsequently, the scenario where the pre-trained model, as configured without fine-tuning, is examined (DTL Case 3). Finally, the MTDL strategy, involving the simultaneous training of each DL model on the datasets of the three islands, is presented. The four proposed forecasting strategies implemented are analyzed and considered in detail below.

3.2.1. Deep Transfer Learning Case 1

In this methodology a DNN pre-trained on a source task is adjusted to perform a related target task. Initially trained on a large dataset for a general task, such as load forecasting, the pre-trained model captures broad features. This knowledge is then transferred to a target task, which is a foreign dataset. During fine-tuning, the model's weights, especially in the deeper layers, are adjusted based on the target task's data, allowing the model to adapt its learned representations to task-specific characteristics. This approach is particularly advantageous when the target task also has limited labeled data, enabling the model to leverage the knowledge gained from the source task and enhance its performance on the target task.

In DTL Case 1, the testing dataset of Chios is used, and the training dataset consists only of the time series of Lesvos island. This approach is followed due to a higher correlation between Chios and Lesvos compared to Rhodes, influencing the selection of the training data for better model performance. After, the training period, fine-tuning is performed on the training parts of each model on the dataset of Chios, creating the final fine-tuned models. Finally, these models are used for Chios day-ahead load forecasting. Figure 10 graphically visualizes the DTL Case 1 strategy.

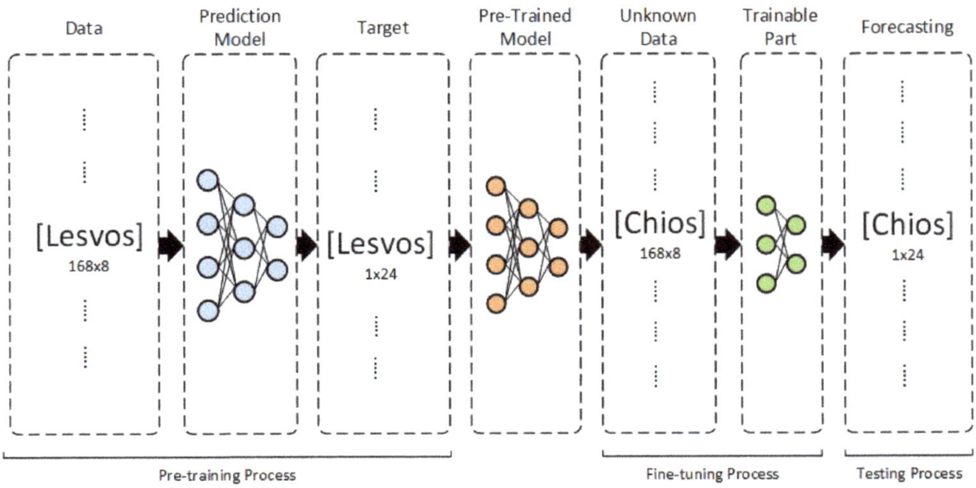

Figure 10. Deep Transfer Learning Case 1.

3.2.2. Deep Transfer Learning Case 2

Similarly with DTL Case 1, for DTL Case 2, the two datasets from Rhodes and Lesvos are merged, and the three models are trained on the combined training dataset. The trainable part of the pre-trained model is fine-tuned on the dataset of Chios. Finally, predictions are made on the unused dataset from Chios island. Figure 11 visualizes this strategy.

3.2.3. Deep Transfer Learning Case 3

Deep transfer learning without fine-tuning involves a two-step process. Initially, a deep neural network is trained on a source task using a substantial dataset, learning hierarchical features relevant to that task. Subsequently, in the transfer phase, the pre-trained model is utilized with the exact same structure it was formed with during the training period in order to create predictions for a target task.The learned features are extracted without further adjusting the model's weights and these fixed representations serve as input to a new task-specific regressor trained on the target task's dataset.

This approach proves advantageous when the target task possesses limited labeled data, as it allows for knowledge transfer from a source task without the computational overhead of fine-tuning the entire model. By utilizing the pre-trained model as a feature extractor, the knowledge encapsulated in the generic representations can be harnessed for tasks that share similar low-level features and structures, promoting effective knowledge transfer across related tasks while mitigating the need for task-specific fine-tuning.

In DTL Case 3 strategy, the three DL models are trained on the dataset of Lesvos, and subsequently, predictions are made directly on the Chios day-ahead load, without a fine-tuning process. This strategy is used in order to draw secure conclusions regarding the approaches with and without fine-tuning. This strategy is presented in Figure 12.

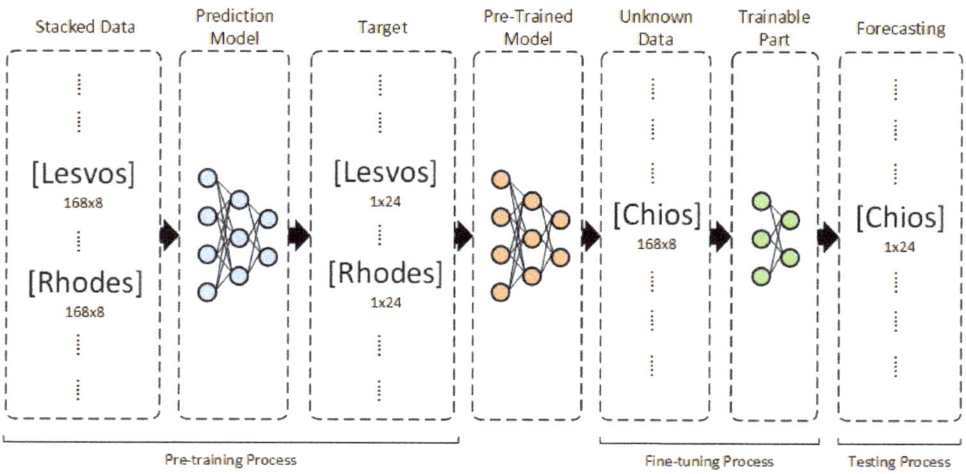

Figure 11. Deep Transfer Learning Case 2.

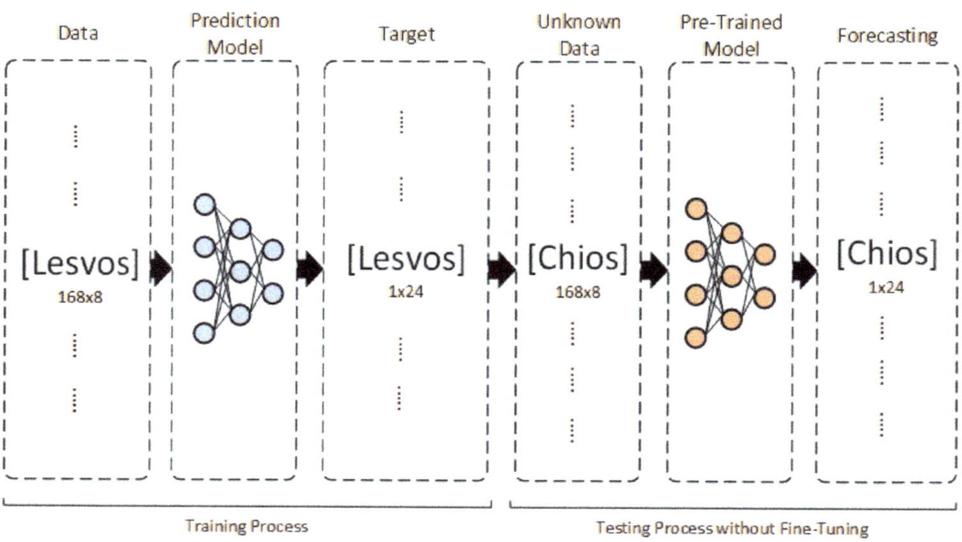

Figure 12. Deep Transfer Learning Case 3.

3.2.4. Multi-Task Deep Learning

Multi-task deep learning (MTDL) is a multi-task learning (MTL) methodology in deep neural networks (DNN), where a DNN model is trained simultaneously on multiple datasets, leveraging the shared knowledge across the different datasets to improve the model's generalization performance. The underlying principle of MTDL is to exploit the relationships and commonalities among distinct but related tasks, allowing the model to learn a shared representation that captures the inherent structure present in the data. Essentially, a unified dataset is created which is a concatenation of different datasets, where each of them corresponds to a specific output. During training, the model optimizes its parameters by jointly minimizing the loss across all tasks. The shared representation learned across tasks facilitates the transfer of knowledge between them, leading to enhanced generalization performance, particularly in scenarios where individual tasks lack sufficient data for robust learning. The success of MTDL lies in its ability to induce a form of regularization, encouraging the model to discover and focus on relevant features that are beneficial for multiple tasks simultaneously.

Each of the training tasks has its own objective function, and the model learns to jointly optimize these objective functions. The general idea is to share information between tasks to improve overall performance. Mathematically, the following applies:

- N is the total number of tasks.
- X are the input data.
- Y_i is the output for task i.
- θ are the parameters of the neural network model.

For each task i, there is an associated loss function $L_i(\theta)$ that measures the error between the predicted output and the true output for that task. The overall loss function for all tasks can be defined as a combination of the individual task loss functions, often using a weighted sum.

$$\mathcal{L}(\theta) = \sum_{i=1}^{N} \alpha_i L_i(\theta), \tag{4}$$

where α_i are optional weighting factors and Li represents the loss for the i-th task.

The goal is to minimize this overall loss function with respect to the model parameters θ. The minimal value of the loss function, θ^*, is given below:

$$\theta^* = \arg\min_{\theta} \mathcal{L}(\theta) \tag{5}$$

The model parameters are then updated using gradient descent or other optimization techniques to minimize this combined loss. The shared representation in the intermediate layers allows the model to discover commonalities and relationships among tasks, promoting a more generalized feature extraction process. By training on diverse datasets simultaneously, MDTL facilitates the development of a model that not only excels in individual tasks but also demonstrates improved performance on new, foreign data.

In the strategy employed in this paper, which is presented in Figure 13, training is conducted simultaneously on all three different datasets of the islands, Rhodes, Chios and Lesvos. The objective is for the model to acquire high generalization capabilities and make predictions on the testing dataset selected of Chios. Due to the distinct variations in the three time series, this approach proves more robust than cases involving singular training, imparting generality in performance across the models.

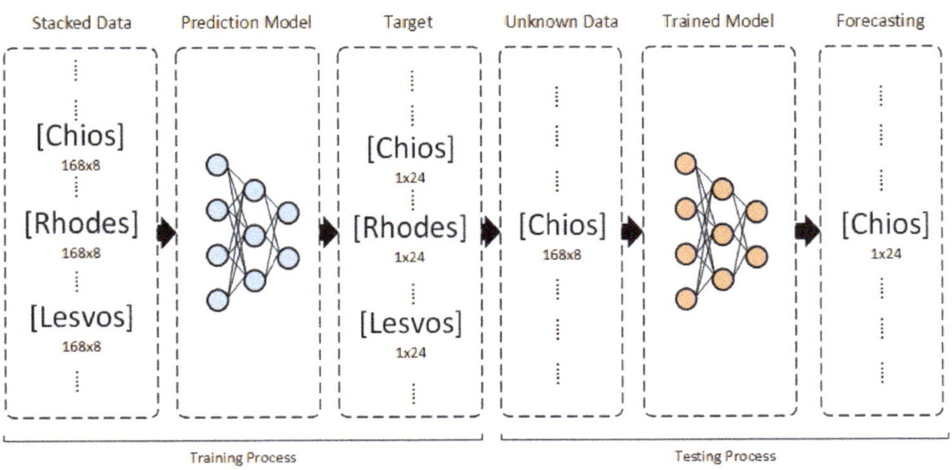

Figure 13. Multi-task Deep Learning.

3.3. *Optimization Framework*

The Bayesian optimization algorithm (BOA) is utilized for each training period and for each model. It is a probabilistic optimization approach designed to tackle complex and computationally expensive objective functions. Central to BOA is the use of a surrogate model, typically a Gaussian process, which provides a probabilistic representation of the unknown objective function. This surrogate model captures both the mean and uncertainty associated with the objective function across the parameter space. The algorithm iteratively refines its understanding of the objective function by selecting points for evaluation based on an acquisition function that balances exploration and exploitation. The chosen points are then used to update the surrogate model through Bayesian inference, adjusting the model's predictions in light of the new information. This iterative process allows BOA to systematically explore the parameter space, adapt to the underlying structure of the objective function, and efficiently converge towards optimal solutions.

BOA excels in making informed decisions using the uncertainty measured by its surrogate model. An acquisition function guides the algorithm to explore areas where the objective function is uncertain or is likely to have optimal values. As the optimization progresses, the surrogate model of BOA improves continuously, enhancing its understanding of the objective function and focusing the search on regions most likely to contain the global optimum. This principled approach makes it particularly well suited for optimization problems in scientific and engineering domains where objective function evaluations are resource-intensive or subject to noise. It efficiently identifies optimal parameter configurations in such scenarios. In this study, the architecture of each DL model was optimized and configured through BOA. For each case, each model was saved in h5 format. Regarding the hyperparameters of each DL model, Table 1 follows, which presents in detail each hyperparameter, the value of which was optimized with the aim of obtaining the most accurate predictions.

Table 1. Models optimization hyperparameters.

Deep Learning Model	Input Sequence Length	Optimization Function	Model Hyperparameters
MLP	168 h	Validation loss of MSE	• Dense layer with search space of Neurons: min_value = 16, max_value = 512, step = 16 • Dropout layer with search space: min_value = 0, max_value = 0.25, step = 0.05 • Optimizer = Adam • Learning rate with search space: min_value = 0.0010, max_value = 0.010, sampling = "log"
CNN	168 h	Validation loss of MSE	• Filters of CNN with range: min_value = 64, max_value = 128, step = 16 • CNN kernel size with search space: min_value = 4, max_value = 8, step = 2 • Neurons of Dense Layer with search space: min_value = 24, max_value = 120, step = 12 • Optimizer = Adam • Learning rate with search space: min_value = 0.0010, max_value = 0.010, sampling = "log"

Table 1. *Cont.*

Deep Learning Model	Input Sequence Length	Optimization Function	Model Hyperparameters
ELM	168 h	Validation loss of MSE	• Dense layer with search space of Neurons: min_value = 16, max_value = 512, step = 16 • Dropout layer with search space: min_value = 0, max_value = 0.25, step = 0.05 • Filters of CNN with range: min_value = 64, max_value = 128, step = 16 • CNN kernel size with search space: min_value = 4, max_value = 8, step = 2 • Neurons of Final Dense Layer with search space: min_value = 24, max_value = 48, step = 4 • Optimizer = Adam • Learning rate with search space: min_value = 0.0010, max_value = 0.010, sampling = "log"

3.4. Evaluation Metrics

For this paper, the following four error prediction metrics are used:

Mean Absolute Error (MAE): In this metric, the average of the absolute differences between the forecast and true values is calculated.

Root Mean Squared Error (RMSE): This metric calculates the square root of the average of the squared differences between the forecast and true values.

Mean Absolute Percentage Error (MAPE): This metric computes the average of the absolute percentage differences between the predicted and actual values.

R-squared (R^2): This is a statistical metric that measures how the independent variable(s) in a forecasting model explain the variation in the dependent variable. It takes values between 0 and 1. A value of 1 implies a satisfactory fit, meaning all variation in the dependent variable is explained by the independent variable(s); 0 demonstrates zero connection between the variables.

The above metrics are defined as follows:

$$MAE = \frac{\sum_{i=1}^{n} |y_i - x_i|}{n} \tag{6}$$

$$RMSE = \sqrt{\sum_{i=1}^{n} \frac{(y_i - x_i)^2}{n}} \tag{7}$$

$$MAPE = \frac{100}{n} \sum_{i=1}^{n} |\frac{y_i - x_i}{x_i}| \tag{8}$$

$$R^2 = 1 - \frac{\sum_{i=1}^{n} (y_i - \hat{x}_i)^2}{\sum_{i=1}^{n} (y_i - \bar{y})^2}, \tag{9}$$

where $x_i, y_i, \hat{x}_i$ and $\bar{y}$ are the forecasting, the actual, the mean of the forecasting values and the mean of the actual values, respectively.

3.5. Software Environment

All the algorithms in this work were developed using the Python 3.10 programming language. The open source software library Tensorflow 2.15.0 and the Keras 2.15.0 high-level API were used to train and test the deep learning algorithms. Furthermore, the Pandas 2.1.0 and Numpy 1.26.0 libraries were used for data analysis. For visualization purposes in exploratory analysis and prediction results, the Seaborn 0.13.2, Plotly 5.19.0, Matplotlib 3.8.3, graphviz 0.20.2 and torchviz 0.0.2 libraries were incorporated. Also, the official Calendar 1.0.1 library was used in order to identify the weekends and the Greek holidays. The research was carried out on the Google Colab Pro environment. For the MLP model, the GPU Tesla T4 with specific characteristics: NVIDIA-SMI 535.104.05 Driver Version: 535.104.05 CUDA Version: 12.2, RAM: 12.7 GB, and disk space: 78.2 GB was used. For the CNN and ensemble model, a cloud TPU 28.6 GB RAM and 107.7 GB disc space was employed.

4. Results Analysis

In this section, the experimental results obtained are presented. First, only the mean variance of the mean absolute error (MAE) on a monthly basis for the MLP, CNN, and ELM, respectively, is presented for economy of space. Then, the MAPE (mean absolute percentage error), the RMSE (root mean square error) and the aggregated results are visualized, analyzed, and compared for each strategy and each model on a monthly basis. Finally, the tables detailing R^2 are presented in the aggregated results.

4.1. Multilayer Perceptron Results

Figure 14 visualizes the variation in MAE on a monthly basis for each of the four strategies for the MLP model. It is observed that May and September exhibit better prediction results, while January, July, and August are the months with poorer performance.

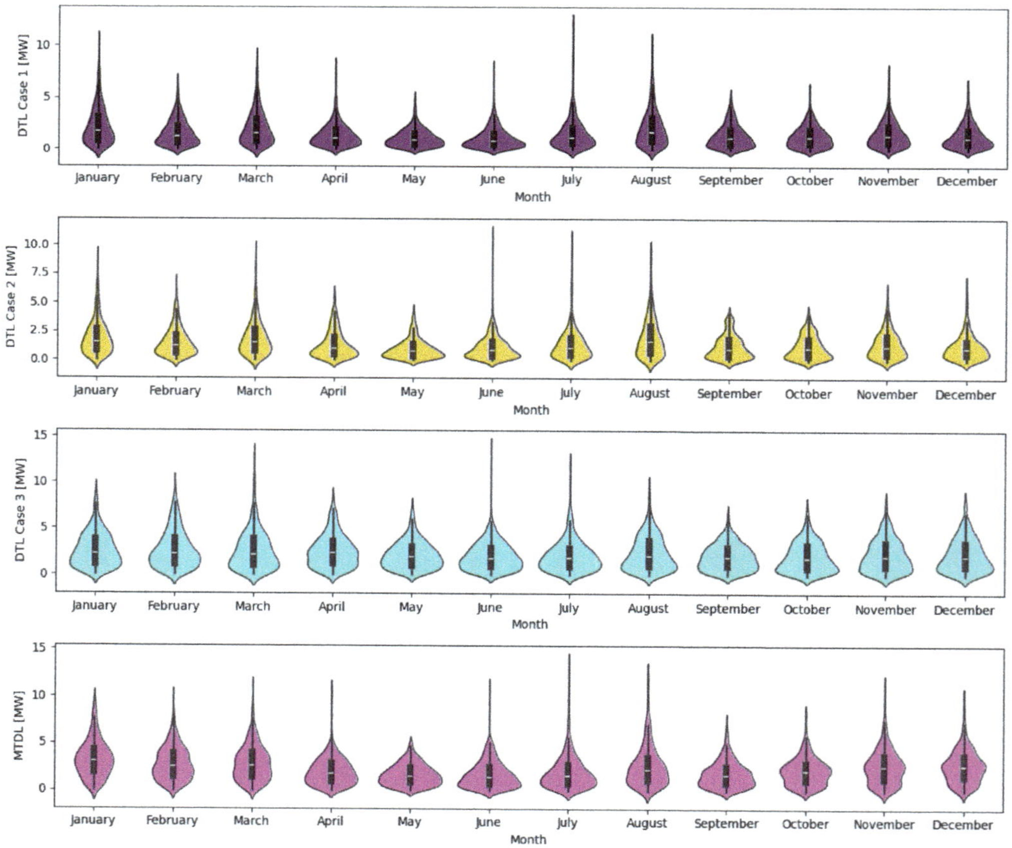

Figure 14. Variation in MAE for MLP model.

4.2. Convolutional Neural Networks Results

Figure 15 presents the variation in MAE on a monthly basis for each of the four strategies for CNN. It is clear that April and May exhibit better average prediction results, while June, July, and August are the months with the worst performance.

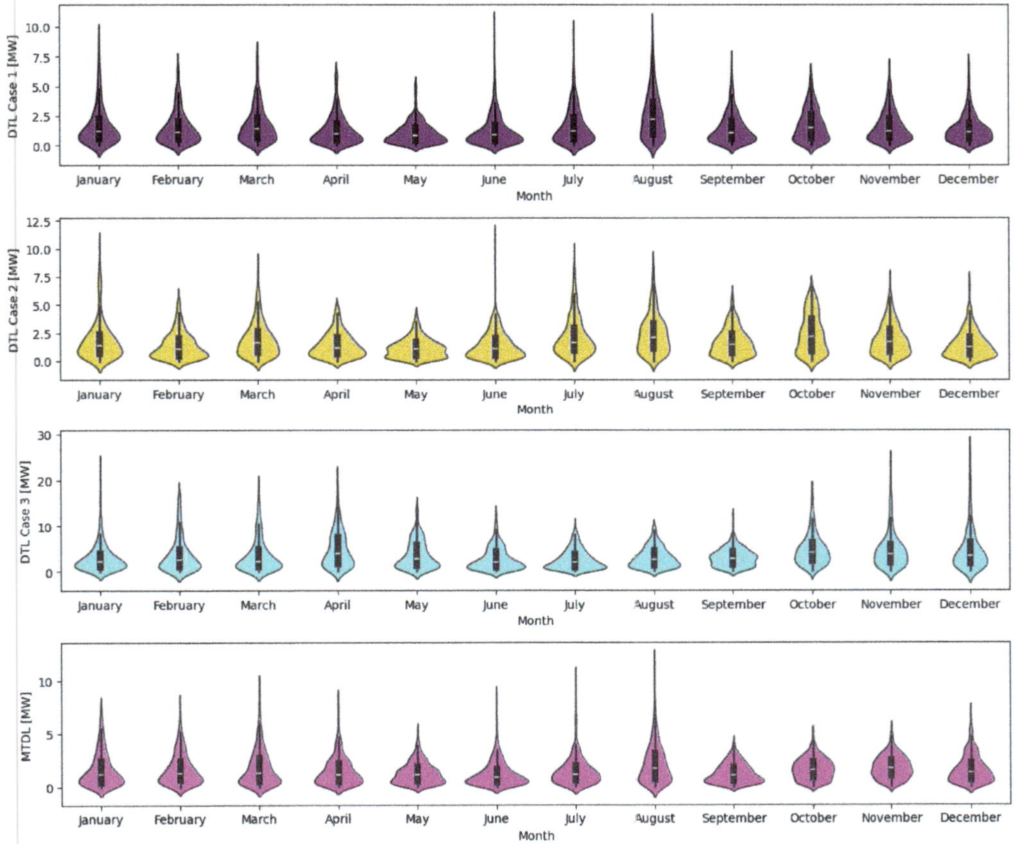

Figure 15. Variation in MAE for CNN model.

4.3. Ensemble Learning Model Results

Figure 16 visualizes the variation in MAE on a monthly basis for each of the four strategies for the ensemble learning model. It is observed that June, July, and August exhibit better average prediction results, while May, September, and October are the months with the lowest prediction accuracy.

4.4. Results Comparison

In order to perform an evaluation of the performance of each model, the results are consolidated and compared for each of the four strategies that were pursued in the following radial graph. This particular graph was chosen because it constitutes an illustrated and clear approach in presenting results for each of the twelve months of the year. In other words, the closer a point is to the center of the circle, the lower the prediction error achieved.

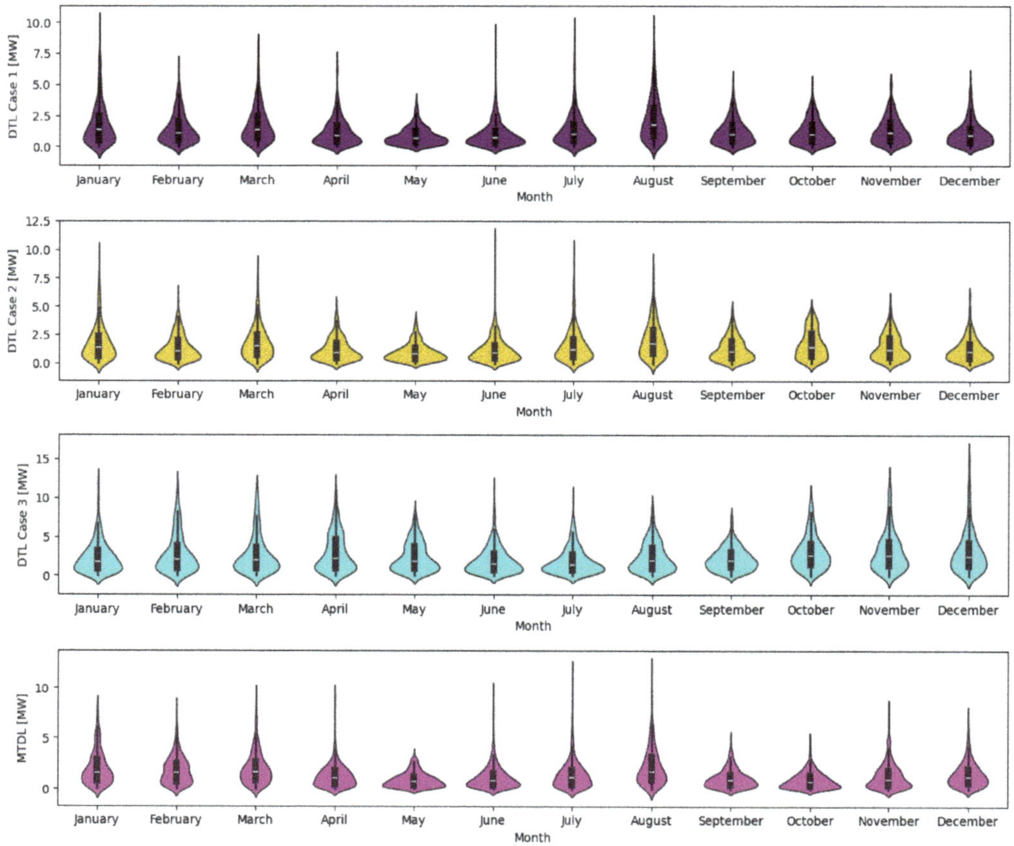

Figure 16. Variation in MAE for ELM.

4.4.1. Variation in MAE for Each Strategy

Figure 17 presents a comparison month-by-month of the mean absolute error (MAE) of the DL models for each strategy.

Based on the comparative results, the following observations emerge:

- For the DTL strategy with fine-tuning and training data from the Lesvos time series, DTL Case 1, it is observed that the MLP and ensemble models outperform the CNN model.
- In the forecasting strategy with fine-tuning and training data from both the Rhodes and Lesvos time series, DTL Case 2, it seems that the MLP and ensemble models exhibit comparable performance, except for October, where MLP outperforms.
- Regarding the use of pre-trained models without fine-tuning, DTL Case 3, the CNN significantly lags behind the other two models, with MLP consistently exhibiting the highest prediction accuracy.
- Finally, regarding the Multi-task Deep Learning application strategy, MTDL, it is observed that the MLP model consistently shows inferior results compared to the other two models for all months. Here, CNN and ensemble achieve similar accuracy.

For additional analysis and understanding of the behavior of the algorithms, Figures 18 and 19 plot the variation in the mean absolute percentage error (MAPE) and the root mean squared error (RMSE) for each month, respectively.

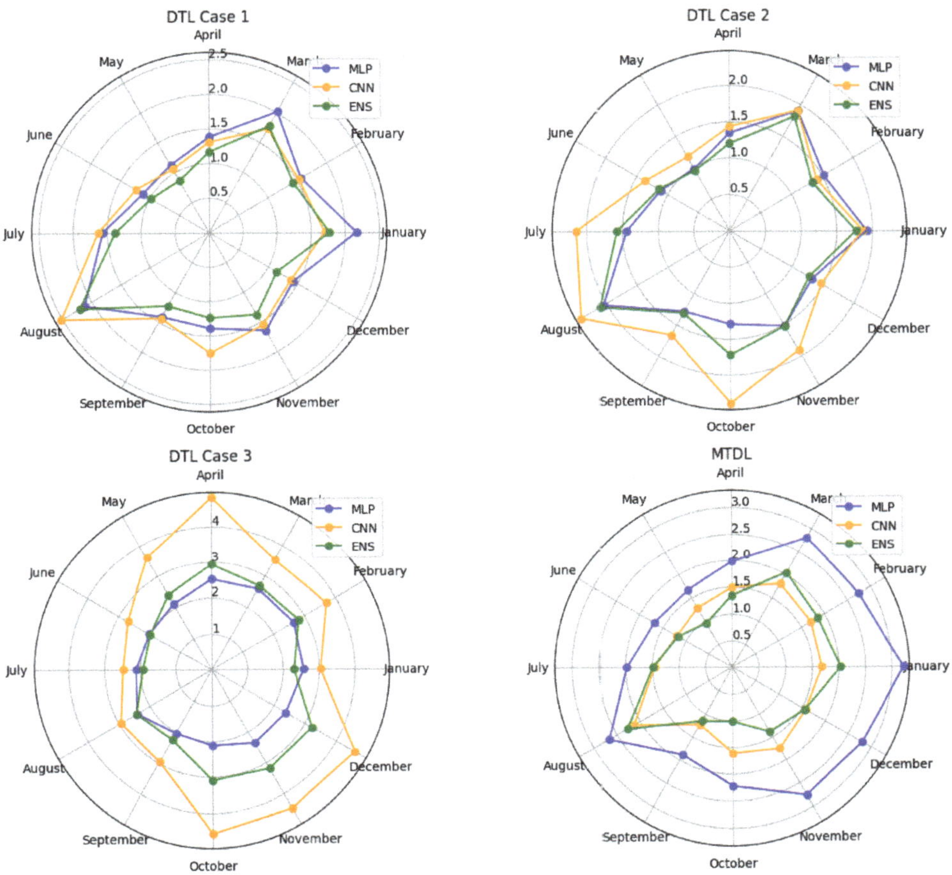

Figure 17. Comparison of monthly MAE for each forecasting strategy.

4.4.2. Variation in MAPE for Each Strategy

Figure 18 presents a comparison of the mean absolute percentage error (MAPE) for each model and strategy on a monthly basis.

It is clearly observed that for the DTL Cases 1 and 2, June exhibits the best prediction accuracy, with the ensemble model achieving a MAPE of 5.29% for Case 1 and the MLP achieving a MAPE of 6.01% for Case 2. In the case of MTDL, the best predictions are observed in the months of June and July, with the ELM presenting predictions of 6.33% and 6.86%, respectively. Finally, for the DTL Case 3, the best prediction is observed in January, corresponding to a MAPE of 7.85%, achieved by the ELM model.

4.4.3. Variation in RMSE for Each Strategy

Figure 19 presents a comparison of the root mean square error (RMSE) for each model and strategy, monthly.

4.5. Aggregated Results

This subsection provides a detailed presentation of the results obtained for each forecasting strategy and each DL model, monthly. Tables 2–5 present all the aggregated results for each forecasting strategy to provide a complete view of the performance of each algorithm and for each forecasting month. The MAE and RMSE metrics are given in MWs, MAPE is given in percent (%), and R^2 takes values between 0 and 1.

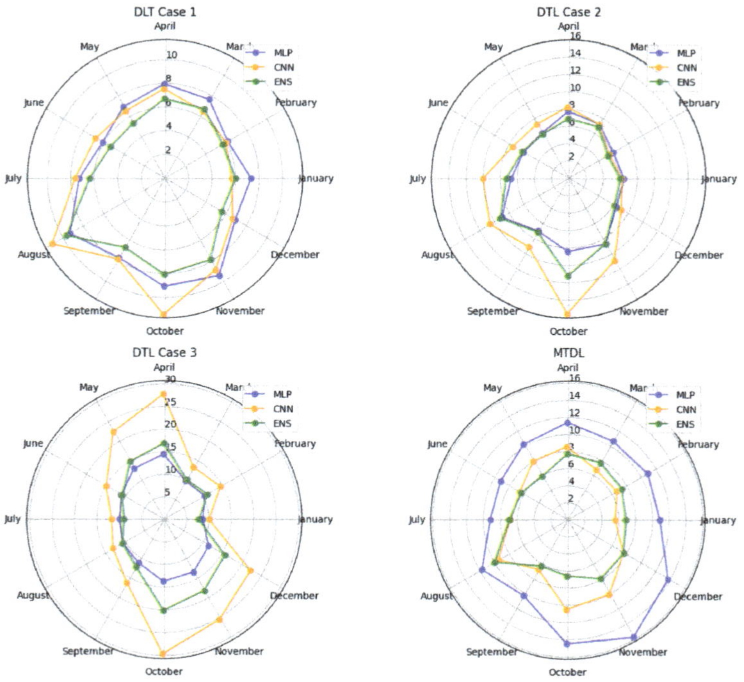

Figure 18. Comparison of monthly MAPE for each forecasting strategy.

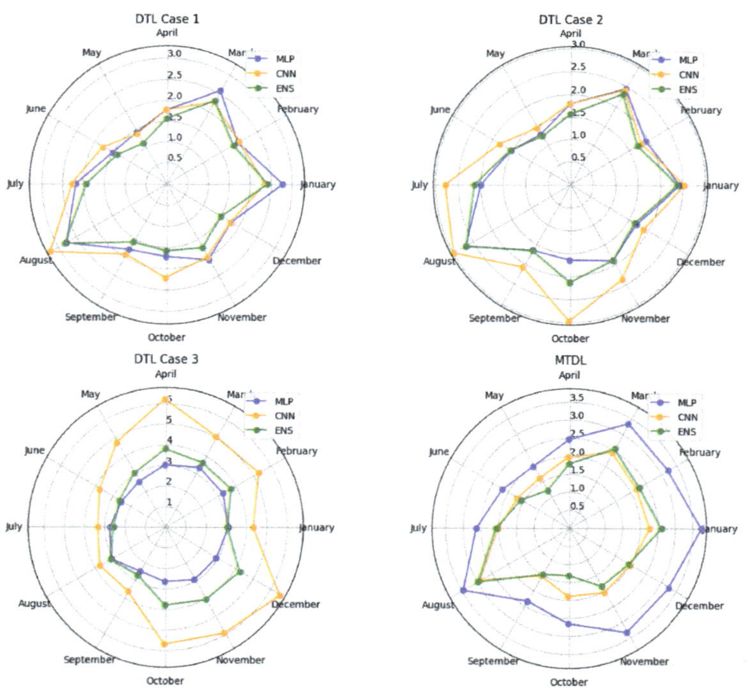

Figure 19. Comparison of monthly RMSE for each forecasting strategy.

Table 2. Deep Transfer Learning Case 1.

Model		Jan	Feb	Mar	Apr	May	June	July	Aug	Sep	Oct	Nov	Dec
							Month						
MLP	MAE	2.16	1.56	2.01	1.38	1.12	1.12	1.57	2.12	1.40	1.39	1.64	1.42
	MAPE	7.38	6.24	7.68	7.98	6.97	6.05	7.01	9.23	7.66	9.04	9.39	7.00
	RMSE	2.79	1.98	2.59	1.78	1.41	1.48	2.15	2.76	1.78	1.70	2.06	1.79
	R^2	0.83	0.87	0.82	0.67	0.73	0.74	0.63	0.08	0.75	0.53	0.70	0.82
CNN	MAE	1.69	1.52	1.73	1.31	1.06	1.24	1.63	2.52	1.43	1.75	1.54	1.38
	MAPE	5.57	6.07	6.60	7.57	6.55	6.69	7.46	10.95	7.79	11.39	8.85	6.79
	RMSE	2.37	2.03	2.28	1.78	1.38	1.74	2.25	3.19	1.90	2.20	1.98	1.79
	R^2	0.88	0.87	0.86	0.68	0.74	0.64	0.59	0.11	0.70	0.21	0.73	0.82
ELM	MAE	1.76	1.42	1.77	1.16	0.86	0.98	1.38	2.20	1.22	1.23	1.37	1.14
	MAPE	5.87	5.72	6.77	6.73	5.36	5.29	6.24	9.57	6.67	8.04	7.82	5.62
	RMSE	2.42	1.84	2.31	1.57	1.12	1.36	1.91	2.80	1.58	1.57	1.73	1.51
	R^2	0.88	0.89	0.86	0.75	0.83	0.78	0.71	6.25	0.80	0.61	0.80	0.87

Table 3. Deep Transfer Learning Case 2.

Model		Jan	Feb	Mar	Apr	May	June	July	Aug	Sep	Oct	Nov	Dec
							Month						
MLP	MAE	1.91	1.51	1.91	1.34	0.98	1.11	1.44	2.04	1.28	1.28	1.51	1.32
	MAPE	6.50	6.07	7.29	7.79	6.24	6.01	6.61	8.87	6.95	8.38	8.66	6.51
	RMSE	2.44	1.94	2.47	1.80	1.29	1.50	1.98	2.67	1.64	1.63	1.89	1.71
	R^2	0.87	0.88	0.84	0.67	0.77	0.73	0.69	0.15	0.78	0.57	0.75	0.84
CNN	MAE	1.85	1.40	1.91	1.43	1.17	1.37	2.14	2.40	1.66	2.39	1.90	1.46
	MAPE	6.26	5.60	7.29	8.30	7.27	7.38	9.85	10.44	9.02	15.51	10.90	7.23
	RMSE	2.56	1.83	2.43	1.80	1.46	1.79	2.76	2.96	2.06	2.96	2.36	1.90
	R^2	0.86	0.89	0.84	0.67	0.71	0.62	0.39	0.02	0.66	0.08	0.61	0.80
ELM	MAE	1.76	1.32	1.82	1.20	0.95	1.14	1.57	2.10	1.30	1.71	1.51	1.27
	MAPE	6.03	5.31	6.94	6.95	5.92	6.17	7.23	9.10	7.09	11.17	8.68	6.26
	RMSE	2.38	1.72	2.33	1.57	1.23	1.52	2.12	2.66	1.65	2.12	1.89	1.65
	R^2	0.88	0.91	0.85	0.75	0.80	0.72	0.64	0.16	0.78	0.27	0.75	0.85

Table 4. Deep Transfer Learning Case 3.

Model		Jan	Feb	Mar	Apr	May	June	July	Aug	Sep	Oct	Nov	Dec
							Month						
MLP	MAE	2.56	2.65	2.58	2.51	2.12	2.01	2.12	2.42	2.03	2.10	2.34	2.37
	MAPE	8.75	10.60	9.87	14.54	13.12	10.86	9.74	10.51	11.05	13.71	13.43	11.69
	RMSE	3.14	3.32	3.38	3.06	2.56	2.48	2.71	3.03	2.44	2.62	2.91	2.94
	R^2	0.79	0.64	0.69	0.05	0.10	0.27	0.41	-0.10	0.52	−0.11	0.41	0.51
CNN	MAE	3.02	3.69	3.53	4.81	3.62	2.71	2.49	2.95	2.96	4.55	4.45	4.59
	MAPE	10.34	14.79	13.48	27.82	22.40	14.63	11.45	12.82	16.09	29.63	25.47	22.62
	RMSE	4.39	5.37	5.10	6.25	4.75	3.70	3.27	3.67	3.58	5.66	5.93	6.59
	R^2	0.58	0.05	0.29	−2.96	−2.08	−0.64	0.14	−0.61	−0.05	−4.18	−1.45	−1.45
ELM	MAE	2.29	2.80	2.68	2.93	2.41	1.98	1.92	2.44	2.23	3.09	3.18	3.21
	MAPE	7.85	11.22	10.22	16.96	14.91	10.68	8.84	10.58	12.16	20.10	18.18	15.84
	RMSE	3.07	3.76	3.65	3.82	3.06	2.61	2.55	3.09	2.67	3.78	4.05	4.27
	R^2	0.80	0.54	0.64	−0.48	−0.28	0.19	0.48	−0.14	0.42	−1.30	−0.14	−0.03

Table 5. Multi-task Deep Learning.

Model		Jan	Feb	Mar	Apr	May	June	July	Aug	Sep	Oct	Nov	Dec
							Month						
MLP	MAE	3.21	2.74	2.81	1.98	1.66	1.69	1.98	2.67	1.88	2.22	2.75	2.80
	MAPE	10.97	10.99	10.73	11.45	10.31	9.10	9.06	11.50	10.23	14.45	15.75	13.80
	RMSE	3.75	3.24	3.36	2.47	2.00	2.16	2.61	3.43	2.32	2.64	3.31	3.27
	R^2	0.70	0.66	0.69	0.38	0.46	0.44	0.45	-0.40	0.56	−0.12	0.24	0.40
CNN	MAE	1.67	1.69	0.80	1.49	1.28	1.20	1.46	2.14	1.22	1.60	1.75	1.57
	MAPE	5.62	6.78	6.86	8.63	7.97	6.47	6.58	9.30	6.66	10.46	10.03	7.75
	RMSE	2.30	2.22	2.45	1.99	1.62	1.67	1.98	2.87	1.51	1.88	2.04	2.02
	R^2	0.89	0.84	0.84	0.60	0.64	0.67	0.69	0.01	0.82	0.43	0.70	0.77
ELM	MAE	2.03	1.85	2.03	1.35	0.96	1.17	1.49	2.28	1.14	1.00	1.38	1.56
	MAPE	6.86	7.42	7.77	7.80	5.96	6.33	6.86	9.89	6.22	6.56	7.92	7.69
	RMSE	2.61	2.28	2.56	1.80	1.21	1.57	2.05	2.96	1.46	1.30	1.84	1.95
	R^2	0.85	0.83	0.82	0.67	0.80	0.71	0.66	-0.05	0.83	0.73	0.77	0.79

Based on the results shown in the Tables, the following summarizations apply:

- In DTL Case 1, it is observed that the ensemble model achieves the best prediction for the month of June, presenting a MAPE of 5.29%.
- In DTL Case 2, again, the ensemble model achieves the best prediction, which pertains to the month of February, exhibiting a MAPE of 5.31%.
- Regarding DTL Case 3, the ensemble model achieves the best prediction in January with a MAPE of 7.85%.
- In MTDL, the CNN model manages the best prediction in January, corresponding to a MAPE of 5.62%.

Generally, based on the Tables 2–5, as well as Figures 17–19, a notable difference is observed between the best and worst prediction months. The best corresponds to the ELM prediction for May, with a MAPE of 5.36% and the worst, corresponds to the CNN model prediction for December, with a MAPE of 22.62%, resulting in a precision difference of 17.26%. This fact further reinforces the high performance of ELM in cases with fine-tuning, which is influenced by the predictions of the two distinct models. On the other hand, CNN, as evidenced and in the case of the worst-performing month, exhibits general instability in the case without fine-tuning, which is related to the inability of its trainable parameters to adapt to foreign data. This fact may reveal the general difficulty of models using convolution mechanisms to adapt to unknown datasets.

4.6. Results Discussion

Based on the above results, it becomes evident that the application of deep learning algorithms in the domain of deep transfer learning (DTL) can yield satisfactory outcomes, reducing the computational power requirements and model training times, due to the fact that, after the initial training of the model, only the fine-tuning needs to take place each time the DL model is applied in a different area. The time required for a DL model to be trained during the fine-tuning period is significantly shorter compared to the time needed for direct training, as in the case of MTDL. Also, the variation in results for each month indicates that the ELM improves predictions for the majority of the forecast months.

In general, it is observed that the two strategies of deep transfer learning with fine-tuning (DTL Case 1 and 2) significantly outperform DTL Case 3 and MTDL. Specifically, in the comparison between fine-tuning strategies and multi-task deep learning, the differences suggest that the utilized models can adapt better when trained separately on different datasets, as opposed to parallel and simultaneous training on multiple datasets together. Both of these cases involve efforts to create models capable of efficiently generalizing to unknown and differently behaving time series.

Additionally, in the case of the direct use of a pre-trained model (DTL Case 3), a poorer performance is achieved compared to other cases. For the ELM, which is influenced by both the MLP and CNN models, the poor performance of the CNN negatively affects the accuracy for most months, with exceptions in January and July.

The variation in results clearly demonstrates that the employment of more than one model in an ensemble combination significantly improves the performance compared to individual algorithms. The reason behind this improvement lies in the weighted average learning, which takes into account the best predictions from both models, MLP and CNN separately. As a result, the final day-ahead load prediction is considerably improved, demonstrating the effectiveness of combining multiple models.

Finally, it is worth noting that the adaptability of the algorithms to the three examined time series relies on both the trainable parameters of each model and the different features and patterns exhibited by each case. Seasonality, the peak demand periods, and the average values of each dataset are some of the characteristics that influence the algorithmic functionalities.

5. Conclusions and Future Study Proposals

In this study, an extensive investigation was conducted regarding Seq2Seq deep transfer learning on time series data. For this reason, a case study of a month-to-month approach was employed with the aim of day-ahead forecasting of electricity load in three islands using the Greek power system. The results obtained provide us with valuable information regarding the application of such methods and their effectiveness. The first major conclusion is that transfer learning outperforms simple learning, even in the case of multi-task deep learning, which is utilized for better model generalization.

Furthermore, another conclusion is that deep transfer learning using ensemble models outperforms simple DL models, as evidenced by the results obtained. More specifically, in the strategies DTL Case 1, DTL Case 2, as well as MTDL, it is observed that, for the majority of months, the ELM enhances the predictions achieved by the two individual DL models. This fact enables particularly optimistic conclusions to be drawn regarding further exploration of ensemble models in the field of prediction in power systems.

DTL strategies are cost-effective, requiring significantly less computational power and time compared to simple prediction methods, due to the fact that the DL models are trained once in a source dataset, after they are saved in appropriate format, and subsequently only their final part is fine-tuned for each specific task. Therefore, DTL minimizes the computational resources required and speeds up training for a specific target task, such as forecasting the day-ahead electrical load. By leveraging the knowledge stored in pre-trained models, DTL efficiently utilizes resources, facilitating swift deployment of effective models across different domains. Thus, it becomes evident that the application of DTL on real-time data can be achieved with high performance due to the aforementioned conditions, as well as the flexibility provided by this specific domain compared to simple transfer learning applications.

More specifically, deploying the models studied in this work in real-time conditions is really challenging. Adapting to fast-changing data streams, maintaining low latency, and balancing model complexity with real-time requirements are key hurdles. Continuous updates of the model to address concept drift and ensure interpretability in real-time settings are also critical. Success requires tailored optimization and infrastructure to handle real-time processing demands effectively.

In the context of future study proposals, it should be emphasized initially that DTL could be applied to other branches of power systems, such as fault prediction in electric power transmission and distribution networks. Furthermore, beyond energy systems, it could be highlighted that a significant challenge lies in the application of DTL to areas like healthcare, where numerous research studies are conducted globally.

As already mentioned, the study of DTL using ELM, both in the field of power systems and in other domains, can help to further improve the results of transfer learning.

Finally, the combination of DTL with reinforcement learning holds promise for future research and offers potential advancements. This could be explored to improve the efficiency of demand forecasting and load management systems. For instance, a pre-trained reinforcement learning agent could learn general patterns and behaviors from historical data across different regions or time periods. This pre-trained agent could then be fine-tuned on a specific locality or timeframe in order to adapt to unique characteristics and changes in electricity demand. This approach may lead to more accurate and adaptable models for load prediction, contributing to improved resource planning and energy efficiency in the electricity grid.

Author Contributions: Conceptualization, V.L. and G.V.; methodology, V.L.; software, V.L. and G.V.; validation, V.L., G.V., A.T., D.B. and L.H.T.; formal analysis, V.L. and G.V.; investigation, V.L. and G.V.; resources, V.L. and G.V.; data curation, V.L. and G.V.; writing—original draft preparation, V.L. and G.V.; writing—review and editing, V.L., G.V., A.T., D.B. and L.H.T.; visualization, V.L. and G.V.; supervision, D.B. and L.H.T.; project administration, D.B. All authors have read and agreed to the published version of the manuscript.

Funding: This research received no external funding.

Data Availability Statement: The load data used in this study are available from the HEDNO, Greece portal in [27].

Conflicts of Interest: The authors declare no conflicts of interest.

Abbreviations

The following abbreviations are used in this manuscript:

ARIMA	Autoregressive Integrated Moving Average
BO	Bayesian Optimization
CNN	Convolutional Neural Network
DNN	Deep Neural Network
DTL	Deep Transfer Learning
DTSF	Dynamic Time Scan Forecasting
EDA	Exploratory Data Analysis
EDL	Ensemble Deep Learning
ELM	Ensemble Learning Model
MAE	Mean Absolute Error
MAPE	Mean Absolute Percentage Error
MDTL	Multi-task Deep Transfer Learning
MLP	Multilayer Perceptron
MMD	Maximum Mean Discrepancy
MSE	Mean Squared Error
NBEATS	Neural Basis Expansion Analysis for Time Series
NILM	Non-Intrusive Load Monitoring
R^2	R-Squared
RMSE	Root Mean Squared Error
Seq2Seq	Sequence-to-Sequence
SVM	Support Vector Machine
TL	Transfer Learning

References

1. Meng, S.; Li, C.; Tian, C.; Peng, W.; Tian, C. Transfer learning based graph convolutional network with self-attention mechanism for abnormal electricity consumption detection. *Energy Rep.* **2023**, *9*, 5647–5658. [CrossRef]
2. Antoniadis, A.; Gaucher, S.; Goude, Y. Hierarchical transfer learning with applications to electricity load forecasting. *Int. J. Forecast.* **2023**, *40*, 641–660. [CrossRef]
3. Yang, C.; Wang, H.; Bai, J.; He, T.; Cheng, H.; Guang, T.; Yao, H.; Qu, L. Transfer learning enhanced water-enabled electricity generation in highly oriented graphene oxide nanochannels. *Nat. Commun.* **2022**, *13*, 6819. [CrossRef] [PubMed]
4. Dong, Y.; Xiao, L. A Transfer Learning Based Deep Model for Electrical Load Prediction. In Proceedings of the 2022 IEEE 8th International Conference on Computer and Communications (ICCC), Chengdu, China, 9–12 December 2022; pp. 2251–2255. [CrossRef]

5. Li, D.; Li, J.; Zeng, X.; Stankovic, V.; Stankovic, L.; Xiao, C.; Shi, Q. Transfer learning for multi-objective non-intrusive load monitoring in smart building. *Appl. Energy* **2023**, *329*, 120223. [CrossRef]
6. Peirelinck, T.; Kazmi, H.; Mbuwir, B.V.; Hermans, C.; Spiessens, F.; Suykens, J.; Deconinck, G. Transfer learning in demand response: A review of algorithms for data-efficient modelling and control. *Energy AI* **2022**, *7*, 100126. [CrossRef]
7. Laitsos, V.; Vontzos, G.; Bargiotas, D. Investigation of Transfer Learning for Electricity Load Forecasting. In Proceedings of the 2023 14th International Conference on Information, Intelligence, Systems & Applications (IISA), Volos, Greece, 10–12 July 2023; pp. 1–7. [CrossRef]
8. Wu, D.; Lin, W. Efficient Residential Electric Load Forecasting via Transfer Learning and Graph Neural Networks. *IEEE Trans. Smart Grid* **2023**, *14*, 2423–2431. [CrossRef]
9. Kamalov, F.; Sulieman, H.; Moussa, S.; Avante Reyes, J.; Safaraliev, M. Powering Electricity Forecasting with Transfer Learning. *Energies* **2024**, *17*, 626. [CrossRef]
10. Syed, D.; Zainab, A.; Refaat, S.S.; Abu-Rub, H.; Bouhali, O.; Ghrayeb, A.; Houchati, M.; Bañales, S. Inductive Transfer and Deep Neural Network Learning-Based Cross-Model Method for Short-Term Load Forecasting in Smarts Grids. *IEEE Can. J. Electr. Comput. Eng.* **2023**, *46*, 157–169. [CrossRef]
11. Laitsos, V.; Vontzos, G.; Bargiotas, D.; Daskalopulu, A.; Tsoukalas, L.H. Enhanced Automated Deep Learning Application for Short-Term Load Forecasting. *Mathematics* **2023**, *11*, 2912. [CrossRef]
12. Santos, M.L.; García, S.D.; García-Santiago, X.; Ogando-Martínez, A.; Camarero, F.E.; Gil, G.B.; Ortega, P.C. Deep learning and transfer learning techniques applied to short-term load forecasting of data-poor buildings in local energy communities. *Energy Build.* **2023**, *292*, 113164. [CrossRef]
13. Arvanitidis, A.I.; Bargiotas, D.; Daskalopulu, A.; Kontogiannis, D.; Panapakidis, I.P.; Tsoukalas, L.H. Clustering informed MLP models for fast and accurate short-term load forecasting. *Energies* **2022**, *15*, 1295. [CrossRef]
14. Luo, T.; Tang, Z.; Liu, J.; Zhou, B. A Review of Transfer Learning Approaches for Load, Solar and Wind Power Predictions. In Proceedings of the 2023 Panda Forum on Power and Energy (PandaFPE), Chengdu, China, 27–30 April 2023; pp. 1580–1584. [CrossRef]
15. Li, S.; Wu, H.; Wang, X.; Xu, B.; Yang, L.; Bi, R. Short-term load forecasting based on AM-CIF-LSTM method adopting transfer learning. *Front. Energy Res.* **2023**, *11*, 1162040. [CrossRef]
16. Chan, S.; Oktavianti, I.; Puspita, V. A deep learning cnn and ai-tuned svm for electricity consumption forecasting: Multivariate time series data. In Proceedings of the 2019 IEEE 10th Annual Information Technology, Electronics and Mobile Communication Conference (IEMCON), Vancouver, BC, Canada, 17–19 October 2019; pp. 488–494. [CrossRef]
17. Silveira Gontijo, T.; Azevedo Costa, M. Forecasting Hierarchical Time Series in Power Generation. *Energies* **2020**, *13*, 3722. [CrossRef]
18. Jung, S.M.; Park, S.; Jung, S.W.; Hwang, E. Monthly electric load forecasting using transfer learning for smart cities. *Sustainability* **2020**, *12*, 6364. [CrossRef]
19. Al-Hajj, R.; Assi, A.; Neji, B.; Ghandour, R.; Al Barakeh, Z. Transfer Learning for Renewable Energy Systems: A Survey. *Sustainability* **2023**, *15*, 9131. [CrossRef]
20. Nivarthi, C.P. Transfer Learning as an Essential Tool for Digital Twins in Renewable Energy Systems. *arXiv* **2022**, arXiv:2203.05026. [CrossRef]
21. Miraftabzadeh, S.M.; Colombo, C.G.; Longo, M.; Foiadelli, F. A Day-Ahead Photovoltaic Power Prediction via Transfer Learning and Deep Neural Networks. *Forecasting* **2023**, *5*, 213–228. [CrossRef]
22. Đaković, D.; Kljajić, M.; Milivojević, N.; Doder, Đ.; Anđelković, A.S. Review of Energy-Related Machine Learning Applications in Drying Processes. *Energies* **2024**, *17*, 224. [CrossRef]
23. Vontzos, G.; Laitsos, V.; Bargiotas, D. Data-Driven Airport Multi-Step Very Short-Term Load Forecasting. In Proceedings of the 2023 14th International Conference on Information, Intelligence, Systems & Applications (IISA), Volos, Greece, 10–12 July 2023; pp. 1–6. [CrossRef]
24. Yang, M.; Liu, Y.; Liu, Q. Nonintrusive residential electricity load decomposition based on transfer learning. *Sustainability* **2021**, *13*, 6546. [CrossRef]
25. Li, K.; Wei, B.; Tang, Q.; Liu, Y. A Data-Efficient Building Electricity Load Forecasting Method Based on Maximum Mean Discrepancy and Improved TrAdaBoost Algorithm. *Energies* **2022**, *15*, 8780. [CrossRef]
26. Silveira Gontijo, T.; Barbosa de Santis, R.; Azevedo Costa, M. Application of a data-driven DTSF and benchmark models for the prediction of electricity prices in Brazil: A time-series case. *J. Renew. Sustain. Energy* **2023**, *15*, 036101. [CrossRef]
27. Publication of NII Daily Energy Planning Data | HEDNO. Available online: https://deddie.gr/en/themata-tou-diaxeiristi-mi-diasundedemenwn-nisiwn/leitourgia-mdn/dimosieusi-imerisiou-energeiakou-programmatismou/ (accessed on 10 January 2023).

electronics

MDPI

Article

A Fast Repetitive Control Strategy for a Power Conversion System

Jinghua Zhou [1,2], Yifei Sun [1,2,*], Shasha Chen [1] and Tianfeng Lan [3]

1 School of Electrical and Control Engineering, North China University of Technology, 5 Jinyuanzhuang Road, Shijingshan District, Beijing 100043, China; zjh@ncut.edu.cn (J.Z.); chenshasha77@126.com (S.C.)
2 Beijing Laboratory of New Energy Storage Technology, Beijing Municipal Education Commission, Beijing 100083, China
3 Shanghai Electrical Apparatus Research Institute, Building 1, No. 505 Wuning Road, Putuo District, Shanghai 200333, China; lantf@seari.com.cn
* Correspondence: spsyf@outlook.com; Tel.: +86-175-3831-8127

Abstract: With the expansion of renewable energy sources, the stable and high-quality operation of microgrids composed of new energy sources has attracted widespread attention. Among them, the power conversion system (PCS), as an important part of microgrids, plays a crucial role in their operation and management. The PCS operation modes are classified into grid-connected and off-grid modes. However, in off-grid mode, due to the access of nonlinear and unbalanced loads, the output voltage quality of a PCS is worse, and the voltage waveform distortion is serious. To solve these problems, a fast repetitive control (FRC) strategy is proposed for a power conversion system with an Active Neutral Point Clamped (ANPC) architecture of three levels. The voltage loop control strategy can be applied to the voltage/frequency (V/f) mode and the grid-forming mode. The control strategy can effectively realize the suppression of the harmonics of the output voltage and has a 100% capability to carry unbalanced loads. Finally, a 1725 kVA PCS prototype is developed, and the proposed control strategy is verified using the MT3200 HIL semiphysical simulator of ModelingTech in the V/f mode as an example. This practically verifies the feasibility and validity of the proposed control strategy, which has a certain degree of engineering practicability and reference due to the simplicity of the design and the ease of realization.

Keywords: power conversion system; repetitive control; harmonic suppression; StarSim MT3200HIL

1. Introduction

In order to foster an eco-friendly society and ensure sustained and stable growth of the national economy, China has embarked on a robust initiative to promote clean energy. This initiative serves as a key measure to bolster the nation's energy competitiveness and addresses the pressing imperatives of climate change mitigation, ecological preservation, and the attainment of sustainable economic and social development in China. Concurrently, as part of the overarching blueprint for ecological civilization construction, China has set forth the ambitious "2030 carbon peak" and "2060 carbon neutral" dual carbon objectives [1]. This strategic framework aims to cultivate a novel power system wherein new energy sources take precedence [2].

A microgrid, comprised of PCS energy storage inverters, is capable of grid-connected operation with the main power grid and off-grid operation to independently supply power to loads. Due to its distinct characteristics of generation ("source"), consumption ("load"), and storage ("storage"), the microgrid stands as a vital component of a novel power system dominated by renewable energy sources [3].

In the off-grid operation mode of PCS, there are primarily two control modes: (1) Constant Voltage–Frequency (V/f) Control and (2) Grid-Forming (GFM) Control. Both droop control [4] and virtual synchronous machine control [5,6] fall under the category of GFM control [7]. Regardless of whether it is V/f control or GFM control, voltage regulation is essential to control the output voltage of a PCS. Different voltage control strategies play a crucial role in determining the quality of the PCS output voltage.

Currently, common voltage control strategies applicable to PCS include the following:

1. Proportional–Integral (PI) Control: Using dq coordinate systems, PI control achieves voltage regulation but lacks harmonic suppression and the capability of operating with unbalanced loads. If it is necessary to add harmonic suppression, and in order to cope with three-phase unbalanced loads on top of PI control, an additional VPI controller is required [8,9]. This approach requires the design of additional parameters, greatly increasing the design complexity, which is not conducive to engineering implementation.
2. Proportional Resonant (PR)/Quasi-Proportional Resonant (QPR) Control [10]: Based on internal model principles and utilizing abc or $\alpha\beta$ coordinate systems, PR/QPR control regulates output voltage without harmonic suppression. Its ability to independently control three-phase voltages allows operation with unbalanced loads.
3. Proportional Multiresonant Control (PMR) [11]: As an enhanced version of PR control, PMR control consists of multiple PR controllers in parallel, providing both harmonic suppression and the capability of operating with unbalanced loads. However, its structure is complex, requiring the tuning of multiple control parameters.
4. Repetitive control (RC) [12]: Also based on internal model principles, RC control is simple in structure and capable of harmonic suppression and operation with unbalanced loads. However, its dynamic performance may be less favorable. To address the dynamic performance issues of repetitive control, some scholars have proposed odd repetitive control [13], which speeds up the dynamic performance but lacks the suppression of even harmonics.
5. Model Predictive Control (MPC) [14,15]: Carrier-based modulated MPC strategies have demonstrated potential for harmonic suppression and operation with unbalanced loads. However, effective MPC controller design requires careful consideration of dynamic system characteristics, constraints, performance metrics, and computational complexity.
6. Robust Control [16,17]: Exhibiting strong robustness, robust control requires a good understanding of system uncertainties and the careful selection of weighting functions. Experimental and simulation methods may be necessary to validate and adjust robust control strategies for reliable application.
7. Sliding Mode Control (SMC) [18,19]: Leveraging strong robustness and nonlinear characteristics, the SMC exhibits capabilities for harmonic suppression and operation with unbalanced loads. However, in practical applications, precise modeling of system dynamics and careful adjustment of controller parameters are essential. Additionally, SMC may introduce high-frequency oscillations, necessitating appropriate design and tuning to balance system performance and stability.

This paper focuses on the ANPC three-level energy storage inverter and proposes a rapid repetitive control strategy. The study adopts a composite control structure that combines proportional control with repetitive control. The proposed control strategy is characterized by its simplicity in design and ease of engineering implementation. Experimental validation on a 1725 kVA power conversion system confirms that the proposed control strategy can achieve optimized output voltage waveforms under both V/f control and grid-forming control.

2. ANPC Three-Level PCS Modeling

With a two-level topology, each IGBT/MOSFET is required to withstand the entire DC bus voltage stress. This leads to the process of selecting the half-bridge module, which must have a module that withstands a voltage value of U_{dc} above the voltage level (at least 1.2 times). As a result, the DC side voltage of the two-level is generally lower than 1000 V. Furthermore, in the chain reaction brought about by voltage stress, the dv/dt becomes larger, causing a serious problem. The large dv/dt introduces serious electromagnetic interference (EMI) problems, which make hardware design difficult, with large switching losses leading to low efficiency.

Compared to the two-level structure, the three-level structure offers several advantages [20]:

1. Enhanced power quality and increased power density: With higher output levels, the output voltage waveform is closer to sinusoidal, improving the power quality of the output waveform. This design also reduces the size of the filter and increases the power density of the system, especially under the same switching frequency [21].
2. Improved efficiency: The three-level structure primarily utilizes the Neutral Point Clamped (NPC) topology, which includes I-type NPC, T-type NPC, and ANPC (Active Neutral Point Clamped) [22].

In this paper, we will focus on ANPC [23]. The relevant parameters of the PCS system in this paper are shown in Table 1, with a rated capacity of 1725 kVA.

Table 1. 1725 kVA PCS system parameters.

DC Characteristics	
Maximum DC voltage	1500 Vdc
Minimum DC voltage	1000 Vdc
Full load DC operating voltage range	1000–1500 Vdc
Maximum DC current	1935 A
AC Characteristics (Off-Grid)	
Nominal output power	1725 kVA
Maximum output current	1578 A
Nominal AC voltage	690 Vac
AC voltage range	−15%–10%
AC voltage harmonics	<3% (Linear loads)
DC voltage components	<0.5% × Un
Nominal frequency/frequency range	50 Hz/45~55 Hz
Overload capacity	120% (20 s)
Number of phases at the output	3 phases/3 lines

Figure 1 provides the basic topology of the 1725 kVA PCS based on an ANPC type three level. The QS1 is the internal DC circuit breaker of a PCS, and the IPM-A/B/C is the switching device (IGBT) composed of the ANPC topology. L represents the three-phase filter inductors, R is the phase-to-phase DC resistance of the filter inductors, and the filter capacitor C_f takes a delta-type connection. The QF1 is the internal AC circuit breaker of the PCS. The grid switch is at the point of common coupling (PCC), which is the external switch of the PCS.

From Figure 1 and combined with Kirchhoff's voltage law, the mathematical model of the three-level converter in the three-phase stationary coordinate system can be obtained as follows:

$$\begin{cases} e_x = L\frac{di_{Lx}}{dt} + Ri_x + u_x \\ i_{Lx} = C\frac{du_x}{dt} + i_x \end{cases}, \tag{1}$$

where $x = a, b, c$, e_x denotes the bridge arm voltage, and C is the filter capacitor ($C = 3C_f$) after the angle-star connection equivalent. i_{Lx} indicates the bridge arm current flowing through phases A, B, and C. u_x denotes the load terminal voltage, and i_x is the PCS output current.

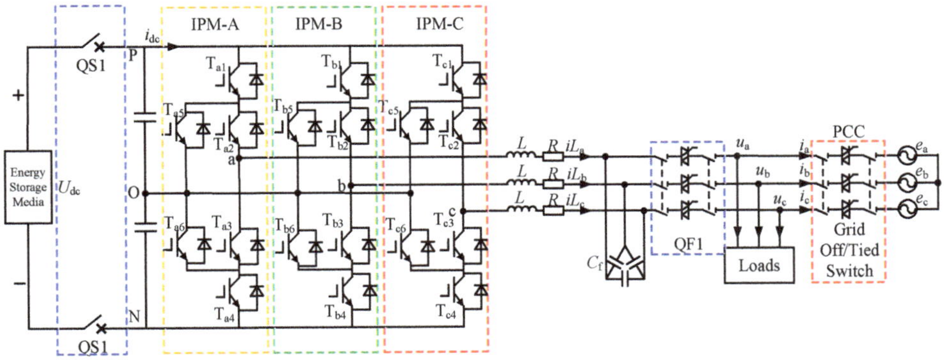

Figure 1. Single-stage ANPC three-level circuit topology.

After modeling the three-level ANPC inverter, the design of the FRC is carried out. To simplify the design of the FRC, the Clarke transformation is conducted, moving from a three-phase stationary coordinate system to a two-phase static coordinate system as follows:

$$\begin{cases} L\frac{di_{L\alpha}}{dt} = e_\alpha - Ri_{L\alpha} - u_\alpha \\ C\frac{du_\alpha}{dt} = i_{L\alpha} - i_\alpha \end{cases} \tag{2}$$

$$\begin{cases} L\frac{di_{L\beta}}{dt} = e_\beta - Ri_{L\beta} - u_\beta \\ C\frac{du_\beta}{dt} = i_{L\beta} - i_\beta \end{cases} . \tag{3}$$

From (2) and (3), there is no coupling on the axis. Further Laplace transform of (2) and (3) gives the relationship between the input and output of the system as

$$e_\alpha = \frac{1}{LCs^2 + RCs + 1}u_\alpha - \frac{Ls + R}{LCs^2 + RCs + 1}i_\alpha \tag{4}$$

$$e_\beta = \frac{1}{LCs^2 + RCs + 1}u_\beta - \frac{Ls + R}{LCs^2 + RCs + 1}i_\beta. \tag{5}$$

The transfer function $G(s)$ of the output voltage and input voltage of the PCS is obtained from (4) and (5):

$$G(s) = \frac{1}{LCs^2 + RCs + 1}. \tag{6}$$

The equivalent output impedance of the PCS is

$$Z_0 = \frac{Ls + R}{LCs^2 + RCs + 1}. \tag{7}$$

In the 1725 kVA PCS, the LC forms a resonant network. If the system is unloaded, there will be a resonance peak. At this time, the system stability is the worst; so, the FRC should be designed assuming the no-load case. From (4) and (5), it can be seen that in the no-load case of the PCS, the transfer function of the controlled object $P(s)$ is

$$P(s) = \frac{1}{LCs^2 + RCs + 1}. \tag{8}$$

The main circuit parameters of the 1725 kVA PCS are shown in Table 2.

Table 2. Parameters of the main circuit of a PCS.

Parameters	Value
Filter inductor DC resistance: R	0.35 Ω
Filter inductors: L	0.07 mH
Filter capacitor: C*	240 μF

C* is the equivalent filtering capacitor after series–parallel connection (C* = 3Cf).

From Table 2 and (8), the discrete time is taken as 1/3600 s, and the discrete transfer function of the controlled object is equal to

$$P(z) = \frac{0.4511z + 0.279}{z^2 - 0.5192z + 0.2494}. \tag{9}$$

The Bode diagram of $P(z)$ is given in Figure 2. In the 1kHz band, the LC resonates, and the phase frequency characteristic has a jump of $-180°$ to $180°$ at this frequency. Due to the filter inductor L in the LC filter, there is a DC resistance; so, the resonance peak of the amplitude–frequency characteristics of $P(z)$ is not apparent. Based on this 1725 kVA PCS, no damping strategy was incorporated. Therefore, in the subsequent design of the FRC, this resonance point should be avoided or attenuated as much as possible.

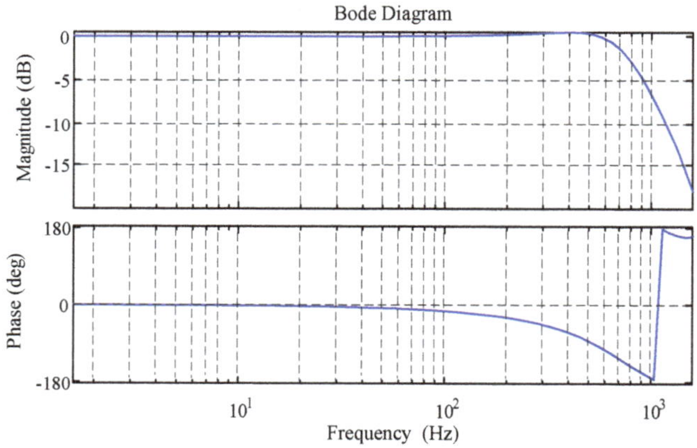

Figure 2. Bode diagram of the controlled object $P(z)$.

3. The Proposed Scheme

Proportional Integral (PI) control is less effective for an AC input signal; so, the conventional method is converted to the *dq* axis. However, this method does not have harmonic suppression capability and cannot deal with unbalanced loads. For this reason, an FRC scheme is proposed in this paper. The block diagram of the proposed FRC is shown in Figure 3.

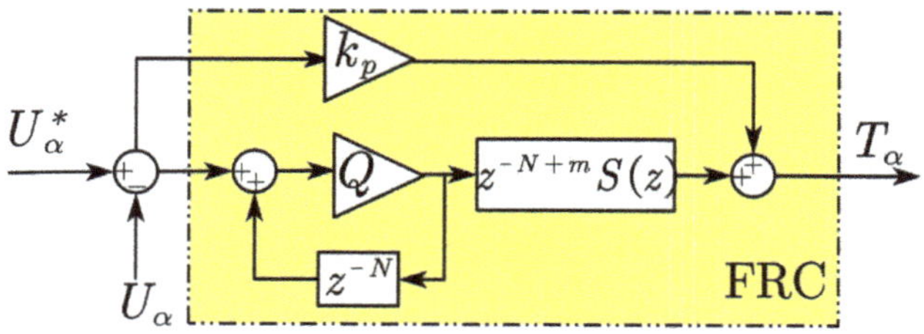

Figure 3. Block diagram of the proposed FRC.

In Figure 3, U_{α}^{*} is the reference voltage signal; U_{α} is the instantaneous voltage value of the PCS output voltage after coordinate transformation to the axis α; k_p is the gain of the FRC; Q is the internal mode (can choose a constant slightly less than 1 or a zero-phase-shift low-pass filter); N is the modulation ratio (f_s/f_n), where f_s is the switching frequency, and f_n is the normal voltage frequency; m is the phase compensation coefficient and $S(z)$ is the low-pass filter. The transfer function of the FRC can be obtained from Figure 3, as follows:

$$G_{\text{FRC}} = \frac{Qz^{-N+m}S(z)}{1 - Qz^{-N}} + k_p. \tag{10}$$

The whole control block of the system for the 1725 kVA PCS in V/f mode is shown in Figure 4.

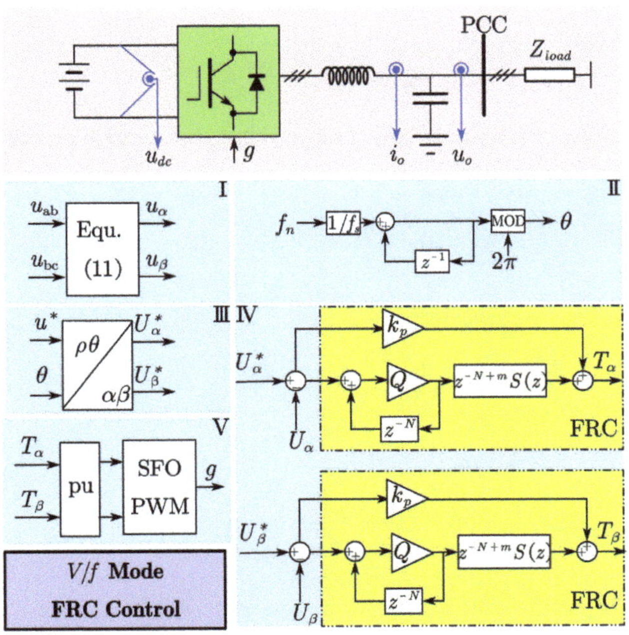

Figure 4. Control block of the inverter system.

In Figure 4, u_{dc} represents the system's DC voltage, i_0 represents the system's output current, and u_o represents the system's voltage. The blue lines represent sampling; the components are as follows:

I is the 3/2 transformation: The output voltages U_{ab} and U_{bc} are sampled from the system and transformed from a three-phase *abc* stationary coordinate system to a two-phase static coordinate system with the following transformation (11). The specific derivation is given in Appendix A.1.

$$\begin{cases} U_{\alpha} = u_{ab} \\ U_{\beta} = \frac{1}{\sqrt{3}}(u_{ab} + 2u_{bc}) \end{cases} ; \tag{11}$$

II is the generation of the output voltage angle θ of PCS, f_n is the rated frequency,f_s is the sampling frequency;

III is the generation of the reference voltage amplitude, the output voltage amplitude can be set via u^*;

IV are the two FRC voltage controllers, output T_{alpha} and T_{beta}, respectively;

V is the generation of six drive signals per phase to the IPM A/B/C, shown in Figure 1, using SFO-PWM.

4. Parameter Design and Analysis of the Proposed FRC

From Figure 4, it can be seen that the design of an FRC is the same in the three-phase stationary coordinate system as in the two-phase static coordinate system. Here, the design of the α-axis is taken as an example, and the β-axis is the same.

4.1. Stability Analysis

As shown in Figure 4, the error transfer function of the system is

$$E(z) = \frac{U_\alpha^* - U_\alpha}{1 + G_{FRC}(z)P(z)}. \tag{12}$$

The polynomial of the error transfer function of the FRC control system is

$$1 + G_{FRC}P(z) = 1 + \left(\frac{Qz^{-N+m}S(z)}{1 - Qz^{-N}} + k_p\right)P(z). \tag{13}$$

To ensure stability, it should be guaranteed that the roots of the characteristic equations of the closed-loop system are inside the unit circle. Observing (13), the equation does not help design the parameters, due to the coupling. Let

$$G_{rc}(z) = \frac{Qz^{-N+m}S(z)}{1 - Qz^{-N}}. \tag{14}$$

Then, Equation (13) after decoupling becomes

$$1 + [G_{rc}(z) + k_p] \cdot P(z) = [1 + k_pP(z)] \cdot [1 + \frac{G_{rc}(z)P(z)}{1 + k_pP(z)}] = [1 + k_pP(z)] \cdot [1 + G_{rc}(z) \cdot P_0(z)]. \tag{15}$$

Among these,

$$P_0(z) = \frac{P(z)}{1 + k_pP(z)}. \tag{16}$$

The stability criterion of the FRC can be obtained from (15), as follows:

Condition 1: The roots of $[1 + k_pP(z)] = 0$ are inside the unit circle.

Condition 2: The roots of $|1 + G_{rc}(z) \cdot P_0(z)| = 0$ are inside the unit circle.

According to Condition 1, a suitable k_p can be selected. From Condition 2, (14), and (16), we obtain

$$\left|1 + \frac{Qz^{-N+m}S(z)}{1 - Qz^{-N}}P_0(z)\right| < 0. \tag{17}$$

Since $1 - Qz^{-N} \neq 0$, the left and right sides of (17) are simultaneously multiplied by $1 - Qz^{-N}$.

$$\left|Qz^{-N}[1 - z^m S(z)P_0(z)]\right| < 1 \tag{18}$$

When the frequency is the fundamental wave or an integer multiple of the fundamental wave frequency, $|Qz^{-N}| \leqslant 1$.

$$|1 - z^m S(z)P_0(z)| < 1 \tag{19}$$

Denote the amplitude and phase characteristics of $S(z)$ and $P_0(z)$ as

$$S(j\omega) = N_S(\omega)e^{-j\theta_S(\omega)}, \tag{20}$$

$$P_0(j\omega) = N_{P_0}(\omega)e^{-j\theta_{P_0}(\omega)}. \tag{21}$$

Substituting (20) and (21) into (19) yields

$$\left|1 - N_S(\omega)N_{P_0}(\omega)e^{-j[\theta_s(\omega)+\theta_{P_0}(\omega)]+m\omega}\right| < 1. \tag{22}$$

According to Euler's formula,

$$e^{-j[\theta_s(\omega)+\theta_{P_0}(\omega)]+m\omega} = \cos\{[\theta_s(\omega) + \theta_{P_0}(\omega)] + m\omega\} - j\sin\{[\theta_s(\omega) + \theta_{P_0}(\omega)] + m\omega\}. \tag{23}$$

Let

$$\varphi = \theta_s(\omega) + \theta_{P_0}(\omega) + m\omega. \tag{24}$$

According to (22)–(24), we can obtain

$$|(1 - N_s(\omega)N_{P_0}(\omega)\cos\varphi) - jN_s(\omega)N_{P_0}(\omega)\sin\varphi| < 1. \tag{25}$$

Taking the square of (25) gives

$$[N_s(\omega)N_{P_0}(\omega)]^2 + 1 + 2N_s(\omega)N_{P_0}(\omega)\cos\varphi < 1. \tag{26}$$

Simplify (26) to obtain

$$N_s(\omega)N_{P_0}(\omega) < 2\cos\varphi. \tag{27}$$

Since $N_s(\omega)N_{P_0}(\omega) > 0$, the final judgment is obtained as follows:

$$-90^\circ < \varphi < 90^\circ. \tag{28}$$

This section started from the system error transfer function and ended with Condition 1 and Condition 2. Condition 1 provides a reference for the design of k_p, and Condition 2 is simplified to obtain the creation of the FRC phase compensation coefficient m. In the following subsection, how to use these two Conditions for the design of the FRC parameters will be explained in detail.

4.2. *Parameter Design of an FRC*

4.2.1. FRC Gain Coefficient k_p

From Condition 1, the k_p of the FRC can be designed. Due to the structural characteristics of the FRC itself, the output signal will lag behind the input signal by N beats, and the dynamic performance of the entire control system is poor at tracking the input signal quickly. Therefore, the proportional coefficient k_p is connected in parallel. k_p will provide a fast channel for the error, and the output of the proportional coefficient k_p will directly change the size of the regulation system to realize the quick adjustment function.

A smaller k_p will make the system's dynamic performance worse, and a larger k_p will make the system's stability worse. Meanwhile, according to (16), there is a coupling relationship between k_p and $P_0(z)$. Different k_ps will lead to different amplitude and phase frequency curves of $P_0(z)$. In the design of phase compensation z^m, the actual compensation object is $P_0(z)$. Therefore, k_p should not be too large, because it will fluctuate the phase of $P_0(z)$. If the phase lag of $P_0(z)$ is too significant, the phase compensation coefficient m will be excessive, and according to (24) and (28), too large an m or too long a phase lag of $P_0(z)$ will cause the stability margin of the system to be small or unstable.

Based on Condition 1, a graph of the root trajectory of $P(z)$ is given under different k_ps. The maximum k_p = 2.68 can be obtained according to Figure 5.

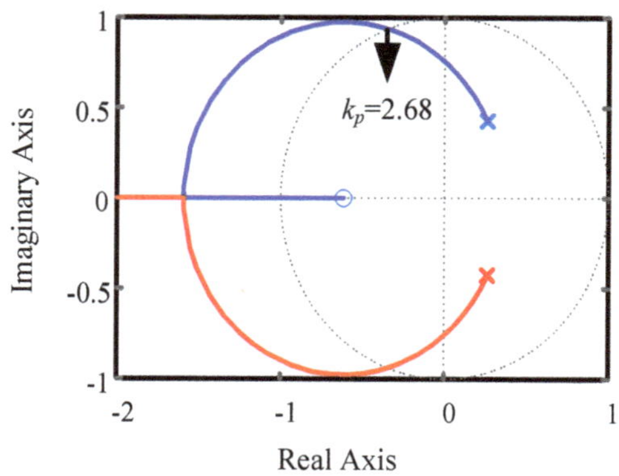

Figure 5. $P(z)$ root trajectory diagram.

According to (15), the FRC controls $P_0(z)$ as a generalized controlled object, and the Bode diagram of $P_0(z)$ is given in Figure 6. With increasing k_p, the phase change in $P_0(z)$ in the middle− and low−frequency bands becomes smaller; however, at the same time, the stability of $P_0(z)$ decreases. When k_p exceeds 2, the system's resonance has started to seriously affect the system's stability.

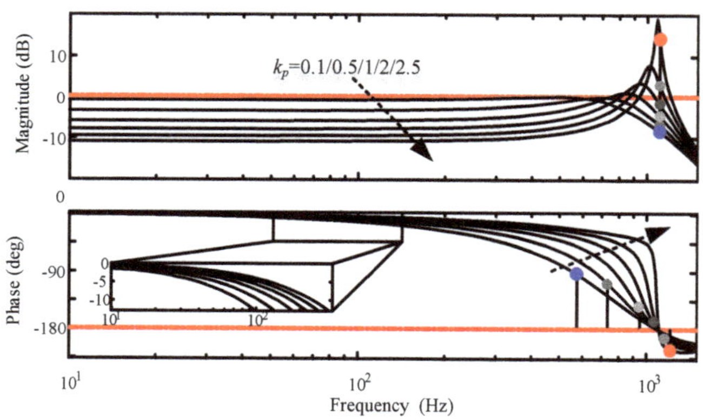

Figure 6. Bode diagram of $P_0(z)$ with different k_p.

With the comprehensive consideration of the above two factors, the FRC phase compensation z^m and the dynamic performance of the system. k_p is chosen as 0.2; at this time, the GM = 7.93 dB, and the PM = ∞deg.

4.2.2. FRC Internal Model Coefficient Q

Q is a key parameter used by the FRC to enhance system stability and achieve the static-free tracking of a reference signal. Q is usually taken as a constant or a low-pass filter slightly less than 1. When Q is a constant, the closer it is to 1, the smaller the steady-state error of the system, and the stronger the harmonic suppression ability. In this work, we take Q as a zero-phase-shift low-pass filter which has no effect on low-frequency signals and quickly attenuates high-frequency signals to improve system stability. The expression of a zero-phase-shift low-pass filter is as follows:

$$Q(z) = \sum_{i=0}^{I} \alpha_i z^i + \sum_{i=1}^{I} \alpha_i z^{-i}. \tag{29}$$

In (29), we have $\alpha_0 + 2\sum_{i=1}^{m} \alpha_i = 1, \alpha_i > 0$, and the first-order zero-phase-shift low-pass filter can be expressed as

$$Q(z) = \frac{z + \alpha_0 + z^{-1}}{2 + \alpha_0}. \tag{30}$$

We can see from Figure 7 that with the small value of α_0, Q attenuates the high-frequency signal more strongly; so, we choose α_0 equal to 2. When α_0 is equal to 2, Q has basically no effect on the base frequency signal (50 Hz), which means that we can realize the non-differential tracking, and for the high-frequency signal (greater than 1 kHz), Q has begun to attenuate the high-frequency signal, which is beneficial for the stability of the system.

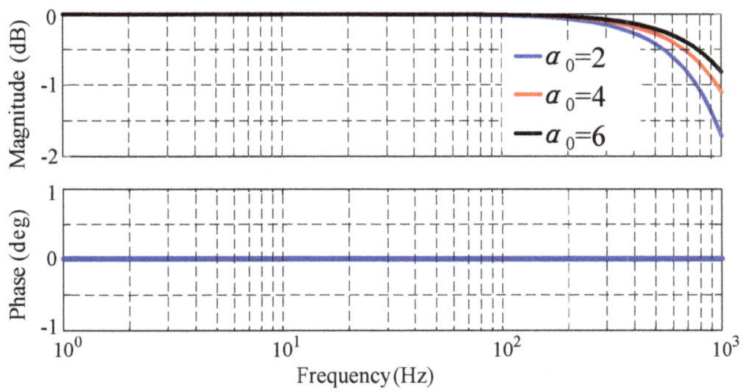

Figure 7. Bode plot of zero-phase low-pass filter Q when α_0 is taken as 2/4/6.

4.2.3. FRC Compensator $S(z)$

Compensator $S(z)$ mainly attenuates the gain brought about by the FRC in the high-frequency band, as well as the strong attenuation of high-frequency disturbing signals in the control system, due to the existence of the compensator $S(z)$, and the system stability performance can be significantly increased. The Butterworth low-pass filter is used in this work because it has zero phase shift in the low-frequency band. As the order of the Butterworth low-pass filter increases, the Butterworth low-pass filter attenuates the signals after the cutoff frequency more strongly, and the filtering effect will be better. However, at the same time, it will also bring about the problem of phase lag. In addition, in the digital control system, a too-high order level will cause a certain computational delay and increase the computational burden of the DSP.

The design of this filter is simple and can be obtained by writing a few lines of the m-language using MATLAB, which is given in Appendix E. The resonant frequency of the LC filter in Figure 2 is around 1 kHz; so, we set the cutoff frequency of the second-order Butterworth low-pass filter to be 1 kHz. The discrete time is 1/3600 s, and the transfer function is obtained as

$$S(z) = \frac{0.3459z^2 + 0.6919z + 0.3459}{z^2 + 0.2047z + 0.179}. \tag{31}$$

4.2.4. FRC Phase Lead Compensation z^m

z^m is mainly used to compensate for the phase lag introduced by $S(z)$ and $P_0(z)$. A suitable z^m should compensate for phase $S(z)$ and $P_0(z)$ at about 0 deg. The phase frequency characteristic curves of $S(z)$ and $P_0(z)$ with different ms are given in Figure 8.

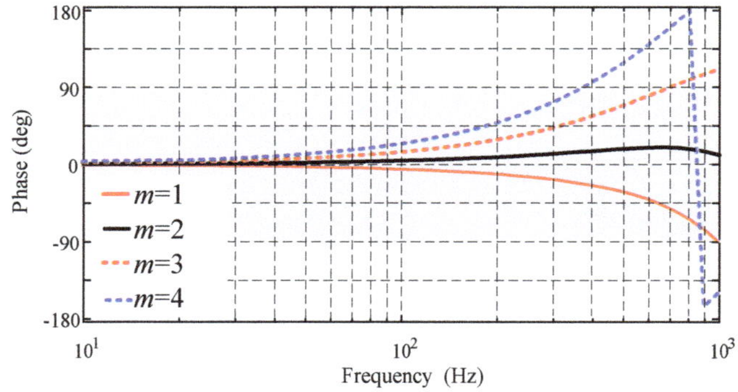

Figure 8. $z^m S(z) P_0(z)$ Phase frequency characteristic curve.

From (28), the stable region in Figure 8 is the shaded part. When $m = 1$, the system is under−compensated, and the phase lag of the system in the low−frequency and mid−frequency bands is quite severe. In the frequency band of 1 kHz, it is about to go beyond the shaded part, and the stability performance is weak. When $m = 2$, the system can meet the stability conditions, while in the frequency band below 1 kHz, there is no significant phase lag. When $m = 3$, the system is in overcompensation, and the phase angle of compensation has exceeded 90°, beyond the stability region. Similarly, when $m = 4$, the steady state condition is not satisfied; so, m is chosen as 2.

The parameters of the FRC are obtained from the above analysis, as shown in Table 3.

Table 3. Parameters of the FRC.

Parameters	Value
N	72
z^m	z^2
Q	$\frac{z + 2 + z^{-1}}{4}$
$S(z)$	$\frac{0.3459z^2 + 0.6919z + 0.3459}{z^2 + 0.2047z + 0.179}$

According to Table 3, the open-loop Bode plot of the FRC in the discrete domain (with a discrete time of 1/3600 s) is obtained, as shown in Figure 9.

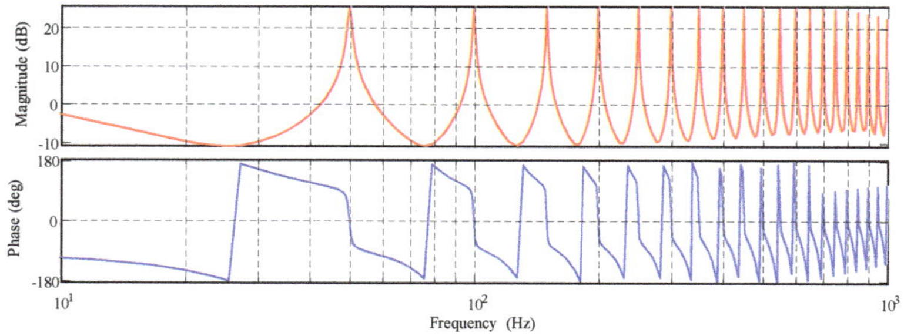

Figure 9. Open−loop Bode plot of G_{FRC}.

5. Experimental Results

The experiments used ModelingTech MT 3200 HIL to simulate the main power circuit, and the control part used the 1725 kVA PCS control board. Host computer 1 is connected to the 1725 kVA PCS via Modbus, through which commands, such as start/stop/set reference voltage values, can be sent to the PCS. The MT3200 HIL is connected to the PCS via DI/DO and AI/AO connection cables to read values such as PCS voltage, current, and switch signals. Host computer 2 serves as the upper computer of the MT3200 HIL, allowing for the modification of different load types and the monitoring of system voltage and current. The experimental hardware setup is shown in Figure 10. For the details of the PI control in the comparison experiments, see Appendix B.

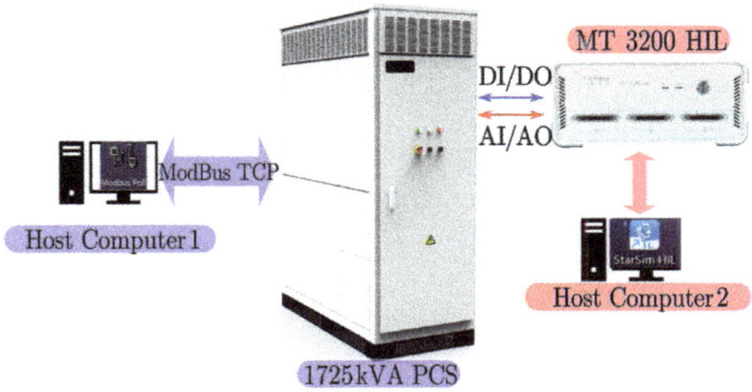

Figure 10. Experimental environment.

5.1. No-Load Experiment

Figures 11–13 show the output voltage waveforms and the THD analysis under PI control and the proposed FRC control, respectively. Analyzing the THD, it can be seen that the proposed control strategy has a very low DC component (less than 0.05%), whereas the DC component using PI control is as high as about 1%; at the same time, the THD of the proposed controller's output voltage is only half that of the PI control.

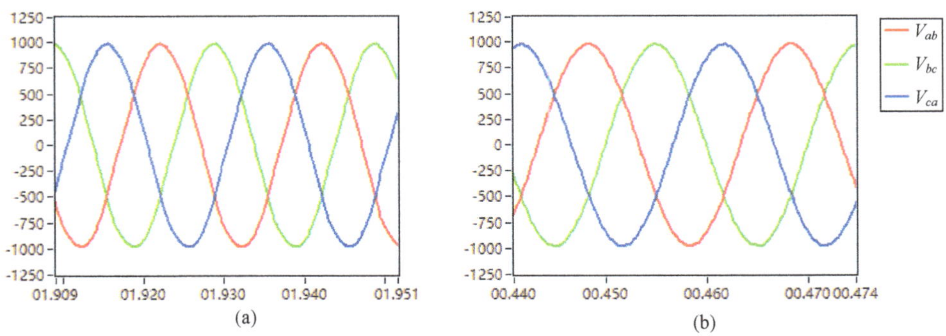

Figure 11. (**a**) PI control no−load output voltage waveform; (**b**) proposed FRC control no−load output voltage waveform.

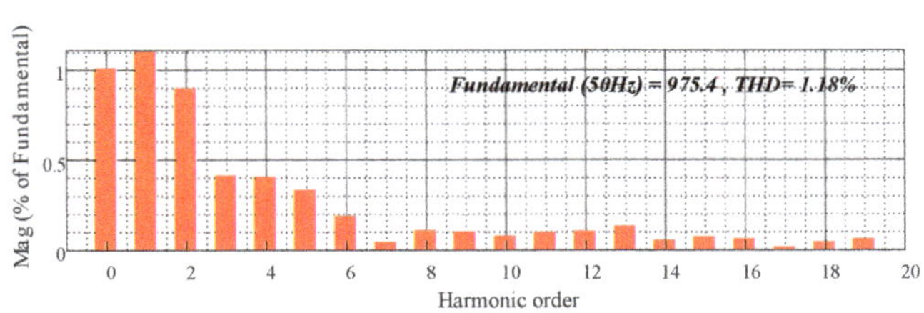

Figure 12. PI control no−load output voltage THD analysis.

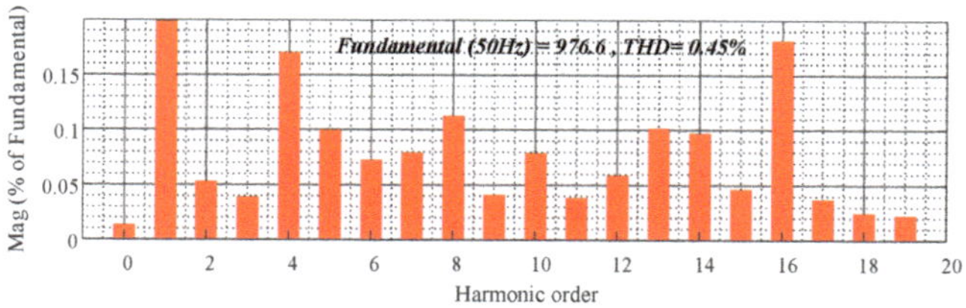

Figure 13. Proposed FRC control no−load output voltage THD analysis.

5.2. Full-Load Experiment

Figures 14–16 show the output voltage waveforms and the THD analysis under the PI control and the proposed FRC control, respectively.

At full load, it is difficult to distinguish the output voltage waveforms of the PI control and the proposed FRC control by the naked eye. This is because the experiments were conducted under very standard conditions. However, in practice, quantitative comparisons can be made through THD analysis. Analyzing the THD from Figures 15 and 16, the experimental results fully demonstrate that the proposed FRC control has a lower DC component (the DC component is close to zero, while the DC component of the PI control is about 1%) and a lower THD (0.38%) compared to the PI control (0.8%).

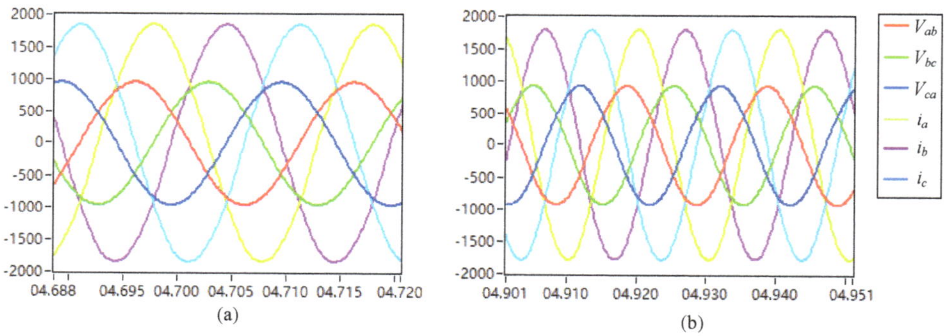

Figure 14. (**a**) PI control full−load output voltage waveform; (**b**) proposed FRC control full−load output voltage waveform.

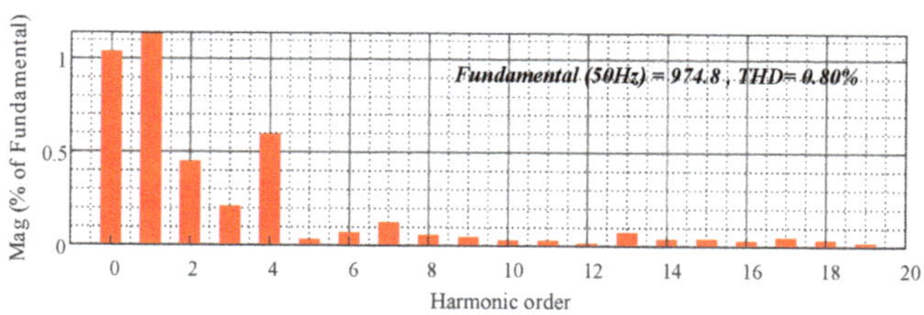

Figure 15. PI control full-load output voltage THD analysis.

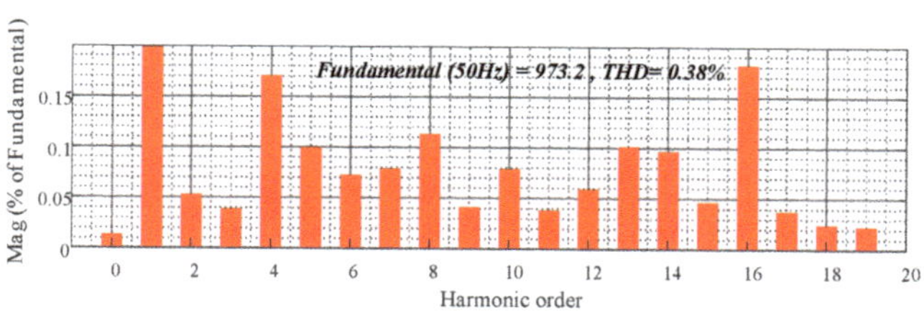

Figure 16. Proposed FRC control full-load output voltage THD analysis.

5.3. Nonlinear Load Experiment

The above two sections discussed resistive loads; in order to further explore the harmonic suppression capability of the proposed FRC control, we accessed a 300 kW nonlinear load (which consists of a three-phase uncontrolled rectifier; refer to Appendix D for details), and the output voltages of the PI control and the proposed FRC control are shown in Figure 17. The output voltage THD was analyzed, as shown in Figures 18 and 19.

Observing Figure 17, it is easy to see that the output voltage waveform of the proposed FRC control is smoother and closer to the sinusoidal waveform.

Observing Figures 18 and 19, the THD of the output voltage obtained by FFT analysis is also smaller; the THD of the output voltage of the proposed FRC control is only 3.26%, while the THD of the output voltage using PI is as high as 7.79%. We focused on the THD and found that the output voltage has the fifth harmonic (PI control > 4%, proposed FRC < 1%); seventh harmonic (PI control > 3%, proposed FRC < 0.6%); eleventh harmonic (PI control > 2.5%, proposed FRC < 1.5%) and thirteenth harmonic (PI control > 2%, proposed FRC < 0.7%), and by observing the content of these harmonics, we again prove that the proposed FRC control has better harmonic suppression capability.

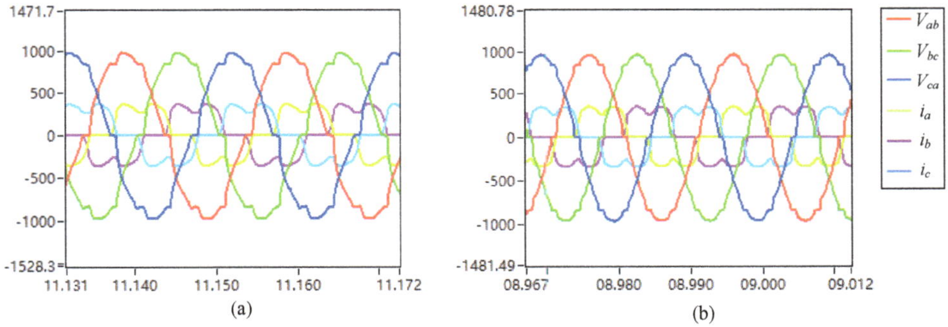

Figure 17. (**a**) PI control with 300 kW nonlinear load output voltage waveform; (**b**) proposed FRC control with 300 kW nonlinear load output voltage waveform.

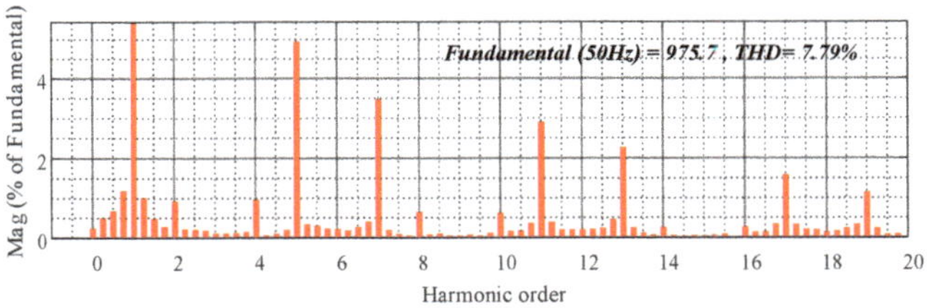

Figure 18. PI control with 300 kW nonlinear load output voltage THD analysis.

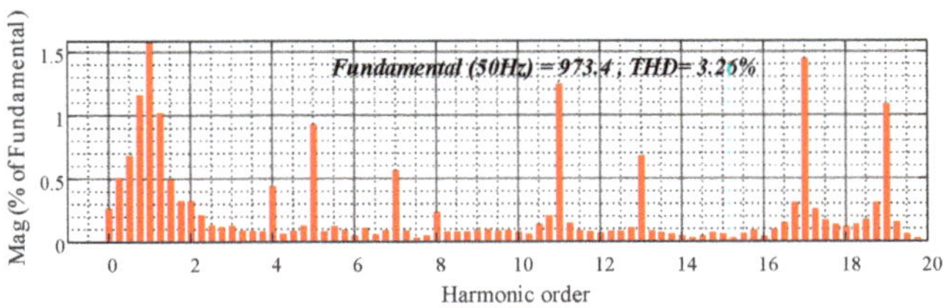

Figure 19. Proposed FRC control with 300 kW nonlinear load output voltage THD analysis.

5.4. Output-Voltage Harmonics Comparison

- To visualize the harmonic suppression capability, the output voltage THD (with resistive loads) at different power levels is shown in Figure 20. The red color represents the PI control, and the blue color represents the FRC control. We take the rated power of 0.1 pu as the step point from zero load to full load. No matter what power level, the proposed output voltage THD content of the FRC control is lower than that of the PI. The THD of the FRC control is approximately 0.5%, which is much lower than the international standard requirements (3%).

- To further demonstrate the harmonic rejection capability of the FRC, its output voltage THD (nonlinear load) is shown in Figure 21. With an increasing nonlinear load power (the step unit is 50 kW), the output voltage waveforms of the PI control and FRC control have a certain degree of distortion. The proposed FRC control is capable of suppressing the voltage harmonics to less than 4%, while the PI control is already close to 8%.

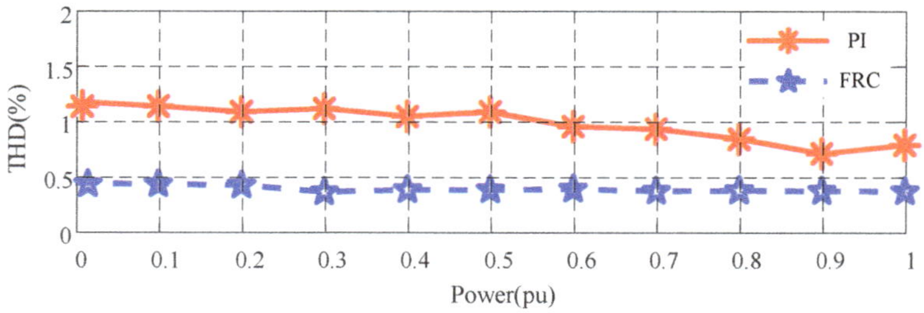

Figure 20. Output voltage THD at different power levels (with resistive loads).

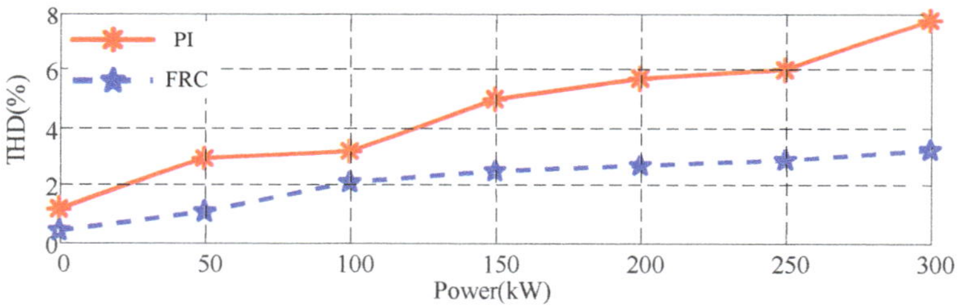

Figure 21. Output voltage THD at different power levels (with nonlinear loads).

5.5. Three-Phase Unbalanced Load Experiment

In order to verify that the proposed FRC control has the ability to carry 100% of the unbalanced load, we set the three-phase unbalanced load conditions: A-phase heavy load, B-phase light load, and C-phase without load.

Figure 22 shows the PI control and FRC control output voltage waveforms. It can be clearly seen that, by using the FRC control, under the unbalanced load conditions, the output voltage is still in a good balance. However, with PI control, the output voltages show a huge imbalance (one of the peak voltages is over 1000 V, and the other peak voltage is less than 900 V).

To further verify that the proposed FRC control has the ability to carry an unbalanced load, the A/B phase load is set as a heavy load, the C phase load is gradually reduced from a heavy load to a light load, and finally, the C phase is unloaded.

Figure 23 shows the phase voltage unbalance ratio of the PI control and the proposed FRC control as the load changes. The definition and calculation of load unbalance and voltage unbalance are provided in Appendix A.2.

According to Figure 23, we can see that the voltage unbalance ratio (PI control) increases as the load unbalance rate increases, while in the proposed FRC control, the voltage unbalance is not affected by the load change. This again shows that the proposed FRC control has the ability to carry a 100% unbalanced load.

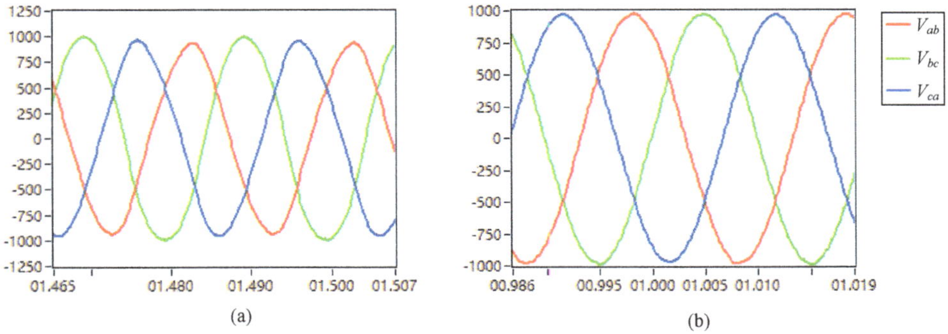

Figure 22. (**a**) PI control with unbalanced load output voltage waveform; (**b**) proposed FRC control with unbalanced load output voltage waveform.

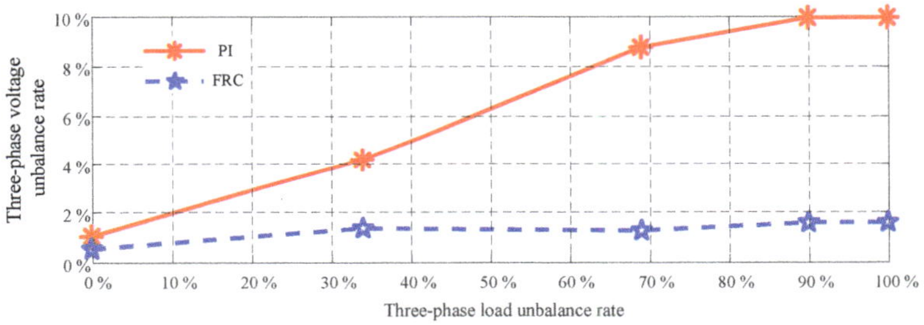

Figure 23. Three-phase voltage unbalance ratio at different load unbalance ratios.

6. Discussion

This paper proposes a voltage loop-control technique for a PCS, which can be applied to the V/f mode and the GFM mode (Refer to Appendix C for details). The advantages of the proposed control are its simple design and easy engineering implementation. The experimental results demonstrate that the proposed FRC control has better harmonic suppression capability and a stronger ability to handle unbalanced loads. Under resistive load conditions, the output voltage THD is always less than 0.5%, which is approximately a 50% reduction in harmonics compared to PI control. Under nonlinear load conditions, the maximum output harmonic of the proposed FRC control is about 3%, while PI control results in about 8% harmonics. Under nonlinear load conditions, the proposed FRC control achieves a maximum output voltage imbalance of less than 2%, while PI control results in a maximum output voltage imbalance of up to 10%.

The proposed control strategy can achieve output voltage harmonic suppression at 100% with unbalanced loads in three phases. However, there are several issues that need to be discussed.

1. In this work, both the V/f mode and GFM mode are based on the voltage single loop to design the controller without considering the current inner loop. Although this design is simple and convenient for engineers to design and debug, the disadvantage is also obvious: in the case of overcurrent, due to the lack of a current inner loop, it is not possible to effectively limit the fault current. However, we believe that this is not an obvious disadvantage for GFM control, and it has been confirmed that the stability of GFM with single-voltage closed-loop control is higher than that of GFM with dual closed-loop control under a robust power grid.
2. In this work, using the V/f control as an example, it has been clearly demonstrated that the proposed controller has excellent control performance. However, in the GFM mode, more tests are needed. The focus of the test should be on whether the proposed

controller has a faster voltage response than the PI control. This response speed directly determines the reactive current of the PCS at LVRT/HVRT.
Theoretically, the proposed voltage controller provides real-time control, while the PI control needs to convert V_{abc} to V_d and V_q, and this link usually adds filters to smooth V_d and V_q. Therefore, the proposed FRC controller should have a faster control speed in GFM mode, and the response speed should be faster than the PI control in the LVRT/HVRT test.

3. The proposed FRC control can be redesigned into a current controller that can be used for the current control of a PCS (e.g., *PQ* mode or AC constant current mode), and this control can achieve very low grid-connected harmonic currents.
4. If it is desired to add a current inner loop to the FRC control discussed in this work, we strongly do not recommend adding a controller with a similar structure of repetitive control, because it will exacerbate the computational burden of the DSP control chip (for the FRC control, we used 50 kHz; however, we found that the control algorithm could not be calculated in an interrupt cycle in DSP28335, so, we later changed to 40 kHz or adopted an ORC kernel, which solved the problem). PR or QPR control can be used as the current inner loop; however, the stability would need to be re-evaluated.

Author Contributions: Conceptualization, Y.S.; methodology, Y.S.; software, Y.S.; validation, Y.S., S.C., and J.Z.; formal analysis, Y.S. and T.L.; investigation, J.Z.; resources, J.Z.; data curation, S.C. and T.L.; writing—original draft preparation, Y.S.; writing—review and editing, S.C. and T.L.; visualization, Y.S.; supervision, J.Z.; project administration, J.Z.; funding acquisition, J.Z. All authors have read and agreed to the published version of the manuscript.

Funding: This paper was supported by the National Key R&D Program of China (No. 2021YFE0103800).

Data Availability Statement: The experimental data are subject to a confidentiality agreement with the company; however, we have open sourced our simulations as well as code generation related files, which can be accessed at the following website: https://ww2.mathworks.cn/matlabcentral/profile/authors/17710614 (accessed on 10 March 2024).

Acknowledgments: We thank ModelingTech, who provided hardware in the loop test equipment MT3200 HIL.

Conflicts of Interest: The authors declare no conflicts of interest.

Abbreviations

The following abbreviations are used in this manuscript:

PCS	power conversion system
ANPC	Active Neutral Point Clamped
RC	repetitive control
PI	Proportional–Integral
FRC	fast repetitive control
VSG	Virtual Synchronous Generator
IPM	Intelligent Power Module
SFO-PWM	Switching Frequency Optimal-PWM
THD	Total Harmonic Distortion
pu	per unit
GFM	Grid Forming
LVRT	Low-Voltage Ride Through
HVRT	High-Voltage Ride Through
ORC	odd repetitive control
PR	Proportional Resonant
QPR	Quasi-Proportional Resonance
DSP	Digital Signal Processing
3P3L	3 Phase 3 Line
LCUR	Line Current Unbalance Ratio

PVUR	phase voltage unbalance ratio
RMS	Root Mean Square
LPF	low-pass filter
PCC	point of common coupling
HIL	hardware in the loop
DI	Digital Input
DO	Digital Output
AI	Analog Input
AO	Analog Output
V/f	voltage/frequency

Appendix A

Appendix A.1. 3P3L System with 3/2 Constant Amplitude Transformation Derivation

Since this PCS is a 3P3L system, the 3/2 constant amplitude transformation is selected to carry out the transformation. The equation below can be obtained.

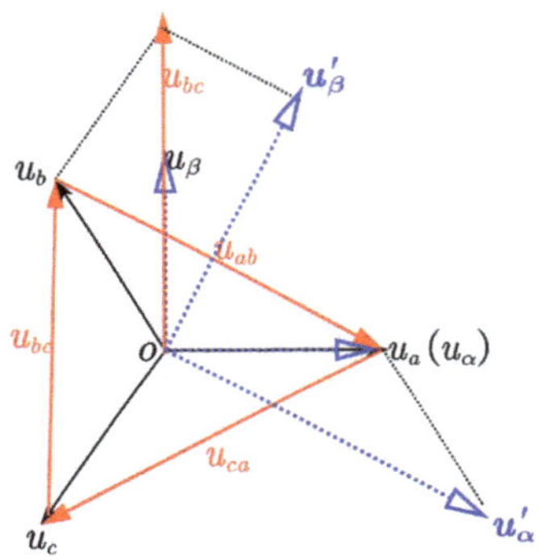

Figure A1. The 3/2 constant amplitude transformation schematic.

$$\begin{cases} u_{ab} = U'_\alpha \\ u_{bc} = \frac{\sqrt{3}}{2}U'_\beta - \frac{1}{2}U'_\alpha \end{cases} \quad \text{(A1)}$$

By sorting (A1), (A2) can be obtained:

$$\begin{cases} U'_\alpha = u_{ab} \\ U'_\beta = \frac{1}{\sqrt{3}}(u_{ab} + 2u_{bc}) \end{cases} \quad \text{(A2)}$$

Appendix A.2. Three-Phase Load Unbalance Rate Calculation Formula

Three-phase load unbalance rate calculation formula: Assume that the three-phase currents (RMS value) are I_a, I_b, and I_c. The calculation formula for the load unbalance rate is

$$\text{LCUR} = \frac{\max[I_a, I_b, I_c] - \min[I_a, I_b, I_c]}{\max[I_a, I_b, I_c]} \times 100\%. \quad \text{(A3)}$$

Assume that the three-line voltages (RMS value) are V_{ab}, V_{bc}, and V_{ca}. The voltage unbalance rate calculation formula is

$$\text{PVUR} = \frac{\max\left[\left|U_{\text{ab}} - U_{\text{avg}}\right|, \left|U_{\text{bc}} - U_{\text{avg}}\right|, \left|U_{\text{ca}} - U_{\text{avg}}\right|\right]}{U_{\text{avg}}} \times 100\%. \quad \text{(A4)}$$

Appendix B

This appendix contains control diagrams showing the PI control. The PI control was set up as a comparison group for the experiment.

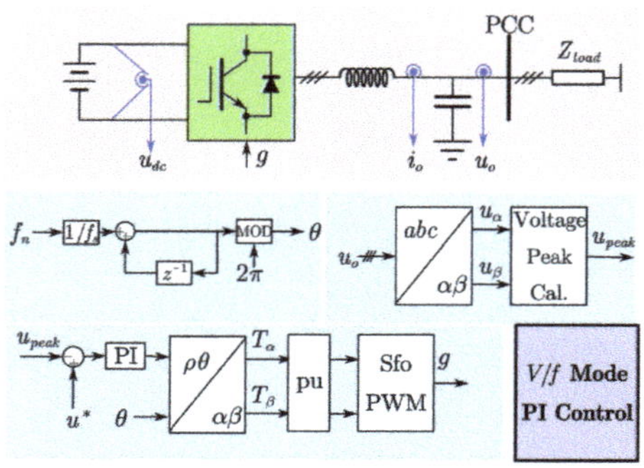

Figure A2. Block diagram using PI control in V/f mode.

The formula for the voltage peak calculation is

$$v_{peak} = \sqrt{u_\alpha^2 + u_\beta^2}. \tag{A5}$$

The formula for the calculation of the pu block is $2/sqrt(3)/u_{dc}$.

Appendix C

Here, we show the application of the FRC control applied to the GFM mode; the experiments showed that the proposed FRC control can be applied to the GFM mode, can work normally in both off-grid and grid-connected conditions, and has the same ability to regulate the voltage amplitude and frequency of the PCC point, as well as the related inertia support. For the relevant test report, the authors can be contacted via email.

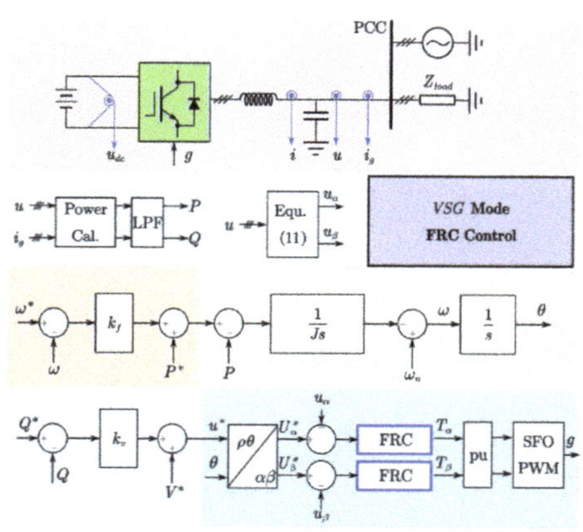

Figure A3. Block diagram using FRC control in GFM mode (VSG).

Appendix D

The nonlinear load applied in the paper is shown in the figure.

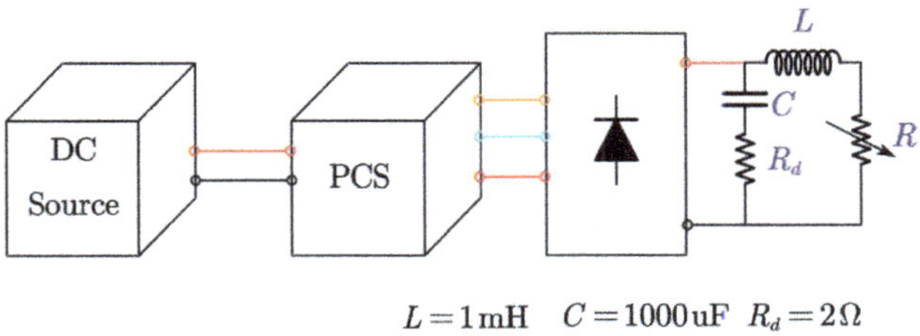

Figure A4. Nonlinear load.

Appendix E

The code for writing the Butterworth low-pass filter using MATLAB is given below:

```
% System input;
n = 2; % Filter Order;
wc = 800; % Set Cutoff Frequency;.
fsw = 3600; % Set Switch Frequency;
wn = wc*2/fsw;
[B,A] = butter(n,wn);
S1 = tf(B,A,1/fsw); % A is Denominator; B is Numerator.
```

References

1. Guo, J.; Ma, S.; Wang, T.; Jing, Y.; Hou, W.; Xu, H. Challenges of developing a power system with a high renewable energy proportion under China's carbon targets. *iEnergy* **2022**, *1*, 12–18. [CrossRef]
2. Breyer, C.; Khalili, S.; Bogdanov, D.; Ram, M.; Oyewo, A.S.; Aghahosseini, A.; Gulagi, A.; Solomon, A.A.; Keiner, D.; Lopez, G.; et al. On the History and Future of 100% Renewable Energy Systems Research. *IEEE Access* **2022**, *10*, 78176–78218. [CrossRef]
3. Zhang, D.; Du, M.; Zhang, Z.; Li, J.; Li, C.; Liu, H.; Shao, L. Effect of multi-energy storage systems on improving the synergy of integrated energy system. In Proceedings of the 2023 3rd New Energy and Energy Storage System Control Summit Forum (NEESSC), Mianyang, China, 26–28 September 2023 ; pp. 90–93.
4. Chen, X.; Chen, Y.; Lin, Z.; Mao, X.; Chen, J.; Zhang, Z. Design of High-Power Energy Storage Bidirectional Power Conversion System. In Proceedings of the 2020 24th International Conference Electronics, Palanga, Lithuania, 15–17 June 2020; pp. 1–6.
5. Zhai, H.; Yang, F.; Xu, Y.; Liu, J.; Huang, L.; Liu, T.; Hao, X. Performance analysis of VSG controlled PCS in Island mode. In Proceedings of the The 16th IET International Conference on AC and DC Power Transmission (ACDC 2020), Online, 2–3 July 2020; pp. 1042–1045.
6. Zhong, Q.-C.; Weiss, G. Synchronverters: Inverters That Mimic Synchronous Generators. *IEEE Trans. Ind. Electron.* **2011**, *58*, 1259–1267. [CrossRef]
7. Rosso, R.; Wang, X.; Liserre, M.; Lu, X.; Engelken, S. Grid-Forming Converters: Control Approaches, Grid-Synchronization, and Future Trends—A Review. *IEEE Open J. Ind. Appl.* **2021**, *2* , 93–109. [CrossRef]
8. Zhou, J.; Zhang, R.; Zhang, X.; Wu, J.; Huang, W. A Compound Control Strategy for Inverter Output Voltage in Micro-grid System. In Proceedings of the 2018 IEEE 4th Southern Power Electronics Conference (SPEC), Singapore, 10–13 December 2018; pp. 1–5.
9. Zhang, Z.; Zhang, D.; Ma, J.; Lu, H.; Zhou, Y.; Xi, Z. Based on Vector Proportional Integral (VPI) Control of Brushless Doubly Fed Induction Generator under Load Imbalance. In Proceedings of the 2022 4th Asia Energy and Electrical Engineering Symposium (AEEES), Chengdu, China, 25–28 March 2022; pp. 64–70.
10. Fantino, R.A.; Busada, C.A.; Solsona, J.A. Optimum PR Control Applied to LCL Filters with Low Resonance Frequency. *IEEE Trans. Power Electron.* **2018**, *33*, 793–801. [CrossRef]
11. Nazeri, A.A.; Saeidi, M.; Zacharias, P. Design and Performance Analysis of a Modified Proportional Multi-Resonant (PMR) Controller for Three-Phase Voltage-Source Inverters. In Proceedings of the 2022 24th European Conference on Power Electronics and Applications, Hanover, Germany, 5–9 September 2022; pp. 1–12.
12. Rahman, M.M.; Biswas, S.P.; Islam, M.R.; Muttaqi, K.M.; Muyeen, S.M. Design of a High Performance Resonant Controller for Improved Stability and Robustness of Islanded Three-Phase Microgrids. *IEEE Access* **2022**, *10*, 119206–119220. [CrossRef]
13. Sun, Y.; Zhou, J. Grid-Tied Inverter Fractional-Order Phase Compensation via Odd Repetition Control. In Proceedings of the International Conference on Electrical Engineering and Green Energy (CEEGE), Grimstad, Norway, 6–9 June 2023; Volume 10, pp. 62–68.
14. Xu, J.; Soeiro, T.B.; Gao, F.; Chen, L.; Tang, H.; Bauer, P.; Dragičević, T. Carrier-Based Modulated Model Predictive Control Strategy for Three-Phase Two-Level VSIs. *IEEE Trans. Energy Convers.* **2021**, *36*, 1673–1687. [CrossRef]
15. Andino, J.; Ayala, P.; Llanos-Proaño, J.; Naunay, D.; Martinez, W.; Arcos-Aviles, D. Constrained Modulated Model Predictive Control for a Three-Phase Three-Level Voltage Source Inverter. *IEEE Access* **2022**, *10*, 10673–10687. [CrossRef]

16. Pardhi, P.K.; Sharma, S. Implementation of Robust Controller for VSI Feeding Stand-alone Loads. In Proceedings of the 2019 IEEE 1st International Conference on Energy, Systems and Information Processing, Chennai, India, 4–6 July 2019; pp. 1–5.
17. Sun, Y.; Zhao, Q.; Liao, W.; Jiang, K.; Tan, W. Robust Repetitive Controller Design Based on S/SK Problem for Single-phase Grid-connected Inverters. In Proceedings of the International Conference on Robotics and Automation Engineering (ICRAE), Guangzhou, China, 19–22 November 2021; pp. 223–230.
18. Vieira, R.P.; Martins, L.T.; Massing, J.R.; Stefanello, M. Sliding Mode Controller in a Multiloop Framework for a Grid-Connected VSI With LCL Filter. *IEEE Trans. Ind. Electron.* **2018**, *65*, 4714–4723. [CrossRef]
19. Shen, X.; Wu, C.; Liu, Z.; Wang, Y.; Leon, J.I.; Liu, J.; Franquelo, L.G. Adaptive-Gain Second-Order Sliding-Mode Control of NPC Converters Via Super-Twisting Technique. *IEEE Trans. Ind. Electron.* **2023**, *38*, 15406–15418. [CrossRef]
20. Rodrguez, J.; Lai, J.; Peng, F. Multilevel inverters: A survey of topologies controls and applications. *IEEE Trans. Ind. Electron.* **2002**, *49*, 724–738. [CrossRef]
21. Barater, D.; Concari, C.; Buticchi, G.; Gurpinar, E.; De, D.; Castellazzi, A. Performance Evaluation of a Three-Level ANPC Photovoltaic Grid-Connected Inverter With 650-V SiC Devices and Optimized PWM. *IEEE Trans. Ind. Appl.* **2016**, *52*, 2475–2485. [CrossRef]
22. Barbosa, P.; Steimer, P.; Meysenc, L.; Winkelnkemper, M.; Steinke, J.; Celanovic, N. Active Neutral-Point-Clamped Multilevel Converters. In Proceedings of the IEEE 36th Power Electronics Specialists Conference, Dresden, Germany, 16 November 2005; pp. 2296–2301.
23. Taallah, A.; Mekhilef, S. Active neutral point clamped converter for equal loss distribution. *IET Power Electron.* **2014**, *7*, 1859–1867. [CrossRef]

electronics

MDPI

Article

A Novel Fractional Delay Proportional–Integral Multi-Resonant-Type Repetitive Control Based on a Farrow-Structure Filter for Grid-Tied Inverters

Fen Liang [1,2], Ho-Joon Lee [1,*] and Qiangsong Zhao [3]

1 Department of Electrical & Control Engineering, Cheongju University, Cheongju 28503, Republic of Korea; lf01023354@cju.ac.kr
2 School of Mechanical and Electrical Engineering, Henan Industry and Trade Vocational College, Zhengzhou 451191, China
3 School of Electronic and Information, Zhongyuan University of Technology, Zhengzhou 451191, China; zhaoqiangsong@126.com
* Correspondence: hjlee@cju.ac.kr; Tel.: +82-043-229-8436

Abstract: The integer-order delay of proportional–integral multi-resonant-type repetitive control (PIMR-RC) cannot provide excellent control performance for grid-tied inverters when the grid frequency fluctuates. To address this issue and reduce control errors, a fractional delay PIMR-RC (FD-PIMR-RC) scheme is proposed. In addition, to reduce the computational load and memory consumption, a Farrow-structure fractional delay (FFD) filter is adopted. The digital filter with the Farrow structure is flexibly and efficiently used for fractional delay. For each new fractional delay, a large number of calculations and storage for the FFD filter coefficients are avoided, which significantly reduces the computational load and memory consumption. The parameter design of the FD-PIMR-RC scheme is provided in detail, including the implementation of fractional delay based on the Farrow structure. Then, a system stability analysis and parameter optimization are presented. Finally, simulations for the steady-state and dynamic responses are presented, and the validity of the proposed method is demonstrated.

Keywords: repetitive control; fractional delay; grid-tied inverter; Farrow structure

1. Introduction

As a result of the energy crisis and environmental pollution, distributed power generation systems (DPGS) based on renewable energy have attracted more and more attention in recent decades [1]. A grid-tied inverter with nonlinear factors such as dead time, as an important interface between the distributed generation system and the grid, will generate a large number of harmonics, which may cause additional power losses or even result in system instability [2]. The total harmonic distortion (THD) of a grid-tied current should be less than 5% according to the IEEE standard. Thus, to suppress harmonics and smooth the grid-tied current, many control schemes have been proposed for grid-tied inverters, such as proportional–integral (PI) control [3], proportional–resonant (PR) control [4], model predictive control (MPC) [5], repetitive control (RC) [6], and so on.

PI control theory is mature and its design is simple. Nevertheless, PI control cannot achieve static zero steady-state error control performance for single-phase or three-phase asymmetric grid-tied inverters due to its limited control-loop gain. To address this problem, some scholars have proposed an ideal PR control [7], which provides an infinite gain and can track references at the selected resonant frequency while maintaining zero steady-state errors. However, the frequency bandwidth of the PR controller is too narrow, which may result in the poor robustness of a system. Subsequently, quasi-proportional–resonant (QPR) controllers have been introduced to increase the bandwidth at the resonant frequency, but the QPR controller can only achieve zero steady-state errors for a single-frequency signal. Several controllers need to operate in parallel to form a proportional multi-resonant (PMR) controller [8], but this results in an increased computational load, a significant reduction in the phase margin, and even system instability.

RC, which is based on the internal model principle, has been widely used in active power filters (APF) [9], grid-tied inverters [10], etc., due to its outstanding performance in static zero-error tracking control and disturbance elimination for the periodic signals (e.g., sine wave signal or harmonics). It not only has high steady-state accuracy and can achieve zero-error tracking but also exhibits excellent performance in suppressing harmonics whose frequencies are integer multiples of the fundamental frequency [11]. However, the ideal RC has a slow dynamic response due to the delay of one fundamental period. To achieve better control performance, it is necessary to combine other feedback controllers such as PI to improve the dynamic response. There are two common structural forms: RC and PI in series or in parallel. For instance, a PIMR-RC adopting an improved RC and a proportional gain in parallel was proposed in [12]. These improved schemes exhibit an excellent dynamic response and performance in suppressing harmonics by accommodating a large gain at resonant frequencies.

The conventional RC (CRC) can be expressed by $z^{-N}/(1-z^{-N})$, where $N = f_s/f_0$, N is the number of samples in one period of the repetitive signal, f_s is the sampling frequency, and f_0 is the grid fundamental frequency. Theoretically, when the sampling frequency is fixed, the delay unit N is an integer. Nevertheless, the grid fundamental frequency in DPGS may vary within a certain range [13]. For instance, f_s = 10 kHz, f_0 = 50 Hz, and the delay unit N is 200; however, when f_0 = 49.6 Hz, N is 201.6. Hence, N may not be an integer when the grid frequency fluctuates. The CRC can only be implemented digitally when N is an integer. Thus, if the value of the fraction N is rounded to the nearest integer, it may result in approximation errors, as well as worse harmonic suppression performance and a higher THD. Given the above issues, adopting fractional delay RC becomes necessary to reduce approximation errors and ensure effective control. Therefore, scholars have proposed frequency adaptive RC schemes for grid-tied inverters [14,15]. In these schemes, a fractional-order RC (FORC) based on a fractional delay (FD) filter is adopted, and the FD filter is used to approximate the fractional part of N. By adjusting the coefficients of the digital filter in real time, the resonant frequency of the FORC is almost equal to the actual values of the grid fundamental and harmonic frequencies, thus achieving current harmonic suppression. Therefore, FORC can achieve good performance in suppressing harmonics while the grid frequency fluctuates.

Generally, there are two kinds of FD filters adopted in FORC: the finite impulse response (FIR) filter [16,17] and the infinite impulse response (IIR) filter [18]. However, both of these filters require recalculating the sub-filter coefficients during coefficient updates, which consumes a significant amount of computational resources. The Farrow-structure filter, based on Taylor series expansion, is more flexible and efficient due to its time-varying nature [19,20]. The Farrow-structure filter can cope with time-varying fractional delays. For each new delay, the polynomial coefficients of the filter remain unchanged, which greatly decreases the computational load of the controller. Moreover, the Farrow-structure filter is easier to implement in hardware and reduces hardware complexity [21,22]. Another benefit of the Farrow-structure filter is that the fractional delay item d is independent of the specific filter coefficients [23]. Thus, when the delay changes, there is no need to reload the filter coefficients, which avoids storing a large number of filter coefficients, thereby greatly reducing the memory. Furthermore, according to [24], the error of the Farrow-structure design is smaller than that of the conventional Lagrange-based design with the same filter order. To date, the Farrow-structure filter has been adopted in many applications [25,26]. A comparison of the FIR filter, conventional IIR filter, and Farrow-structure filter is provided in Table 1. In addition, there is no fractional-order PIMR-RC based on a Farrow-structure filter in the published literature, so the idea of adopting an FFD filter in PIMR-RC is novel.

In summary, to further improve harmonic suppression performance and control accuracy when the grid frequency fluctuates, based on PIMR-RC and an FFD filter, an FD-PIMR-RC scheme is proposed in this paper. Item z^{-N} with a fractional N can be approximated using an FFD filter, where each sub-filter of the FFD filter is designed offline in advance, which simplifies the calculations. The proposed FD-PIMR-RC scheme not only decreases the computational load and memory consumption but also exhibits excellent harmonic suppression performance and smaller tracking errors when the grid frequency fluctuates.

Table 1. Comparison of the different fractional delay filters.

	Computational Load	Memory Consumption	Performance
Conventional fractional delay FIR filter	High	High	High approximation accuracy but needs sufficient bandwidth of the amplitude-frequency.
Fractional delay IIR filter	High	High	Excellent dynamic performance and accurate frequency adaptation capability; the difficulty of hardware implementation is increased due to the existence of zeros and poles.
Fractional delay Farrow-structure filter	Low	Low	Wide stability region, small tracking error, and low THD; simple hardware implementation.

The contributions of this paper are as follows:

(1) To reduce the computational load and memory consumption of the grid-tied inverter, an FFD filter is adopted to achieve fractional delay for the PIMR-RC system.
(2) Based on the FFD filter, an FD-PIMR-RC scheme is proposed to effectively improve the quality of the grid current against frequency variations.
(3) From both the steady-state and dynamic response aspects of the system, a synthesis analysis is carried out on the comparison of the control effects between the conventional integer-order PRMR-RC and the proposed FD-PIMR-RC scheme when the grid frequency fluctuates.

2. Model of the LCL-Type Grid-Tied Inverter

Figure 1 shows the model of the single-phase LCL-type grid-tied inverter, including the single-phase inverter bridge, LCL filter, and controller [27]. E_{dc} is the bus voltage; u_{inv} is the grid-tied inverter output voltage; and L_1, L_2, and C are nominal values of the LCL filter, respectively. I_{ref} is the tracked reference current amplitude; i_g is the grid current; u_g is the grid voltage; i_c is the capacitance current; and ZOH is the zero-order holder. The phase-locked loop (PLL) is adopted to provide the phase angle θ of the point common coupling (PCC) voltage u_{PCC}. Moreover, the reference current i_{ref} is produced by combining the phase θ with I_{ref}; k_{ic} is the capacitance current feedback coefficient, which is used to suppress the LCL filter resonance peak; and $G_i(z)$ is the FD-PIMR-RC current controller.

In practice, high-frequency SPWM modulation technology is mainly adopted in grid-tied inverters, with a switching frequency of more than 10 kHz. Thus, the inverter unit can be regarded as a link with a gain of k_{PWM}, which can be equivalent to 1 [28]. To suppress the LCL resonance peak, capacitive current feedback active damping is adopted in this system. Then, the open-loop transfer function from the inverter output voltage $u_{inv}(s)$ to the grid current $i_g(s)$ can be expressed as:

$$P(s) = \frac{i_g(s)}{u_{inv}(s)} = \frac{1}{L_1 L_2 C s^3 + L_2 C k_{ic} s^2 + (L_1 + L_2)s} \tag{1}$$

where $P(s)$ is the controlled object transfer function in the s-domain. Accordingly, a block diagram of an LCL-type grid-tied inverter control with capacitive current feedback active damping is shown in Figure 2.

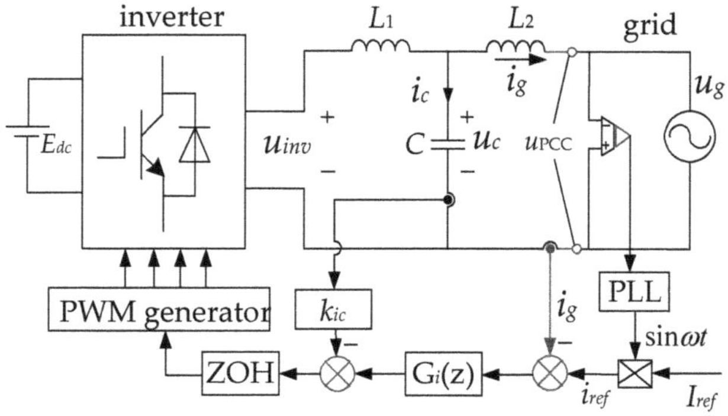

Figure 1. The model of the single-phase LCL-type grid-tied inverter.

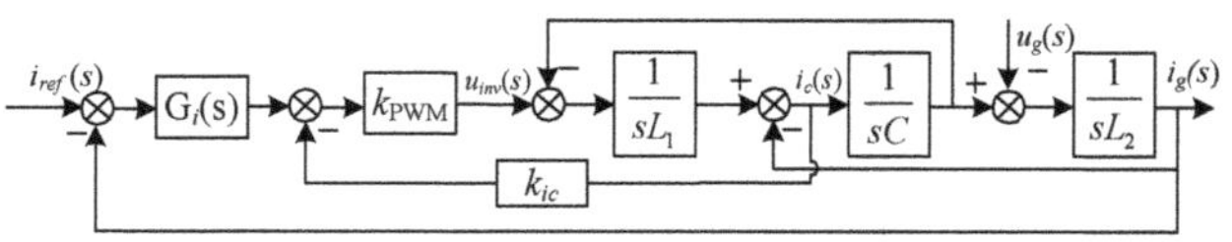

Figure 2. Block diagram of an inverter control with active damping.

According to the LCL filter parameters in Table 2, by selecting $k_{ic} = 18$ (in Section 5), $P(s)$ with active damping can be derived as

$$P(s) = \frac{1}{8.36 \times 10^{-11}s^3 + 3.96 \times 10^{-7}s^2 + 0.006s} \quad (2)$$

The discrete transfer function $P(z)$ with f_s = 10 kHz is given by

$$P(z) = \frac{0.001718z^2 + 0.005903z + 0.001352}{z^3 - 2.084z^2 + 1.707z - 0.6227} \quad (3)$$

Table 2. Parameters of the LCL-type single-phase grid-tied inverter.

Parameter	Value
Inverter-side inductor: L_1	3.8 mH
Grid-side inductor: L_2	2.2 mH
Filter capacitor: C	10 μF
DC-link voltage: E_{dc}	380 V
Grid frequency: f_g	50 Hz
Switching frequency: f_{sw}	10 kHz
Sampling frequency: f_s	10 kHz

Figure 3 shows the frequency characteristics of $P(s)$ with active damping and $P_{LCL}(s)$ without damping ($k_{ic} = 0$), respectively. As shown in Figure 3, f_r is the resonance frequency. Due to a lack of damping, the magnitude characteristics of $P_{LCL}(s)$ exhibit a resonance spike at the resonance frequency (1348 Hz). The phase characteristics of $P_{LCL}(s)$ exhibit a $-180°$ jump, which can lead to system instability. Thus, it is necessary to adopt a damping method to suppress the resonance peak and stabilize the system. In this paper, the method of capacitive current feedback active damping is adopted because of its simplicity and losslessness [29].

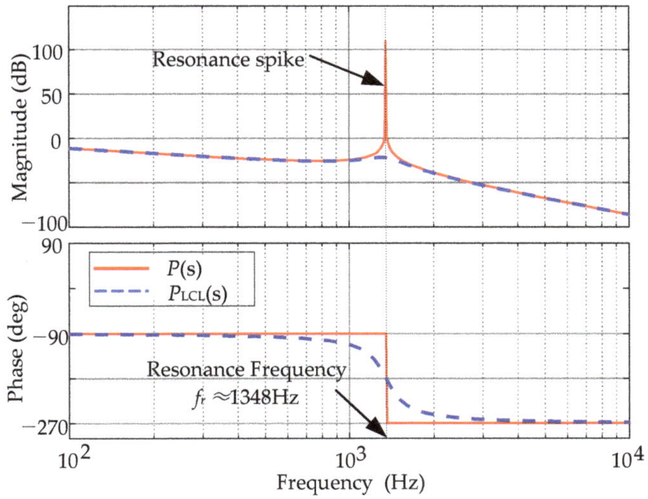

Figure 3. Frequency characteristics of $P(s)$ and $P_{LCL}(s)$.

3. PIMR-RC

The transfer function of CRC in the s-domain can be described as:

$$G_{rc}(s) = k_r \frac{e^{-sT_0}}{1 - e^{-sT_0}} \tag{4}$$

where k_r is the RC gain and T_0 is the fundamental period of the reference signal. Based on the Taylor series expansion, Expression (4) can be derived as

$$G_{rc}(s) = -\frac{k_r}{2} + \frac{k_r}{T_0 s} + \frac{2k_r}{T_0} \sum_{n=1}^{\infty} \frac{s}{s^2 + (n\omega_0)^2} \tag{5}$$

By adding a positive gain, proportional link k_p, on both sides of Equation (5), $k_p > \frac{k_r}{2}$, a composite controller $G_i(s)$ composed of RC+PI can be obtained:

$$G_i(s) = G_{rc}(s) + k_p = (k_p - \frac{k_r}{2}) + \frac{k_r}{T_0 s} + \frac{2k_r}{T_0} \sum_{n=1}^{\infty} \frac{s}{s^2 + (n\omega_0)^2} \tag{6}$$

Thus, $G_i(s)$ contains a proportional term $(k_p - k_r/2)$, an integral term $k_r/T_0 s$, and a multi-resonant term, indicating that the composite controller $G_i(s)$ is a PIMR controller [27,30]. ω_0 is the resonance frequency.

As a digital control system, the repetitive controller needs to be discretized. The discretized RC transfer function in the z-domain is

$$G_{rc}(z) = \frac{Q(z)z^{-N}}{1 - Q(z)z^{-N}} z^m k_r S(z) \tag{7}$$

where $Q(z)$ is usually a low-pass filter [31], which is employed to improve system stability; z^{-N} is the period delay; $S(z)$ is a compensator, which is adopted to further attenuate high frequencies; and z^m is a phase-lead compensator.

A block diagram of a PIMR-RC current controller, which comprises RC and P components in parallel, after discretization is shown in Figure 4. k_p is the proportional gain and $E(z)$ is the tracking error of the system.

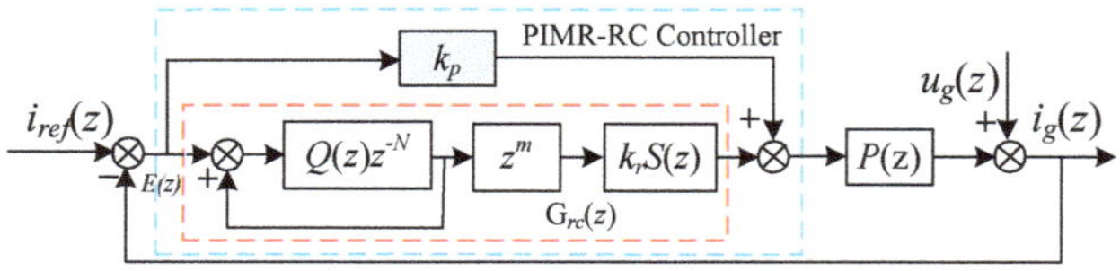

Figure 4. Block diagram of the PIMR-RC system.

The PIMR-RC controller, based on CRC, is also sensitive to grid frequency fluctuations. In addition, N may not be an integer, which can reduce tracking accuracy and stability. Here, to address this issue, an FD-PIMR-RC scheme based on an FFD filter is proposed to improve robustness against varying grid frequencies.

4. Design of FD-PIMR-RC

4.1. Realization of FFD Filter

When the grid frequency fluctuates, N is a fraction, and it can be separated into an integer N_i and a fraction d, as follows:

$$N = N_i + d \tag{8}$$

A block diagram of FD-PIMR-RC based on an FFD filter is shown in Figure 5.

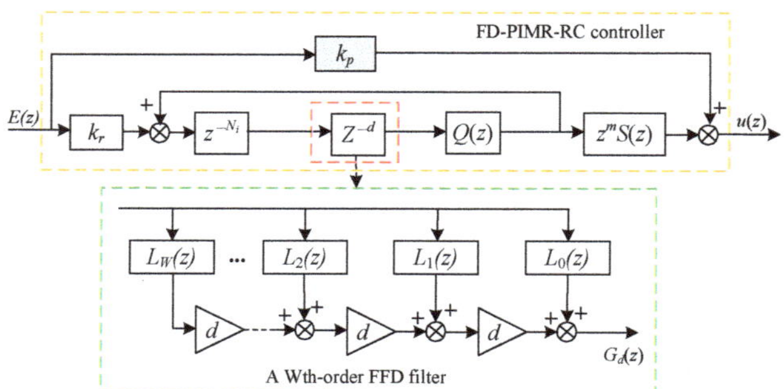

Figure 5. Structure of the FD-PIMR-RC based on an FFD filter.

The transfer function of a Wth-order FFD filter in the z-domain can be described as

$$G_d(z) = \sum_{k=0}^{W} L_k(z) d^k \tag{9}$$

where W is the order of the filter and $L_k(z)$ is the k-th sub-filter ($k = 0, 1, 2, \ldots, W$). $L_k(z)$ can be calculated as follows:

$$f_{sub} = U^{-1} z_{sub} = \begin{bmatrix} L_0(z) & L_1(z) & L_2(z) & \cdots & L_W(z) \end{bmatrix}^{\mathrm{T}} \tag{10}$$

where z_{sub} is the delay operator matrix

$$z_{sub} = \begin{bmatrix} 1 & z^{-1} & z^{-2} & \cdots & z^{-W} \end{bmatrix}^{\mathrm{T}} \tag{11}$$

$$U = \begin{bmatrix} 0^0 & 0^1 & 0^2 & \cdots & 0^S \\ 1^0 & 1^1 & 1^2 & \cdots & 1^S \\ 2^0 & 2^1 & 2^2 & \cdots & 2^S \\ \vdots & \vdots & \vdots & \vdots & \vdots \\ W^0 & W^1 & W^2 & \cdots & W^S \end{bmatrix} \tag{12}$$

where U is a Wth-order Vandermonde matrix ($S = W$). For instance, the second-order FFD filter is derived as follows:

$$\begin{aligned} G_d(z) &= \sum_{k=0}^{2} L_k(z)d^k = L_0(z) + L_1(z)d + L_2(z)d^2 \\ &= 1 + (-1.5 + 2z^{-1} - 0.5z^{-2})d + (0.5 - z^{-1} + 0.5z^{-2})d^2 \end{aligned} \tag{13}$$

When $W = 3$, $z^{-198.4}$ can be expressed as $z^{-198}z^{-0.4}$, according to Equations (9)–(12)

$$z^{-0.4} = G_d(z) = \sum_{k=0}^{3} L_k(z)0.4^k = 0.416 + 0.832z^{-1} - 0.312z^{-2} + 0.064z^{-3} \tag{14}$$

Hence,

$$z^{-198.4} = z^{-198}z^{-0.4} = z^{-198}(0.416 + 0.832z^{-1} - 0.312z^{-2} + 0.064z^{-3}) \tag{15}$$

Figure 6 shows the frequency characteristics of $z^{-0.4}$ based on an FFD filter with different W-th order. The bandwidths of the first- and second-order FFD filters are 51% and 66% of the Nyquist frequency, respectively, which is almost the same as the FIR filter. Additionally, the FFD filter has a wider bandwidth and better phase response linearity when $W = 3$. Therefore, using a third-order FFD filter can effectively approximate fractional delay.

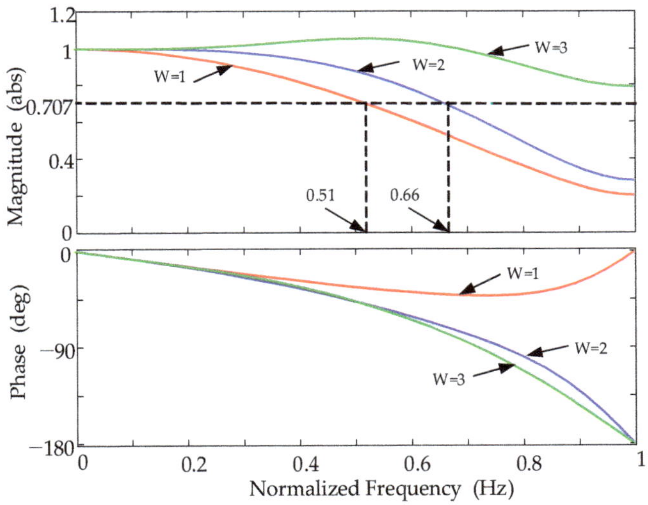

Figure 6. Frequency characteristics of the FFD filter with different W-th order.

4.2. Stability of FD-PIMR-RC

According to Figure 4, the error transfer function $E(z)$ can be derived as

$$E(z) = \frac{1}{1 + (G_{rc}(z) + k_p)P(z)}(i_{ref}(z) - u_g(z)) \tag{16}$$

The characteristic polynomial of $E(z)$ is:

$$1 + (G_{rc}(z) + k_p)P(z) = (1 + k_pP(z))(1 + G_{rc}(z)P_0(z)) \tag{17}$$

where $P_0(z)$ is a new equivalent controlled object of RC, and its expression is:

$$P_0(z) = \frac{P(z)}{1 + k_p P(z)} \tag{18}$$

There are two stability conditions for the system [12]: ① The roots of the polynomial $1 + k_p P(z) = 0$ are all inside a unit circle. ② $|1 + G_{rc}(z)P_0(z)| \neq 0$. By selecting an appropriate proportional gain k_p, condition ① can be met. Subsequently, by substituting (7) into condition ②, condition ② can be written as

$$\left|1 - Q(z)z^{-N} + Q(z)z^{-N+m}k_r S(z)P_0(z)\right| \neq 0 \tag{19}$$

$$i.e., \quad \left|Q(z)z^{-N}(1 - z^m k_r S(z)P_0(z))\right| < 1, \forall z = e^{j\omega T_s}, 0 < \omega < \pi/T_s \tag{20}$$

Substitute (8) into (20)

$$|Q(z)(1 - z^m k_r S(z)P_0(z))| < \left|z^{-(N_i+d)}\right|^{-1} \tag{21}$$

When both the frequency of the reference signal and disturbance signal are equal to the fundamental frequency or its integer multiples within the bandwidth range of the fractional delay filter, $\left|z^{-N_i}z^{-d}\right|^{-1} \to 1$, $|Q(z)| < 1$, then, condition ② can be derived as

$$|1 - z^m k_r S(z)P_0(z))| < 1 \tag{22}$$

Expression (22) indicates that to maintain the stability of the PIMR-RC system while ω varies from zero to the Nyquist frequency, the vector $(1 - z^m k_r S(z)P_0(z))$ should fall within the unit circle. Accordingly, the system stability condition is independent of the FFD filters. Thus, according to the literature [12], Equation (22) can be derived as

$$\left|\theta_s(\omega) + \theta_{P_0}(\omega) + m\omega T_s\right| < 90^\circ \tag{23}$$

$$0 < k_r < \min_{\omega} \frac{2\cos(\theta_s(\omega) + \theta_{P_0}(\omega) + m\omega T_s)}{N_s(\omega)N_{P_0}(\omega)} \tag{24}$$

where $N_s(\omega)$ and $\theta_s(\omega)$ are the magnitude characteristics and the phase characteristics of $S(z)$, respectively. $N_{P_0}(\omega)$ and $\theta_{P_0}(\omega)$ are the magnitude characteristics and the phase characteristics of $P_0(z)$, respectively.

5. Parameter Design of FD-PIMR-RC

According to Figure 5, there are five parameters to be designed for the proposed FD-PIMR-RC. They are the proportional gain k_p, internal mode filter $Q(z)$, phase-lead compensator z_m, RC controller gain k_r, and compensator $S(z)$. The design processes of these parameters are described below.

5.1. Proportional Gain k_p

It is necessary to select the appropriate k_p to meet condition ①, which is equivalent to having all the poles of $P_0(z)$ inside a unit circle. According to [12], when k_p varies from 10 to 25, the poles of $P_0(z)$ are within a unit circle. As shown in Figure 7a, the pole plots of $P_0(z)$ with the proportional gain k_p are 10, 13, 16, 19, respectively. Then, by selecting an appropriate capacitance current feedback coefficient k_{ic}, condition ① is satisfied. Figure 7b shows the pole map of $P_0(z)$ when k_p = 15 and k_{ic} = 18. The poles of $P_0(z)$ are all inside the unit circle on the z-plane.

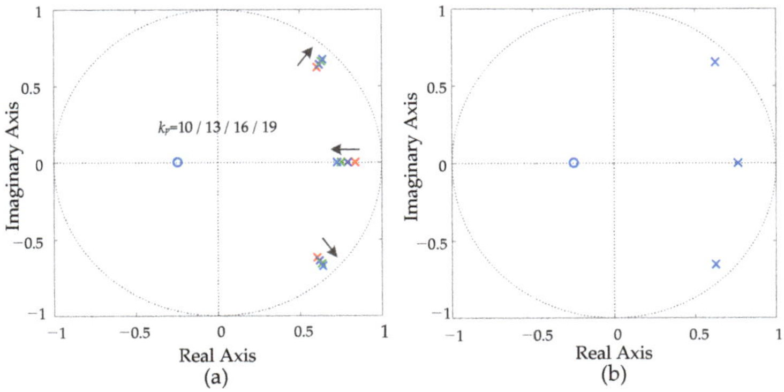

Figure 7. Pole plots of $P_0(z)$. (**a**) Pole plots of the $P_0(z)$ with different k_p. (**b**) Pole plots of the $P_0(z)$ with k_p and k_{ic} is 15 and 18, respectively.

5.2. Design the Compensator $S(z)$

The compensator can use a fourth-order Butterworth low-pass filter with a cut-off frequency of 850 Hz. The discrete expression of $S(z)$ is as follows:

$$S(z) = \frac{0.00276z^4 + 0.01104z^3 + 0.01656z^2 + 0.01104 + 0.00276}{z^4 - 2.612z^3 + 2.72z^2 - 1.308z + 0.2428} \tag{25}$$

5.3. Design the Phase-Lead Compensator z^m and RC Gain k_r

The bode diagram of $P_0(z)S(z)z^m$ is illustrated in Figure 8 with four typical values of m (m = 7, 8, 9, 10, respectively). It is clear that within the 1 kHz frequency range, the compensated phase-frequency curve is closest to the 0 dB line when m is equal to 9. Therefore, m = 9 can be selected as the lead beat.

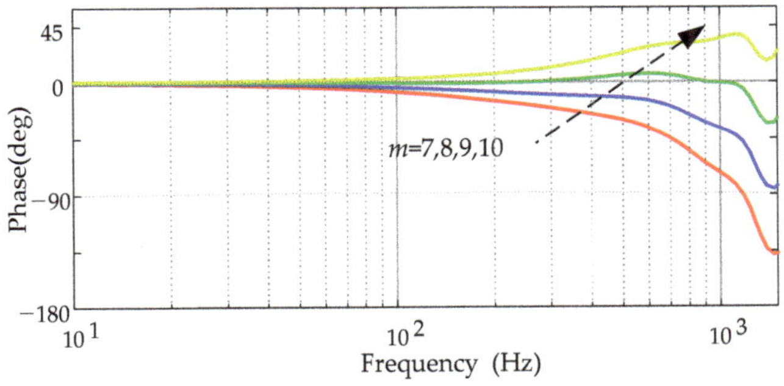

Figure 8. Bode of $P_0(z)S(z)z^m$.

It is necessary to find a balance between harmonic suppression capability and stability margins when selecting an appropriate value for k_r. According to Expression (24), the maximum k_r can be selected as

$$k_r = \min_{0 \le \omega \le \omega^N} \frac{2\cos(\theta_s(\omega) + \theta_{P_0}(\omega) + m\omega T_s)}{N_s(\omega)N_{P_0}(\omega)} = \frac{2\min\left[\cos(\theta_s(\omega) + \theta_{P_0}(\omega) + m\omega T_s)\right]}{\max\left[N_s(\omega)N_{P_0}(\omega)\right]} \tag{26}$$

Define the error convergence function for the system by

$$H(e^{j\omega T_s}) = 1 - k_r e^{j\omega m T_s} S(e^{j\omega T_s}) P_0(e^{j\omega T_s}) \tag{27}$$

According to the system stability Formula (22), the Nyquist graph of the $H(e^{j\omega T_s})$ should be within the unit circle to ensure system stability. Figure 9 shows the trajectories $H(e^{j\omega T_s})$, with k_r varying from 14 to 18. It can be seen that when $k_r = 18, m = 9$, and the curve is closest to the center of the unit circle.

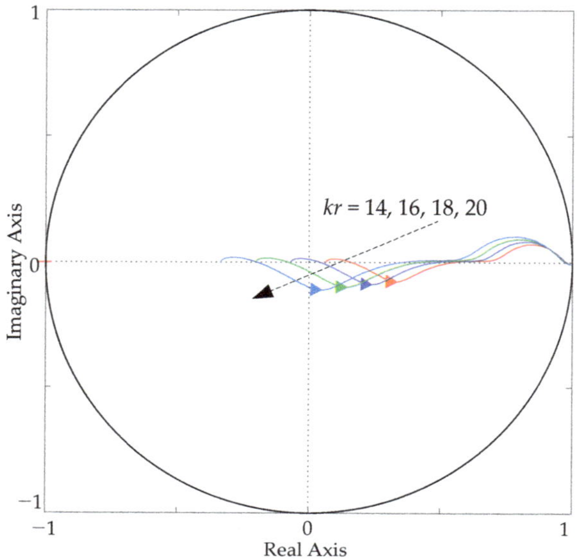

Figure 9. Nyquist graph of the $H(e^{j\omega T_s})$ with different values of k_r.

5.4. Internal Mode Filter Q(z)

The internal mode $Q(z)$ is an important component for keeping the system stable. $Q(z)$ can be selected as a constant less than 1 (such as 0.95, 0.98, etc.). However, to further improve system control performance and increase harmonic suppression capability, a zero-phase low-pass filter is usually employed. The expression for the zero-phase low-pass filter can be selected as

$$Q(z) = 0.25z^{-1} + 0.5 + 0.25z \tag{28}$$

5.5. Harmonic Suppression Analysis

When the sampling frequency is 10 kHz and the grid frequency varies from 49.5 Hz to 50.5 Hz, the delay N correspondingly fluctuates from 202 to 198. Furthermore, N may be a fraction; using an integer to approximate the fraction N will result in control errors. Consequently, the proposed FD-PIMR-RC based on an FFD filter is adopted to reduce control errors against grid frequency variations.

The frequency characteristics of the PIMR-RC (N = 200) and the FD-PIMR-RC (N = 198.4 and N = 201.6) are shown in Figure 10. The PIMR-RC exhibits excellent harmonic suppression performance at 50 Hz, 100 Hz, ..., etc. However, when the grid frequency varies within a certain range, the harmonic suppression performance deteriorates due to this deviation. i.e., a large tracking error and high THD of the grid current are produced. When the grid fundamental frequency fluctuates (50 ± 0.4 Hz) at the 7th harmonic frequency (347.2 Hz or 352.8 Hz) as an example, the open-loop gain of PIMR-RC decreases from 38 dB to 9 dB, as shown in Figure 11. Conversely, the open-loop gain of FD-PIMR-RC can maintain 38 dB at the same resonant frequency. Therefore, the newly proposed control scheme exhibits excellent harmonic suppression performance when the grid frequency fluctuates.

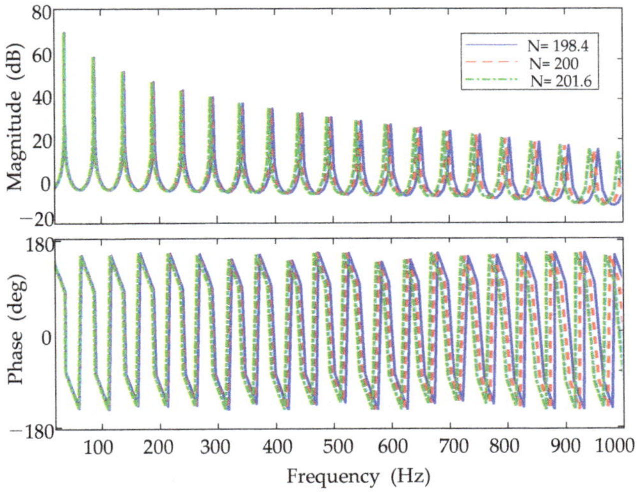

Figure 10. Bode diagrams of the PIMR-RC (N = 200) and FD-PIMR-RC (N = 198.4 and N = 201.6).

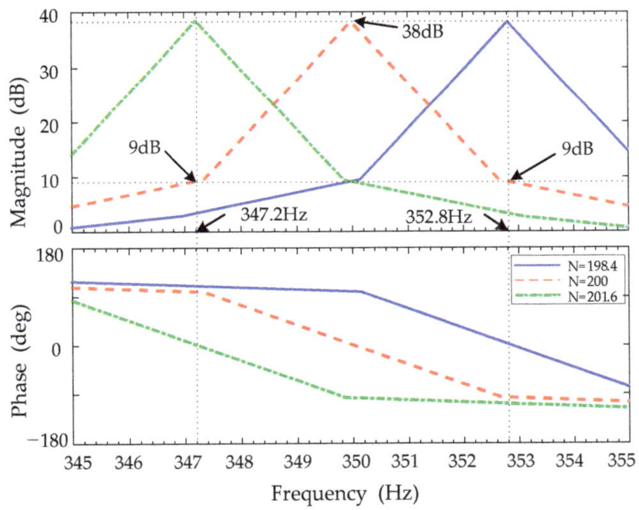

Figure 11. Frequency characteristics of PIMR-RC (N = 200) and FD-PIMR-RC (N = 198.4 and N = 201.6) at 7th harmonic frequency.

6. Simulation Analysis

As illustrated in Figures 4 and 5, a model of the 2.2 kW single-phase grid-tied inverter system adopting the PIMR-RC and FD-PIMR-RC schemes was built by MATLAB/Simulink. The rated value of the reference current was 10 A, and the other parameters were set based on the above analysis and the calculation parameter values presented in this paper.

6.1. Steady-State Response

- Grid frequency: the grid fundamental frequency (50 Hz)

When the grid frequency was the grid fundamental frequency, f_g = 50 Hz and N = 200. The grid voltage u_g and grid current i_g waveforms are shown in Figures 12 and 13. The THDs of PIMR-RC and FD-PIMR-RC were 0.5% and 0.74%, respectively. Both schemes demonstrated excellent performance in suppressing harmonics at the grid fundamental frequency.

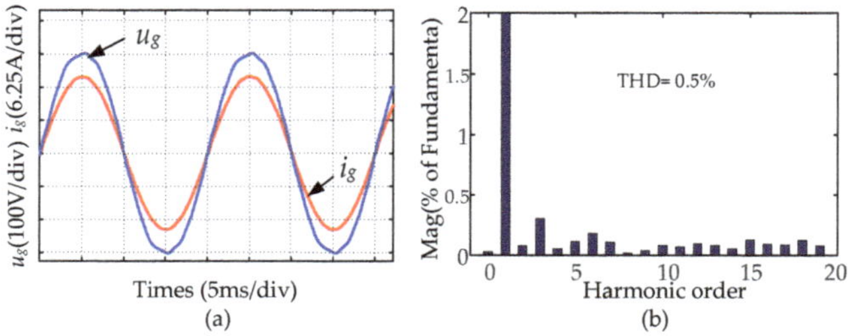

Figure 12. Steady-state response of the PIMR-RC system when f_g = 50 Hz. (**a**) Output voltage and current of grid. (**b**) Spectrum analysis of output current.

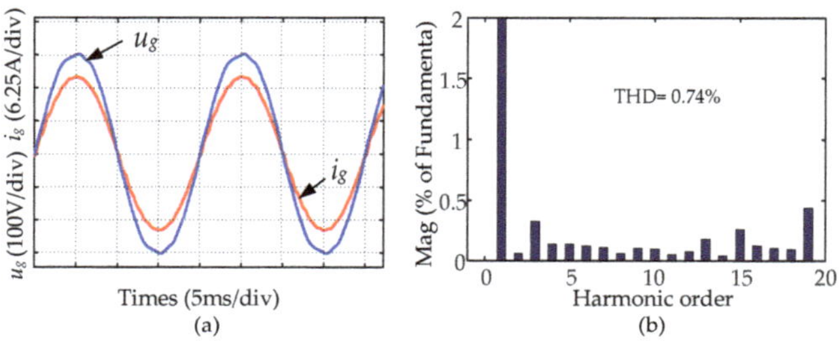

Figure 13. Steady-state response of the FD-PIMR-RC system when f_g = 50 Hz. (**a**) Output voltage and current of grid. (**b**) Spectrum analysis of output current.

- Grid frequency fluctuations (50 ± 0.4 Hz)

When the grid frequency fluctuates, the grid fundamental frequency is not 50 Hz but 50 ± 0.4 Hz. The conventional integer order delayed PIMR-RC system still takes the N as an integer 200, the control errors will be produced. As shown in Figure 14, if the conventional PIMR-RC scheme is adopted, when the grid frequency fluctuates from 50 Hz to 49.6 Hz, the THD of the i_g increases from 0.5% to 2.37%. However, the proposed FD-PIMR-RC can effectively cope with frequency fluctuations. Figure 15 shows that when the grid frequency fluctuates from 50 Hz to 49.6 Hz, N is 201.6, compared with the conventional PIMR-RC scheme, the THD of i_g decreases from 2.37% to 1.38% by the proposed FD-PIMR-RC.

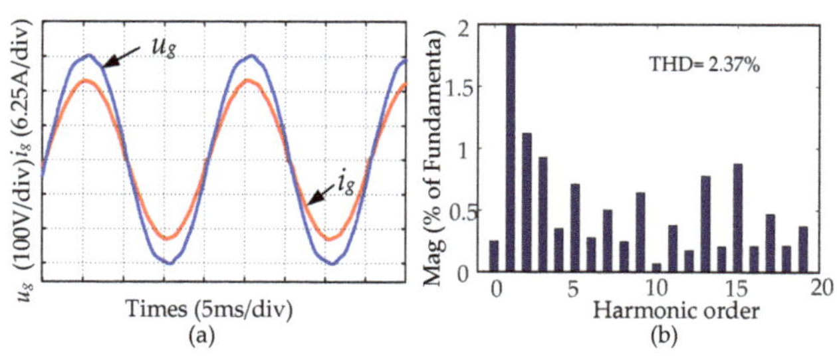

Figure 14. Steady-state response of the PIMR-RC system when f_g = 49.6 Hz. (**a**) Output voltage and current of grid. (**b**) Spectrum analysis of output current.

Similarly, Figure 16 shows that when the grid frequency fluctuates from 50 Hz to 50.4 Hz, the THD of i_g increases from 0.5% to 3.43% by the conventional PIMR-RC system. However, while the proposed FD-PIMR-RC is adopted in Figure 17, N is 198.4, and the THD of i_g decreases from 3.43% to 1.3%.

Therefore, the simulation results of the steady-state response illustrate that compared to the conventional PIMR-RC scheme, the proposed FD-PIMR-RC system is more robust to grid frequency fluctuations.

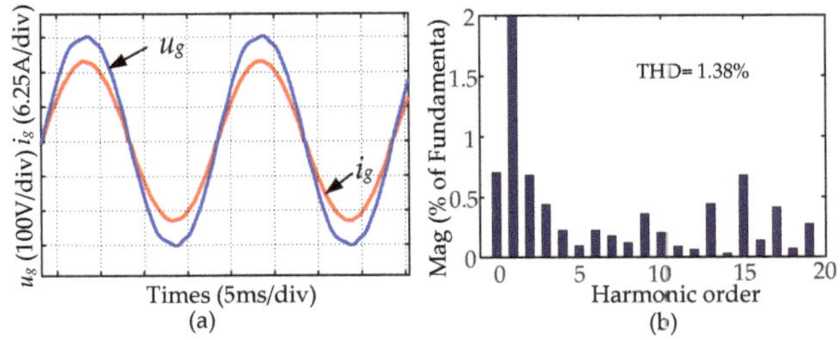

Figure 15. Steady-state response of the FD-PIMR-RC system when f_g = 49.6Hz. (**a**) Output voltage and current of grid. (**b**) Spectrum analysis of output current.

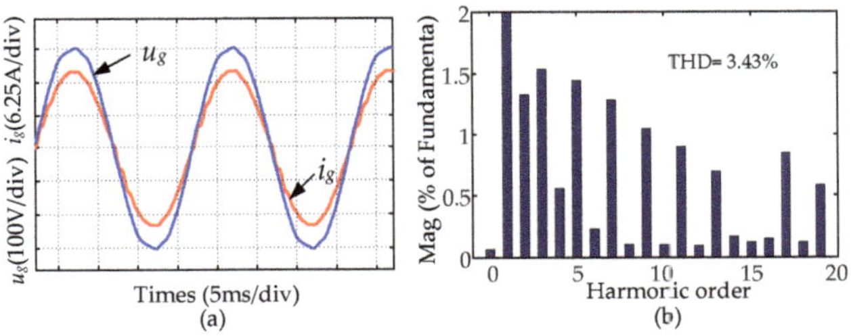

Figure 16. Steady-state response of the PIMR-RC system when f_g = 50.4 Hz. (**a**) Output voltage and current of grid. (**b**) Spectrum analysis of output current.

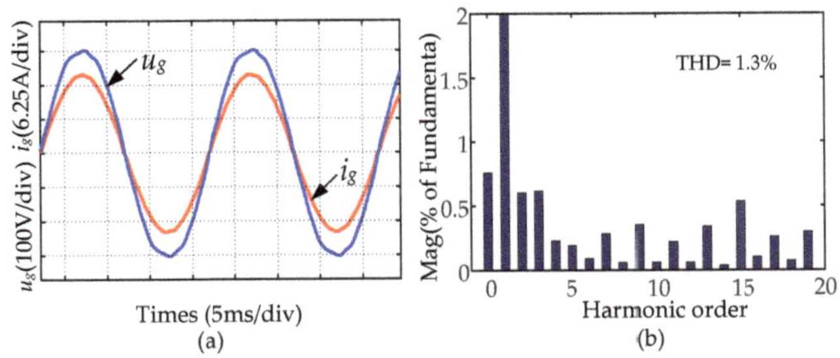

Figure 17. Steady-state response of the FD-PIMR-RC system when f_g = 50.4 Hz. (**a**) Output voltage and current of grid. (**b**) Spectrum analysis of output current.

In addition, to further verify the harmonic suppression performance when the grid frequency fluctuates (f_g = 50 ± 0.4 Hz), the PIMR-RC, PIMR-RC+FIR, and proposed FD-PIMR-RC schemes were adopted. The THD values of output current i_g are shown in Figure 18. Due to the absence of fractional delay, the greater the grid frequency fluctuation, the more severe the current distortion of PIMR-RC. The PIMR-RC+FIR scheme, based on a third-order FIR to achieve fractional delay, exhibited excellent harmonic suppression performance when the grid frequency fluctuated. However, for each new fractional delay, the FIR filter coefficients needed to be recalculated, which increased the computation load of the system.

The proposed FD-PIMR-RC scheme is based on an FFD filter to achieve fractional delay. Because the fractional delay item d is independent of the coefficients of the FFD filter, when the fractional delay item d changes, the coefficients of the FFD filter remain unchanged. That is, for each new fractional delay, a large number of calculations and storage for the FFD filter coefficients are avoided, which effectively reduces the computational load and memory consumption. Clearly, the proposed FD-PIMR-RC scheme not only retains the excellent harmonic suppression performance of the PIMR-RC+FIR but also avoids recalculating the filter coefficients for each new fractional delay. Therefore, the proposed FD-PIMR-RC scheme is more efficient and flexible.

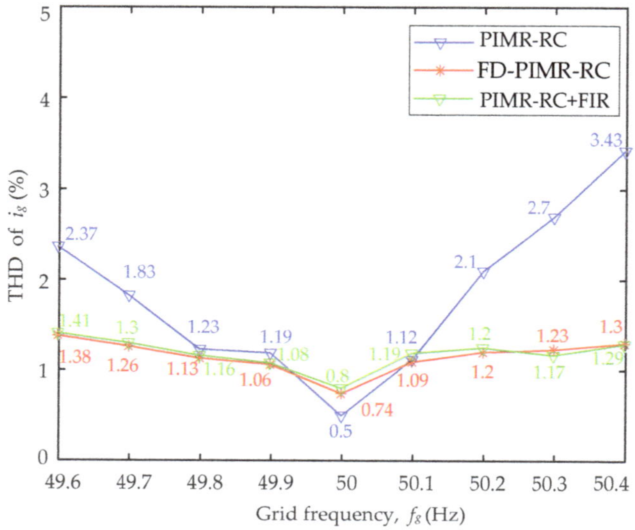

Figure 18. THD values of i_g with different control schemes when the grid frequency fluctuates.

6.2. Dynamic Response

Figures 19 and 20 show the grid current i_g and error current i_{error} transient waveforms of the PIMR-RC and FD-PIMR-RC systems when the reference current i_{ref} drops from 10 A to 5 A. As can be seen, the two schemes exhibited excellent error convergence rates when the reference current changed at different frequencies. Both of them could recover stability after about two fundamental periods. Nevertheless, compared to the PIMR-RC, the FD-PIMR-RC more effectively reduced the steady-state error.

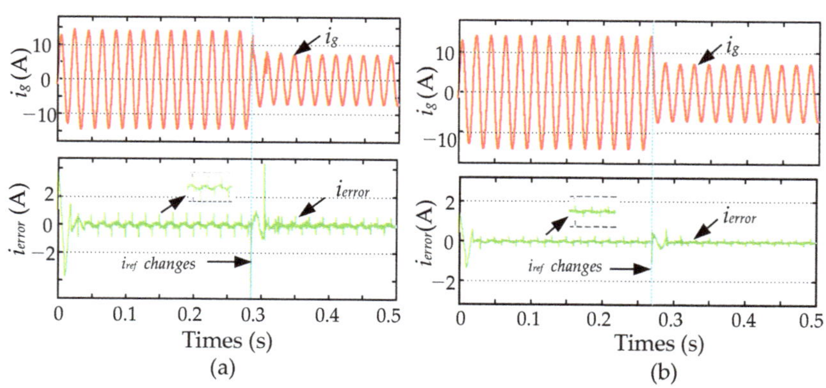

Figure 19. Transient waveforms of i_g and i_{error} when the reference current changes with f_g = 49.6 Hz. (**a**) PIMR-RC. (**b**) FD-PIMR-RC.

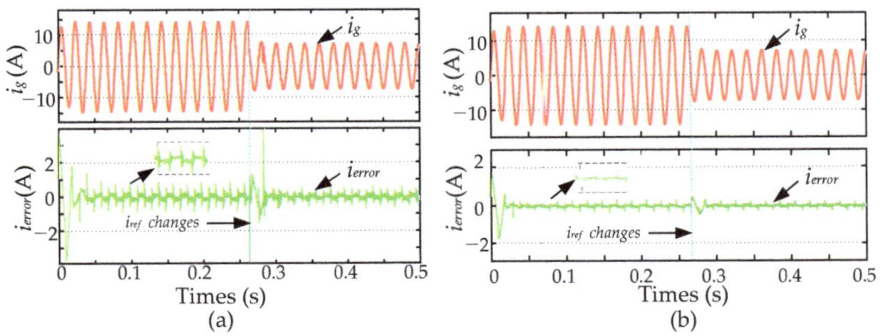

Figure 20. Transient waveforms of i_g and i_{error} when the reference current changes with f_g = 50.4 Hz. (**a**) PIMR-RC. (**b**) FD-PIMR-RC.

7. Conclusions

In this paper, a frequency-adaptive FD-PIMR-RC scheme is proposed for grid-tied inverters to resist grid frequency variations. The proposed scheme, based on an FFD filter, can flexibly and efficiently achieve fractional delay with reduced computational load and memory consumption. Furthermore, the parameter design and stability analysis of the FD-PIMR-RC system are presented. Then, the new controller based on the FFD filter is established, and the steady-state and dynamic responses of the conventional PIMR-RC and FD-PIMR-RC systems are compared and analyzed. The simulation results show that the proposed FD-PIMR-RC scheme is robust to grid frequency fluctuations, fast error convergence rates, and small tracking errors, thereby improving the quality of the grid current.

However, there are still some limitations of the proposed scheme that need to be studied in future works. Firstly, the research object is the ideal grid-tied inverter model, without considering the 1.5-period digital delay that exists in the practical grid-tied inverter. The impact of grid voltage and grid impedance fluctuations is also ignored. Secondly, the detailed mathematical analysis and simulation verification of the specific computational load and memory consumption required for the controller are not provided in this paper. Moreover, other methods used to reduce a system's computational load, such as multi-bandwidth RC, multi-rate RC, etc., need to be further researched.

Author Contributions: Conceptualization, F.L.; methodology, F.L. and H.-J.L.; software, F.L.; validation, F.L.; formal analysis, F.L. and H.-J.L.; writing—original draft preparation, F.L.; writing—review and editing, F.L., H.-J.L. and Q.Z.; visualization, F.L.; supervision, H.-J.L. and Q.Z.; project administration, H.-J.L.; funding acquisition, H.-J.L. All authors have read and agreed to the published version of the manuscript.

Funding: This work was supported in part by the National Research Foundation of Korea (NRF) grant, funded by the Korean government (MSIT) (NRF-RS-2022-00143178); in part by the Basic Science Research Program through the National Research Foundation of Korea (NRF), funded by the Ministry of Education (NRF-2020R1I1A3069569); in part by the Incubation Program for Young Master Supervisors under the grant of China (No. D202213); and in part by the Foundation for University Youth Key Teachers by Henan province of China (No. 2018GGJS-213).

Data Availability Statement: The authors confirm that the data supporting the findings of this study are available within the article.

Conflicts of Interest: The authors declare no conflict of interest.

Abbreviations

The following abbreviations are used in this manuscript:

RC	Repetitive control
PI	Proportional integral
PIMR-RC	Proportional–integral multi-resonant-type repetitive control
FFD	Farrow-structure fractional-order delay
FD-PIMR-RC	fractional delay proportional–integral multi-resonant-type repetitive control
DPGS	Distributed power generation systems
THD	Total harmonic distortion
PR	Proportional resonant
MPC	Model predictive control
QPR	Quasi-proportional resonant
PMR	Proportional multi-resonant
APF	Active power filters
CRC	Conventional repetitive control
FORC	Fractional-order repetitive control
FIR	Finite impulse response
IIR	Infinite impulse response
PLL	Phase-locked loop
PCC	Point common coupling

References

1. Blaabjerg, F.; Yang, Y.; Yang, D.; Wang, X. Distributed Power-Generation Systems and Protection. *Proc. IEEE* **2017**, *105*, 1311–1331. [CrossRef]
2. Yang, Y.; Zhou, K.; Wang, H.; Blaabjerg, F.; Wang, D.; Zhang, B. Frequency Adaptive Selective Harmonic Control for Grid-Connected Inverters. *IEEE Trans. Power Electron.* **2015**, *30*, 3912–3924. [CrossRef]
3. Dannehl, J.; Wessels, C.; Fuchs, F.W. Limitations of Voltage-Oriented PI Current Control of Grid-Connected PWM Rectifiers With LCL Filters. *IEEE Trans. Ind. Electron.* **2009**, *56*, 380–388. [CrossRef]
4. Quan, X. Improved Dynamic Response Design for Proportional Resonant Control Applied to Three-Phase Grid-Forming Inverter. *IEEE Trans. Ind. Electron.* **2021**, *68*, 9919–9930. [CrossRef]
5. Zhou, L.; Preindl, M. Decentralized Circulating Current Attenuation With Model Predictive Control for Distributed/Shunted Single/Three-Phase Grid-Tied Inverters. *IEEE Trans. Power Electron.* **2022**, *37*, 11534–11539. [CrossRef]
6. Zhou, K.; Wang, D.; Zhang, B.; Wang, Y. Plug-In Dual-Mode-Structure Repetitive Controller for CVCF PWM Inverters. *IEEE Trans. Ind. Electron.* **2009**, *56*, 784–791. [CrossRef]
7. Parvez, M.; Elias, M.F.M.; Rahim, N.A.; Blaabjerg, F.; Abbott, D.; Al-Sarawi, S.F. Comparative Study of Discrete PI and PR Controls for Single-Phase UPS Inverter. *IEEE Access* **2020**, *8*, 45584–45595. [CrossRef]
8. Sou, W.-K.; Chao, C.-W.; Gong, C.; Lam, C.-S.; Wong, C.-K. Analysis, Design, and Implementation of Multi-Quasi-Proportional-Resonant Controller for Thyristor-Controlled LC-Coupling Hybrid Active Power Filter (TCLC-HAPF). *IEEE Trans. Ind. Electron.* **2022**, *69*, 29–40. [CrossRef]
9. Lin, H.; Guo, X.; Chen, D.; Wu, S.; Chen, G. A Frequency Adaptive Repetitive Control for Active Power Filter With 380V/75A SiC-Inverter. *IEEE Trans. Ind. Appl.* **2022**, *58*, 5469–5479. [CrossRef]
10. Hornik, T.; Zhong, Q.-C. A Current-Control Strategy for Voltage-Source Inverters in Microgrids Based on H∞ and Repetitive Control. *IEEE Trans. Power Electron.* **2011**, *26*, 943–952. [CrossRef]
11. Lu, W.; Zhou, K.; Wang, D.; Cheng, M. A general parallel structure repetitive control scheme for multiphase DC–AC PWM converters. *IEEE Trans. Power Electron.* **2013**, *28*, 3980–3987. [CrossRef]
12. Zhao, Q.; Ye, Y. A PIMR-Type Repetitive Control for a Grid-Tied Inverter: Structure, Analysis, and Design. *IEEE Trans. Power Electron.* **2018**, *33*, 2730–2739. [CrossRef]
13. Yang, Y.; Zhou, K.; Blaabjerg, F. Enhancing the frequency adaptability of periodic current controllers with a fixed sampling rate for grid-connected power converters. *IEEE Trans. Power Electron.* **2016**, *31*, 7273–7285. [CrossRef]
14. di Benedetto, M.; Faro, A.; Bigarelli, L.; Lidozzi, A.; Solero, L. Constant Delay-Line Frequency Adaptive Repetitive-Resonant Control for Grid-Tied and Intentional Islanding Operations. *IEEE Trans. Ind. Appl.* **2023**, *59*, 1944–1955. [CrossRef]
15. Ye, J.; Liu, L.; Xu, J.; Shen, A. Frequency Adaptive Proportional-Repetitive Control for Grid-Connected Inverters. *IEEE Trans. Ind. Electron.* **2021**, *68*, 7965–7974. [CrossRef]
16. Liu, Z.; Zhang, B.; Zhou, K. Universal Fractional-Order Design of Linear Phase Lead Compensation Multirate Repetitive Control for PWM Inverters. *IEEE Trans. Ind. Electron.* **2017**, *64*, 7132–7140. [CrossRef]
17. Xiong, Y.; Ye, Y.; Cao, Y.; Wu, Y. Separate-Structure UDE-Based Current Resonant Control Strategy on LCL-Type Grid-Tied Inverters With Weighted Average Current Method for Improved Injected Current Quality and Robustness. *IEEE Trans. Power Electron.* **2020**, *35*, 13641–13651. [CrossRef]

18. Zhu, M.; Ye, Y.; Xiong, Y.; Zhao, Q. Parameter Robustness Improvement for Repetitive Control in Grid-Tied Inverters Using an IIR Filter. *IEEE Trans. Power Electron.* **2021**, *36*, 8454–8463. [CrossRef]
19. Johansson, H. Farrow-Structure-Based Reconfigurable Bandpass Linear-Phase FIR Filters for Integer Sampling Rate Conversion. *IEEE Trans. Circuits Syst. II Express Briefs* **2011**, *58*, 46–50. [CrossRef]
20. Tertinek, S.; Vogel, C. Reconstruction of Two-Periodic Nonuniformly Sampled Band-Limited Signals Using a Discrete-Time Differentiator and a Time-Varying Multiplier. *IEEE Trans. Circuits Syst. II Express Briefs* **2007**, *54*, 616–620. [CrossRef]
21. Bindima, T.; Elias, E. Low-Complexity 2-D Digital FIR Filters Using Polyphase Decomposition and Farrow Structure. *IEEE Trans. Circuits Syst. I Regul. Pap.* **2019**, *66*, 2298–2308. [CrossRef]
22. Wang, W.; Lu, W.; Zhou, K.; Fan, Q. Fractional-Order New Generation of nk ± m-Order Harmonic Repetitive Control for PWM Converters. *IEEE Access* **2020**, *8*, 180706–180721. [CrossRef]
23. Deng, T.-B. Hybrid Structures for Low-Complexity Variable Fractional-Delay FIR Filters. *IEEE Trans. Circuits Syst. I Regul. Pap.* **2010**, *57*, 897–910. [CrossRef]
24. Tseng, C.-C. Digital differentiator design using fractional delay filter and limit computation. *IEEE Trans. Circuits Syst. I Regul. Pap.* **2005**, *52*, 2248–2259. [CrossRef]
25. Canese, L.; Cardarilli, G.C.; Di Nunzio, L.; Fazzolari, R.; Giardino, D.; Re, M.; Spanò, S. Efficient Digital Implementation of a Multirate-Based Variable Fractional Delay Filter for Wideband Beamforming. *IEEE Trans. Circuits Syst. II Express Briefs* **2023**, *70*, 2231–2235. [CrossRef]
26. Roy, S.; Chandra, A. On the Design of Variable Filtered-OFDM Based LDACS for Future Generation Air-to-Ground Communication System. *IEEE Trans. Circuits Syst. II Express Briefs* **2022**, *69*, 644–648. [CrossRef]
27. Xiong, Y.; Ye, Y.; Zhu, M. UDE-Based Current Controller With Enhanced Grid Frequency Fluctuation Adaptability for LCL-Type Grid-Tied Inverters. *IEEE Trans. Power Electron.* **2023**, *38*, 1683–1691. [CrossRef]
28. Li, H.; Wu, W.; Huang, M.; Chung, H.S.-H.; Liserre, M.; Blaabjerg, F. Design of PWM-SMC Controller Using Linearized Model for Grid-Connected Inverter With LCL Filter. *IEEE Trans. Power Electron.* **2020**, *35*, 12773–12786. [CrossRef]
29. Miao, Z.; Yao, W.; Lu, Z. Single-Cycle-Lag Compensator-Based Active Damping for Digitally Controlled LCL/LLCL-Type Grid-Connected Inverters. *IEEE Trans. Ind. Electron.* **2020**, *67*, 1980–1990. [CrossRef]
30. Chen, S.; Zhao, Q.; Ye, Y.; Qu, B. Using IIR Filter in Fractional Order Phase Lead Compensation PIMR-RC for Grid-Tied Inverters. *IEEE Trans. Ind. Electron.* **2023**, *70*, 9399–9409. [CrossRef]
31. Zhu, M.; Ye, Y.; Xiong, Y.; Zhao, Q. Multibandwidth Repetitive Control Resisting Frequency Variation in Grid-Tied Inverters. *IEEE Trans. Emerg. Sel. Top. Power Electron.* **2022**, *10*, 446–454. [CrossRef]

electronics

MDPI

Article

Hardware-in-the-Loop Simulation of Flywheel Energy Storage Systems for Power Control in Wind Farms

Li Yang [1,2] and Qiaoni Zhao [1,*]

1 Intelligent Control Institute, Hunan Railway Professional Technology College, Zhuzhou 412001, China; justdoit850928@163.com

2 National Engineering Research Center for Robot Vision Perception and Control, College of Electrical and Information Engineering, Hunan University, Changsha 410082, China

* Correspondence: zqn1402@163.com; Tel.: +86-18932129397

Abstract: Flywheel energy storage systems (FESSs) are widely used for power regulation in wind farms as they can balance the wind farms' output power and improve the wind power grid connection rate. Due to the complex environment of wind farms, it is costly and time-consuming to repeatedly debug the system on-site. To save research costs and shorten research cycles, a hardware-in-the-loop (HIL) testing system was built to provide a convenient testing environment for the research of FESSs on wind farms. The focus of this study is the construction of mathematical models in the HIL testing system. Firstly, a mathematical model of the FESS main circuit is established using a hierarchical method. Secondly, the principle of the permanent magnet synchronous motor (PMSM) is analyzed, and a nonlinear dq mathematical model of the PMSM is established by referring to the relationship among d-axis inductance, q-axis inductance, and permanent magnet flux change with respect to the motor's current. Then, the power grid and wind farm test models are established. Finally, the established mathematical models are applied to the HIL testing system. The experimental results indicated that the HIL testing system can provide a convenient testing environment for the optimization of FESS control algorithms.

Keywords: power regulation in wind farm; flywheel energy storage system (FESS); hardware-in-the-loop (HIL) testing system; mathematical model

1. Introduction

Wind energy is a clean and pollution-free renewable energy source, and there is a substantial amount of wind energy. The use of wind energy for power generation is very environmentally friendly; therefore, it is increasingly receiving attention from the Chinese government. At the end of 2023, the installed capacity of wind power generation systems in China was 2919.65 million kilowatts, an increase of 13.9% compared to the end of the previous year [1]. The installed capacity of wind power connected to the power grid is 44.134 million kilowatts, an increase of 20.7% compared to the end of the previous year [1]. It can be observed that the installed capacity of wind power connected to the power grid is much lower than the installed capacity of power generation. Abandoning wind and limiting electricity is an important reason. A further reason is that the instability of wind energy results in fluctuations and intermittency in the output power of wind farms [2,3]. When the output power of wind farms cannot accommodate power grid connection conditions, the traditional approach is to abandon wind power and limit electricity, which results in substantial energy waste. Fortunately, the output power of wind farms can be balanced when using energy storage methods, and wind power abandonment and limitations to electricity can also be reduced. Chemical, electromagnetic, and mechanical energy storage methods are common [2]. Among them, FESSs are one of the most common mechanical energy storage methods, where electric energy is stored in a high-speed rotating flywheel rotor [3]. During the acceleration or deceleration of the flywheel rotor, electrical energy and mechanical energy are converted into one another [4,5]. The outstanding advantages of FESSs include long service life, low charging cost, and environmental friendliness. FESS is currently a popular method for regulating the power of wind farms. Due to the particularity and complexity of the operating environment, the on-site debugging of FESSs for wind farms is difficult [5,6]. Therefore, it is necessary to build a simulation testing environment in the laboratory so that the FESS algorithm can be repeatedly verified before on-site debugging. In this manner, the development cycle is shortened, and development costs are also saved.

In general, PCs are used for simulation testing systems. However, due to the difference between PC and actual hardware conditions, traditional simulation testing systems inevitably have some deviation from the actual system. HIL is a more advanced simulation test method that generally comprises a real-time simulator and a physical controller [5]. Among them, one part of the system runs in the real-time simulator via mathematical models, and the other parts still run within the physical controller [5]. Finally, the two parts are connected for debugging and experimentation. An HIL testing system reproduces the field conditions to the greatest extent in a laboratory environment; thus, the test results are closer to the field test results than traditional simulation test systems; this is more conducive to system optimization [5].

This article outlines the construction of an HIL testing system to assist in the optimization of the control algorithm of FESSs in wind farms. This study involves the realization of an FESS control algorithm, the construction of mathematical models, and the construction of an HIL testing system. Classic FESS control algorithms include field-oriented control, direct torque control, artificial intelligence control, etc. [6]. For example, the authors of [7] presented an energy management and control system designed with an integrated FESS for residential users. The power/current tracking of the machine-side and the network-side converter was the key to the proposed FESS control system. The authors of [8] presented a high-speed FESS for DC1500V transit transportation traction grids, and a three-level neutral point clamped control method for high-speed maglev permanent magnet motors based on square wave modulation–two-phase conduction was proposed. The authors of [9] presented an FESS controlled via a modeled predictive control algorithm for short-term high-frequency power smoothing in wind farms. The high-frequency components of a wind farm's output power were extracted via a wavelet packet decomposition algorithm, and it was optimized using mathematical interpolation. The authors of [10] presented a new adaptive droop controller for FESSs in order to maximize the contribution of the FESS during the first instances following a frequency deviation. The droop coefficient was altered in real time according to the grid's frequency. The authors of [11] presented 3D-SVPWM technology in order to operate a phase-loss PMSM in FESSs. A dual closed-loop control strategy relative to speed and current control was proposed.

The establishment of mathematical models is key to simulations, and the scholarly research provides references. For example, the authors of [12] presented nonlinear PMSM modeling and a traditional model reference adaptive system of FESSs. The authors of [13] presented linear frequency control modeling and a primary frequency modulation control model of FESS-assisted wind power. The authors of [5] presented models of dynamic flywheels and wind farms. The authors of [14] presented an HIL simulation of wind power, including mathematical models of wind farms, concentrated wind turbines, PMSMs, rectifiers, inverters, and batteries. The authors of [15] presented models of PMSMs, power compensation control strategies, and wind farms. Among these, the actual wind power output of a wind turbine was measured every minute, and 51-min variable data were used for the wind farm models. The authors of [16] presented six different numerical modeling methods for horizontal-axis wind turbine arrays in large wind farms. The authors of [17] presented effective windage loss modeling in FESSs. The proposed model was based on several analytical and semi-empirical windage loss solutions for cylindrical and planar surface interactions.

The HIL testing system can provide convenient simulation conditions for algorithm optimization. For example, the authors of [18] presented an HIL testing system to verify the correctness of the proposed fault-tolerant control strategy for offshore wind farms. The system consisted of a wind park computer simulator, interface circuits, and an onboard processor. The authors of [19] presented an HIL testing system for FESS-based microgrid controllers. Experiments were performed with real equipment, and the FESS was connected to a digital real-time simulator. Several frequency events were triggered, and the frequency profile was observed to compare the microgrid's performance with and without the developed frequency control. The authors of [20] presented an HIL testing system to validate the performance of a proposed adaptive droop control strategy. The controller was simulated in real time via an Opal-RT OP5700 real-time simulator, and it was implemented on a real commercial 60 kW high-speed FESS. The simulation time steps only took 24 μs. The authors of [21] presented an HIL platform in an NI myRIO 1900 to evaluate the performance of control algorithms in a small wind system that serves as a distributed generator. The authors of [22] presented a wind tunnel HIL testing system to make up for the shortages of traditional HILs with respect to flight control system evaluation, and it further reduced the risks of flight tests. The authors of [23,24] presented a testing system in which converters and controllers were simulated using MATLAB 2018, and the models were simulated in the HIL testing platform.

This article is organized into seven sections. Section 1 comprises the introduction; Section 2 describes the structure of wind farms with FESSs; Section 3 describes the structure of HIL testing systems; Section 4 describes the modeling of FESSs, including the main circuit, PMSMs, and FESS dynamic models; Section 5 describes the modeling of wind farms and power grids; and Section 6 describes the experiments on HIL testing systems. Finally, Section 7 provides conclusions.

2. Structure of Wind Farms with FESSs

A structural diagram of wind farms with FESSs is shown in Figure 1. An FESS is mainly composed of a flywheel, a PMSM, a main circuit, a controller, etc. [3]. The main circuit is mainly composed of a motor-side inverter, a network-side inverter, and a filter circuit. The operation of FESSs is divided into three states: energy charging, energy discharging, and energy retention [2–5]. The energy charging state comprises the following: When the output power of the wind turbine is higher than the power grid, the energy is output from the network side to the motor side. The motor is an electric motor, the flywheel rotates rapidly, and electrical energy is converted into kinetic energy [5]. The 50 Hz alternating current (AC) is tuned to the rated frequency via the network-side and motor-side inverters in order to control the movement of the flywheel. The energy retention state comprises the following: When the flywheel's speed approaches the limit, it no longer increases. The energy-discharging state is as follows: When the output power of the wind turbine is lower than the power grid, energy is outputted from the motor side to the network side. The motor is an engine, and kinetic energy is converted into electrical energy [5]. In this case, the output energy of the flywheel's rotor is fed back to the power grid via the motor-side and network-side inverters. During the discharging process, the flywheel's speed decreases continuously. When the preset minimum working speed is reached, the system stops releasing energy. The role of the controller is to run the FESS control algorithm. The total power of the power grid is the output power P_d of the wind farm plus the output power P_f of the FESS.

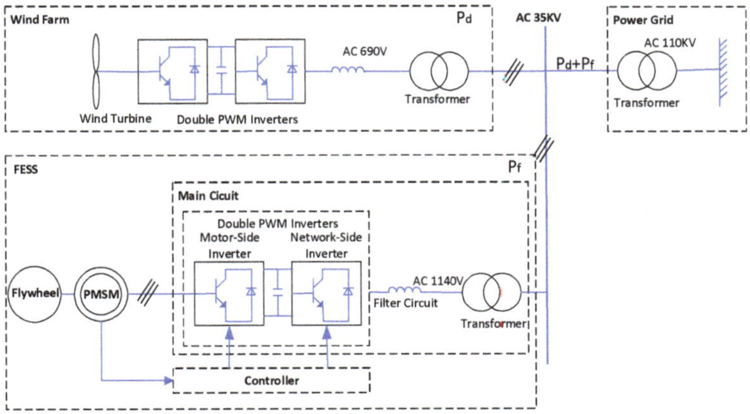

Figure 1. Structural diagram of wind farms with FESSs.

3. The Structure of HIL Testing Systems

According to the system structure described in Figure 1, an HIL testing system was built, as shown in Figure 2. The testing system consists of three parts: a PC, a real-time simulator, and a physical controller. Among these, the function of the PC is model building, testing, and downloading, in addition to the real-time monitoring of simulation processes. The function of the real-time simulator is to carry out real-time calculations of the models. This system uses a real-time simulator developed by the German dSPACE company. The dSPACE simulator is built based on CPU+FPGA architecture and high-speed IO technology [24]. Among them, FPGA is mainly used for the calculation of models with complex structures and high operation rates and to maintain calculation accuracy [24]. FPGA can greatly reduce the CPU's calculation burden and improve the working efficiency of real-time simulators [24]. As shown in Figure 1, the flywheel dynamics model, PMSM model, main circuit model, power grid model, and other models in FESSs are run in the real-time simulator. Among these, the flywheel dynamics, the power grid, and wind farm models are run using the CPU board; the PMSM and main circuit models are run using the FPGA board. This article focuses on the construction of these mathematical models.

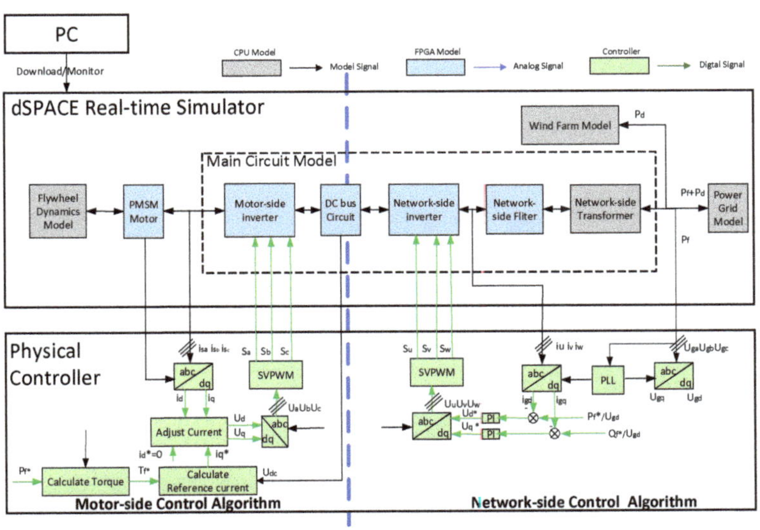

Figure 2. The structure of HIL testing systems.

The function of the physical controller is to run the FESS control algorithm. The physical controller is actually the FESS controller shown in Figure 1. The control algorithm is divided into motor-side control algorithms and network-side control algorithms. The function of the motor-side control algorithm is to collect the angular velocity of PMSMs and the three-phase input stator current of PMSMs given the power and other signals sent by the real-time simulator, and it outputs PWM signals to drive the motor-side converter in order to control the rotation of the PMSM [5]. The function of the network-side control algorithm is to collect the three-phase voltage of the power grid and the three-phase current of network-side converters sent by the real-time simulator, and it outputs PWM signals to drive the network-side converter according to the given active and reactive power; finally, it compensates the output power of the wind farm [5]. The specific definitions of the parameters in Figure 2 are defined in the Nomenclature Section.

4. Modeling of FESSs

4.1. Modeling of the Main Circuit

As shown in Figure 2, the main circuit model includes a motor-side inverter, a DC bus circuit, a network-side inverter, a network-side filter, and a network-side transformer. The most critical part is the motor-side and network-side inverters. The circuit of the network-side inverter is shown in Figure 3. The structure of the motor-side inverter is basically the same as that of the network-side inverter; thus, the establishment of the motor-side inverter model is not discussed in this paper. As shown in Figure 3, S_{C1}~S_{C6}, S_{B1}~S_{B6}, and S_{A1}~S_{A6} are IGBT (insulated gate bipolar transistor) devices; C_{dc1} and C_{dc2} are DC support capacitors; L_1, L_2, and L_3 are AC filter inductors; C_1, C_2, and C_3 are AC filter capacitors. The main circuit model is divided into IGBT, inductor, capacitor, resistor, and other models via hierarchical modeling. The main circuit model of FESS is formed via superposition and a combination of each part's model.

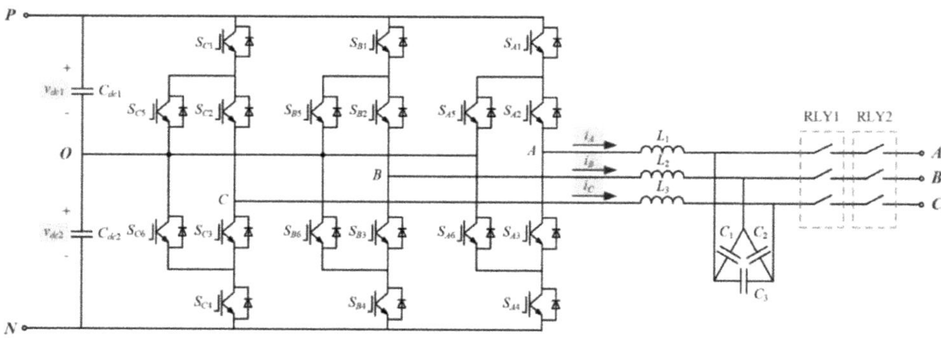

Figure 3. Circuit of the network-side inverter.

4.1.1. Modeling of ANPC Topology

As shown in Figure 3, the IGBTs in the circuit comprise three identical bridge arms A, B, and C, which form the active neutral point clamped (ANPC) topology. A single bridge arm consists of six IGBTs, as shown in Figure 4. The A-phase bridge arm is taken as an example to illustrate the modeling of ANPC topology. IGBTs in the inverter circuit are switching elements with operating states that vary at high frequencies. Smaller simulation steps are needed to reflect the dynamic characteristics, and the modeling method will affect overall accuracies and efficiencies in simulations. The common modeling methods of inverter circuits comprise detailed modeling, average modeling, adjoint network modeling, etc. Among these, the adjoint network modeling method is an innovative circuit analysis method, and it simplifies the transient analysis process of complex circuits by discretizing dynamic components and constructing adjoint networks. This method improves analysis efficiency and accuracy, and its models are easier to program.

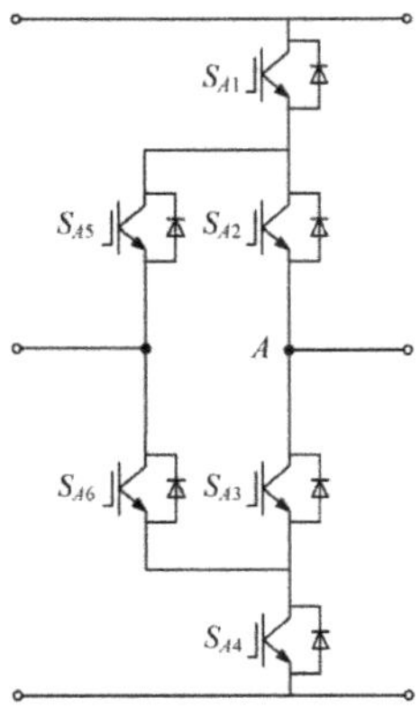

Figure 4. Circuit of A-phase bridge arm.

Now, adjoint network modeling methods are used to model the A-phase bridge arm:

- Step 1: Let the circuit time interval T (simulation step) be 1×10^{-8} s.
- Step 2: At each time interval, a single IGBT is replaced with its discrete model.

The circuit of a single IGBT is shown in Figure 5a, where $u_s(k+1)$ represents the voltage at both ends of the IGBT at $k+1$ simulation step. $i_s(k+1)$ represents the current flowing through the IGBT at the $k+1$ simulation step. s is a Boolean quantity that indicates the IGBT's switch status. IGBT is a fully controlled device with a switching state that is determined via an external control signal. Ideally, the relationship between s and $u_s(k+1)$, $i_s(k+1)$ is shown in Equation (1).

$$\begin{array}{l} open : s(k+1) = 1, u_s(k+1) = 0 \\ close : s(k+1) = 0, i_s(k+1) = 0 \end{array} \tag{1}$$

The IGBT is equivalent to a small inductor when it is switched on, and it is equivalent to a small capacitor when it is off. The mathematical discretization model of inductance is shown in Equation (2), and the mathematical discretization model of capacitance is shown in Equation (3). Here, T is the simulation step, G_l is the admittance of the equivalent inductance, G_c is the admittance of the equivalent capacitance, L_s is the equivalent inductance, and C_s is the equivalent capacitance. j_l is the inductively equivalent current source, and j_c is the capacitance equivalent current source.

$$i_l(k+1) \approx \frac{T}{L_s} u_l(k+1) + i_l(k) = G_l u_l(k+1) - j_l(k+1) \tag{2}$$

$$i_c(k+1) \approx \frac{C_s}{T} u_c(k+1) - \frac{C_s}{T} u_c(k) = G_c u_c(k+1) - j_c(k+1) \tag{3}$$

- Step 3: According to the discrete model, the adjoint network corresponding to the original circuit is constructed.

In Equations (2) and (3), let $G_s = G_l = G_c$ and $j_s(k+1) = j_l(k+1) = j_c(k+1)$. Figure 5a is equivalent to the parallel connection of G_s and j_s, as shown in Figure 5b. G_s denotes universal admittance, and j_s denotes the controlled current source. The value of j_s is shown in Equation (4). If G_s remains a constant and j_s changes according to Equation (4), the model of IGBT is obtained. In order to maintain the convergence of the system, G_s is generally selected between 0 and 10.

$$j_s(k+1) = \begin{cases} -i_s(k), (open : s(k+1) = 1) \\ G_s u_s(k), (close : s(k+1) = 0) \end{cases} \tag{4}$$

Figure 5. Circuit of a single IGBT: (a) original circuit; (b) equivalent circuit.

- Step 4: According to the single IGBT equivalent circuit, the equivalent circuit of the A-phase bridge arm is obtained, as shown in Figure 6.

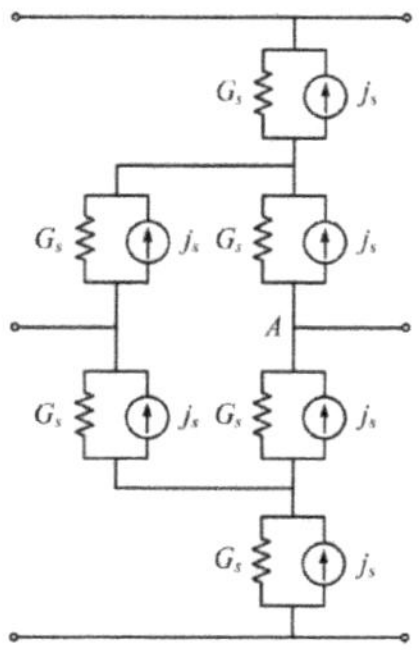

Figure 6. Equivalent circuit of A-phase bridge arm.

4.1.2. Modeling of Inductive Filter

As shown in Figure 3, the inductive filter consists of L_1, L_2, and L_3. The equivalent circuit of the inductive filter is shown in Figure 7. i_A denotes the current flowing through L_1, i_B denotes the current flowing through L_2, i_C denotes the current flowing through L_3, and R denotes the resistance value of the inductors in series. u_1, u_2, u_3, and u_4 denote the terminal voltage of adjacent inductors. The specific meaning is shown in Figure 7. When $L_1 = L_2 = L_3 = L$, the inductance current is shown in Equation (5).

$$\begin{cases} i_A = \int \left(\frac{2 \times (u_1 - u_3) + u_2 - u_4 - 3R \cdot i_A}{3L}\right)dt \\ i_B = \int \left(\frac{(u_2 + u_3) - u_2 - u_4 - 3R \cdot i_B}{3L}\right)dt \end{cases} \quad (5)$$

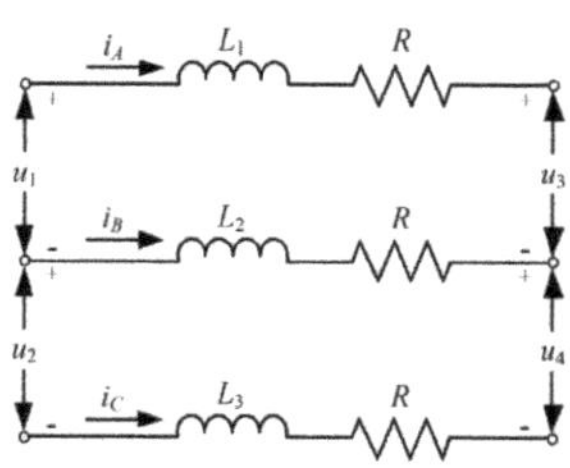

Figure 7. Equivalent circuit of the inductive filter.

4.1.3. Modeling of the Capacitive Filter

As shown in Figure 3, the capacitive filter consists of C_1, C_2, and C_3. The equivalent circuit of the capacitive filter is shown in Figure 8. i_a, i_b, i_c, $i_a{}'$, $i_b{}'$, and $i_c{}'$ denote the current at different places. R is the resistance value of the capacitor in parallel. u_1 is the voltage at both ends of capacitor C_1. u_2 is the voltage at both ends of capacitor C_2. u_3 is the voltage at both ends after capacitors C_1 and C_2 are connected in series. When $C_1 = C_2 = C_3 = C$, the capacitance voltage is as shown in Equation (6).

$$\begin{cases} u_1 = \int (\frac{i_a - i_a' - i_b + i_b'}{3C} - \frac{u_1}{RC})dt \\ u_2 = \int (\frac{2 \times (i_b - i_b') + i_a - i_a'}{3C} - \frac{u_2}{RC})dt \end{cases} \quad (6)$$

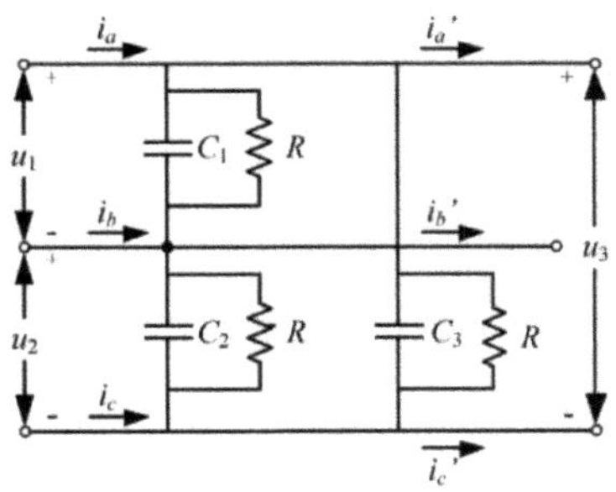

Figure 8. Equivalent circuit of the capacitive filter.

4.1.4. Modeling of the DC Support Capacitor

As shown in Figure 3, C_{dc1} and C_{dc2} are the DC support capacitors. An equivalent circuit of a DC support capacitor is shown in Figure 9 C is the DC support capacitor C_{dc1} or C_{dc2}. i_c is the current flowing through the support capacitor, and R_C is the resistance value of the capacitor in parallel. During the test, the change in the DC bus load is simulated by modifying the value of R_C. u_C is the voltage at both ends of C. The calculation formula of capacitor voltage is shown in Equation (7).

$$u_C = \frac{1}{C}\int (i_C - \frac{u_C}{R_C})dt \quad (7)$$

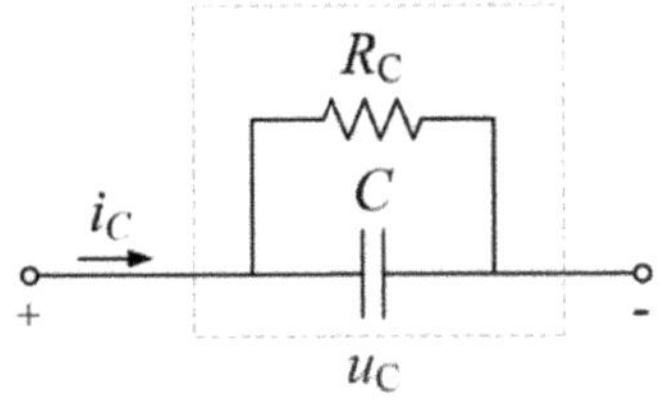

Figure 9. Equivalent circuit of the support capacitor.

4.2. Modeling of PMSMs

The PMSM model is shown in Figure 10, where U_a, U_b, and U_c denote the three-phase input phase voltages of the motor, and i_a, i_b, and i_c denote the three-phase output stator currents of the motor. The PMSM modeling process finds the relationship between the output stator current and the input phase voltage. The derivation of this relation is described below.

Figure 10. PMSM model.

After the coordinate transformation, U_a, U_b, and U_c can be converted to the stator voltage U_α of α-axes and the stator voltage U_β of β-axes in the static coordinate system, as shown in Equation (8):

$$\begin{pmatrix} U_\alpha \\ U_\beta \end{pmatrix} = \begin{bmatrix} \frac{2}{3} & -\frac{1}{3} & -\frac{1}{3} \\ 0 & \frac{\sqrt{3}}{3} & -\frac{\sqrt{3}}{3} \end{bmatrix} \begin{pmatrix} U_a \\ U_b \\ U_c \end{pmatrix} \tag{8}$$

U_α and U_β are converted to the stator voltage U_d of the d-axis and the stator voltage U_q of the q-axis in the rotating coordinate system, as shown in Equation (9):

$$\begin{pmatrix} U_d \\ U_q \end{pmatrix} = \begin{bmatrix} \sin\theta - \cos\theta \\ \cos\theta \ \sin\theta \end{bmatrix} \begin{pmatrix} U_\alpha \\ U_\beta \end{pmatrix} \tag{9}$$

By synchronously rotating Equation (9) in the coordinate system, Equations (10) and (11) are obtained:

$$\begin{bmatrix} U_d \\ U_q \end{bmatrix} = R\begin{bmatrix} i_d \\ i_q \end{bmatrix} + \frac{d}{dt}\begin{bmatrix} \varphi_d \\ \varphi_q \end{bmatrix} + w_e\begin{bmatrix} -\varphi_q \\ \varphi_d \end{bmatrix} \tag{10}$$

$$\begin{bmatrix} \varphi_d \\ \varphi_q \end{bmatrix} = \begin{bmatrix} L_d & 0 \\ 0 & L_q \end{bmatrix}\begin{bmatrix} i_d \\ i_q \end{bmatrix} + \begin{bmatrix} \varphi_f \\ 0 \end{bmatrix} \tag{11}$$

Here, R denotes the motor's resistance, i_d denotes the d-axis's stator current, i_q denotes the q-axis's stator current, L_d denotes the d-axis's inductance, L_q denotes the q-axis's inductance, φ_d denotes the d-axis's stator flux, φ_q denotes the q-axis's stator flux, and φ_f denotes the permanent magnet flux. The final current formula is as follows:

$$\begin{bmatrix} i_d \\ i_q \end{bmatrix} = \int \frac{1}{L_d L_q}\begin{bmatrix} -i_d L_q R + L_q U_d + \varphi_q L_q w_e \\ -i_q L_d R + L_d U_q - \varphi_d L_d w_e \end{bmatrix} \tag{12}$$

The i_d and i_q of the rotating coordinate system are converted to i_α and i_β:

$$\begin{pmatrix} i_\alpha \\ i_\beta \end{pmatrix} = \begin{bmatrix} \sin\theta & \cos\theta \\ -\cos\theta & \sin\theta \end{bmatrix} \cdot \begin{pmatrix} i_d \\ i_q \end{pmatrix} \tag{13}$$

After the coordinate transformation of i_α and i_β in the stationary coordinate system, three-phase stator currents i_a, i_b, and i_c are obtained.

$$\begin{pmatrix} i_a \\ i_b \\ i_c \end{pmatrix} = \begin{bmatrix} 1 & 0 & 1 \\ -\frac{1}{2} & \frac{\sqrt{3}}{2} & 1 \end{bmatrix} \cdot \begin{pmatrix} i_\alpha \\ i_\beta \\ i_0 \end{pmatrix} \tag{14}$$

In practice, the performance of PMSMs is greatly affected by the rotor's magnetic field space harmonics, magnetic circuit saturation, cross-saturation, and cross-coupling effects. In order to simulate the running state and performance of PMSMs more accurately, a general dq nonlinear model of PMSMs should be derived.

The steps of establishing the PMSM's nonlinear model are as follows:

- Step 1: The finite element method is used to obtain the relation curve of L_d, L_q, and φ_f, changing with i_d and i_q, respectively. It should be noted that the finite element calculation method is the work of other members of the team, and the specific process is not described in this paper.
- Step 2: Using the relationship curve obtained in step 1, the three parameters L_d, L_q, and φ_f are obtained by looking up the table according to the current values i_d and i_q.
- Step 3: The dq nonlinear model of the PMSM is obtained by substituting parameters L_d, L_q, and φ_f into Equation (12).

4.3. Modeling of Flywheel

The mechanical equation of the flywheel is shown in Equation (15). Here, ω_m denotes the rotor mechanical angular velocity of the PMSM, and B denotes the friction coefficient. T_e denotes the electromagnetic torque of the PMSM, and T_L denotes the drive torque of the PMSM.

$$\begin{array}{l} T_e - T_L = J\frac{d\omega_m}{dt} + B\omega_m \\ \frac{d\theta_m}{dt} = \omega_m \end{array} \tag{15}$$

The calculation formula for T_e and T_L is shown in Equation (16). Here, ψ_f is the permanent magnet flux. i_q is the d-axis current of the stator. P_m is the number of poles of the PMSM.

$$\begin{array}{l} T_L = -J\frac{d\omega_m}{dt} \\ T_e = P_m\psi_f i_q \end{array} \tag{16}$$

In the case of ignoring mechanical loss, the output active power P_f of the flywheel is equal to the electromagnetic power P_e, as shown in Equation (17). It can be observed that the output active power of the flywheel is directly related to i_q.

$$P_f = P_e = T_e\omega_m = P_m\psi_f i_q\omega_m \tag{17}$$

5. Modeling of Power Grids and Wind Farms

- Modeling of power grids

According to the HIL testing system's structure, as shown in Figure 2, a power grid test model was built. As shown in Figure 11, the power grid test model is equivalent to three AC voltage sources in parallel, and the output voltage is 35 kV before transformer processing. Here, U_{sA}, U_{sB}, and U_{sC} denote three-phase AC voltage sources, R_s denotes the equivalent resistance, and L_s denotes the equivalent inductance. i_{sA}, i_{sB}, and i_{sC} denote the phase current.

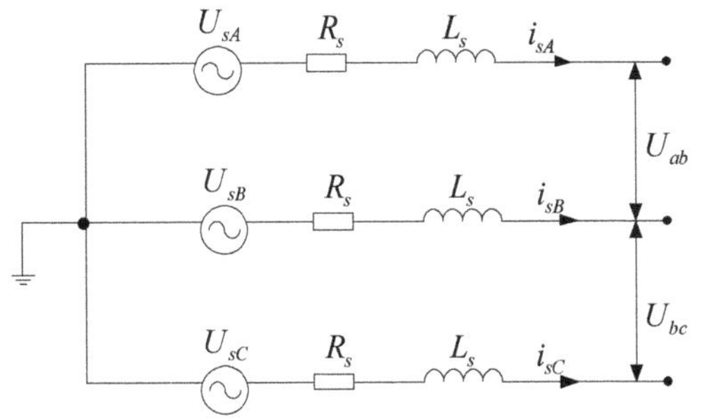

Figure 11. Topological structure of the power grid test model.

- Modeling of wind farms

As shown in Figure 12, the wind farm test model is equivalent to three controllable current sources in parallel. Here, i_{A_WF}, i_{B_WF}, and i_{C_WF} denote three-phase controllable current sources, and the output value of the current source is shown in Equation (18). V_{nom} denotes the rated voltage of the power grid, and its value is 35 kV. θ_g denotes the voltage angle of the power grid. As shown in Figure 2, θ_g can be obtained from the three-phase voltage of the power grid through a PLL (phase-locked loop) operation. P_{WF} is the current source power, simulating the active power output of the wind farm under wind speed fluctuations.

$$\begin{array}{ll} i_{A\_WF} & = \frac{P_{WF}}{V_{nom}}\sin(\theta_g) \\ i_{B\_WF} & = \frac{P_{WF}}{V_{nom}}\sin(\theta_g - 2*\pi/3) \\ i_{c\_WF} & = -i_{A\_WF} - i_{B\_WF} \end{array} \tag{18}$$

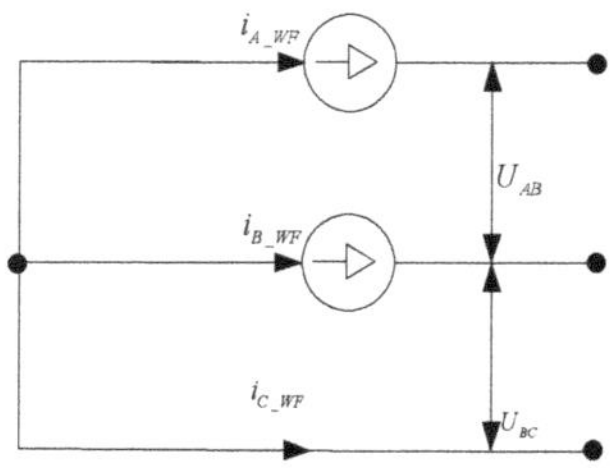

Figure 12. Topological structure of wind farm test models.

6. HIL Testing System Experiment

As mentioned above, the research focus of this article is the establishment of an FESS simulation model, rather than the development of FESS control algorithms. Therefore, in order to verify the correctness of the established FESS simulation models, the same FESS control algorithm developed by other members of the team was used in all following experiments. The aim of the experiments conducted with respect to the HIL testing system is to verify the following three points:

- The correctness of the motor-side models in FESSs;
- The correctness of the network-side models in FESSs;
- The effectiveness of the power control function of FESSs for wind farms.

The experimental conditions are as follows. The dSPACE real-time simulator consists of four CPU boards and six FPGA boards. Among them, the CPU boards comprise DS1007, and the calculation steps are 1×10^{-8} s. The FPGA boards are DC5203, and they contain 12 high-resolution 14-bit analog-to-digital and digital-to-analog converters and 32 high-speed digital IO channels. The processing chip of DC5203 is Xilinx's Kintex7 xc7k325t-1 fbg900, with a data rate of 6.6 Gb/s, which can be used for high-speed model calculations and simulations. Each board is connected through a peripheral high-speed bus, and the data transmission speed is 1.25 Gb/s. Because the electrical interface of the real-time simulator cannot be directly connected to the physical controller due to the voltage specification, a dSPACE signal conditioning unit is used to realize the signal's conversion. The signal conditioning unit is equipped with various plug-ins to complete the conversion between different signals. Three 4 MW/125 kWh flywheel units are adopted in the system, and the selected parameters are shown in Table 1.

Table 1. Parameters of the HIL testing system.

Number	Parameter	Value
1	Rated frequency of power grid	50 Hz
2	Rated voltage of power grid	35 kV
3	Rated power of PMSM	4 MW
4	Rated voltage of PMSM	3AC 850 V
5	Rated current of PMSM	2900 A
6	Rated frequency of PMSM	300 Hz
7	Rated speed of PMSM	5400 rpm
8	Rated voltage of FESS	1140 V
9	Electric storage capacity of FESS	125 kWh

6.1. Verification of Motor-Side Models

In order to verify the correctness of motor-side models, the FESS process from the energy-charging state to the energy retention state and then to the energy-discharging state was simulated. As shown in Figure 13, at 0.6 s, the FESS is in the energy charging state, and the PMSM output power is rated at 4 MW. At 0.8 s, the FESS is in the energy retention state, and the PMSM output power is 0 kW. At 1 s, the FESS is in the energy-discharging state, and the PMSM output power is −4 MW. The DC bus voltage waveform is shown in Figure 13. When the FESS is in the energy charging state, energy flows out of the power grid. When the FESS is in the energy-discharging state, energy flows to the power grid. When the FESS switches from the charging to the discharging state, the resulting change in the direction of energy causes the DC bus voltage to decrease. At 0.8–1 s, when the FESS turns into the discharging state, the system outputs energy to the power grid, and the DC bus voltage also increases. The current waveform of the PMSM is shown in Figure 13. It is obvious that when the FESS completes the state switch at 0.8 s, the motor current is also reversed. As mentioned above, it can be observed that the motor-side models can realize the three-state switching of the FESS, which is consistent with the expected function. This shows the correctness of the motor-side models.

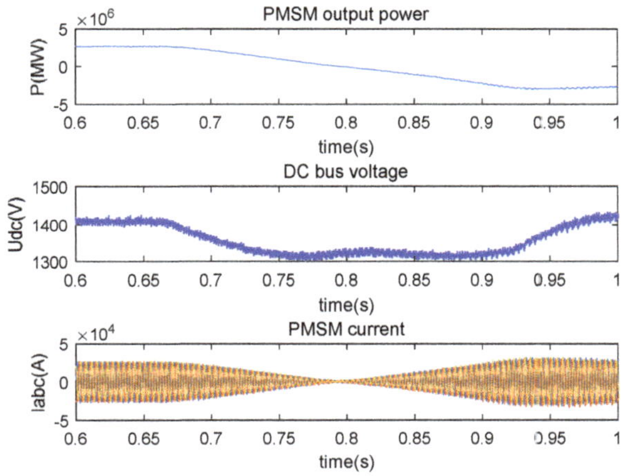

Figure 13. Motor-side waveform.

6.2. Verification of Network-Side Models

The output power of the FESS is adjusted when the voltage frequency of the power grid changes. The specific adjustment strategy is shown in Figure 14. In order to verify the power regulation effect of FESSs on wind farms, the step disturbance can be set in the voltage of the power grid to observe the power changes in the FESS.

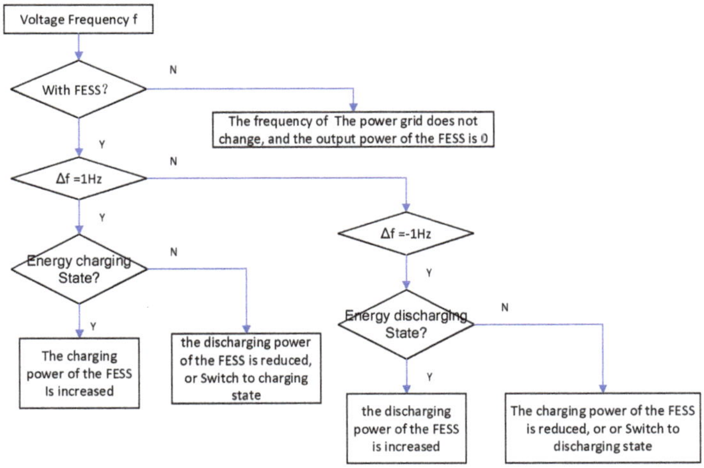

Figure 14. Power adjustment strategy from experience.

- Positive disturbance

When the voltage disturbance signal is positive, if the FESS is in the charging state, the charging power should be increased according to the power adjustment strategy from experience, as shown in Figure 14. It can be observed in Figure 15a that when the frequency of the power grid voltage is disturbed and increased to 51 Hz, the charging output power of the FESS increases. This is in line with the expected effect. When the voltage disturbance signal is positive, if the FESS is in the discharging state, the discharge power should be reduced according to the power adjustment strategy from experience, as shown in Figure 14. It can be observed in Figure 15b that when the frequency of the power grid voltage is disturbed and increased to 51 Hz, the discharge output power of the FESS decreases. This is in line with the expected effect.

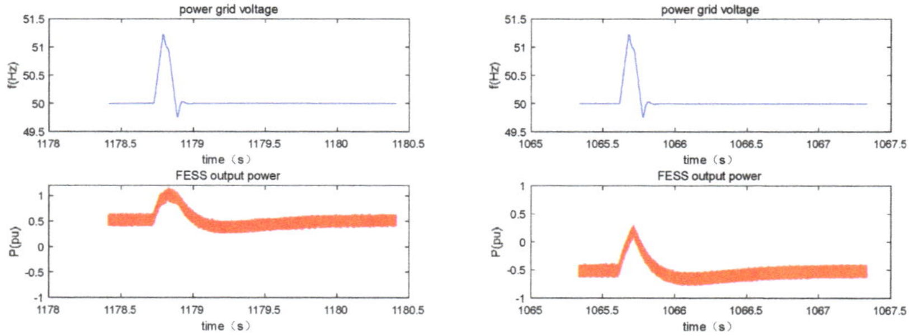

(a) (b)

Figure 15. FESS output power when there is positive disturbance: (**a**) under the charging state; (**b**) under the discharge state.

- Negative disturbance

When the voltage disturbance signal is negative, if the FESS is in the charging state, the charging power should be reduced according to the power adjustment strategy as shown in Figure 14. It can be observed in Figure 16a that when the frequency of the power grid voltage is disturbed and decreases to 49 Hz, the charging output power of the FESS decreases. This is in line with the expected effect. When the voltage disturbance signal is positive, if the FESS is in the discharge state, the discharge power should be increased according to the power adjustment strategy from experience, as shown in Figure 14. As shown in Figure 16b, the discharge output power of the FESS increases when the frequency of the power grid voltage is disturbed and decreases to 49 Hz. This is in line with the expected effect.

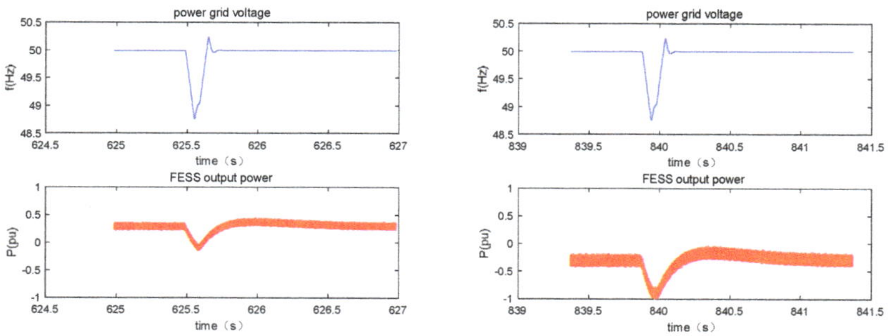

(a) (b)

Figure 16. FESS output power when there is negative disturbance: (**a**) under the charging state and (**b**) under the discharge state.

6.3. Verification of Power Control in Wind Farms

An array of three 4 MW/125 kW·h FESS units is used to match a 120 MW wind farm. As mentioned above, the output power of the wind power system can be simulated via the adjustment of the P_{WF} of the wind farm model in Equation (18). A smoothing simulation was carried out for the wind power output within 180 min, and the results are shown in Figure 17. After the smoothing operation, the fast fluctuations in wind power output were significantly reduced.

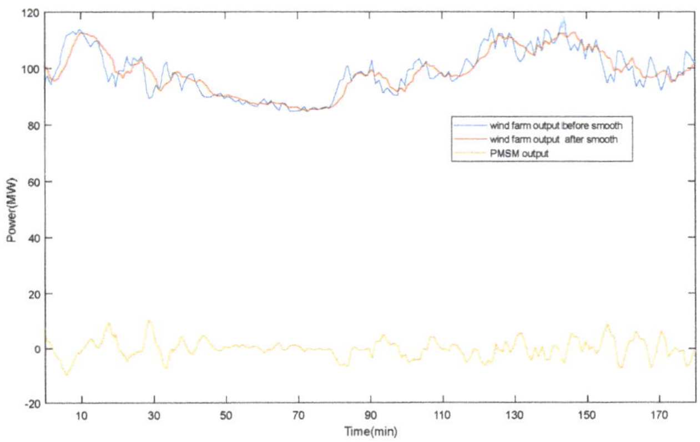

Figure 17. Comparison of wind farm output power.

7. Conclusions

In this paper, an HIL testing system was built for wind farm power control systems with FESSs. Firstly, the principle and structure of the system were analyzed. Then, the FESS, wind farm test, and power grid test models were analyzed and built. The modeling of the FESS main circuit and PMSM was the focus of this paper. Finally, the models were imported into the HIL testing system in order to carry out HIL experiments. The following conclusions can be drawn from the HIL experiments:

- The constructed FESS model can realize the switching of three states—energy charging, energy discharging, and energy retention—which is in agreement with the expected theory.
- The output power of FESS can be adjusted according to the voltage fluctuations of the power grid, and finally, the rapid fluctuations of the output power of the wind farm can be smoothed via the FESS, which is conducive to improving the wind farm's grid connection rate.
- The smoothing effect on wind farm output power indicates the reliability of the FESS algorithm in this experiment.
- In general, the HIL testing system built in this article can provide a more convenient environment for the optimization of the team's FESS control algorithm.

There is one limitation to the present study, which is the lack of a comparison of the current model with field data. Nevertheless, this study still shows the effectiveness of FESSs with respect to wind farm power control. In the future, we will further optimize the models based on field data in order to provide a more realistic simulation environment for the optimization of FESS control algorithms.

Author Contributions: Conceptualization, L.Y. and Q.Z.; methodology, L.Y.; software, L.Y.; validation, Q.Z.; formal analysis, L.Y.; investigation, Q.Z.; resources, L.Y.; data curation, L.Y.; writing—original draft preparation, L.Y.; writing—review and editing, Q.Z.; visualization, Q.Z.; supervision, Q.Z.; project administration, L.Y.; funding acquisition, L.Y. All authors have read and agreed to the published version of the manuscript.

Funding: This research was funded by the Hunan Provincial Natural Science Foundation of China, grant number 2022JJ60073.

Data Availability Statement: Further inquiries can be directed to the corresponding authors.

Conflicts of Interest: The authors declare no conflicts of interest. The funders had no role in the design of the study; in the collection, analyses, or interpretation of data; in the writing of the manuscript; or in the decision to publish the results.

Nomenclature

FESS	Flywheel energy storage system
HIL	Hardware-in-the-loop
PMSM	Permanent magnet synchronous motor
AC	Alternating current
IGBT	Insulate gate bipolar transistor
ANPC	Active neutral point clamped
PLL	Phase-locked loop
ω	Electrical angular velocity of the PMSM
θ	Angular position of the PMSM
θ_g	Voltage angle of the power grid
i_{sa}, i_{sb}, i_{sc}	Three-phase stator current of the PMSM
U_a, U_b, U_c	Three-phase phase voltage of the PMSM
i_u, i_v, i_w	Three-phase current of the network-side inverter
U_u, U_v, U_w	Three-phase voltage of the network-side inverter
U_{ga}, U_{gb}, U_{gc}	Three-phase voltage of the power grid
S_a, S_b, S_c	PWM control signals of the motor-side inverter
S_u, S_v, S_w	PWM control signals of the network-side inverter
i_{gd}	The d-axis current of the network-side inverter
i_{gq}	The q-axis current of the network-side inverter
$U_d{}^*$	The given d-axis voltage of the network-side inverter
$U_q{}^*$	The given q-axis voltage of the network-side inverter
i_d	The d-axis stator current of the PMSM
i_q	The q-axis stator current of the PMSM
$i_d{}^*$	The given d-axis stator current of the PMSM
$i_q{}^*$	The given q-axis stator current of the PMSM
$P_f{}^*$	The given active power of the PMSM
P_d	The given active power of the wind farm
$Q_f{}^*$	The given reactive power of the PMSM
$T_f{}^*$	The given torque of the PMSM
U_{dc}	The voltage of the DC bus

References

1. The Law of the People's Republic of China on Statistics Bulletin of the National Economy and Social Development in 2023. Available online: https://www.gov.cn/lianbo/bumen/202402/content_6934935.htm (accessed on 19 August 2024).
2. Olabi, A.G.; Wilberforce, T.; Abdelkareem, M.A.; Ramadan, M. Critical Review of Flywheel Energy Storage System. *Energies* **2021**, *14*, 2159. [CrossRef]
3. Xu, K.; Guo, Y.; Lei, G.; Zhu, J. A Review of Flywheel Energy Storage System Technologies. *Energies* **2023**, *16*, 6462. [CrossRef]
4. Ji, W.; Hong, F.; Zhao, Y.; Liang, L.; Du, H.; Hao, J.; Fang, F.; Liu, J. Applications of flywheel energy storage system on load frequency regulation combined with various power generations: A review. *Renew. Energy* **2024**, *223*, 119975. [CrossRef]
5. Li, Y.; Liu, S.C.; Yong, C.X. Modeling and Application of a Rectifier Transformer with Primary Winding in Series in A Metallurgical Rolling Mill System. *Mechatron. Syst. Control* **2023**, *51*, 182–192. [CrossRef]
6. Zhang, J.W.; Wang, Y.H.; Liu, G.C.; Tian, G.Z. A review of control strategies for flywheel energy storage system and a case study with matrix converter. *Energy Rep.* **2022**, *8*, 3948–3963. [CrossRef]
7. Floris, A.; Porru, M.; Damiano, A.; Serpi, A. Energy Management and Control System Design of an Integrated Flywheel Energy Storage System for Residential Users. *Appl. Sci.* **2021**, *11*, 4615. [CrossRef]
8. Wang, B.; Yu, X.; Yang, X.; Zhang, D. Control strategy for high speed flywheel energy storage system based on voltage threshold of DC1500 V transit transportation traction grid. *Energy Rep.* **2022**, *8*, 640–647. [CrossRef]

9. Lei, M.; Meng, K.; Feng, H.; Bai, J.; Jiang, H.; Zhang, Z. Flywheel energy storage controlled by model predictive control to achieve smooth short-term high-frequency wind power. *J. Energy Storage* **2023**, *63*, 106949. [CrossRef]
10. Wang, K.; Tian, L.; Li, N.; Yin, X.; Han, L.; Jiang, T. Adaptive comprehensive control strategy for primary frequency regulation of coal-fired units assisted by flywheel energy storage system. *J. Phys. Conf. Ser.* **2023**, *1*, 2592. [CrossRef]
11. Zheng, X.; Wu, Z.; Jia, Y.; Zhang, J.; Yang, P.; Zhang, Z. Fault-Tolerant Control Strategy for Phase Loss of the Flywheel Energy Storage Motor. *Electronics* **2023**, *12*, 3076. [CrossRef]
12. Song, G.; Wu, Z.; Zheng, X.; Zhang, J.; Yang, P.; Zhang, Z. Control Strategy of Flywheel Energy Storage System for Improved Model Reference Adaptive System Based on Tent-Sparrow Search Algorithm. *Electronics* **2024**, *13*, 2699. [CrossRef]
13. Qin, R.; Chen, J.; Li, Z.; Teng, W.; Liu, Y. Simulation of Secondary Frequency Modulation Process of Wind Power with Auxiliary of Flywheel Energy Storage. *Sustainability* **2023**, *15*, 11832. [CrossRef]
14. Song, H.H.; Tian, D. Study on Hardware-in-the-loop-simulation of Wind Power Based on dSPACE. In Proceedings of the 2010 Second International Conference on Computational Intelligence and Natural Computing (CINC), Wuhan, China, 13–14 September 2010.
15. Jia, Y.; Wu, Z.; Zhang, J.; Yang, P.; Zhang, Z. Control Strategy of Flywheel Energy Storage System Based on Primary Frequency Modulation of Wind Power. *Energies* **2022**, *15*, 1850. [CrossRef]
16. Young, L.; Zheng, X.; Gao, E. Numerical Modeling and Application of Horizontal-Axis Wind Turbine Arrays in Large Wind Farms. *Wind* **2023**, *3*, 459–484. [CrossRef]
17. Simani, S.; Farsoni, S.; Turhan, C. Hardware-In-The-Loop Assessment of a Fault Tolerant Fuzzy Control Scheme for an Offshore Wind Farm Simulator. *IFAC PapersOnLine* **2022**, *55*, 390–395. [CrossRef]
18. Kikusato, H.; Ustun, T.S.; Suzuki, M.; Sugahara, S.; Hashimoto, J.; Otani, K.; Ikeda, N.; Komuro, I.; Yokoi, H.; Takahashi, K. Flywheel energy storage system based microgrid controller design and PHIL testing. *Energy Rep.* **2022**, *8*, 470–475. [CrossRef]
19. Karrari, S.; De Carne, G.; Noe, M. Adaptive droop control strategy for Flywheel Energy Storage Systems: A Power Hardware-in-the-Loop validation. *Electr. Power Syst. Res.* **2022**, *212*, 108300. [CrossRef]
20. Martínez-Nolasco, J.; Sámano-Ortega, V.; Botello-Álvarez, J.; Padilla-Medina, J.; Martínez-Nolasco, C.; Bravo-Sánchez, M. Development of a Hardware-in-the-Loop Platform for the Validation of a Small-Scale Wind System Control Strategy. *Energies* **2023**, *16*, 7813. [CrossRef]
21. Venturini, S.; Cavallaro, S.P.; Vigliani, A. Windage loss characterisation for flywheel energy storage system: Model and experimental validation. *Energy* **2024**, *307*, 132641. [CrossRef]
22. Huang, M.; Wang, Z.; Guo, Z.; Zeng, Q.; Niu, Y. Wind Tunnel Hardware-in-the-loop Simulation Techniques for Flight Control System Evaluation. In Proceedings of the 2017 2nd International Conference on Control Automation and Artificial Intelligence (CAAI), Sanya, China, 25–26 June 2017.
23. Mathivanan, V.; Ramabadran, R.; Nagappan, B.; Devarajan, Y. Assessment of photovoltaic powered flywheel energy storage system for power generation and conditioning. *Sol. Energy* **2023**, *264*, 112045. [CrossRef]
24. Wu, X.; Meng, Y.; Hu, S.; Song, B. Hardware-In-the-Loop Simulation of Wind Turbine Based on 6 Degree-of-Freedom Load. In Proceedings of the 2017 International Electrical and Energy Conference (CIEEC), Beijing, China, 25–27 October 2017.

electronics

MDPI

Article

Research on Fast Frequency Response Control Strategy of Hydrogen Production Systems

Tao Shi [1,2,*], Zeyan Xu [1], Libo Gu [1] and Hangyu Zhou [1]

[1] College of Automation, Nanjing University of Posts and Telecommunications, Nanjing 210023, China; 1222056624@njupt.edu.cn (Z.X.); 1222056621@njupt.edu.cn (L.G.); 1022051527@njupt.edu.cn (H.Z.)

[2] Institute of Advance Technology for Carbon Neutrality, Nanjing University of Posts and Telecommunications, Nanjing 210023, China

* Correspondence: shitao@njupt.edu.cn

Abstract: With the large-scale integration of intermittent renewable energy generation presented by wind and photovoltaic power, the security and stability of power system operations have been challenged. Therefore, this article proposes a control strategy of a hydrogen production system based on renewable energy power generation to enable the fast frequency response of a grid. Firstly, based on the idea of virtual synchronous control, a fast frequency response control transformation strategy for the grid-connected interface of hydrogen production systems for renewable energy power generation is proposed to provide active power support when the grid frequency is disturbed. Secondly, based on the influence of VSG's inertia and damping coefficient on the dynamic characteristics of the system, a VSG adaptive control model based on particle swarm optimization is designed. Finally, based on the Matlab/Simulink platform, a grid-connected simulation model of hydrogen production systems for renewable energy power generation is established. The results show that the interface-transformed electrolytic hydrogen production device can actively respond to the frequency disturbances of the power system and participate in primary frequency control, providing active support for the frequency stability of the power system under high-percentage renewable energy generation integration. Moreover, the system with parameter optimization has better fast frequency response control characteristics.

Keywords: hydrogen production system; fast frequency response; virtual synchronous control; VSG adaptive control; particle swarm optimization

1. Introduction

With the large-scale grid connection of intermittent renewable energy generation presented by wind power and photovoltaics, system operation risks have emerged, along with the frequent absence of wind and sunlight [1,2]. On the one hand, the intermittent, random, and volatile nature of renewable energy power generation increases the flexible regulation demand of the power system. The system needs to configure additional regulatory resources to increase the absorption capacity of renewable energy. On the other hand, renewable energy power generation serves as a power electronic interface for power supply, which does not possess the inertia, damping, and primary frequency control support capabilities of traditional synchronous generators under conditions of disturbance.

Hydrogen energy, as a widely available, clean, carbon-free, flexible, and application-rich secondary energy source, is an essential carrier for supporting energy transformation and in building a modern energy system. We can realize the full utilization of renewable energy and alleviate the phenomenon of abandoned wind and abandoned light by closely integrating the hydrogen production strategy through electrolysis with the fluctuation of renewable energy output. Secondly, electrolytic hydrogen production serves as a flexible load regulation and storage solution, providing high-quality resources with flexible adjustments for new power systems. The hydrogen production device can serve as an energy storage carrier for renewable energy conversion. Combining electrolytic water with renewable energy generation can store excess electrical energy as chemical energy in hydrogen, thereby stabilizing renewable energy power fluctuations, increasing absorption capacity, and promoting clean energy.

There has been preliminary progress in the research of hydrogen generation from renewable energy power generation, both domestically and internationally. In ref. [3], to address the issues of wind turbine and photovoltaic power fluctuation, a grid-connected structure with an electrolyzer and supercapacitor assembled on a DC bus is constructed. A coordinated control model for the hybrid system is established, and the accuracy and effectiveness of the wind/PV/hydrogen production/supercapacitor grid-connected system model and control strategies are verified through Simulink simulations. In ref. [4], a refined model of electrolyzer participation in grid frequency control isestablished, and the fast frequency response capability of the electrolyzer is studied. In ref. [5], combining wind turbines with batteries and alkaline electrolyzer, a comprehensive control scheme for effectively managing the operation of the hydrogen generation system is proposed. In ref. [6], a virtual inertia and droop characteristic mechanism of the wind-storing hydrogen station is proposed to simulate a synchronous unit, followed by the establishment of a mathematical model for an alkaline electrolyzer, analyzing the matching between different energy storage devices and their synergistic effects in grid frequency control. In ref. [7], the authors compare the regulatory characteristics of alkaline electrolyzers and proton-exchange membrane (PEM) electrolyzers, and propose a dynamic power-allocation-control strategy for a hybrid system of both types of electrolyzers. In ref. [8], a comparative evaluation of fast active power-regulation-control strategies implemented on megawatt-scale controllable electrolyzers is presented, with the goal of achieving enhanced frequency support during large active power imbalances that lead to main low-frequency deviations. In ref. [9], the authors studied the coupling structure of wind turbines and electrolyzers, and compared the response characteristics of four different electrolyzer models under variable wind speed and grid fluctuation conditions. In ref. [10], a control strategy for a photovoltaic energy-storing grid-connected system based on a virtual synchronous generator has been proposed, which effectively suppresses the output power fluctuations of the photovoltaic inverter by coordinating the control of the energy storage unit. The above examples of the literature have demonstrated that controlling the electrolytic hydrogen system can stabilize the grid-connected power fluctuations of renewable energy sources. However, most of these studies consider using dispatch instructions to make the electrolytic hydrogen production device operate passively, without fully utilizing the inherent characteristics of the system. There is therefore a lack of active regulation capability for grid frequency.

In order to enable a renewable energy power-generation hydrogen production system to have a primary frequency-control active-response characteristic, the VSG control strategy is combined with the renewable energy power-generation hydrogen production system. The VSG control technology simulates the rotor motion characteristics of traditional synchronous generators and introduces a virtual inertia coefficient and damping coefficient into the control system through algorithms, thereby improving the system's frequency and voltage-regulation capabilities [11]. Compared with traditional synchronous generators, virtual synchronous generators are more flexible because they can change control parameters. During operation, changing control parameters to achieve a more stable performance and response speed greatly increases the flexibility of VSGs. The adaptive control method of rotor inertia and damping coefficient is based on this idea [12]. In ref. [13], a control strategy based on adaptive rotational inertia for VSGs has been proposed, which, compared with the traditional VSG method, has better stability, a faster response speed, and a smaller overshoot. In ref. [14], a VSG-coordination adaptive control strategy for rotational inertia and damping coefficient has been proposed, which effectively improves the dynamic regulation performance of the entire system. However, the above articles do not consider the impact of initial rotational inertia and damping coefficient on the system.

This paper uses the particle swarm optimization to optimize the initial damping coefficient and rotational inertia of the VSG, thus improving the dynamic response characteristics of the VSG. When combining the renewable energy power-generation hydrogen production system with it, the entire system can not only alleviate the phenomenon of abandoned wind and sunlight, but also actively provide frequency support for the grid.

2. Model of Electrolytic Hydrogen Production Systems

2.1. Principle of Electrolytic Hydrogen Production

Electrolytic hydrogen production is a mature technology for hydrogen production within the industry. It involves the supply of electrical energy to enable the electrochemical reaction of water molecules in the electrolytic cell, resulting in the decomposition of hydrogen and oxygen. The water decomposition reaction primarily consists of two half-reactions: hydrogen evolution at the cathode and oxygen precipitation at the anode. The reaction equations are presented as follows [15]:

Anode: $2H_2O - \rightarrow 4H^+ + O_2 + 4e$.

Cathode: $4H^+ + 4e \rightarrow 2H_2$.

Total: $2H_2O - \rightarrow 2H_2 + O_2$.

Hydrogen production systems that utilize renewable energy for power generation not only makes the electrolytic hydrogen production process more energy-efficient and environmentally friendly, achieving "zero carbon emissions", but also transform intermittent and unstable renewable energy into stable and easily stored chemical energy. This approach solves the imbalance between long-term power supply and consumption, alleviates the limitation of grid output capacity on renewable energy power-generation systems, and realizes more effective development and utilization of renewable energy.

2.2. Simulation Model

2.2.1. Model of PEM Electrolyzer

The electrolyzer is a device that uses electrical energy to split water into hydrogen and oxygen through electrochemical reactions. The structural principle is illustrated in Figure 1. The electrolyzer comprises multiple independent electrolyzer cells connected in series or parallel. This article initially calculates the working voltage required for each electrolyzer cell, as well as the hydrogen and oxygen production rates generated by the electrolyzer [16].

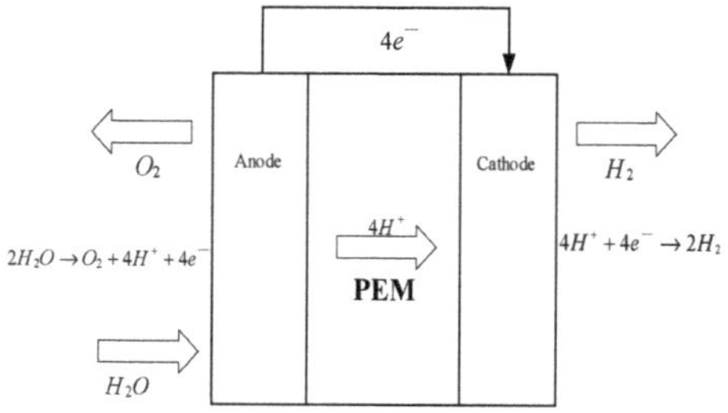

Figure 1. Electrolyzer basic structure.

Studying the I–V characteristics of an electrolyzer can significantly enhance the efficiency of gas production. From an electrical circuit perspective, an electrolyzer is equivalent to a voltage-sensitive, nonlinear DC load. After connecting to a power supply, the electrolyzer is linked to the DC bus via a DC/DC converter. The current and voltage characteristics of the electrolytic device are determined by the working temperature of the electrolyzer. Given the high nonlinearity of the current and voltage characteristics, curve fitting methods are commonly employed to simulate them. Considering its applicability in the field of electrical engineering, the voltage equation of a single electrolyzer is as follows [17]:

$$V_{cell} = E + V_{el,act} + V_{el,ohm} \tag{1}$$

where E is open circuit voltage, $V_{el,act}$ is activation polarization voltage, and $V_{el,ohm}$ is ohmic polarization voltage.

Open circuit voltage is defined as the Nernst Equation:

$$E = E_0 + \frac{RT_{el}}{2F}\left[\ln\left(\frac{P_{H_2}P_{0.5\cdot O_2}}{\alpha H_2O}\right)\right] \tag{2}$$

$$E_0 = 1.229 - 0.009(T - 298.15) \tag{3}$$

where E_0 is the standard cell potential, R is the universal gas constant, T_{el} is the cell temperature, α_{H_2O} is the water activity between anode and electrolyzer (for simplicity, it is assumed here to be 1), P_{H_2} is the cathode hydrogen pressure, $P_{0.5\cdot O_2}$ is the anode oxygen pressure, and F is the Faraday constant.

The activation polarization voltage is obtained as follows:

$$V_{el,act} = \frac{RT_{el}}{2\alpha F}\ln\left(\frac{i}{i_0}\right) \tag{4}$$

where α is the charge transfer coefficient, i is the current density, and i_0 is the exchange current density.

The ohmic polarization voltage is calculated as follows:

$$V_{el,ohm} = iR_{el,ohm} \tag{5}$$

where $R_{el,ohm}$ is the membrane resistance.

When combining multiple electrolysis cells in series and parallel to form an electrolysis tank stack, the total power consumption and terminal voltage of the entire system are calculated as follows:

$$\begin{cases} P_{el} = N_b \cdot V_{el} \cdot I_{cell} \\ V_{el} = N_c \cdot V_{cell} \end{cases} \tag{6}$$

where I_{cell} is the electrolytic current; N_b and N_c are the parallel and series numbers of the electrolytic unit, respectively.

The current and voltage characteristics of an electrolyzer are dependent on its operating temperature. According to Faraday's law, the generation rate of hydrogen in the electrolyzer is directly proportional to the rate of electron transfer at the electrode.

$$n_{H_2} = \frac{n_F \cdot I_{cell} \cdot N_c \cdot N_b}{2F} \tag{7}$$

where n_{H_2} is the output hydrogen flow; n_F is the Faraday efficiency.

The ratio between the actual and theoretical maximum hydrogen gas amounts in an electrolyzer is called the Faraday efficiency. At a working temperature of 80 °C for the electrolyzer, the Faraday efficiency is as follows:

$$n_F = 96.5 \cdot \exp(0.09/I_{cell} - 75.5/I_{cell}^2) \tag{8}$$

2.2.2. Model of Hydrogen Storage Tank

When temperature changes during hydrogen charging and discharging are not considered, the storage capacity can be expressed as follows:

$$n_{sto}(t_1) = n_{sto}(t_0) + \int_{t_0}^{t_1} v_{sto}(t)dt \tag{9}$$

where $n_{sto}(t_0)$ and $n_{sto}(t_1)$ are the hydrogen storage capacities of the hydrogen storage tank at moments t_0 and t_1, respectively; $v_{sto}(t)$ is the net intake rate of the hydrogen storage tank at time t.

The hydrogen storage tank pressure is as follows:

$$P_{sto} = \frac{RT_{sto}}{V_{sto}} n_{sto} \quad (10)$$

where P_{sto} is the hydrogen storage tank pressure, R is the gas constant, T_{sto} is the gas thermodynamic, and V_{sto} is the hydrogen storage tank capacity.

2.2.3. Electrolyzer Dynamic Characteristics

Figures 2 and 3 illustrate the dynamic response characteristics of the electrolyzer, demonstrating that hydrogen production varies with the changes in input current.

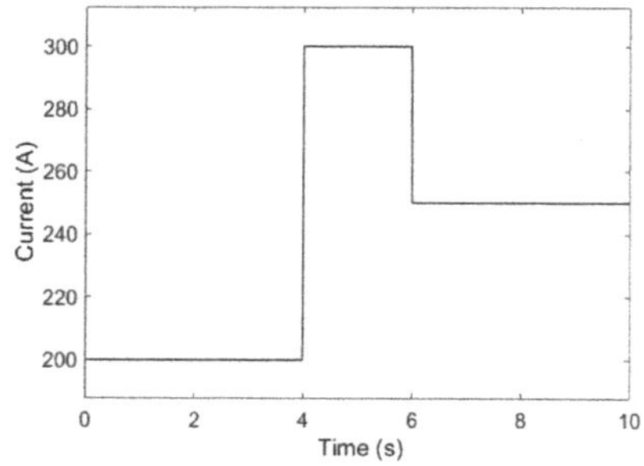

Figure 2. Input current of electrolyzer.

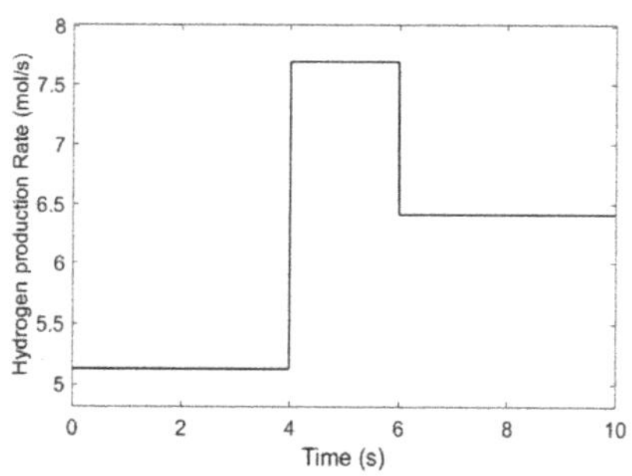

Figure 3. Hydrogen production rate of electrolyzer.

3. Objective of Fast Frequency Response Control

3.1. The Principle of Frequency Response

The principle of primary frequency control in power systems is illustrated in Figure 4. When the balance of active power within the grid becomes disturbed, leading to a deviation in the grid's frequency from its rated value, the regulation systems of the parallel operating generating units or other frequency adjustment devices rapidly adjust their active power output according to their individual speed-regulated static and dynamic characteristics to swiftly restore the grid's active power balance.

As illustrated in Figure 4, $P_G(f)$ and $P_D(f)$ represent the frequency characteristics of the generator set and the total load, respectively. The system initially operates at point A, where at $P_G = P_D$, a power balance is maintained, and the frequency stays stable at the rated frequency f_1. When the system load increases by ΔP_{D0}, the load's frequency characteristic shifts to $P'_D(f)$, resulting in a disruption of power balance. If left uncontrolled, the grid frequency will continue to decline. To mitigate the frequency drop, the generator set increases its output of active power ΔP_G to maintain the system's power balance, ultimately stabilizing the grid frequency at point f_2.

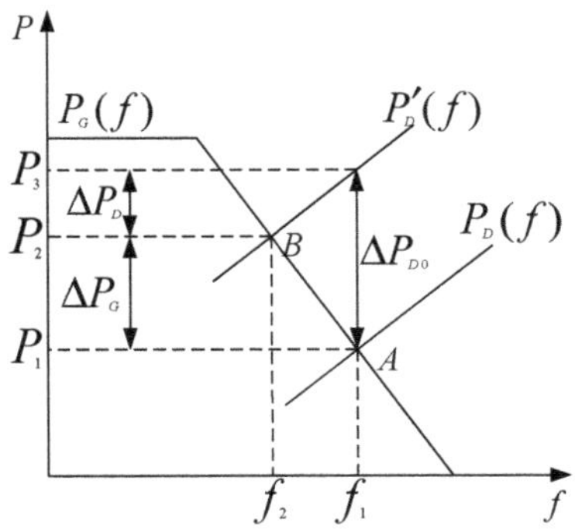

Figure 4. Primary frequency control.

3.2. Control Objectives of Primary Frequency Control

Figures 5 and 6 depict schematic diagrams illustrating the fast frequency response and active-frequency droop characteristics of wind farms and photovoltaic power stations, respectively. The renewable energy power generation system utilizes corresponding active control strategies and adds independent control devices to complete the active-frequency droop characteristic control, enabling it to have fast adjustment capabilities for participating in grid frequency regulation at the grid connection point. When the frequency at the grid connection point exceeds the frequency regulation dead zone, the frequency regulation control function is activated. Based on the active-frequency droop characteristic of the VSG, the active target value is calculated, and the control instructions are distributed to the DC/DC converter connected to the electrolytic hydrogen production device according to the active control strategy. Through the adjustment of the duty cycle, the voltage and current of the hydrogen production device are altered, thus changing the power consumption and efficiency of hydrogen production, and ensuring the grid frequency remains stable near the rated value.

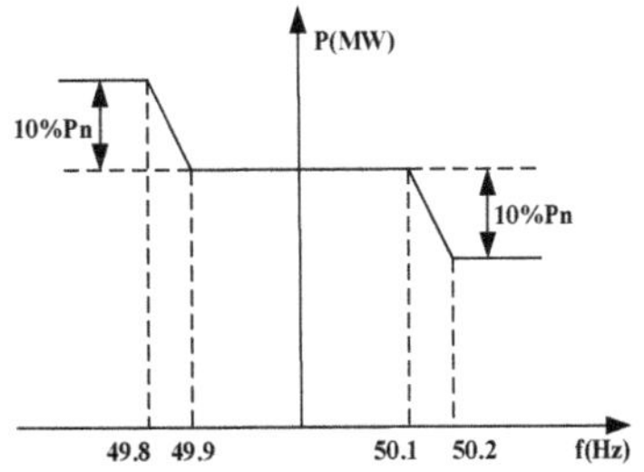

Figure 5. Fast frequency response and characteristics of wind farms.

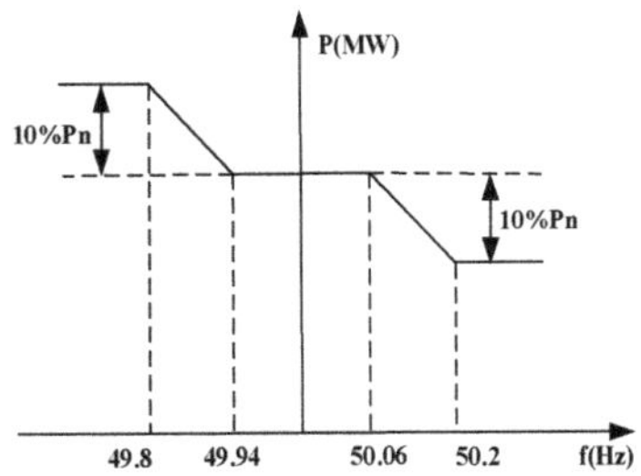

Figure 6. Fast frequency response and characteristics of photovoltaic power plants.

4. Strategy of Fast Frequency Response Control

4.1. System Structure

The typical structure of a renewable energy power-generation hydrogen production system is illustrated in Figure 7. This system utilizes a hybrid AC/DC structure, as described herein, with wind turbines and photovoltaic modules functioning as power generation units that operate in MPPT mode by default. These units are connected to the DC bus via AC/DC and DC/DC converters. The electrolysis hydrogen production module serves as the energy storage unit, connected to the DC bus via a DC/DC converter and linked to the load and the grid via the grid-side converter. The grid-side converter employs VSG control to impart inertia and fast frequency response characteristics to the entire system, facilitating friendly grid connection. The electrolysis hydrogen production module serves as power support for VSG's primary frequency response, aiming to maintain the stability of the DC bus voltage.

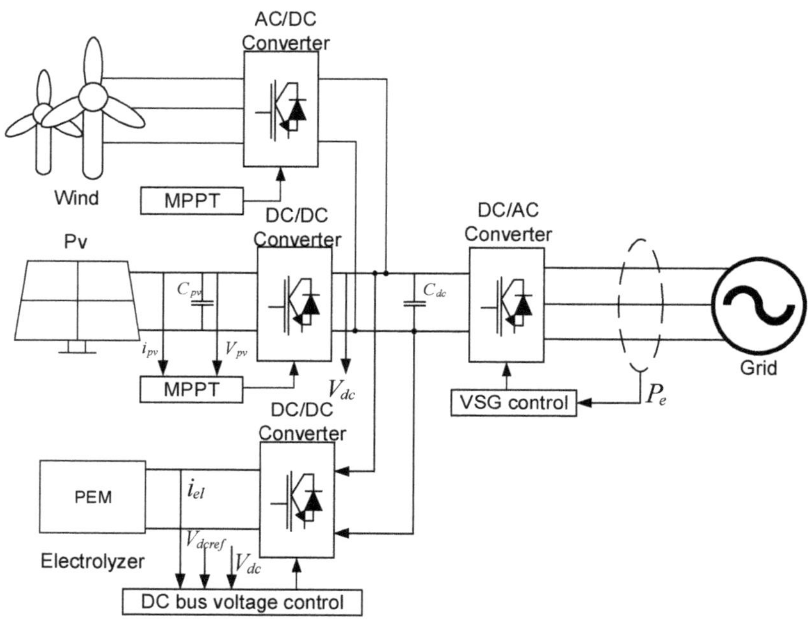

Figure 7. Renewable energy power-generation hydrogen production system.

4.2. DC/DC Converter Control Strategy on the Electrolyzer Side

The excellent regulatory capabilities of the hydrogen production equipment by electrolysis provides favorable conditions for its participation in grid frequency regulation and improvement of the phenomenon of abandoned wind and solar energy generation systems. The electrolysis hydrogen production equipment is connected to the DC bus through a DC/DC converter. On the one hand, a stable power supply is essential for the smooth progress of the electrolysis process; on the other hand, the electrolysis hydrogen production equipment ensures system power balance via the converter, thereby maintaining DC bus voltage stability. Figure 8 illustrates the control strategy of the electrolysis hydrogen production equipment. The renewable energy generation system and VSG provide inertia and damping support for the grid. When the balance of DC side power and inverter output power changes, the DC bus voltage is disturbed. The electrolysis hydrogen production equipment promptly adjusts its own operating power to maintain a stable DC bus voltage, close to the reference voltage.

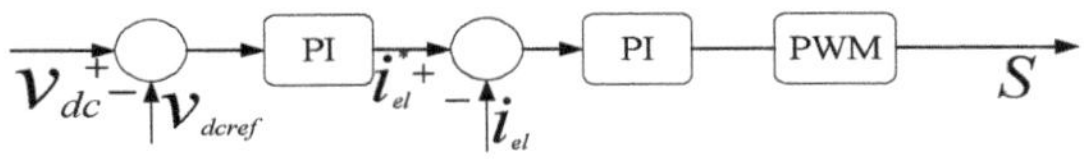

Figure 8. Control strategy for electrolyzer.

4.3. DC/AC Converter Control Strategy on the Power Grid Side

The grid-side converter adopts VSG control, which involves embedding the mathematical model of a synchronous generator into the control algorithm of the converter. This technology emulates a rotating motor by considering the static power electronic device to be a rotating machine. By emulating the primary frequency and voltage regulation of a synchronous generator, VSG equips the device with functions such as damping voltage and frequency fluctuations, automatic power allocation, and synchronous grid operation.

Figure 9 illustrates the structure and control block diagram of the virtual synchronous generator [18]. By embedding the synchronous generator equations in the converter control system, VSG can achieve power exchange between the DC power supply and the system according to the characteristics of the synchronous generator. When viewed from a system perspective and disregarding the high-frequency components caused by the switching action of power electronic devices, VSG is analogous to a synchronous generator. In a synchronous generator, both the mechanical shaft and winding are crucial in providing the necessary inertia and damping for the system's stable operation. For VSG, it is imperative to consider the establishment of virtual inertia and damping coefficients.

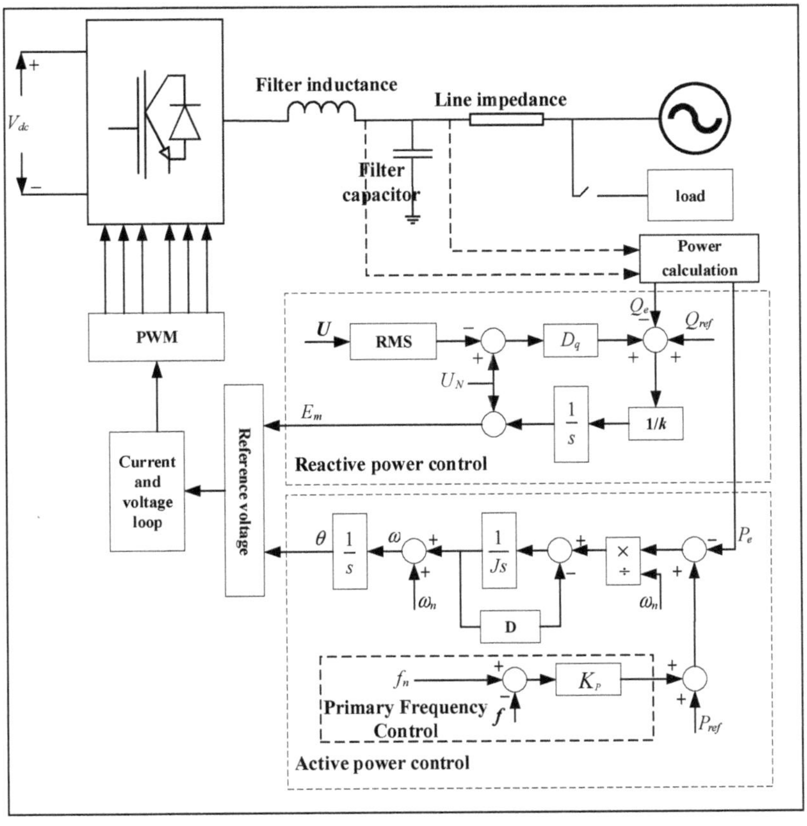

Figure 9. VSG structure and control block diagram.

The VSG rotor motion equation incorporates the second-order model of a traditional synchronous generator into the control mechanism, and Equation (11) demonstrates the inertial and damping characteristics of VSG [19].

$$\begin{cases} \omega = \frac{d\theta}{dt} \\ \frac{P_m - P_e}{\omega_n} - D(\omega - \omega_n) = J\frac{d\omega}{dt} \end{cases} \tag{11}$$

where J is the virtual inertia coefficient, D is the virtual damping coefficient, ω is the rotor angular velocity, ω_n is the rated rotor angular velocity, P_m is the VSG theoretical output active power, and P_e is the VSG actual output active power.

VSG emulates the *P-f* droop characteristic of synchronous generators, thereby enabling the grid-side converter to achieve primary frequency control performance in grid-connected mode. Upon a disturbance in the grid frequency, the actual grid frequency calculated is compared with the reference value of the grid frequency, and the output power needed by the converter side is obtained according to the *P-f* droop characteristic to regulate the frequency disturbance, thus fulfilling the primary frequency control function. The expression is as follows:

$$P_m = -K_p(f - f_n) + P_{ref} \tag{12}$$

where P_{ref} is the active power reference value, P_m is the VSG theoretical output active power, K_p is the primary frequency modulation coefficient, f is the grid actual frequency, and f_n is the grid reference frequency.

5. Fast Frequency Response Control Parameters on Regulation Performance

5.1. The Impact of System Control Parameters on Regulation Performance

Virtual rotational inertia and virtual damping coefficient constitute the core control parameters of the VSG control system. Therefore, based on the small-signal model of VSG active power, the transfer function is established as Equation (13) to study the influence of these two parameters on the system's dynamic characteristics [20].

$$G(s) = \frac{\frac{1}{J\omega_0}\frac{EU}{Z}}{s^2 + \left(\frac{D}{J} + \frac{K_\omega}{J\omega_0}\right)s + \frac{1}{J\omega_0}\frac{EU}{Z}} \tag{13}$$

The natural oscillation frequency ω_n and damping coefficient ξ of the corresponding second-order model can be obtained according to Equation (13).

$$\begin{cases} \omega_n = \sqrt{\frac{EU}{J\omega_0 Z}} \\ \xi = D\sqrt{\frac{\omega_0 Z}{4JEU}} + K_\omega\sqrt{\frac{Z}{4J\omega_0 EU}} \end{cases} \tag{14}$$

Assuming the system is in an underdamped status, and selecting an error band $\Delta = 2\%$, the overshoot and adjustment time of the system are, respectively, as follows:

$$\sigma\% = e^{-\pi\xi/\sqrt{1-\xi^2}} \times 100\% \tag{15}$$

$$t_s = \frac{3.5}{\xi\omega_n} = 3.5/\left(\frac{D}{2J} + \frac{K_\omega}{2J\omega_0}\right) \tag{16}$$

Figure 10 illustrates the output power and frequency waveforms of the converter under different virtual inertia coefficients; Figure 11 illustrates the output active power and frequency waveforms of the converter under different damping coefficients. Upon analyzing the simulation diagrams in conjunction with Equations (15) and (16), it is observed that when the virtual damping coefficient is constant, an increase in the virtual rotational inertia results in greater output overshoot and a more pronounced oscillation; with the virtual rotational inertia remaining constant, a decrease in the virtual damping coefficient leads to greater overshoot, extended regulation time, and a more pronounced system oscillation.

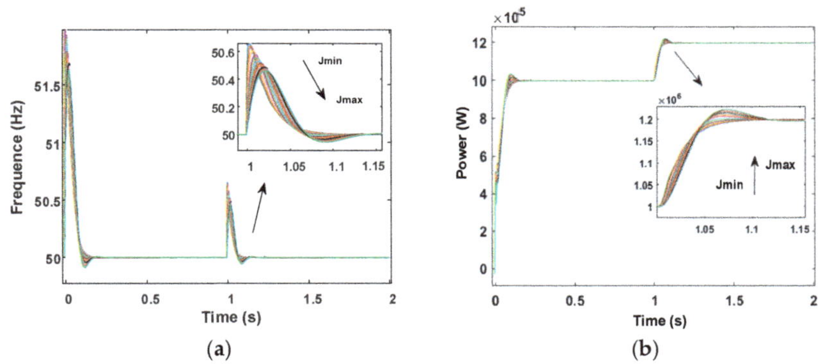

Figure 10. Output power and frequency waveforms of converter under different virtual inertia coefficients. (**a**) Frequency. (**b**) Power.

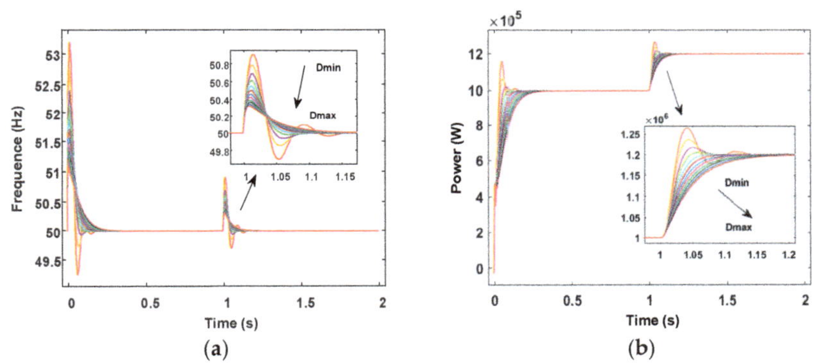

Figure 11. Output power and frequency waveform converter under different damping coefficients. (**a**) Frequency. (**b**) Power.

5.2. Establishment of Control Parameter Optimization Model

The core of VSG technology lies in the introduction of a simplified synchronous generator model into the converter control, enabling it to mimic synchronous generator characteristics. However, in comparison with actual synchronous generators, VSG is more flexible due to its ability to alter control parameters. Based on the analysis in the previous section, utilizing the particle swarm optimization algorithm to optimize the rotational inertia and damping coefficient of VSG facilitates more stable performance and faster response speed. The ITAE (Integral Time Absolute Error) index serves to establish the optimization objective function, as shown in Equation (17). The ITAE index may serve as a representative parameter for overshoot and regulation time, exhibiting minimal transient response oscillation.

$$ITAE = \int_0^t t \cdot |f - f_n| dt \tag{17}$$

where t is the simulation time, f is the grid actual frequency, and f_n is the grid reference frequency.

5.3. Establishment of Adaptive Control Parameter Model

When power fluctuations occur during the grid-connection process, the system frequency experiences decaying oscillations at the moment of fluctuations, and the overshoot and regulation time of the oscillations are key indicators for assessing the stability of the system frequency. The power angle characteristics and the angular velocity curve of synchronous generators are illustrated in Figures 12 and 13. The decaying oscillation process can be divided into four stages. The first stage system power increases from P_1 to P_2, resulting in a sudden increase in frequency. The angular velocity of the VSG is higher than the rated angular velocity of the grid, which is $\Delta\omega > 0$, and will continue to increase, $d\omega/dt > 0$. Therefore, it is necessary to increase the virtual inertia and damping coefficient to suppress $d\omega/dt$ and reduce the deviation of the angular frequency. In the second stage, system power increases from P_2 to P, during which $\Delta\omega > 0$, $d\omega/dt < 0$ need to reduce the virtual inertia to accelerate the adjustment of ω, stabilizing its curve to the rated angular velocity, while appropriately reducing the damping coefficient. The third and fourth stages are similar to the first and second stages [21].

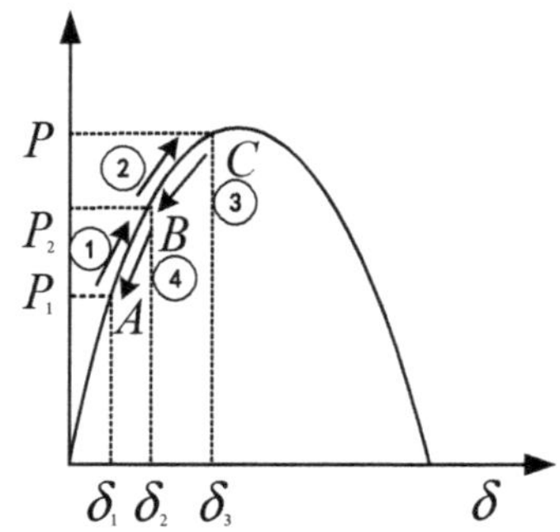

Figure 12. Synchronous generator power angle characteristic curve.

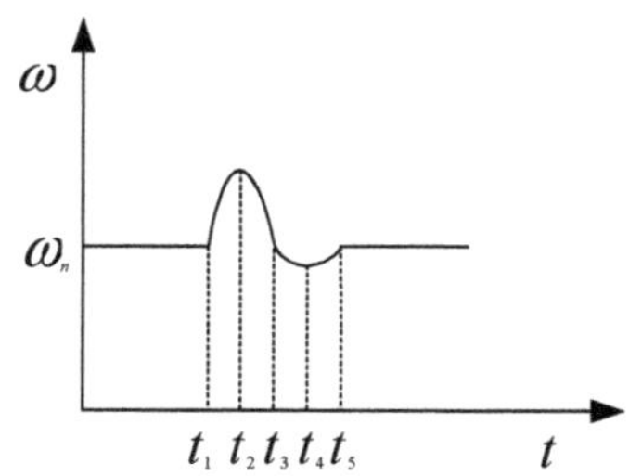

Figure 13. Synchronous generator angular velocity curve.

According to the above analysis, the selection principles of inertia and damping coefficients under different conditions are shown in Table 1. Combining the optimized initial inertia and damping coefficients based on the particle swarm optimization, an adaptive control strategy is designed as Equations (18) and (19).

$$J = \begin{cases} J_0 & \left|\frac{d\omega}{dt}\right| \le M_J \\ \max(J_0 + A\left|\frac{d\omega}{dt}\right| + B|\Delta\omega|, J_0 + J_1) & \Delta\omega \cdot \frac{d\omega}{dt} \ge 0 \cap \left|\frac{d\omega}{dt}\right| > M_J \\ \min(J_0 + C\left|\frac{d\omega}{dt}\right| + D|\Delta\omega|, J_0 - J_2) & \Delta\omega \cdot \frac{d\omega}{dt} < 0 \cap \left|\frac{d\omega}{dt}\right| > M_J \end{cases} \tag{18}$$

$$D = \begin{cases} D_0 & \left|\frac{d\omega}{dt}\right| \le M_J \\ \max(D_0 + E\left|\frac{d\omega}{dt}\right| + F|\Delta\omega|, D_0 + D_1) & \Delta\omega \cdot \frac{d\omega}{dt} \ge 0 \cap \left|\frac{d\omega}{dt}\right| > M_J \\ \min(D_0 + G\left|\frac{d\omega}{dt}\right| + H|\Delta\omega|, D_0 - D_2) & \Delta\omega \cdot \frac{d\omega}{dt} < 0 \cap \left|\frac{d\omega}{dt}\right| > M_J \end{cases} \tag{19}$$

where J_0 and D_0 represent the initial inertia coefficient and damping coefficient of the VSG after particle swarm optimization, M_J is the $d\omega/dt$ transformation threshold, *A-D* and *E-H* are the adjustment coefficients for the moment of inertia and damping coefficient, and J_1, J_2, D_1, and D_2 are the support coefficients for the moment of inertia and damping coefficient.

Table 1. Principles for selecting J and D in different situations.

Phase	Variation	Change Rate	J	D
A	$\Delta\omega > 0$	$\frac{d\omega}{dt} > 0$	Increase	Increase
B	$\Delta\omega > 0$	$\frac{d\omega}{dt} < 0$	Reduce	Reduce
C	$\Delta\omega < 0$	$\frac{d\omega}{dt} < 0$	Increase	Increase
D	$\Delta\omega < 0$	$\frac{d\omega}{dt} > 0$	Reduce	Reduce

5.4. The Process of Parameter Optimization

Through the introduction of an optimization model for control parameters and an adaptive control parameter model, we have established a method for optimizing the parameters of a combined particle swarm optimization and adaptive VSG control strategy. The specific optimization process proceeds as follows:

(a) Initialize particle parameters.
(b) Calculate the ITAE value based on the VSG output frequency.
(c) Update the individual best values and the global best solution based on the individual and overall ITAE values, finding the most suitable J and D for this iteration.
(d) Modify the J and D of the VSG based on the particle swarm's individual update formula.
(e) After 20 iterations, calculate the ITAE value to complete optimization, and display the optimized J and D.
(f) Conduct a grid-connected simulation using the optimized J_0 and D_0.
(g) Monitor the VSG output angular frequency and angular frequency rate changes during operation.
(h) Determine if the VSG output angular frequency rate exceeds the threshold.
(i) If the threshold is exceeded, adaptively adjust the VSG's moment of inertia and damping coefficient online, based on the designed VSG control parameters.

6. Case Study

6.1. Case Description

To verify the effectiveness of the control strategy of hydrogen production systems based on renewable energy power, a system simulation model was built on the Matlab/Simulink platform, as illustrated in the structure in Figure 7. In this system, the rated capacity of the photovoltaic power generation is 1.22 MW, the rated capacity of the wind power generation is 1.06 MW, the rated capacity of electrolytic hydrogen production equipment is 1.2 MW, and the DC bus voltage is 2500 V. The system connects to a public grid with a rated frequency of 50 Hz.

6.2. Simulation of Coordinated Control Strategy

Based on the above system, two disturbance events were established to analyze the electrolytic hydrogen production device's ability to actively adjust the hydrogen production power when there is an excess of electricity in renewable energy. The specific process is as follows:

(1) When t = 1.5 s, the light intensity increased from 800 W/m^2 to 1000 W/m^2.
(2) When t = 3 s, the wind speed increased from 10 m/s to 12 m/s.

Figures 14 and 15 illustrate the results of the grid-connected simulation of the hydrogen production system based on renewable energy power. It can be seen that at t = 1.5 s, the illumination intensity increased from 800 W/m^2 to 1000 W/m^2, and the photovoltaic output power increased from 0.98 MW to 1.22 MW; at t = 3 s, the wind speed increased from 10 m/s to 12 m/s, and the wind turbine output power increased from 0.61 MW to 1.05 MW. The output power limit at the grid connection point is 1 MW. When there was excess electricity in renewable energy, the electrolysis hydrogen production power increased by 0.24 MW at t = 1.5 s and 0.39 MW at t = 3 s, respectively.

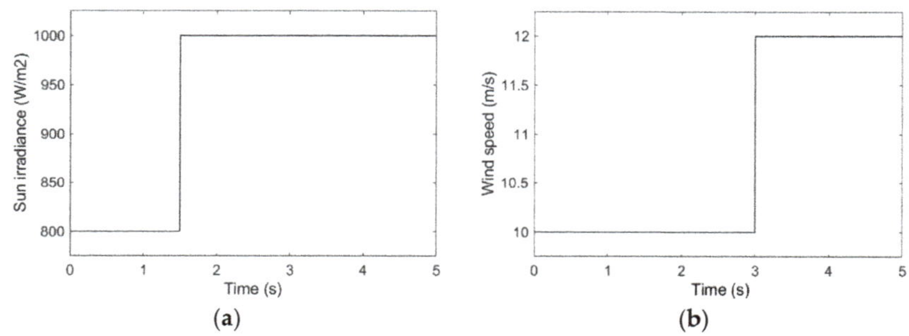

Figure 14. (**a**) Sun irradiance. (**b**) Wind speed.

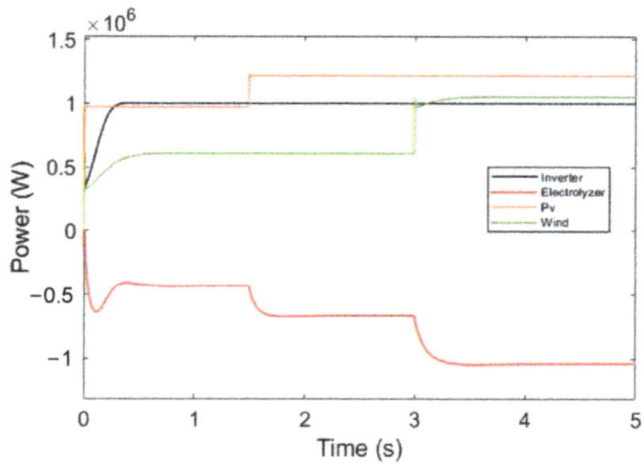

Figure 15. Wind, Pv, electrolyzer, and inverter output power.

6.3. Simulation of Fast Frequency Response Control Strategy

Based on the above system settings, we analyzed the rapid frequency response capability of the grid connection interface of hydrogen production system based on renewable energy power generation. The specific process is as follows:

(1) When t = 1.5 s, disturbance in the external power grid leads to an increase in system frequency of 0.15 Hz;
(2) When t = 3 s, restore the system frequency to its rated value;
(3) When t = 4 s, the output power limit of the grid connection point has been increased from 1 MW to 1.2 MW.

Figures 16 and 17 illustrate the simulation results of the renewable energy power-generation hydrogen production system participating in the primary frequency control of the power grid. Figure 16a illustrates the output frequency of VSG, and Figure 16b presents the hydrogen production rate. Figure 17 presents the output power of wind turbines, photovoltaic cells, electrolyzers, and VSGs. It can be seen that the output power of photovoltaic cells and wind turbines remains constant at 1.22 MW and 0.61 MW, respectively. At t = 1.5 s, due to the disturbance of the external power grid, the system frequency increases by 0.15 Hz. VSG controls the DC/AC converter on the grid side according to its active power droop characteristic, reducing the output power from 1 MW to 0.855 MW. The electrolytic hydrogen production system quickly increases the hydrogen production power to 0.823 MW. At t = 3 s, the system frequency returns to the rated frequency of 50 Hz, and the output power of VSG also returns to 1 MW. The electrolytic hydrogen production system quickly reduces the hydrogen production power to 0.66 MW. At t = 4 s, the output power limit of the parallel point increases from 1 MW to 1.2 MW, resulting in a 0.142 Hz output frequency disturbance of VSG. However, under the action of adaptive VSG control parameters, the frequency quickly returns to the rated frequency of 50 Hz, avoiding a long frequency recovery event. The electrolytic hydrogen production device adjusts the hydrogen production power to 0.43 MW. Throughout the process, the electrolytic hydrogen production system actively responds to the frequency disturbance of the power system and participates in the primary frequency control of the power system. The hydrogen production rate also continuously varies with the electrolytic power.

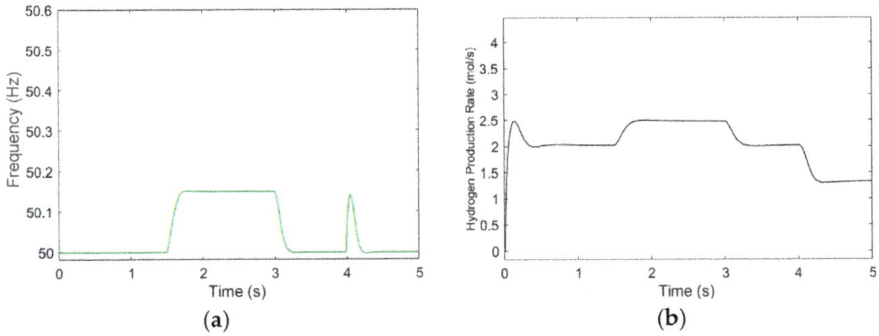

Figure 16. (**a**) VSG output frequency. (**b**) Hydrogen production rate.

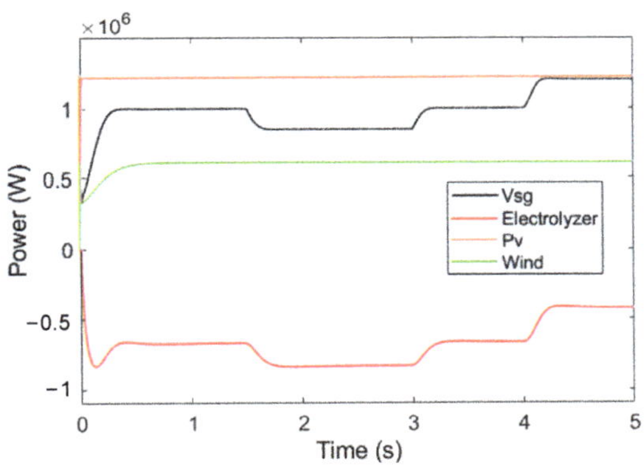

Figure 17. Wind, Pv, electrolyzer, and inverter output power.

Figure 18 illustrates the DC bus voltage waveform. It can be seen that at t = 0 s, t = 1.5 s, t = 3 s, and t = 4 s, the system experiences power imbalance and brief disturbances. However, due to the power support provided by the electrolytic hydrogen production system, the DC bus voltage remains close to the set value of 2500 V.

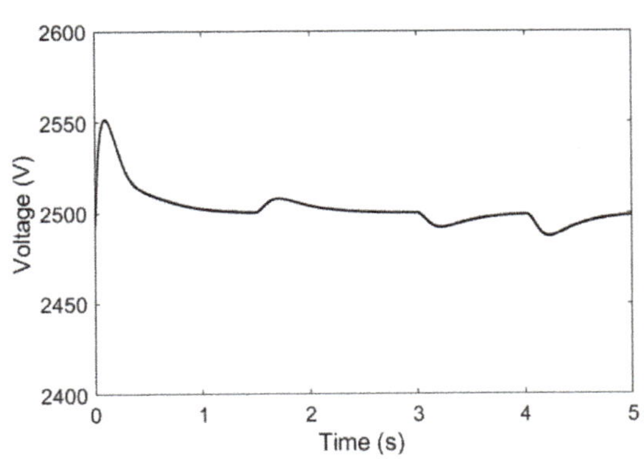

Figure 18. DC bus voltage.

6.4. Comparative Analysis of Parameter Optimization

To demonstrate the superiority of combining the particle swarm optimization algorithm with adaptive VSG control strategy, frequency and power curves obtained from simulations under identical working conditions for varying inertia and damping coefficients are illustrated in Figure 19. Mode 1 adopts the traditional VSG control strategy; Mode 2 employs the VSG adaptive control strategy without particle swarm optimization; Mode 3 initializes the rotational inertia and damping coefficients via the particle swarm optimization algorithm, without the use of an adaptive algorithm; Mode 4 combines the particle swarm optimization algorithm with adaptive control parameters.

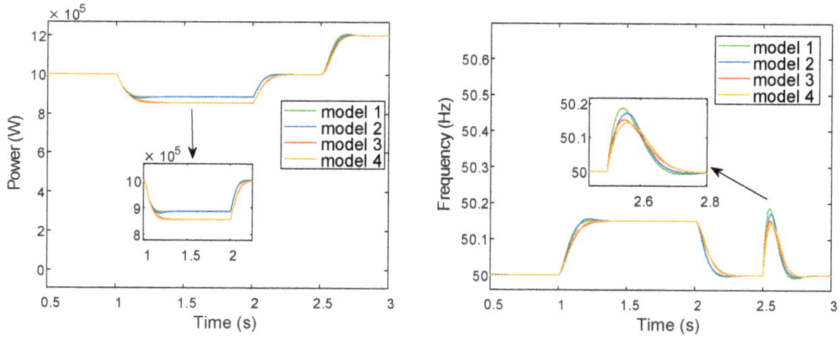

Figure 19. Frequency and power curve under different control parameter modes. (**a**) Frequency curve comparison chart. (**b**) Power curve comparison chart.

To study the advantages of the control strategy proposed herein, a comparison was conducted with traditional control strategies based on a fixed moment of inertia and damping coefficients. As illustrated in Figure 19a, at t = 2.5 s, a fluctuation in the VSG output power was observed. Under the traditional control strategy with fixed moment of inertia and damping coefficients, the overshoot of VSG output power was 1.6%, with a regulation time of 0.246 s. As illustrated in Figure 19b, the frequency fluctuation was 50.186 Hz. In contrast, under the adaptive control strategy proposed in this paper, the overshoot of VSG output power was reduced to 0.58%, with a regulation time of 0.187 s. Additionally, from Figure 19b, the frequency fluctuation was measured to be 50.143 Hz.

To highlight the importance of particle swarm optimization in determining the initial moment of inertia in this paper, a comparison was made with the VSG adaptive control strategy proposed in this paper alone. As illustrated in Figure 19, it can be observed that from t = 1 s to t = 2 s, the VSG participated in primary frequency regulation of the grid. Compared to the control strategy before optimization, the application of particle swarm algorithm to optimize the initial damping coefficient and moment of inertia resulted in an increase of 0.31 MW in the primary frequency regulation power, indicating a stronger primary frequency regulation capability.

At t = 2.5 s, fluctuations in VSG output power were observed. Under the VSG adaptive control strategy without particle swarm optimization, the overshoot of VSG output power was 0.88%, with a regulation time of 0.225 s, and a frequency fluctuation of 50.171 Hz. In the adaptive control strategy proposed in this paper that incorporates particle swarm optimization, the overshoot of VSG output power decreased by 0.3%, the regulation time reduced by 0.038 s, and the frequency fluctuation decreased by 0.028 Hz.

7. Conclusions

This article addressed the issue of the lack of autonomous frequency regulation capabilities in renewable energy power-generation systems. A structural model of coupling renewable energy power-generation systems with electrolytic hydrogen production systems to participate in rapid frequency response of a grid was established, and the dynamic regulation characteristics of the electrolysis hydrogen production system were verified through a Simulink simulation. When the power conversion of the renewable energy system and the grid frequency fluctuated, the electrolysis hydrogen production system was able to adjust its operating power accordingly, thus actively maintaining the system power balance and the stability of the DC bus voltage as well as participating in primary frequency control of the grid. Additionally, aiming at solving the problem of fixed inertia and damping coefficient in traditional VSG control strategies, a Simulink simulation was built to compare the influence of different control parameters on a system's active power and frequency. A VSG control strategy that combines particle swarm optimization and adaptive control parameters was proposed to enhance the primary frequency control capability of the entire system and improve the response characteristics of system frequency and output active power. This paper, by incorporating control strategies, transformed the electrolytic hydrogen production device from a simple power load into a flexible regulation resource in the power system, supporting grid frequency disturbances at the millisecond timescale. It also addressed the issue that renewable energy generation, serving as a power electronic interface, lacks the inertia, damping, and primary frequency control capabilities of traditional synchronous generators under disturbance conditions. However, current research only focuses on single-machine systems, and future studies will investigate the consistency of fast frequency response control for multiple renewable energy generation and hydrogen production systems, considering the uncertainty of communication environments.

Author Contributions: Conceptualization, T.S.; formal analysis, T.S.; writing—review and editing, T.S.; supervision, T.S.; project administration, T.S.; funding acquisition, T.S.; methodology, T.S., Z.X. and L.G.; software, Z.X.; data curation, Z.X.; writing—original draft preparation, Z.X.; visualization, Z.X.; validation, Z.X. and H.Z.; investigation, L.G. and H.Z. All authors have read and agreed to the published version of the manuscript.

Funding: This work is supported by the major project of National Natural Science Foundation of China: "Theory and method of coordinated operation optimization of multi energy supply and demand system containing hydrogen" (No. 62192753).

Data Availability Statement: Derived data supporting the findings of this study are available from the corresponding author on request.

Conflicts of Interest: The authors declare no conflicts of interest.

References

1. Liu, D.; Zhao, N.; Li, P.; Liu, M.; Xu, X.; Shao, P.; Cao, X. Market-Oriented Consumption Model Based on the Joint Tracking of Renewable Energy Generation Curve of "Shared Energy Storage & Demand Side Resources". *Dianwang Jishu/Power Syst. Technol.* **2021**, *45*, 2791–2800. [CrossRef]
2. Sheng, W.; Wu, M.; Ji, Y.; Kou, L.; Pan, J.; Shi, H.; Niu, G.; Wang, Z. Key Techniques and Engineering Practice of Distributed Renewable Generation Clusters Integration. *Proc. CSEE* **2019**, *39*, 2175–2186. [CrossRef]
3. Cai, G.; Chong, C.; Kong, L.; Long, P.; Hao, Z. Modeling and Control of Grid-Connected System of Wind/PV/Electrolyzer and SC. *Power Syst. Technol.* **2016**, *40*, 2982–2990.
4. Albarghot, M.; Rolland, L. MATLAB/Simulink Modelling and Experimental Results of a PEM Electrolyzer Powered by a Solar Panel. In Proceedings of the 2016 IEEE Electrical Power and Energy Conference (EPEC), Ottawa, ON, Canada, 12–14 October 2016; IEEE: Piscataway, NJ, USA, 2016; pp. 1–6. [CrossRef]
5. Valenciaga, F.; Evangelista, C.A. Control Design for an Autonomous Wind Based Hydrogen Production System. *Int. J. Hydrogen Energy* **2010**, *35*, 5799–5807. [CrossRef]
6. Sun, D.; Zheng, W.; Yu, J.; Li, J. Research on the Primary Frequency Regulation Control Strategy of a Wind Storage Hydrogen-Generating Power Station. *Electronics* **2022**, *11*, 3669. [CrossRef]
7. Agredano-Torres, M.; Xu, Q.; Zhang, M.; Söder, L.; Cornell, A. Dynamic Power Allocation Control for Frequency Regulation Using Hybrid Electrolyzer Systems. In Proceedings of the Conference Proceedings—IEEE Applied Power Electronics Conference and Exposition—APEC, Orlando, FL, USA, 19–23 March 2023; Institute of Electrical and Electronics Engineers Inc.: Piscataway, NJ, USA, 2023; Volume 2023, pp. 2991–2998. [CrossRef]
8. Veerakumar, N.; Ahmad, Z.; Adabi, M.E.; Torres, J.R.; Palensky, P.; van der Meijden, M.; Gonzalez-Longatt, F. Fast Active Power-Frequency Support Methods by Large Scale Electrolyzers for Multi-Energy Systems. In Proceedings of the 2020 IEEE PES Innovative Smart Grid Technologies Europe (ISGT-Europe), The Hague, The Netherlands, 26–28 October 2020; IEEE: Piscataway, NJ, USA, 2020; pp. 151–155. [CrossRef]

9. Sarrias-Mena, R.; Fernández-Ramírez, L.M.; García-Vázquez, C.A.; Jurado, F. Electrolyzer Models for Hydrogen Production from Wind Energy Systems. *Int. J. Hydrogen Energy* **2015**, *40*, 2927–2938. [CrossRef]
10. Zhang, X.; Gao, Q.; Guo, Z.; Zhang, H.; Li, M.; Li, F. Coordinated Control Strategy for a PV-Storage Grid-Connected System Based on a Virtual Synchronous Generator. *Glob. Energy Interconnect.* **2020**, *3*, 51–59. [CrossRef]
11. Zhong, Q. Virtual Synchronous Machines and Autonomous Power Systems. *Proc. CSEE* **2017**, *37*, 336–348. [CrossRef]
12. Cheng, Q.; Yu, D.; Cheng, Y.; Gao, J.; Zhang, Y.; Tan, F. Control Strategy of Virtual Synchronous Generator Based on Adaptive Rotational Inertia. *Electr. Power Autom. Equip.* **2018**, *38*, 79–85. [CrossRef]
13. Gao, F.; Qian, K.; Li, M.; Jiang, Y.; Li, T.; Wen, B. Control Strategy of Virtual Synchronous Generator Based on Self-Adaptive Damping Control Algorithm. *Power Capacit. React. Power Compens.* **2021**, *42*, 109–114. [CrossRef]
14. Yang, Y.; Mei, F.; Zhang, C.; Liao, H.; Chen, H.; Zheng, J. Collaborative Adaptive Control Strategy for Virtual Synchronous Generator's Rotational Inertia and Damping Coefficient. *Electr. Power Autom. Equip.* **2019**, *39*, 125–131. [CrossRef]
15. Falcão, D.S.; Pinto, A.M.F.R. A Review on PEM Electrolyzer Modelling: Guidelines for Beginners. *J. Clean. Prod.* **2020**, *261*, 121184. [CrossRef]
16. Ma, Z.; Witteman, L.; Wrubel, J.A.; Bender, G. A Comprehensive Modeling Method for Proton Exchange Membrane Electrolyzer Development. *Int. J. Hydrogen Energy* **2021**, *46*, 17627–17643. [CrossRef]
17. Görgün, H. Dynamic Modelling of a Proton Exchange Membrane (PEM) Electrolyzer. *Int. J. Hydrogen Energy* **2006**, *31*, 29–38. [CrossRef]
18. Ling, Z.; Liu, Y.; Wang, Z.; Yin, J. Adaptive Inertia and Damping of Grid-Connected Inverter with Improved VSG Control. *IET Power Electron.* **2023**, *16*, 2769–2781. [CrossRef]
19. Ding, L.; Ke, S.; Yang, J.; Shi, X. Control Strategy of Virtual Synchronous Generator Based on Adaptive Control Parameter Setting. *Integr. Intell. Energy* **2023**. Available online: https://kns.cnki.net/kcms/detail/41.1461.TK.20230823.0909.002.html (accessed on 27 February 2024).
20. Shen, Z.; Liu, Y.; Zheng, B.; Qi, Y.; Li, J. Microgrid Optical Storage Charge and Discharge Control Technology Based on Adaptive Virtual Synchronous Generator. *Distrib. Energy* **2021**, *6*, 18–25. [CrossRef]
21. Zheng, T.; Chen, L.; Wang, R.; Li, C.; Mei, S. Adaptive Damping Control Strategy of Virtual Synchronous Generator for Frequency Oscillation Suppression. In Proceedings of the 12th IET International Conference on AC and DC Power Transmission (ACDC 2016), Beijing, China, 28–29 May 2016; Institution of Engineering and Technology: Stevenage, UK, 2016; pp. 1–5. [CrossRef]

electronics MDPI

Article

Total Power Factor Smart Contract with Cyber Grid Guard Using Distributed Ledger Technology for Electrical Utility Grid with Customer-Owned Wind Farm

Emilio C. Piesciorovsky [1,*], Gary Hahn [1], Raymond Borges Hink [1] and Aaron Werth [2]

[1] Oak Ridge National Laboratory, Electrification and Energy Infrastructures Division, One Bethel Valley Road, Oak Ridge, TN 37831, USA; hahng@ornl.gov (G.H.); borgesrc@ornl.gov (R.B.H.)
[2] Oak Ridge National Laboratory, Cyber Resilience and Intelligence Division, One Bethel Valley Road, Oak Ridge, TN 37831, USA; werthaw@ornl.gov
* Correspondence: piesciorovec@ornl.gov

Abstract: In modern electrical grids, the numbers of customer-owned distributed energy resources (DERs) have increased, and consequently, so have the numbers of points of common coupling (PCC) between the electrical grid and customer-owned DERs. The disruptive operation of and out-of-tolerance outputs from DERs, especially owned DERs, present a risk to power system operations. A common protective measure is to use relays located at the PCC to isolate poorly behaving or out-of-tolerance DERs from the grid. Ensuring the integrity of the data from these relays at the PCC is vital, and blockchain technology could enhance the security of modern electrical grids by providing an accurate means to translate operational constraints into actions/commands for relays. This study demonstrates an advanced power system application solution using distributed ledger technology (DLT) with smart contracts to manage the relay operation at the PCC. The smart contract defines the allowable total power factor (TPF) of the DER output, and the terms of the smart contract are implemented using DLT with a Cyber Grid Guard (CGG) system for a customer-owned DER (wind farm). This article presents flowcharts for the TPF smart contract implemented by the CGG using DLT. The test scenarios were implemented using a real-time simulator containing a CGG system and relay in-the-loop. The data collected from the CGG system were used to execute the TPF smart contract. The desired TPF limits on the grid-side were between +0.9 and +1.0, and the operation of the breakers in the electrical grid and DER sides was controlled by the relay consistent with the provisions of the smart contract. The events from the real-time simulator, CGG, and relay showed a successful implementation of the TPF smart contract with CGG using DLT, proving the efficacy of this approach in general for implementing electrical grid applications for utilities with connections to customer-owned DERs.

Keywords: distributed ledger technology; smart contract; relays; power system; cyber security

1. Introduction

The number of distributed energy resources (DERs) interconnected with utility systems, as well as the complexity of the configurations of these interconnections, has significantly increased [1]. The characteristics of electric power produced by customer-owned DERs using renewable energy sources (wind or photovoltaic) are often not as stable. Therefore, DER-produced electricity may exhibit undesirable variations from the norm in frequency, voltage, harmonic content, reliability/continuity, power factor, and stability. One approach to safely incorporating DERs into the power system is to install a relay at the point of common coupling (PCC) that can isolate the DER from the power system if the characteristics of the DER's electricity production become unstable or cannot be tolerated. The approach presented in this article defines the rules of engagement for the DER (customer-owned wind farm) to connect to the electrical utility grid through a smart contract and transmit the operating and status data securely using distributed ledger technology (DLT), which are implemented using a Cyber Grid Guard (CGG) hardware/firmware system using a relay at the PCC when the DER must be disconnected from the utility in order to maintain power system stability.

Smart contracts are digital contracts stored on a distributed ledger that are automatically executed when predetermined terms and conditions are met [2] between an electrical grid utility and customer-owned DERs. The large and continuing increase in customer-owned DERs is accompanied by a corresponding increase in the number of relays deployed at the PCCs between the electrical grid utility and customer-owned DERs. Each renewable DER and relay require communications and data management at the PCC. The integrity of the data communicated among the DERs, the PCC relays, and the bulk power system is vital. DLTs are decentralized storage platforms that maintain data integrity without requiring mutual trust among participants [3].

Security, scalability, and governance are prominent subjects of recent advancements in DLT and smart contracts. Security is still a critical issue, with ongoing efforts to improve smart contract safety using libraries like OpenZeppelin and formal verification methods [4]. For example, composite smart contracts, which are smart contracts that call upon other smart contracts, introduce additional security challenges. These could be addressed using finite-state machine models and checking techniques. Other relevant formal methods include the Observe-based Statistical Model Checking (OSM) framework, which has been designed for the verification of complex software systems [5]. The BlockASP framework leverages Aspect-Oriented Programming (AOP) to address the dynamic nature of blockchain systems, including smart contracts, by enabling flexible instrumentation for monitoring runtime events and facilitating analysis [6]. Scalability is addressed through techniques like sharding [7,8] and rollups [9,10], particularly in the context of public DLTs, and off-chain storage strategies [11]. DLT governance encompasses the policies and procedures that help DLT development and usage adhere to both legal requirements and ethical responsibilities. Transitioning to consensus mechanisms like Proof-of-Stake (PoS) from Proof-of-Work (PoW) reduces resource consumption and enhances scalability [12], which is a significant trend pertinent to incorporating DLT into the regulated energy sector to meet regulatory standards [13].

Traditional cybersecurity solutions in protective relays and centralized monitoring systems managing DERs lack the decentralization, transparency, and immutability of smart contracts and DLT [14]. Smart contracts execute predefined agreements automatically and independently [15]. Cryptographic primitives such as hashing and digital signatures are used by DLTs to protect data integrity and secure sensitive information against unauthorized access [16]. These features make smart contracts and DLT well-suited for managing customer-owned DERs and their relays, as they can provide a secure, transparent, and tamper-evident means of translating operational constraints into relay actions/commands. The use of DLT to solve cybersecurity problems and improve resilience within electrical grids has been examined in recent studies, particularly for distributed generation systems [17]. For example, one study implemented a prototype using Hyperledger Fabric DLT and IoT for metering and billing small-scale consumers, emphasizing its scalability, low energy requirements, and reduced vulnerability to cyberattacks due to its data encryption capabilities and decentralized architecture [18]. A fault location and traceability system based on DLT was developed to protect IoT sensor data integrity, which is vital for grid security and reliability [19]. A realistic electric grid substation testbed was developed for researching fault detection and cyberattack scenarios involving DERs [20], which was used for executing the total power factor experiments described in this study.

Most potential energy applications of DLT have been based on software simulations [21–25]. However, software simulation algorithms cannot always be directly integrated with intelligent electronic devices (IEDs) like relays and meters because of measurement equipment boundaries. For example, problems could arise when IEDs have different measurement condition behaviors. When the breakers are opened, IEDs could measure a frequency of 60 Hz and/or power factor of 1.00 [26] or have different power factor sign conventions depending on IEEE and IEC standards [27]. This study used an advanced real-time simulator with a hardware in-the-loop (HIL) testbed, integrating the CGG system with DLT and a smart contract for a Schweitzer Engineering Laboratories (SEL) 700GT relay. This advanced testbed was essential for studying blockchain applications because it is based on using real relays and CGG with DLT and smart contracts. This testbed was established at Oak Ridge National Laboratory [28–30].

In this research, the CGG with DLT was combined with smart contracts to study the control application for improving the total power factor (TPF) between the electrical utility grid side and customer-owned wind farm side to reduce the power line losses generated by the reactive energy. This TPF smart contract was applied with the CGG using DLT and one SEL 700GT relay that controls the breakers at the grid and customer-owned DER sides of the PCC. The CGG with DLT and smart contract demonstrated a way to improve the integrity of data received from the breakers at the grid and customer-owned DER (wind farm) sides. The TPF smart contract was activated with a time window of 450 s because the frequency protection elements could have a maximum time setting of 400 s [26], and traditional power quality improvement methods (like load shedding, capacitor banks or transformer/load tap changers) could be applied on the utility grid side. Also, this 450 s time window avoids having the TPF smart contract operating breakers during anomalous events like disturbances or electrical faults. The TPF smart contract flowchart was developed by using the power factor explanations [31] and setting the condition of the good power factor between +0.90 and +1.00 because usually electrical utilities adjust customer bills for power factors smaller than +0.90 [32].

An HIL real-time simulator testbed platform was utilized in this smart contract experimental model, with a real CGG using DLT and one relay in the PCC, instead of using smart contract applications with only software simulations [33–36]. Therefore, this testbed platform allowed us to create a more realistic TPF smart contract algorithm, based on the relay TPF measurement observed conditions. This experimental model permitted us to compare and measure the different results of the relay versus the simulation in the PCC when the breaker was opened. In the simulation, the TPF depends on the "true power/apparent power" ratio, and when the breaker was opened and phase current magnitudes were zero, the numerator and denominator of the TPF resulted in an indeterminate TPF. However, the relay measured a TPF of +1.00 when the breaker was opened based on the relay. Another consideration is that the signs for measuring the TPF in a relay or meter could have different conventions, while the IEC standard defines the positive and negative power factor when the true power is positive and negative, respectively. On the other side, the IEEE standard defines the positive and negative power factor for leading current (capacitive effect) and lagging current (inductive effect), respectively. These TPF measurement considerations were observed when an HIL testbed platform was implemented. In this case, the relay's behavior had shown the use of the IEC standard for the power factor sign convention based on its instruction manual [26]. The boundary condition of the opened breakers for the SEL 700GT relay, power factor limits, breaker states, and operation times were successfully integrated into the TPF smart contract with CGG using DLT at an electrical utility grid with a customer-owned DER. The novelties of this study are focused on the use of an (1) advanced power quality measurement method for operating a customer-owned DER to reduce the reactive energy on the grid-side power lines using a TPF smart contract with DLT as backup method after using capacitor banks, load shedding, or transformer/load tap changers on the grid side, (2) the development of a data security platform to use the CGG with DLT and smart contracts for securing data from the relay located at the PCC between the electrical utility grid and the customer-owned DER, and (3) an assessment of the HIL testbed platform to evaluate the TPF smart contract in an advanced substation grid testbed with a real-time simulator using DLT nodes and a protective relay instead of using only a software simulation platform. Table 1 shows the novelties of this study.

Table 1. Novelties of this study.

Field	Novelties
Application	Operating a power quality method with a customer-owned DER, based on reducing the reactive energy in the grid-side power lines using a TPF smart contract with CGG and DLT, following the use of traditional power quality methods like capacitor banks, load shedding, or transformer/load tap changers.
Data Security	Using the CGG with DLT and smart contracts for securing data from the relay located at the PCC between electrical utility grid and customer-owned DER.
Assessment	Evaluating the TPF smart contract in an advanced substation grid testbed with a real-time simulator using DLT nodes and relay in-the-loop instead of using software simulations.

DER: distributed energy resource; TPF: total power factor; CGG: cyber grid guard; DLT: distributed ledger technology; PCC: point of common coupling.

In this research article, the experimental model, power line diagrams, architecture, and equipment are described first. Then, the phase and TPF descriptions and equations with the smart contract flowcharts are described in detail. The test event plots from the real-time simulator, CGG system, and relay are presented and assessed. Finally, the main findings of this research are given, along with the possible future implementation of new smart contract applications.

2. Test Model

2.1. Grid and Equipment

The TPF smart contract was applied to a substation grid testbed with customer-owned wind farm, using IEDs that managed, monitored, and controlled the smart contract-based application. This testbed used a software model and a simulated electrical grid with a trigger-event system and synchronized time source. The time source enables synchronization between the relays and meters and, consequently, the CGG system. The trigger-event system was implemented in the testbed for recording the events of the relay at the desired time.

The objective of this testbed was to assess whether the DLT architecture was successful in executing the TPF smart contract between the electrical utility grid and customer-owned wind farm. A diagram of the equipment and electrical grid is shown in Figure 1. Figure 1a represents a sectionalized bus substation configuration [37] and electrical grid with one customer-owned DER (wind farm). Utility A has one substation with two transformers, using primary/secondary voltages of 34.5/12.47 kV, and feeders that are connected to power lines. The loads can be fed by the wind farm (Utility B) or substation (Utility A). The customer-owned DER has six 1.5 MW wind turbines (Utility B), and Utility C represents a fossil fuel power plant. In Figure 1b, the equipment rack includes the meters and relays. These IEDs (meters and relays) are wired to communication devices and a real-time simulator that are linked to a synchronized-time system. All meters [38] and relays [39–41] use the Generic Object-Oriented Substation Event (GOOSE) IEC 61850 protocol. In this experimental model, the test scenarios were run in the red dashed-line square area (Figure 1a) to evaluate the CGG system for the TPF smart contract, based on operating the breakers BKY (grid-side) and BKX (wind farm-side). The SEL 700GT relay measured the TPFs and states of the BKY (Utility A) and BKX (Utility B) breakers.

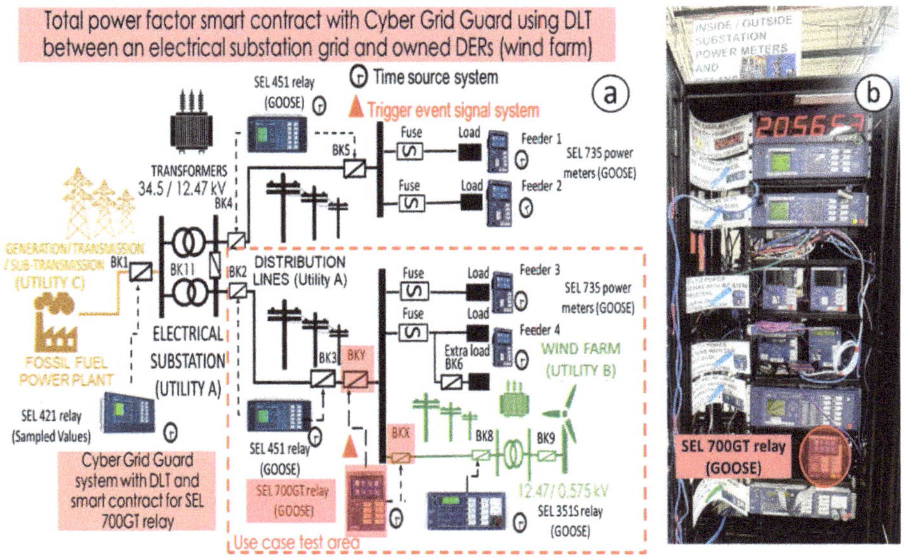

Figure 1. (**a**) Electrical grid diagram and (**b**) equipment rack.

2.2. Testing Platform

In this study, the testbed was implemented with the CGG using DLT and smart contracts. This testbed had IEDs (meters and relays) wired to a real-time simulator. This testbed has a real-time simulator rack, relay/meter rack, and CGG rack (Figure 2). The four workstation computers in Figure 2 are the host laptop, monitoring laptop, human machine interface (HMI) laptop, and CGG laptop. The synchronized time system was implemented using a time source clock used as a master and clock displays as slaves, which were connected to the meters and relays.

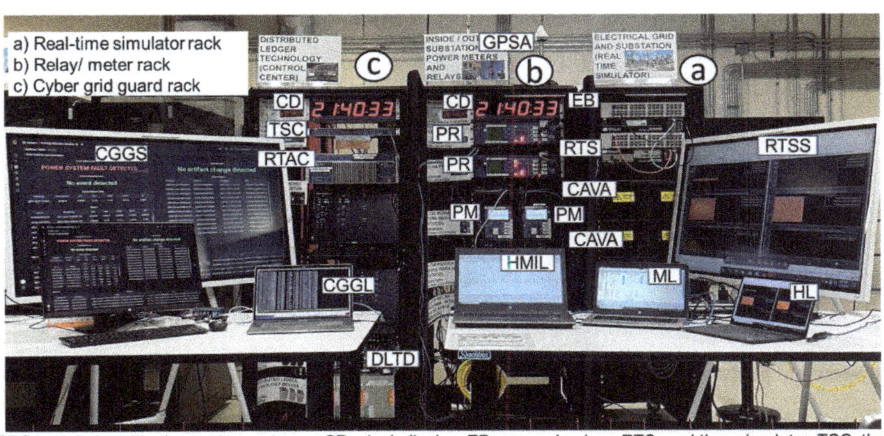

GPSA: global positioning system antenna; CD: clock display; EB: expansion box; RTS: real time simulator; TSC: time source clock; PR: protective relay; PM: power meter; RTAC: real time automation controller; CAVA: current and voltage amplifier; DLTD: distributed ledger technology devices; RTSS: real time simulation screen; CGGS: cyber grid guard screen; HL: host laptop; ML: monitoring laptop; HMIL: human machine interface laptop; CGGL: cyber grid guard laptop.

Figure 2. Testbed with relays, meters, and workstation computers.

Figure 3 shows the architecture of the testbed with CGG using DLT and smart contracts. In the control center, the settings of the HMI local substation, HMI control center, EmSense high-speed smart visu, and virtual machine Blueframe servers/computers are set. This architecture has four layers:

- The first layer (the physical layer) includes the breakers, power lines, and other grid elements simulated by the real-time simulator.
- The second layer (protection and metering layer) includes the HIL, represented by the relays and meters.
- The third layer (the automation layer) includes the Ethernet switches and remote terminal units.
- The fourth layer includes the CGG system, synchronized time system, trigger-event system, supervisory control and data acquisition, and HMI.

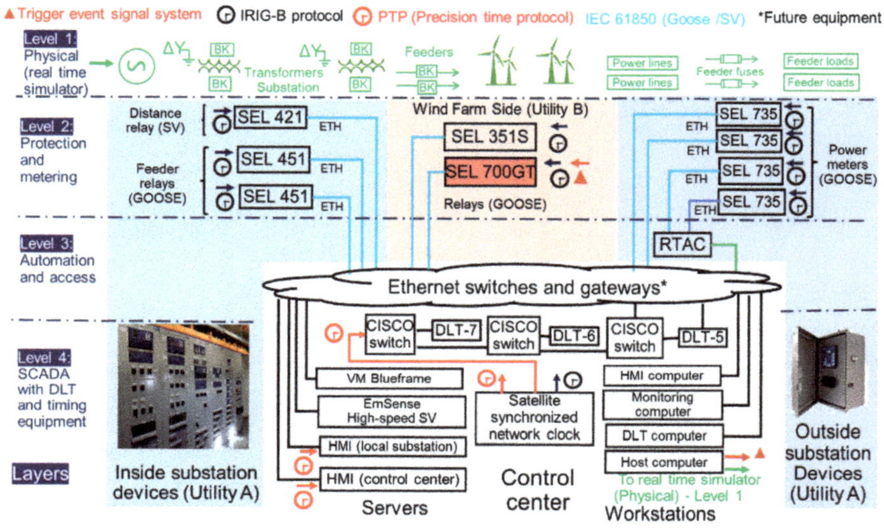

Figure 3. Architecture of testbed.

The SEL 700GT relay was connected to the trigger-event trigger system and synchronized time system. The relay recorded the test events by receiving a digital signal from the trigger-event system. In the testbed architecture, the synchronized time protocol implemented was based on the inter-range instrumentation group time code B (IRIG-B) signals for relays and meters. However, the CGG system through the Ethernet network used precision time protocol (PTP) communication. The relays transmitted IEC 61850 GOOSE messages.

2.3. Three-Phase Diagram and Real-Time Simulation Properties

In this experimental model, a three-phase diagram of the power system was created with MATLAB/Simulink (2015b) software models, which were then turned into an RT-LAB (V2020.2.2.82) software project which was integrated into a real-time simulator with meters and relays in-the-loop. The RT-LAB software is fully integrated with MATLAB/Simulink, and it enables the Simulink models to interact with the real-time simulator in real time. The RT-LAB software is a platform used for real-time simulation to develop and validate the power system application. The RT-LAB software is a multi-domain platform that provides a flexible and scalable solution for power system test scenarios, like in other research applications [42–45].

This advanced testbed platform simulated different use case scenarios for normal operation and temporary electrical faults tests. However, anomalous events related to electrical faults are not likely to be present in a real electrical grid on the field. This is because electrical fault tests are non-desired testing situations that can damage electrical equipment (power transformers, capacitor banks, power lines, generators, etc.) and/or personal staff, and/or can generate electrical grid blackouts. As such, the real-time simulator with relays and meters in-the-loop was the best testbed platform to perform tests on the TPF smart contract with the CGG using DLT. Figure 4 shows the power grid and substation with the customer-owned wind farm. A 34.5/12.47 kV voltage system was used in the substation. The wind farm used a 0.575/12.47 kV voltage system. The wind farm included six 1.5 MW wind turbines. In the substation, two power transformers were connected in parallel with two power lines and load feeders. The load feeders had 50 T and 100 T fuses [46]. In the PCC between the grid-side and wind farm-side, the SEL 700GT relay controlled the BKY (grid-side) and BKX (wind farm-side) breakers.

In the CGG with DLT and the smart contract, the BKY/BKX breaker states and TPF values were measured from the SEL 700GT relay. Then, the BKY/BKX breakers were operated based on the TPF boundary conditions. The desired TPF limits on the grid-side were between +0.90 and +1.00. The main goal was to minimize the power losses in the grid-side power line by reducing the reactive power. The BKY/BKX breakers were operated by the SEL 700GT relay through the CGG using the DLT and smart contract if the TPF at the grid-side was less than +0.90. The CGG using DLT and the smart contract were used in the PCC between the electrical utility grid and customer-owned wind farm. The integrity of data from the SEL 700GT relay at the PCC is vital, and DLT enhances the resilience of the electrical grid by securing the shared data.

Figure 4 shows a diagram of the three-phase power system. The substation grid (Utility A) can be connected to the customer-owned wind farm (Utility B). Utility C represents a large fossil power plant with a transmission and sub-transmission model (Figure 4a). Utility A has a substation (Figure 4b), power line/s (Figure 4c), and feeder loads (Figure 4d). Utility B represents a customer-owned wind farm (Figure 4e). The BKY breaker on the grid side (Figure 4f) and BKX breaker on the farm side (Figure 4g) have three-phase poles. The temporary fault block (Figure 4j) is controlled by the signals from the breaker operation circuits in Figure 4h. The trigger-event circuit (Figure 4i) generates a 24 VDC signal that is sent to relays and meters for recording the events during the simulation tests at a specific time or when the breakers are operated. The temporary fault block (Figure 4j) can generate no permanent electrical faults during 60 cycles at a specific time in the grid-side power line when the fault signal is received from the temporary fault block signal circuit (Figure 4h).

The real-time simulation tests were run with the real-time simulator and relays/meters in the loop, using MATLAB/Simulink (2015b) and the RT-LAB (V2020.2.2.82) software. In the RT-LAB project, a simulation powergui block was set up with a Discrete simulation, Tustin/Backward Euler solver, and sample time of 50 micro-seconds. Then, the real time simulation tests were started from the host computer before each simulation was run. The real-time properties were set for the target platform (OPAL-RT Linux, x86-based), real-time simulation mode (Hardware synchronized), real time communication link type (UDP/IP), time factor (1.0), and stop/pause time (600 s).

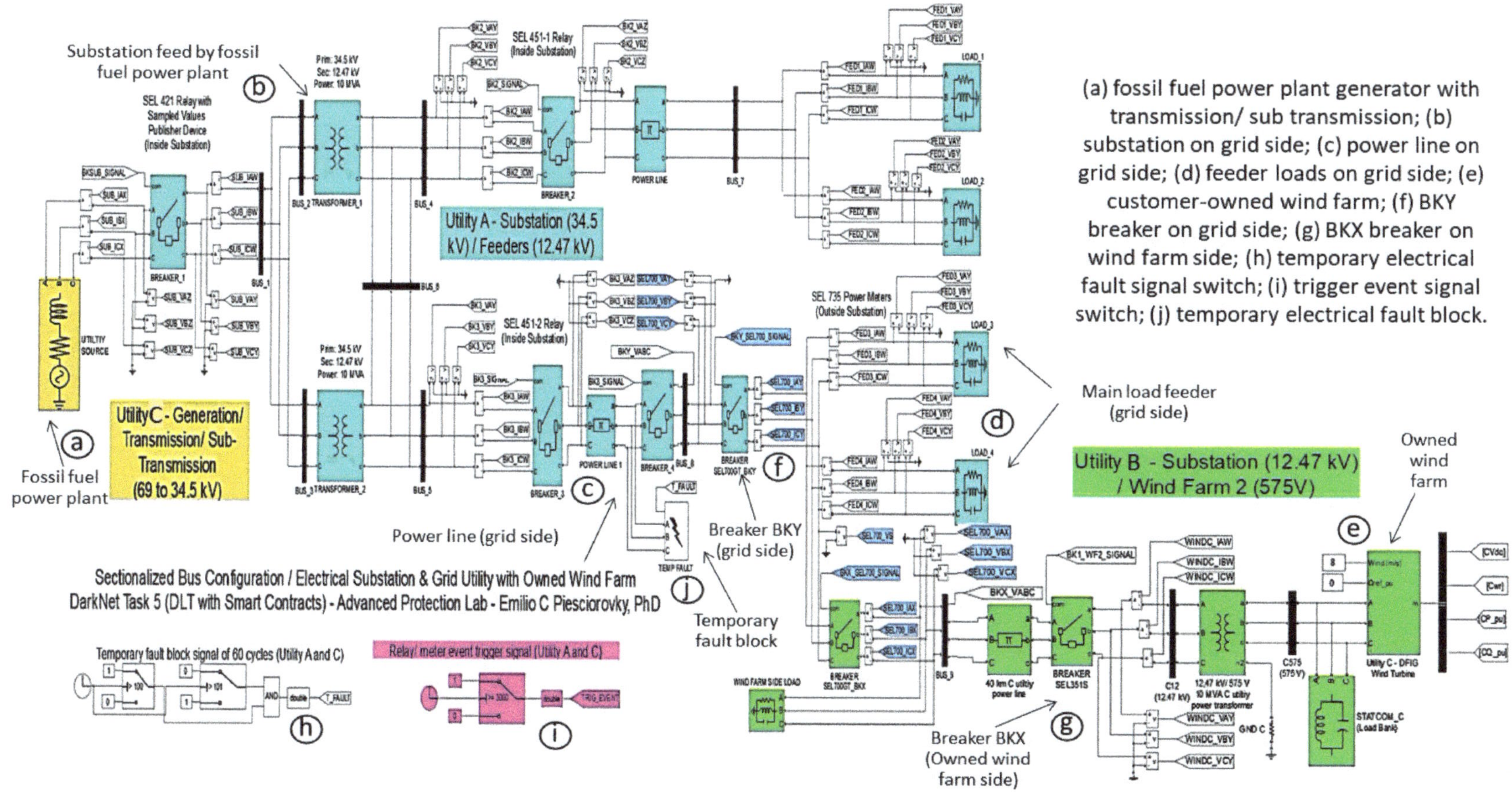

Figure 4. Three-phase power system diagram of the substation and grid with customer-owned wind farm.

3. Theory and Equations

3.1. Power Factor

The power factor [47] is represented by Equation (1) as the ratio between the true power and the apparent power. The magnitude of the voltage and current multiplied by the cosine angle between the voltage and current is the true power. However, the apparent power is the product of the current and voltage. A power factor magnitude of less than one indicates that the voltage and current are not in phase. A negative power factor occurs when the true power flows back toward the source.

$$PF = \frac{P}{S} = \frac{V \times I \times cos(\theta_V - \theta_I)}{V \times I} = cos(\theta_V - \theta_I), \tag{1}$$

where PF is the power factor measured between ± 1.00, P is the true power in Watts, S is the apparent power in volt-amperes, V is the phase voltage magnitude in volts, I is the phase current magnitude in amps, θ_V is the phase voltage angle in degrees, θ_I is the phase current angle in degrees.

3.2. Average versus Total Power Factor

The average power factor represents the average of the measured power factors for phases A, B, and C. The average power factor can be estimated by calculating the power factor for each phase using Equation (1), measuring the phase current/voltage magnitudes and angles. Then, the average power factor is calculated as the sum of the power factors for phases A, B, and C divided by three. The average power factor is represented by Equation (2):

$$APF = \frac{PF_A + PF_B + PF_C}{3}, \tag{2}$$

where APF is the average power factor, PF_A is the power factor of phase A, PF_B is the power factor of phase B, and PF_C is the power factor of phase C.

Meters and relays usually measure the TPF instead of the average power factor because the average power factor does not represent the power factor of a three-phase power system effectively, especially when phases A, B, and C loads are unbalanced. Therefore, the TPF represents a better way of measuring the efficiency of three-phase power systems. The TPF is estimated using the ratio between the true power sum and apparent power sum for phases A, B, and C. The TPF can be calculated by using Equation (3).

$$TPF = \frac{P_A + P_B + P_C}{S_A + S_B + S_C}, \tag{3}$$

where TPF is the total power factor, P_A is the true power of phase A in watts, P_B is the true power of phase B in watts, P_C is the true power of phase C in watts, S_A is the apparent power of phase A in volt-amperes, S_B is the apparent power of phase B in volt-amperes, and S_C is the apparent power of phase C in volt-amperes.

If the relay-recorded events do not provide the measured true and apparent power for phases A, B, and C, calculate t the TPF using Equation (3), and then the TPF can be estimated directly using the measured current/voltage magnitudes and angles for phases A, B, and C. Then, by combining Equations (1) and (3), the TPF can be calculated using Equation (4):

$$TPF = \frac{\sum_{m=A}^{C} V_m \times I_m \times cos(\theta_{Vm} - \theta_{Im})}{\sum_{m=A}^{C} V_m \times I_m}, \tag{4}$$

where TPF is the total power factor, V_m is the voltage magnitude of the generic phase m (A, B, C) in volts, I_m is the current magnitude of the generic phase m (A, B, C) in amperes, θ_{Vm} is the voltage angle of the generic phase m (A, B or C) in degrees, and θ_{Im} is the current angle of the generic phase m (A, B or C) in degrees.

3.3. Power Factor Convention Signs

The power factor is measured by relays or meters between ± 1.00. The power factor can be measured as a percentage or unit. However, a unit measurement is usually the more common method. The power factor represents the efficiency of the power system with respect to losses generated by the reactive energy flowing in an electrical grid point. The maximum power factor is +1.00, and the minimum power factor limits are usually between +0.80 and +0.98 [47].

Based on an industrial meter manual [27], the measured power factor sign convention could be defined for the IEEE or IEC standards, and sometimes meters allow for setting the power factor sign convention that is used on the display to either IEC or IEEE [27]. The conventional power factor signs based on the IEC and IEEE standards are shown in Figure 5a *and* Figure 5b, respectively, based on a meter instruction manual [27].

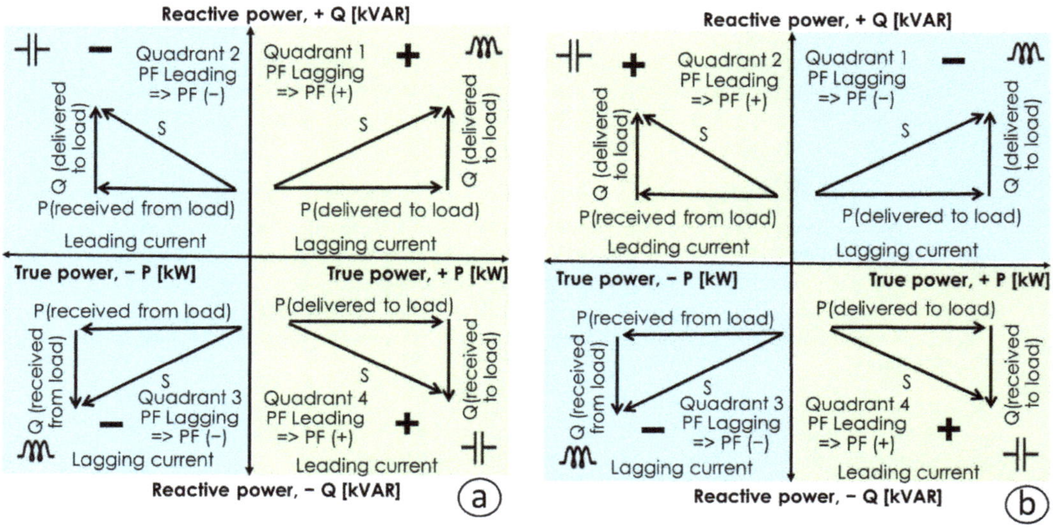

Figure 5. Power factor sign convention based on IEC (**a**) and IEEE (**b**) standards.

The IEC standard (Figure 5a) defines the positive and negative power factor when the true power is positive and negative, respectively. However, the IEEE standard (Figure 5b) defines the positive and negative power factor for leading current (capacitive effect) and lagging current (inductive effect), respectively. For the IEC standard [27], in Figure 5a, quadrants 1 and 4 with positive true powers have a positive power factor. However, quadrants 2 and 3 with negative true powers have a negative power factor. For the IEEE standard [27], in Figure 5b, quadrants 2 and 4 with capacitive loads (power factor leading) have a positive power factor. However, quadrants 1 and 3 with inductive loads (power factor lagging) have a negative power factor. In Figure 5, the green and blue quadrants represent the positive and negative power factors, respectively. In this study, the TPF smart contract with CGG using DLT for the electrical utility grid with a customer-owned wind farm was used with an SEL 700GT relay [26] that uses the power factor sign convention, based on the IEEE standard in Figure 5b.

4. Methodology

4.1. Total Power Factor Smart Contract

The smart contract is defined as a digital contract stored on a distributed ledger that is automatically performed when the predetermined terms and conditions are reached. Smart contracts are commonly used to automate the execution of transactions consistent with agreed-upon set rules or according to specific conditions so that all members (electrical utilities) can operate according to the agreed-upon consensus. The TPF smart contract was applied to a CGG with DLT, to manage an electrical utility grid with a customer-owned wind farm, therefore maintaining the integrity of data from a relay at the PCC. In addition, it becomes more difficult in the future, as the number of DERs and PCCs increases; however, DLT could enhance the resilience of modern electrical grids by effectively securing the shared data. The TPF smart contract was based on using CGG with DLT and a relay that measures the TPF at breaker BKY (grid-side) and breaker BKX (wind farm-side) for the A, B, and C phases. The TPF smart contract application was implemented by integrating a real-time simulator with the CGG system and a relay in-the-loop.

The main goal of the TPF smart contract is to use the CGG with DLT to offer a solution at PCCs for improving the TPF at the grid-side of an electrical utility grid connected to a customer-owned DER (wind turbine farm). The data collected from the CGG system were used for performing the TPF smart contract. The desired TPF limits on the grid-side were set to be between +0.9 and +1.0, and the operation of the breakers in the electrical grid and DER side were controlled by the CGG with DLT and the SEL 700GT relay by applying the terms of the smart contract. The boundaries of the TPF smart contract are defined by the breaker states at the grid side (BKY) and wind farm side (BKX), the TPF measured on the grid side (TPF_{BKY}) and wind farm side (TPF_{BKX}), and the condition that the TPF measured at the grid side must be between +0.9 and +1.0 for the main load feeders.

Another boundary condition is based on the selected period for measuring and keeping the TPF from the CGG. If the TPF measured at the grid-side (TPF_{BKY}) is smaller than +0.9, this condition needs to be measured and kept for more than 400 s to allow for BKX and BKY breaker operations from the TPF smart contract. In the TPF smart contract, the time window was defined at 450 s, but it depends on the implementation of the TPF smart contract for normal operation and electrical fault situations in the electrical grid. In normal operation, this time window allows for the operation of the breakers by the TPF smart contract after the use of other power factor quality techniques (like capacitor banks, load shedding, transformer/load tap changers) from the grid-side utility, without the use of the smart contract and with other relays and breakers. Coordination with load tap changers is commonly used in distribution substations. The load tap changer can raise the voltages but cannot improve the reactive power or power factor [48]. Then, capacitor banks are usually selected instead of overload tap changers at the distribution substation. However, the feed of industrial loads is achieved with a proper time delay at both controllers. Capacitor bank controllers are used on the high-voltage side, but load tap changers controllers are used on the low-voltage side of the transformer, which are regulated by the load tap changer [48]. In transformers, the onload tap changers can regulate the voltages in steps of 0.625% to 2.5% with a time delay of 1–3 min (60–180 s) for each step operation, [48] but switching on a capacitor bank can provide the same or a larger voltage increase much faster [48]. The other power quality method is the load shedding that may occur if there is a shortage of electricity supply, or to keep prevent power lines from becoming overloaded by poor power factors. A load shedding program is successfully implemented when the system frequency has recovered to 60 Hz, but it can take several minutes [49] or hours depending on the electrical grid.

In this study, a time window greater than 400 s (6.66 min) is also related to the TPF smart contract condition that it must not operate the breakers for a poorly measured TPF of less than +0.9 from the grid-side ($TPF_{BKY} < +0.9$) during anomalous events such as electrical fault states. The reason for this is that poor measured TPF could be detected during an electrical fault state for short periods of less than 1 s, and relays first need to clear the electrical fault by opening the breakers before the CGG triggers a low-TPF condition. In addition, a time window greater than 400 s was selected based on the maximum time setting limit for the frequency protection elements, defined as 400 s for frequency protection elements [50] based on the relay manual [26] and shown in Table 2.

Table 2. Time range of frequency protection elements.

Protection Elements	Protection Number	Protection Description	Time Setting	Time Range (s)
X side frequency	81	81: Frequency protection on the BKX breaker side	Frequency delay time	0.00 to 400.00
Y side frequency	81	81: Frequency protection on the BKY breaker side	Frequency delay time	0.00 to 400.00
V/Hz	24	24: Overexcitation protection, ratio of the voltage to frequency	Reset time	0.00 to 400.00

In the SEL 700GT relay, the conventional sign for the power factor is based on the IEEE standard. Based on the power factor sign convention for the IEEE standard, when the current lagging voltage (Figure 6a) or power factor is lagging (inductive behavior), the power factor is negative. However, when the current leading voltage (Figure 6b) or power factor is leading (capacitive behavior), the power factor is positive. Figure 6 shows phasor diagrams (Figure 6a,b) and the IEEE standard power factor sign conventions (Figure 6c) for the SEL 700GT relay.

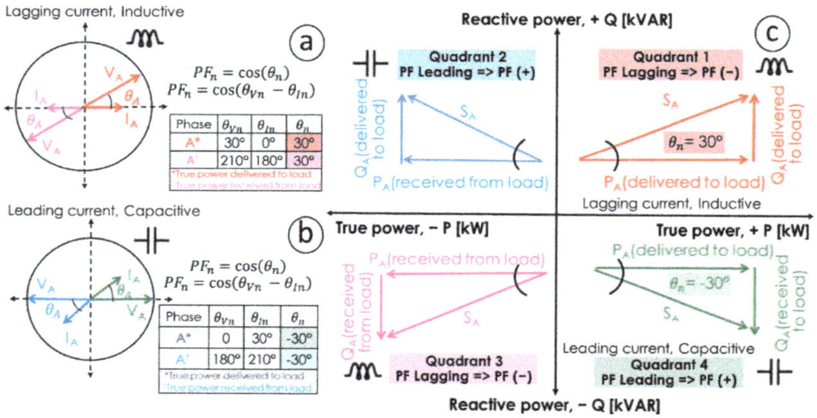

Figure 6. Phasor diagrams (**a**,**b**) and IEEE standard power factor sign convention (**c**).

In the SEL 700GT relay, the measured power factor at phases A, B, and C in the breakers on the grid side and the wind farm side can be calculated as the cosine of the angle between the phase voltage and phase current based on Equation (5),

$$PF_{mBKn} = \cos(\theta_{VmBKn} - \theta_{ImBKn}), \tag{5}$$

where PF_{mBKn} is the power factor of generic phase m (A, B, or C) for the breaker n (BKY or BKX) measured between ±1.00, θ_{VmBKn} is the voltage angle of the generic phase m (A, B, or C) for the breaker n (BKY or BKX) measured in degrees, and θ_{ImBKn} is the current angle of the generic phase m (A, B, or C) for the breaker n (BKY or BKX) measured in degrees.

In the SEL 700GT relay, the TPF or three-phase power factor for the grid side (TPF_{BKY}) and the wind farm side (TPF_{BKX}) depends on the true and apparent power of phases A, B, and C; it can be calculated using Equation (6):

$$TPF_{BKn} = \frac{PA_{BKn} + PB_{BKn} + PC_{BKn}}{SA_{BKn} + SB_{BKn} + SC_{BKn}}, \tag{6}$$

where TPF_{BKn} is the total power factor for the breaker n (BKY or BKX) measured between ±1.00; PA_{BKn}, PB_{BKn}, and PC_{BKn} are the true powers of phases A, B, and C, respectively, for the breaker n (BKY or BKX) measured in kW; SA_{BKn}, SB_{BKn}, and SC_{BKn} are the apparent powers of phases A, B, and C, respectively, for the breaker n (BKY or BKX) measured in kVA.

In the SEL 700GT relay, if the three-phase pole breaker is open (BKY or BKX), the measured TPF is +1.00. This approach is taken to avoid the measurement of indetermined TPF values that result when the breakers (BKY or BKX) are opened and the phase currents are not flowing there. The situation when the breakers are opened could be represented by the generic expression of the TPF definition in Equation (7). When the phase currents are zero, the total true and apparent powers also are zero; consequently, the TPF result is indeterminate, and the SEL 700GT relay measures the TPF as +1.00.

$$If\ TPF = \frac{Total\ true\ power}{Total\ apparent\ power} = \frac{0.00}{0.00} = \nexists \rightarrow Then \text{ measured TPF} = +1.00. \quad (7)$$

The TPF is +1.00 when a breaker is opened; this is a boundary condition that is set by the manufacturer instead of measurement of an indetermined (swing) TPF. Thus, the TPF smart contract should not operate the grid-side (BKY) and wind farm-side (BKX) breakers. Figure 7 shows the measured TPF from the real-time simulation versus the relay. From the real-time simulator (Figure 7a), the measured TPF swings when the breaker BKX is opened. However, in the CGG system (Figure 7b), the measured TPF from the SEL 700GT relay is +1.00, validating the boundary condition calculated for the relay using Equation (7).

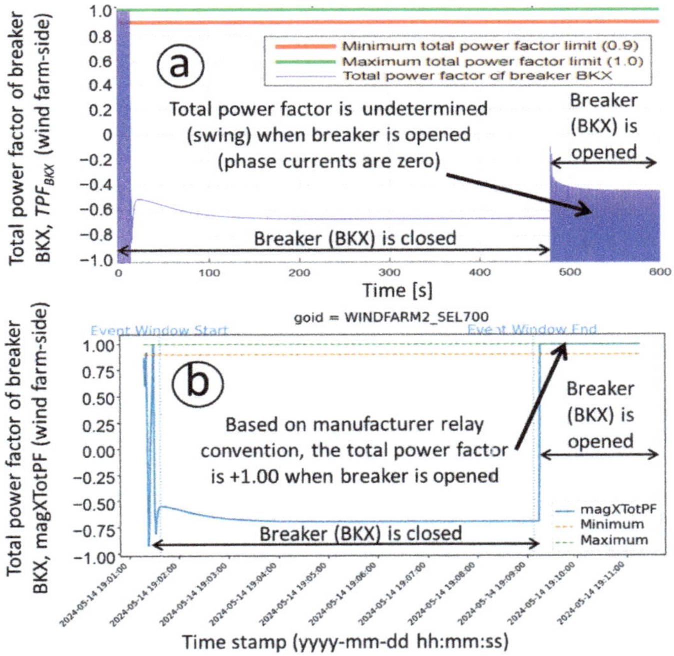

Figure 7. Measured TPF from real-time simulator (**a**) and CGG (**b**).

The objective of the TPF smart contract is to reduce the reactive power flowing along the power line at the grid side when an electrical utility grid is integrated with a customer-owned DER (wind farm); another objective is to maintain the data security and reliability offered by CGG using DLT and smart contracts. The power line losses caused by reactive power need to be kept with a TPF between +1.00 and +0.90. A flowchart of the TPF smart contract for the time window operation is shown in Figure 8.

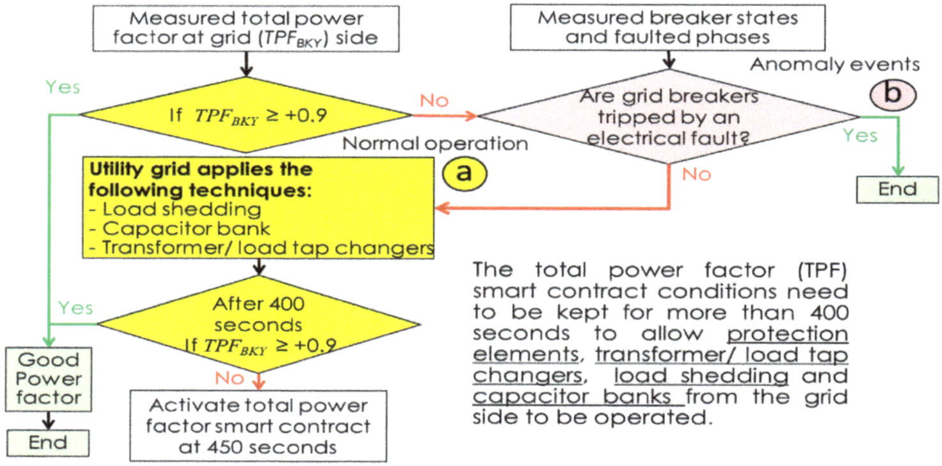

Figure 8. Time window flowchart for TPF smart contract.

4.2. Time Window Flowchart

In this study, the proposed time window operation for the TPF smart contract is defined at 450 s; the time was selected for normal grid operation. This time window allows breaker BKX and BKY to be operated under the TPF smart contract following the power factor improvement techniques (like capacitor banks, load shedding, transformer/load tap changers) used in the grid-side utility (Figure 8a). The time window of 450 s is also chosen so that the TPF smart contract is not operated for a transient anomaly event that generates a poorly measured TPF for only a couple of cycles, such as an electrical fault that is cleared by nearby relays (Figure 8b). The TPF smart contract boundary conditions need to be measured and held for a period of 450 s to allow protection elements, such as transformer/load tap changers, load shedding, and capacitor banks from the grid-side, to be used, before the BKX and BKY breakers will be operated with the CGG system using DLT and smart contracts.

4.3. Total Power Factor Smart Contract Flowchart

The phases and main conditions of the TPF smart contract are shown in the flowchart in Figure 9. The TPF smart contract was defined by a breaker non-operation condition for a measured TPF between +1.00 and +0.90 (good power factor), and a breaker operation condition defined for measured TPFs smaller than +0.90 (poor power factor). Based on the definition of good and poor power factors, the TPF boundaries and time window of 450 s were chosen for the flowchart of the TPF smart contract.

The flowchart (Figure 9) was created based on the power factor measurement behavior of the SEL 700GT relay from Equation (7) and the TPF or three-phase power factors (PF3X and PF3Y) measured from the relay using the CGG with IEC 61850 GOOSE messages. In Figure 9, the phases of the TPF smart contract flowchart are defined by the measurements (I), time state conditions (II), breaker state conditions (III), total power factor conditions (IV), and alarms for breaker operations (V).

In phase (I) of the flowchart (Figure 9a), the time (day, month, year, minutes, and seconds), states, and TPFs of the BKY (grid side) and BKY (wind farm side) breakers are measured. In phase (II) of the flowchart (Figure 9b), the time sate conditions are assessed for the TPF smart contract. This contract must consider periods when disturbances related to low-power-quality scenarios and shut down situations on the wind farm impact the utility grid-side. These scenarios are programmed maintenance operations (November), possible extreme ice accumulation on the blades (January), and excess noise site regulations (from

10 p.m. to 7 a.m.). The time stamps are measured in the "dd/MM/yyyy HH:mm:ss.SS" format from the SEL 700GT relay to activate the TPF smart contract (Figure 9b).

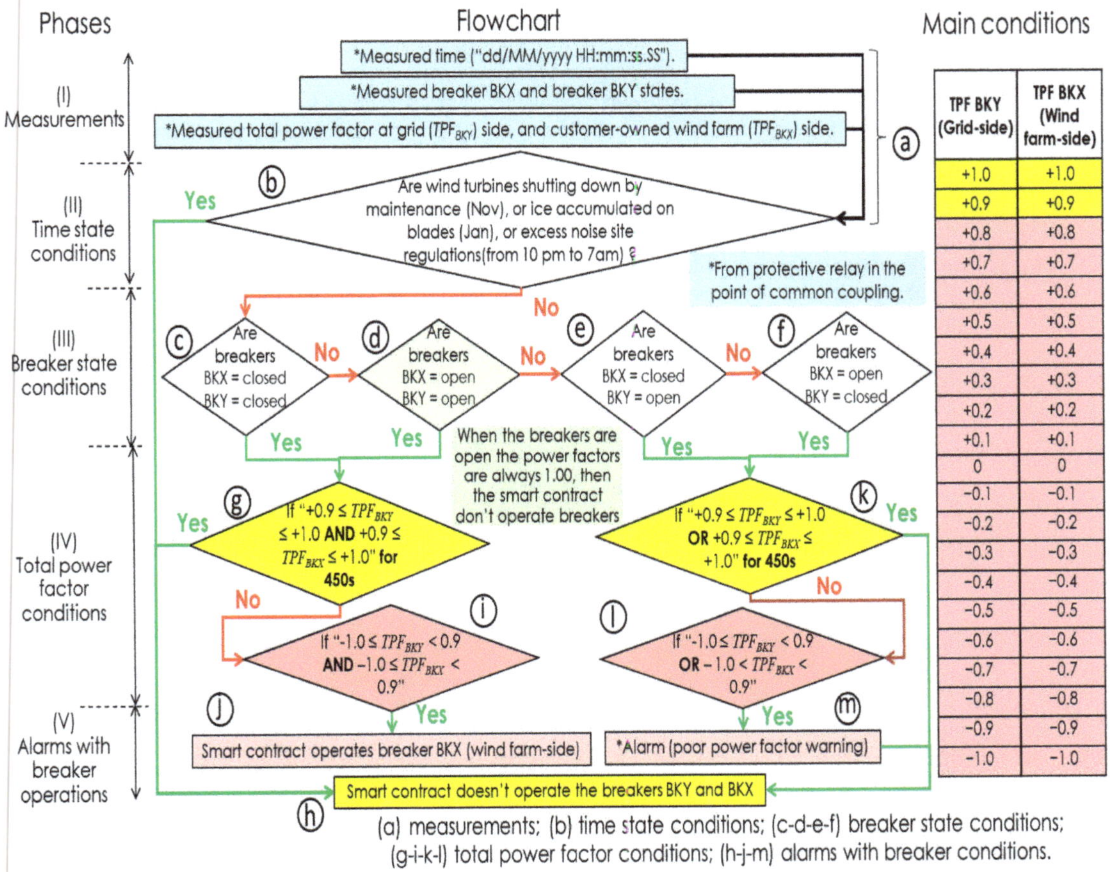

Figure 9. Phases and main conditions of TPF smart contract alongside a flowchart of the operation.

In phase (III) of the flowchart (Figure 9c–f), the breaker state conditions are assessed. The breaker state situations were defined for scenarios for the conditions of the flowchart when the BKX (wind farm side) and BKY (grid side) breakers are closed and open. In phase (IV) of the flowchart (Figure 9g,i,k,l), the AND/OR logic TPF conditions for the BKX (wind farm side) and BKY (grid side) breakers are assessed. In phase (V) of the flowchart (Figure 9h,j,m), the TPF smart contract operates the BKX breaker (wind farm side) for a measured TPF smaller than +0.90 on the grid side (TPF_{BKY}). Also, a warning alarm is activated in case the TPF smart contract does not operate the BKX (wind farm-side) and BKY (grid-side) breakers.

The main goal of the TPF smart contract flowchart (Figure 9) is to prioritize the operation of breaker BKX (wind farm-side) instead of breaker BKY (grid-side), because breaker BKY (grid-side) feeds the main load feeders (Figure 1a), and the main electrical grid (utility A) does not permit the breaker to be controlled by the smart contract of the CGG. In addition, the smart contract does not operate the BKY and BKX breakers for a well-measured TPF between +0.90 and +1.00. In the flowchart of the TPF smart contract (Figure 9) the following conditional sequences are defined:

- If both breakers are closed (Figure 9c), and the measured TPFs on the grid side (TPF_{BKY}) and wind farm side (TPF_{BKX}) are between +0.90 and +1.00 (Figure 9g) for 450 s, then the smart contract does not operate any breakers (Figure 9h). However, if the measured TPFs on the grid side (TPF_{BKY}) and wind farm side (TPF_{BKX}) are smaller than +0.90 (Figure 9i) after 450 s, then the smart contract operates the BKX breaker on the wind farm side (Figure 9j).
- If both breakers are open (Figure 9d), and the measured TPFs on the grid-side (TPF_{BKY}) and wind farm-side (TPF_{BKX}) are +1.00 because of the measurement relay condition in Equation (7) when breakers are open, then the smart contract does not operate any breakers (Figure 9h).
- If only one breaker is closed (Figure 9e,f), and the measured TPF on the grid side (TPF_{BKY}) or wind farm side (TPF_{BKX}) is between +0.90 and +1.00 (Figure 9k) for 450 s, then the smart contract does not operate any breakers (Figure 9h). However, if the measured TPF on the grid side (TPF_{BKY}) or wind farm side (TPF_{BKX}) is smaller than +0.90 (Figure 9l) after 450 s, then a warning alarm is triggered (Figure 9m), to indicate a poor power factor without operating the BKY (grid-side) and BKX (wind farm-side) breakers.

5. Cyber Grid Guard with DLT and Smart Contract

5.1. Cyber Grid Guard and Definitions

The CGG system is a modular software platform created for handling data ingestion and processing tasks related to securing IED data and configuration settings. It is responsible for ingesting data into the storage layer, consisting of on-chain (DLT) and off-chain (SQL database) data stores, and using that data to detect anomalies and other issues. At a high level, the modules are organized around tasks that include data ingestion, processing, and anomaly detection. Data ingestion involves collecting data and configuration settings from devices via the network using protocols such as IEC 61850 GOOSE and FTP. Data are stored on-chain in the DLT by sending transactions using smart contract functions; they are stored off-chain in a PostgreSQL database with the TimescaleDB extension. Data processing includes converting data from wire formats to JSON for more convenient storage and parsing or generating statistics. Anomaly detection is carried out using methods such as analyzing generated statistics and looking at comparisons of on-chain and off-chain data to detect anomalous events. Table 3 shows the definitions of terms related to DLT.

Table 3. Definitions of terms related to Distributed Ledger Technology.

Terms	Definitions
Cyber Grid Guard (CGG)	Cyber Grid Guard is a modular software framework that performs data integrity attestation using a storage layer consisting of a database and DLT.
Distributed Ledger Technology (DLT)	Distributed Ledger Technology (DLT) refers to distributed and decentralized secure database platforms that use consensus mechanisms to reach agreements on state. The most common types of DLT are implemented using blockchains.
Smart Contracts	Smart contracts are automated programs that run on top of DLTs and enforce terms and conditions on transaction data and are executed in a distributed manner.
On-chain	On-chain refers to functionality or data that exist in the distributed ledger.
Off-Chain	Off-Chain refers to functionality or data that exist outside of the distributed ledger (e.g., in a traditional database).

5.2. Architecture of Smart Contract

The TPF smart contract was implemented with the CGG using DLT and an SEL 700GT relay. The smart contract for checking TPF conditions for the BKY (grid side) and BKX (wind farm side) breakers was created using Hyperledger Fabric (HLF) 2.5. HLF is a modular, open-source, and permissioned DLT platform. It is designed for enterprise-grade, general-purpose use test cases. Smart contracts are defined as self-executing contracts in which the contract terms are defined in the code. They operate on DLT platforms, enabling execution and automatic enforcement without the need for intermediaries. In the HLF DLT platform, smart contract programs are referred to as "chaincode" and can be implemented in various programming languages. In this study, the Go language was used to implement the TPF chaincode. The TPF chaincode is used to store IEC 61850 GOOSE-based measurements, breaker states, and the TPFs presented in Table 4 from the WINDFARM2_SEL700 relay in the ledger and initiate the condition checks.

Table 4. GOOSE data set fields used in event checks.

Field Name	Data Type	Data Description
magXTotPF	Float32	Total power factor for X.
magYTotPF	Float32	Total power factor for Y.
breakerXOpen	Boolean	Status of breaker X (open = 1, closed = 0).
breakerYOpen	Boolean	Status of breaker Y (open = 1, closed = 0).

The architecture of the implemented system is shown in Figure 10. The software modules form parts of the CGG system. The IEC 61850 GOOSE measurements in Table 4 are published by the WINDFARM2_SEL700 relay and received by the observer program. This program uses the *libiec61850* library to receive GOOSE messages and process the data, which includes filtering duplicate messages and formatting them as JSON. The TPF ledger storage module handles receiving the formatted GOOSE messages and creating and sending DLT transactions with the relevant data using the TPF chaincode AddOrUpdateRecord function.

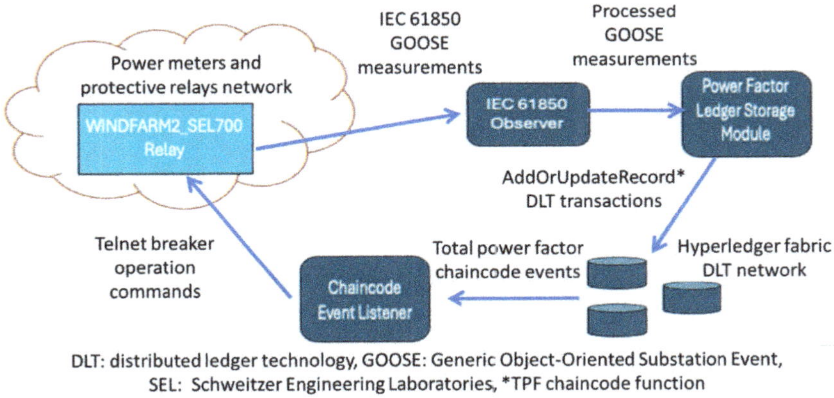

Figure 10. Architecture of the DLT with TPF condition chaincode.

The conditions for the TPF smart contract are created based on the flowcharts in Figures 8 and 9. The TPF smart contract is a backup power quality method that can be applied following the use of load shedding, transformer/load tap changers, and capacitor bank applications on the grid side. Therefore, a time of 450 s is implemented for measuring TPF conditions and operating the BKX (DER side) breaker. These conditions depend on the measured time, breaker state, and TPF, as shown in Figure 11. When a condition is met (based on the conditions and logic detailed in Figure 11), a DLT event is emitted. This event is a JSON message that is received by a listener application that handles operating a breaker. The listener application implemented was a Go program using the HLF Fabric Gateway client API version 1.4.0 to receive the event from the smart contract and handle the communication with the SEL 700GT relay that operates the breakers. This design choice was made because smart contracts are unable to directly interact with external resources. They are executed in a distributed manner and must reach a result that is consistent and deterministic.

(a)

Phase	Measured smart contract conditions (IF)			Smart contract decisions (THEN)
	BKY (Grid-side)	BKX (DER-side)	Total power factor for a period of 450 seconds	
No breaker operation	closed	closed	+0.90 ≤ TPF BKY ≤ +1.00 **AND** +0.90 ≤ TPF BKX ≤ +1.00	**Do not operate the breakers BKY and BKX**
No breaker operation	closed	open	+0.90 ≤ TPF BKY ≤ +1.00 **OR** +0.90 ≤ TPF BKX ≤ +1.00	**Do not operate the breakers BKY and BKX**
No breaker operation	open	closed		
No breaker operation	open	open	When breakers are opened, then the measured power factor is 1.00; therefore, breakers must not be operated by smart contract	**Do not operate the breakers BKY and BKX**
Breaker operation	closed	closed	−1.00 ≤ TPF BKY < +0.90 **AND** −1.00 ≤ TPF BKX < +0.90	**Operate breaker BKX**
Breaker operation	closed	open	−1.00 ≤ TPF BKY < 0.90 **OR** −1.00 ≤ TPF BKX < 0.90	**Alarm* (poor power factor warning)**
Breaker operation	open	closed		

*Do not operate the breakers BKY (Grid-side) and BKX (Wind farm-side)

(b)

TPF BKY (Grid-side)	TPF BKX (Wind farm-side)
+1.0	+1.0
+0.9	+0.9
+0.8	+0.8
+0.7	+0.7
+0.6	+0.6
+0.5	+0.5
+0.4	+0.4
+0.3	+0.3
+0.2	+0.2
+0.1	+0.1
0	0
−0.1	−0.1
−0.2	−0.2
−0.3	−0.3
−0.4	−0.4
−0.5	−0.5
−0.6	−0.6
−0.7	−0.7
−0.8	−0.8
−0.9	−0.9
−1.0	−1.0

Figure 11. Measured smart contract conditions (**a**) and boundaries (**b**).

The smart contract conditions (Figure 11a) that represent the TPF boundaries are shown in yellow and red in Figure 11b. If the measured TPF conditions are reached for a duration of 450 s, then the wind farm-side breaker is operated for a measured TPF smaller than 0.90. The breaker states are based on the "AND/OR" logic conditions in Figure 11a. When a condition check is triggered, a transaction is sent to the DLT using the CheckConditions function of the TPF chaincode; the TPF smart contract conditions (Figure 11) are evaluated for the past 450 s period based on the provided timestamp. If a condition is met, a chaincode event will be generated. The chaincode event listener subscribes to these events and, upon receiving one, will initiate the required action. If the event requires a breaker operation, the listener will trigger the code to send a breaker operation command using the Telnet protocol.

6. Results

6.1. Test Scenarios

In this study, a CGG using DLT and a TPF smart contract was tested under normal and electrical fault situations. In a normal situation, the TPF smart contract conditions are delayed for 450 s to allow for the operation of transformer/load tap changers, load shedding, and/or capacitor banks on the grid side. After that period, the CGG can implement the smart contract using DLT and the relay. As in an electrical fault situation, the TPF smart contract should not operate the breakers in the SEL 700GT relay. Therefore, based on the maximum time limit for the frequency protection elements of 400 s (Table 2), the TPF smart contract time window was set at 450 s. Tests 1 and 2 were based on normal electrical grid situations. In test 1, the main feeder loads were connected only with the grid-side breaker (BKY), which measured TPFs between +0.90 and +1.00; in that case, the smart contract did not operate any breaker. In test 2, the main feeder loads were connected with the grid-side breaker (BKY) and the wind farm-side breaker (BKX). In this case, the grid-side breaker (BKY) measured a poor TPF, and the wind farm-side breaker (BKX) was opened.

Tests 3 and 4 were based on temporary electrical fault situations—such as lightning, windblown tree branches or wires, birds, or rodents [51]—that do not initiate the TPF smart contract because they last only for a few cycles or s [51]. In test 3, the main feeder loads were connected only with the grid-side breaker (BKY) and measured a TPF between +0.90 and +1.00. In this case, a temporary three line-to-ground (3LG) electrical fault of 1 s (60 cycles) at the grid-side power line was set, but the smart contract did not operate a breaker because the electrical fault cleared itself. In test 4, a similar scenario was presented, but a temporary single line-to-ground (SLG) electrical fault of 1 s (60 cycles) in the grid-side power line occurred. Table 5 shows the test scenarios implemented based on the three-phase power system of the substation grid with customer-owned DER (a wind farm) in Figure 4.

Table 5. Test scenarios.

*† TEST 1: Normal situation and no breaker operation.	Title: Main feeder loads are connected only with the grid-side breaker (BKY) that measures a TPF ≥ +0.90; thus, the smart contract does not operate any breakers.
	Description: Initially the grid-side breaker (BKY) is closed and the wind farm-side breaker (BKX) is opened. The wind farm–side load has 2.06 MW, +1000 VAR, −1000 VAR. The fossil fuel power plant (Utility C) and the substation (Utility A) are connected to the main feeder loads (2 × [1.25 MW, +0.5 MVAR, −1000 VAR]). The grid-side breaker (BKY) shows a TPF ≥ +0.90, meaning reactive losses are practically none on the grid-side power line. The TPF smart contract does not operate the breakers or the grid side (BKY) or wind farm (BKX) for a TPF ≥ +0.90.
* TEST 2: Normal situation and wind farm-side breaker (BKX) operation.	Title: Main feeder loads are connected with the grid-side breaker (BKY) and wind farm-side breaker (BKY); then the grid-side breaker (BKY) measures a poor TPF, and the wind farm-side breaker (BKX) is opened.
	Description: Initially the grid-side breaker (BKY) is closed and the wind farm-side breaker (BKX) is closed. The wind farm-side load has 2.06 MW, +1000 VAR, −1000 VAR. The fossil fuel power plant (Utility C) through the substation (Utility A) and the wind farm are connected to the main feeder loads (2 × [1.25 MW, +0.5 MVAR, −1000 VAR]). The grid-side breaker (BKY) shows a TPF < +0.90, meaning that the TPF in the grid-side power line generates reactive losses. The wind farm-side breaker (BKX) is opened after 400 s, and the TPF is ≥ +0.90 at the grid-side breaker (BKY).
*† TEST 3: Temporary 3LG electrical fault situation and no breaker operation.	Title: Main feeder loads are connected only with the grid-side breaker (BKY) and measure a TPF > +0.90. Thus, a temporary 3LG electrical fault of 1 s (60 cycles) at the grid-side power line is set, but the smart contract does not operate any breaker because the electrical fault is cleared.
	Description: Initially the grid-side breaker (BKY) is closed and the wind farm–side breaker (BKX) is opened. The wind farm–side load has 2.06 MW, +1000 VAR, −1000 VAR. The fossil fuel power plant (Utility C) through the substation (Utility A) is connected to the main feeder loads (2 × [1.25 MW, +0.5 MVAR, −1000 VAR]). Then, a temporary 3LG electrical fault of 1 s (60 cycles) at the end of the grid-side power line is set at 100 s. The grid-side breaker (BKY) shows a swing and poor power factor during the fault state, until the fault is clears by itself. However, the TPF smart contract does not operate the grid-side breaker (BKY), and the wind farm–side breaker (BKX) remains open.
*† TEST 4: Temporary SLG electrical fault situation and no breaker operation.	Title: Main feeder loads are connected only with the grid-side breaker (BKY) and measure a TPF > +0.90. Thus, a temporary SLG electrical fault of 1 s (60 cycles) at the grid-side power line is set, but the smart contract does not operate any breaker because the electrical fault is cleared.
	Description: Initially the grid-side breaker (BKY) is closed and the wind farm-side breaker (BKX) is opened. The wind farm-side load has 2.06 MW, +1000 VAR, −1000 VAR. The fossil fuel power plant (Utility C) and the substation (Utility A) are connected to the main feeder loads (2 × [1.25 MW, +0.5 MVAR, −1000 VAR]). Then, a temporary SGL electrical fault of 1 s (60 cycles) at the end of the grid-side power line is set at 100 s. The grid-side breaker (BKY) swings and shows a poor power factor during the fault state, until the fault clears by itself. However, the TPF smart contract does not operate the grid-side breaker (BKY), and the wind farm-side breaker (BKX) remains open.

* Tests 1–4 were run for a total time of 600 s. † Tests 1, 3, and 4 started the simulation with all breakers closed; then, the breaker BKX (wind farm-side) was tripped manually from the SEL 700GT relay to obtain the initial condition, 3LG (three lines to ground), and SLG (single line to ground).

6.2. Measured Total Power Factor

The TPF smart contract was assessed by comparing the measured data from the real-time simulator, CGG, and relay. The data collected from the CGG system were used to operate the TPF smart contract. The desired TPF limits on the grid-side were defined as being between +0.9 and +1.0, and the operation of the breakers on the grid side and wind farm side was controlled by the SEL 700GT relay using the smart contract. The events from the real-time simulator, CGG, and relay showed a successful assessment of the TPF smart contract with the CGG using DLT. The TPF smart contract was assessed for normal situation tests and temporary electrical fault tests; then, the CGG using DLT with the smart contract was evaluated for normal and anomalous event situations, observing operation and non-operation of the smart contract. The normal situation tests (Figures 12 and 13) and temporary electrical fault situation tests (Figures 14 and 15) were run for a total simulation time of 600 s to allow for the implementation of the smart contract for a TPF smaller than +0.9 on the grid-side for more than 400 s. In the temporary electrical fault situation tests, for a couple of cycles, a measured TPF smaller than +0.9 during fault states was observed; however, the TPF smart contract did not operate the SEL 700GT relay breakers in the PCC between the electrical grid side and the wind farm side because anomalous transient events were present.

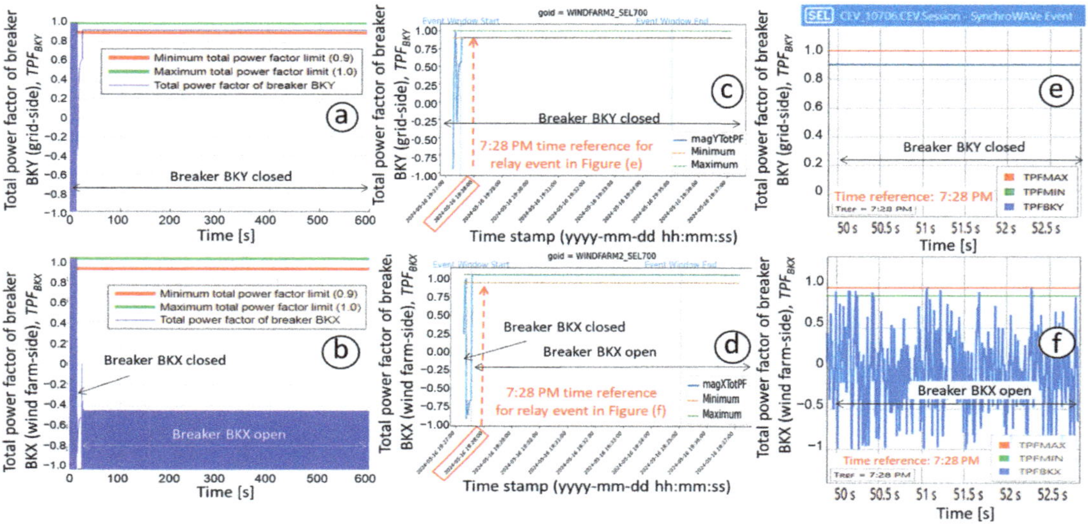

Figure 12. Measured TPF from real-time simulator (**a**,**b**), CGG (**c**,**d**), and relay (**e**,**f**) in test 1 for a normal situation and no breaker operation.

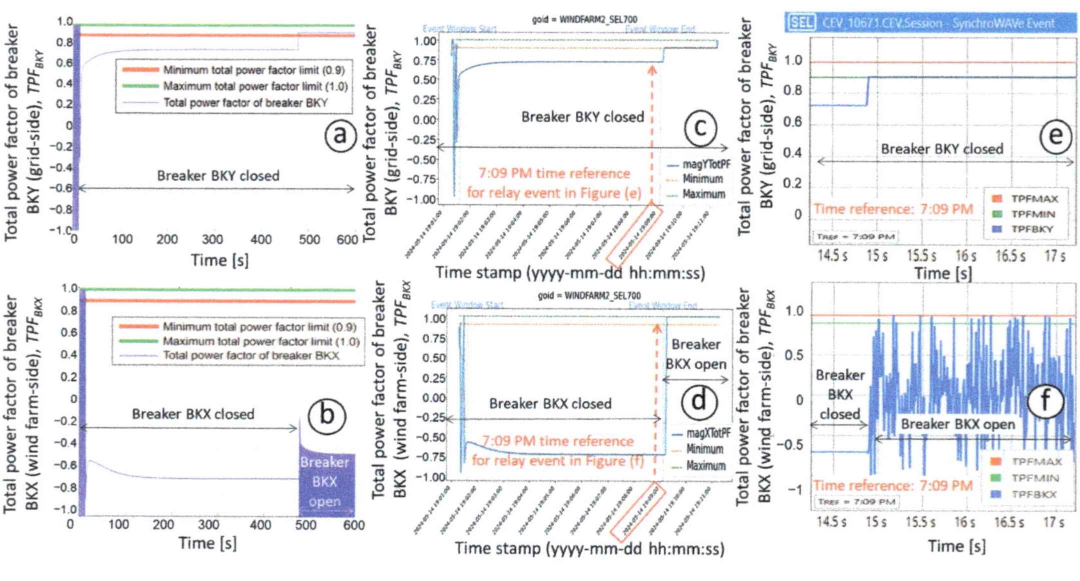

Figure 13. TPF measured using real-time simulator (**a**,**b**), CGG (**c**,**d**), and relay (**e**,**f**) in test 2 for a normal situation and wind farm-side breaker BKX operation.

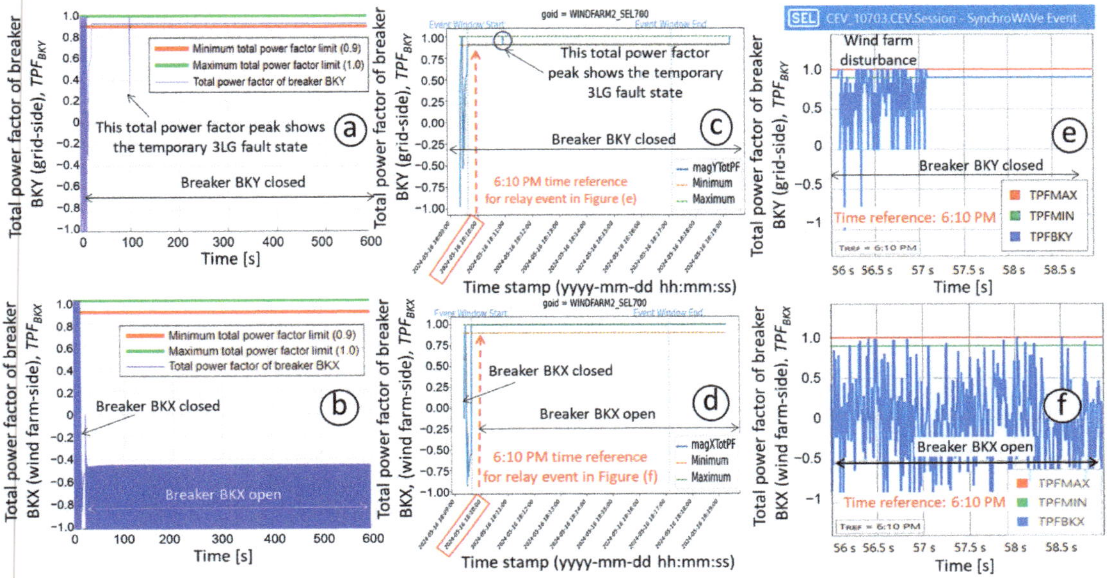

Figure 14. TPF measured using real-time simulator (**a**,**b**), CGG (**c**,**d**), and relay (**e**,**f**) in test 3 for temporary 3LG electrical fault situation and no breaker operation.

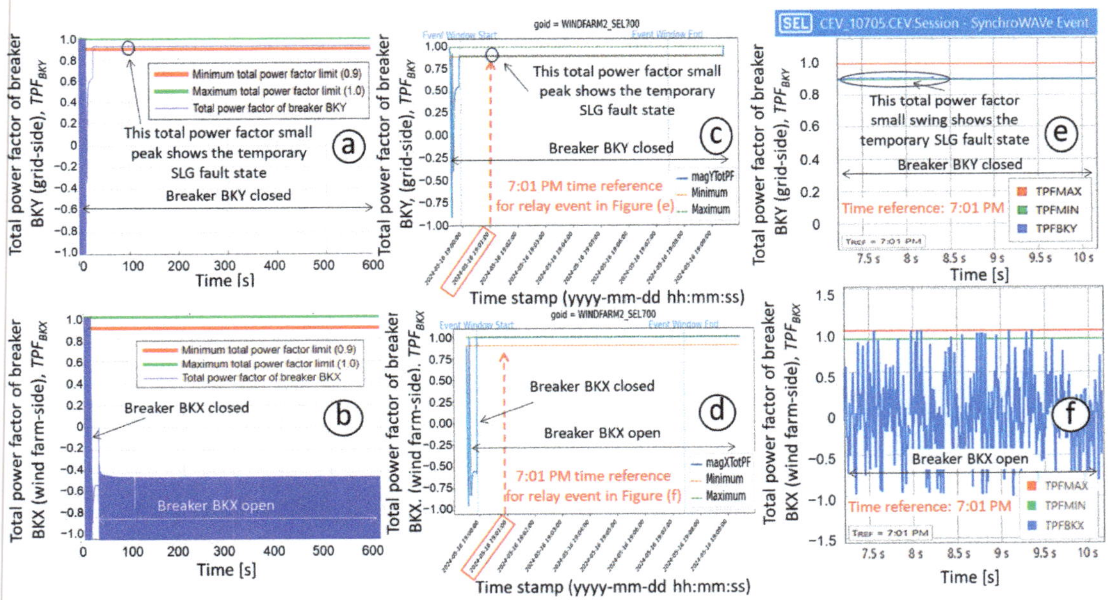

Figure 15. TPF measured using real-time simulator (**a**,**b**), CGG (**c**,**d**), and relay (**e**,**f**) in test 4 for temporary SLG electrical fault situation and no breaker operation.

6.2.1. Normal Situation Tests

The normal situation in test 1 (Table 5) presents a comparison between the measured TPFs of the grid side (TPF_{BKY}) and wind farm side (TPF_{BKX}) obtained from the real-time simulator (Figure 12a,b), the CGG (Figure 12c,d), and the relay (Figure 12e,f). The real-time simulator and CGG recorded the TPF plots for 600 s, and the relay recorded the TPF plots at a specific time (7:28 p.m.). Figure 12a,c,e show the measured TPFs for the grid side (TPF_{BKY}) and Figure 12b,d,f show the measured TPFs of the wind farm side (TPF_{BKX}). In the real-time simulator and CGG (Figure 12a–d), initially, the grid simulation has all breakers closed and a measured TPF swing of around ± 1.00 for a couple of s when the wind farm is connected. Then, the wind farm-side breaker (BKX) is opened to obtain the desired conditions for test 1. From Figure 4, in this test, the fossil fuel power plant (Utility C) and the substation (Utility A) are connected to the main feeder loads. The grid-side breaker (BKY) presents a TPF between +0.90 and +1.00, which means that reactive losses are practically nonexistent on the grid-side power line. The TPF smart contract does not operate the breakers on the grid side (BKY) or the wind farm side (BKX) for a TPF between +0.90 and +1.00. In Figure 12b, as can be seen from the real-time simulator, a TPF swing was measured for the wind farm side because the breaker (BKX) was opened and the phase currents were zero; then, an indetermined value was measured based on Equation (4). However, from the CGG, the wind farm-side TPF (Figure 12d) was measured to be +1.00 because the manufacturer's conditions for the relay state that when the breaker is opened at a measured TPF of +1.00, an indetermined value is not measured. However, from the relay event (Figure 12f) at 7:28 p.m., a wind farm-side TPF swing was seen because the breaker (BKX) was opened and phase currents were zero, and then an indeterminate value was measured based on Equation (4). In this normal situation test, it was concluded that the TPF smart contract did not operate the relay's breakers because the measured TPF of the grid-side (TPF_{BKY}) was between the TPF limits of +0.90 to +1.00.

The normal situation in test 2 (Table 5) presents a comparison between the measured TPFs for the grid side (TPF_{BKY}) and wind farm side (TPF_{BKX}) obtained from the real-time simulator (Figure 13a,b), CGG (Figure 13c,d), and relay (Figure 13e,f). The real-time simulator and CGG record the TPF plots for 600 s, and the relay records the TPF plots at a specific time (7:09 p.m.). Figure 13a,c,e show the measured TPFs for the grid-side (TPF_{BKY}) and Figure 13b,d,f show the measured TPFs for the wind farm side (TPF_{BKX}). In the real-time simulator and CGG (Figure 13a–d), initially, the grid simulation has all breakers closed and the measured TPF swing is around $\pm$ 1.00 for a couple of seconds when the wind farm is connected. Then, the measured TPF for the grid-side (TPF_{BKY}) is smaller than +0.90 for the real-time simulator and CGG plots (Figure 13a–c) for more than 400 s. As seen in Figure 4, in this test, the fossil fuel power plant (Utility C), the substation (Utility A), and the wind farm (Utility B) are connected to the main feeder loads. The grid-side breaker (BKY) shows a TPF smaller than +0.90, meaning the TPF on the grid-side power line generates reactive losses. The wind farm-side breaker (BKX) is opened after 400 s. and a TPF between +0.90 and +1.00 is obtained for the grid side breaker (BKY). In Figure 13b, from the real-time simulator, TPF swing was measured for the wind farm side because the breaker (BKX) was opened and phase currents were zero; then, an indeterminate value was measured based on Equation (4). However, the wind farm-side TPF measured from the CGG (Figure 13d) is +1.00 because the manufacturer's conditions for the relay state that, when the breaker is opened, the measured TPF is +1.00. However, from the relay event (Figure 13f) at 7:09 p.m., TPF swing was measured for the wind farm side because the breaker (BKX) was opened and the phase currents were zero, and then an indeterminate value was measured based on Equation (4). In this normal situation test, it was concluded that the TPF smart contract opened the wind farm-side breaker (BKX) of the relay because the TPF measured on the grid side (TPF_{BKY}) was smaller than +0.90.

6.2.2. Temporary Electrical Fault Situation Tests

The temporary electrical fault in test 3 (Table 5) represents the TPFs measured for the grid side (TPF_{BKY}) and wind farm side (TPF_{BKX}) by the real-time simulator (Figure 14a,b), the CGG (Figure 14c,d), and the relay (Figure 14e,f). The real-time simulator and CGG recorded the TPF plots for 600 s, and the relay recorded the TPF plots at a specific time (6:10 p.m.). Figure 14a,c,e show the TPF measured for the grid side (TPF_{BKY}) and Figure 14b,d,f shows the TPF measured for the wind farm side (TPF_{BKX}). In the real-time simulator and CGG

(Figure 14a–d), initially, the grid simulation has all breakers closed and a measured TPF swing of around ±1.00 for a couple of seconds when the wind farm is connected. Then, the wind farm-side breaker (BKX) is opened to obtain the desired initial conditions for test 3. From Figure 4 it can be seen that, in this test, the fossil fuel power plant (Utility C) and the substation (Utility A) are connected to the main feeder loads. Then, a temporary 3LG electrical fault of 1 s (60 cycles) at the end of the grid-side power line occurs for 100 s. Then, the grid-side breaker (BKY) presents a poor power factor during the fault state (Figure 14a), until a time when the temporary electrical fault clears by itself. However, the TPF smart contract does not operate the grid-side breaker (BKY) and wind farm-side breaker (BKX) because this is a temporary electrical fault situation. In Figure 14a, from the real-time simulator, it can be seen that the measured grid-side TPF presents a peak during a fault state, like the grid-side TPF measured by the CGG (Figure 14c). However, from the relay event (Figure 14f) at 6:10 p.m., the measured wind farm-side TPF swings because the breaker (BKX) is opened and the phase currents are zero, and then an indeterminate value is measured based on Equation (4). In this temporary electrical fault situation test, it was concluded that the TPF smart contract did not operate the relay's breakers because the poorly measured TPF on the grid side (TPF_{BKY}) was measured during a temporary electrical fault.

The temporary electrical fault in test 4 (Table 5) represents the TPFs for the grid side (TPF_{BKY}) and wind farm side (TPF_{BKX}) measured by the real-time simulator (Figure 15a,b), CGG (Figure 15c,d), and relay (Figure 15e,f). The real-time simulator and CGG record the TPF plots for 600 s, and the relay records the TPF plots at a specific time (7:01 p.m.). Figure 15a,c,e show the TPF measured for the grid side (TPF_{BKY}) and Figure 15b,d,f show the TPF measured forthe wind farm side (TPF_{BKX}). In the real-time simulator and CGG (Figure 15a–d), initially, the grid simulation has all breakers closed and a measured TPF swing of around ±1.00 for a couple of seconds when the wind farm is connected. Then, the wind farm side breaker (BKX) is opened to obtain the desired initial conditions for test 4. From Figure 4, in this test, the fossil fuel power plant (Utility C) and the substation (Utility A) are connected to the main feeder loads. Then, a temporary SLG electrical fault of 1 s (60 cycles) at the end of the grid-side power line occurs for 100 s. Then, the grid-side breaker (BKY) presents a small change in TPF during the fault state (Figure 15a,c,e), until the temporary electrical fault clears by itself. However, the TPF smart contract does not operate the grid-side (BKY) and wind farm-side (BKX) breakers because this is a temporary electrical fault situation. In Figure 15a, from the real-time simulator, it can be seen that the measured grid-side TPF shows a small peak during the fault state, like the measured grid-side TPF from the CGG (Figure 14c). However, from the relay event (Figure 15f) at 7:01 p.m., it can be seen that the measured wind farm-side TPF swings because the breaker (BKX) is opened and the phase currents are zero, and then an indeterminate value is measured based on Equation (4). In this temporary electrical fault situation test, it was concluded that the TPF smart contract did not operate the relay's breakers during a temporary electrical fault.

7. Discussion

The experimental model for this research was based on using different case studies (normal situation tests and temporary electrical fault tests) instead of performing statistical studies. As such, the results from the real-time simulator, CGG and relay events were collected to compare the behavior of the TPFs measured on the utility grid side and the customer-owned wind farm side. As the main goal was to assess the TPF smart contract, a statistical analysis was not needed in this study because the results were based on plotting the behavior of measured TPFs (Figures 12–15) without quantifying the results. These measured TPF plots from the real-time simulator, CGG, and relay validated the TPF smart contract flowchart in Figure 9.

In the TPFs measured by the real-time simulator (Figures 12 and 13a,b) in tests 1 and 2 (normal situations) without breaker operation, a measured swing TPF behavior was observed for the breakers on the grid side (TPF_{BKY}) and wind farm side (TPF_{BKX}). In these real-time simulation tests, initially, the electrical grid had all breakers closed and a measured TPF swing of around ±1.00 for a couple of s was assessed when the wind farm was connected. This transient event behavior with a measured swing TPF is related to the impact of connecting the customer-owned wind farm (six 1.5 MW wind turbines) to the load feeders on the grid side. This transient event shows a ±1.00 variation in the measured TPF that is related to a power swing phenomenon that occurs when disturbances in a power system cause oscillations in the active and reactive power flows of the power lines. If these transient events continue for a long period of time, it could affect the electrical grid reliability by triggering the breakers' protective relay for non-desired transient events, consequently generating possible protective relay misoperations because of wrongly calibrated time windows for protection functions (over–under frequency, over–under voltage, etc.).

The TPF smart contract was satisfactorily evaluated by analyzing the test scenarios of normal (Figures 12 and 13) and non-normal (Figures 14 and 15) operations. In the normal operation, a good TPF between +0.90 and +1.00 (grid side) and more than 400 s was observed as the breakers were not operated by the smart contract (Figure 12). In the normal operation, a poor TPF smaller than +0.90 (grid side) and more than 400 s was observed as the wind farm side breaker (BKX) was operated by the smart contract (Figure 13). In non-normal operations like temporary electrical faults that could occur over a period of less than 400 s and could cause a poor TPF smaller than +0.90 on the grid-side (Figure 14a), the smart contract did not operate the breakers of the SEL 700GT relay (Figures 14 and 15). The test scenarios for the temporary electrical faults were performed at the 3LG (Figure 14) and SLG (Figure 15) faults located at the end of the power line on the grid side, and these temporary faults had durations of 60 cycles (1 s).

In the test scenarios of the temporary electrical faults (Figures 14 and 15), the measured TPFs at the fault states on the grid side (TPF_{BKY}) showed a high and low swing value for the 3LG (Figure 14a) and SLG (Figure 15a) fault, respectively. This is because the 3LG fault had a power factor impact greater than the SLG fault. Also, the TPFs measured for the grid side (TPF_{BKY}) and wind farm side (TPF_{BKX}) at the fault states in the real-time simulator (Figures 14a and 15a) had a better resolution than the CGG (Figures 14b and 15b) because the real-time simulator used a time step of 50 micro-seconds to calculate the total power factor values, and the CGG collected the three-phase power factors (PF3X and PF3Y) from the relay using IEC 61850 GOOSE messages. However, if a relay with time domain protocols is connected to the CGG using DLT and smart contracts, the accuracy of the measurements and algorithms could be improved for the CGG. The implementation of sample values or phasor measurement unit protocols in the measurement of voltage/current harmonics on inverter-based photovoltaic (IB-PV) arrays in the PCC could enhance the power system applications of both DLT and smart contracts.

The three-phase power system circuit of the grid substation with the customer-owned wind farm (Figure 4) simulated real devices (power lines, loads, power transformers, wind farm, etc.) because the simulation time step was 50 micro-seconds in the real-time simulation tests, with real IEDs (relays and meters) in the loop. Then, the noise (or harmonics) from the phase current/voltage signals for the wind farm interconnection and/or electrical fault states was simulated because of the time step of 50 micro-seconds that was used, which represents a sampling frequency of 20 kHz that can be used to measure up to 333 samples per cycle (166th harmonics).

In this research, a TPF smart contract flowchart (Figure 9) was designed based on the theory behind and traditional equations for power factors [26,27,31,32,47]. However, these factors were insufficient and so a study of the relay behavior at the PCC between the electrical grid and wind farm sides was necessary to define the boundaries of the TPF smart contract algorithm. This approach was based on the measured unity TPF when the breakers were opened for the relay, and the IEEE or/and IEC conventional power factor signs (IEEE was used for the SEL 700GT relay). Many of the studies of the electrical grid applications of smart contracts used software simulations [33–36] that did not use real IEDs, an approach that could result in issues with verifying their applicability to realistic scenarios, as opposed to when those smart contracts are integrated with HIL. A way of using real relays with substation protocols is critical to properly assess the proposed smart contract algorithms in the CGG system with DLT for an electrical grid utility with a customer-owned wind farm.

The TPF smart contract between the electrical grid and customer-owned wind farm is a backup method to reduce the reactive losses on the grid-side power line before applying the capacitor bank, load shedding, and transformer/load tap changer techniques from the grid side. In the TPF smart contract, the time taken to operate the breakers in the PCC will depend on the time for operating the power quality grid methods (capacitor bank, load shedding and transformer/load tap changers). Also, the time window will depend on the maximum time limit of the frequency-protection elements of the relays, based on the complexity of the electrical grid. The TPF smart contract needs to allow time for the power quality grid methods on the grid side; in this case, it was chosen to be 450 s, but it could be set at a different time depending on the control systems of electrical distribution utilities.

In the TPF smart contract, while DLT is capable of providing a low latency of less than 1000 ms for 10 transactions per second [52], the introduction of DLT may still introduce additional computational overhead and network jitter that could affect real-time performance. This overhead could include the time taken for transaction preparation, consensus algorithms, and data replication across nodes [53]. These factors raise concerns about scalability and real-time performance in large-scale grids. While designed as a backup method, DLT scalability and performance issues could lead to potential delays in time-critical processing commands such as trip commands for breakers in fault states. To mitigate these concerns, optimizations and enhancements can be applied to the DLT platform, such as sharding or partitioning the network to reduce the transaction load on individual nodes [54] or employing lightweight consensus algorithms [55]. For example, Hyperledger Fabric is a permissioned DLT, meaning all participants are known and authenticated. This avoids the need for resource-intensive consensus mechanisms designed to prevent malicious behavior in open, anonymous networks. Additionally, integrating DLT with existing communication protocols like GOOSE messages must be carefully designed to ensure the timely delivery of time-critical commands, as the transfer time for the trip command in the GOOSE messages must be within 3 ms [56]. Therefore, the CGG system with smart contract could be a more appropriate and useful tool in this application for measuring the TPF in a PCC between the grid side and customer-owned wind farm side, notably to operate breakers as a backup alternative through the relay using the smart contract.

Power systems are vulnerable to both cyberattacks and severe weather events, each posing distinct threats to their stability and functionality, imposing severe financial costs on grid operators. Cyberattacks can exploit vulnerabilities in smart grids and DERs, affecting integrity, confidentiality, availability, and accountability. Integrity attacks, such as false data injection (FDI) [57], can lead to unauthorized modifications of field measurement data, potentially causing cascading failures in the electric grid [14]. Severe weather events are a primary cause of large-scale power outages, particularly in distribution systems. These events, such as hurricanes and winter storms, have led to significant economic damages, with up to 90% of power failures being attributed to disruptions in the distribution system [58]. The implementation of DLT with smart contracts for monitoring power factor data at the PCC could offer significant economic benefits due to enhanced cybersecurity and reliability. The immutability of blockchain ensures that all data transactions are secure and tamper-evident, which reduces the risk of fraudulent activities and operational errors [59]. This increased data security can lead to cost savings by minimizing the need for additional manual oversight and reducing potential downtime due to cyber threats [14]. DLT also provides a resilient platform that ensures continuous operation even in the event of individual component failures. This reliability could translate into lower operational costs as it reduces the frequency and impact of outages. While initial investments in setting up a DLT-based system can be substantial in both cost and complexity [18], the long-term benefits in terms of reduced maintenance costs, enhanced security, and improved grid resilience make this approach economically viable for grid operators and utilities.

The TPF application using a CGG with smart contract was satisfactorily implemented on this study. The novelties herein are based on securing data between the utility grid and customer-owned wind farm and reducing the grid-side power line losses at the PCC, using a CGG with DLT and smart contracts. In addition, the TPF smart contract, applied with the CGG and relay—for controlling breakers at the PCC between the grid-side and customer-owned wind farm-side—was based on a smart contract with multiple boundaries represented by the operation time, breaker states, and power quality conditions, to obtain a TPF between 0.90 and 1.00 at the grid-side and consequently reduce power line losses. As a future research topic, the number of use case test scenarios with multiple DERs will be increased, focusing not only on power factor deviations during normal operations and temporary electrical faults, but also covering extreme electrical grid conditions like cyberattacks or protective relay misoperations. It will assess the robustness of the CGG system, evaluating the scalability and real-time performance of the DLT in large-scale grids with numerous PCCs for applications using customer-owned wind farms and IB-PV array farms, and in advanced smart contract applications, like the detection of transient anomaly events with high harmonics using time domain protocols.

8. Conclusions

The novel employment of CGG using DLT with a TPF smart contract is a modern electrical grid application, using a relay at the PCC between the main electrical utility and a customer-owned wind farm and protecting the integrity of shared data from the relay. In this study, the TPF smart contract with CGG using DLT for the electrical grid with a customer-owned wind farm was implemented, and the flowcharts for the TPF smart contract were satisfactorily assessed with a real-time simulator and a CGG system with relays in the loop.

This TPF smart contract was satisfactorily assessed in different test scenarios. In normal operations, a poor TPF smaller than +0.90 (grid-side) lasting for more than 400 s was observed as the breaker (wind farm side) was operated by the smart contract. A good TPF between +0.90 and +1.00 on the grid side for more than 400 s was observed as the breakers were not operated by the smart contract. Also, in abnormal conditions, such as temporary electrical faults that resulted in a poor TPF smaller than +0.90 on the grid side, the smart contract did not operate the breakers of the SEL 700GT relay.

The TPF smart contract improved the TPF on the grid-side, and consequently reduced the power line losses on the grid-side. It did so by eliminating the reactive power fed to the wind farm-side and by implementing a CGG with DLT that secured the data and breaker control commands between the main electrical grid utility and a customer-owned wind farm. The desired TPF limits on the grid-side were defined as being between +0.9 and +1.0, and the operation of the breakers on the grid side and customer-owned wind farm side were controlled by the relay through the smart contract. The data collected from the CGG system were used to enact the TPF smart contract. The events from the real-time simulator, CGG, and relay show a successful assessment of the TPF smart contract with a CGG using DLT. In the future, a CGG with DLT and a smart contract will be used in other electrical grid applications between the main utility grids and customer-owned DER.

Author Contributions: Conceptualization, E.C.P., G.H., R.B.H. and A.W.; methodology, E.C.P., G.H. and A.W.; validation, E.C.P. and G.H.; formal analysis, E.C.P. and G.H.; investigation, E.C.P. and G.H.; resources, E.C.P., G.H. and R.B.H.; data curation, E.C.P. and G.H.; writing—original draft preparation, E.C.P., G.H., A.W. and R.B.H.; writing—review and editing, E.C.P., G.H., A.W. and R.B.H. All authors have read and agreed to the published version of the manuscript.

Funding: This research received no external funding.

Data Availability Statement: Data available on request from the authors.

Acknowledgments: This manuscript has been authored by UT–Battelle, LLC, under contract DE–AC05–00OR22725 with the US Department of Energy (DOE). The US government retains and the publisher, by accepting the article for publication, acknowledges that the US government retains a nonexclusive, paid-up, irrevocable, worldwide license to publish or reproduce the published form of this manuscript, or allow others to do so, for US government purposes. DOE will provide public access to these results of federally sponsored research in accordance with the DOE Public Access Plan (http://energy.gov/downloads/doe-public-access-plan, accessed on 1 January 2023).

Conflicts of Interest: The authors declare no conflicts of interest.

References

1. Rizwan, M.; Hong, L.; Waseem, M.; Ahmad, S.; Sharaf, M.; Shafiq, M. A Robust Adaptive Overcurrent Relay Coordination Scheme for Wind-Farm-Integrated Power Systems Based on Forecasting the Wind Dynamics for Smart Energy Systems. *Appl. Sci.* **2020**, *10*, 6318. [CrossRef]
2. Naheed, K.S.; Louki, S.; Ghedira-Guegan, C.; Benkhelifa, E.; Bani-Hani, A. Blockchain smart contracts: Applications, challenges, and future trends. *Peer-Peer Netw. Appl.* **2021**, *14*, 2901–2925.
3. Foti, M.; Vavalis, M. What blockchain can do for power grids? *Blockchain Res. Appl.* **2021**, *2*, 100008. [CrossRef]
4. Bellaj, B.; Ouaddah, A.; Bertin, E.; Crespi, N.; Mezrioui, A. Drawing the Boundaries Between Blockchain and Blockchain-Like Systems: A Comprehensive Survey on Distributed Ledger Technologies. *Proc. IEEE* **2024**, *112*, 247–299. [CrossRef]
5. AlSobeh, A.M.R. OSM: Leveraging model checking for observing dynamic behaviors in aspect-oriented applications. *Online J. Commun. Media Technol.* **2023**, *13*, e202355. [CrossRef]
6. AlSobeh, A.M.R.; Magableh, A.A. BlockASP: A Framework for AOP-Based Model Checking Blockchain System. *IEEE Access* **2023**, *11*, 115062–115075. [CrossRef]
7. Hong, Z.; Guo, S.; Zhou, E.; Chen, W.; Huang, H.; Zomaya, A. GriDB: Scaling Blockchain Database via Sharding and Off-Chain Cross-Shard Mechanism. *Proc. VLDB Endow.* **2023**, *16*, 1685–1698. [CrossRef]
8. Li, Y.; Wang, J.; Zhang, H. A survey of state-of-the-art sharding blockchains: Models, components, and attack surfaces. *J. Netw. Comput. Appl.* **2023**, *217*, 103686. [CrossRef]
9. Gorzny, J.; Derka, M. A Rollup Comparison Framework. *arXiv* **2024**, arXiv:2404.16150. [CrossRef]
10. Thibault, L.T.; Sarry, T.; Hafid, A.S. Blockchain Scaling using Rollups: A Comprehensive Survey. *IEEE Access* **2022**, *20*, 93039–93054. [CrossRef]
11. Cai, T.; Chen, W.; Psannis, K.E.; Goudos, S.K.; Yu, Y.; Zheng, Z.; Wan, S. On-Chain and Off-Chain Scalability Techniques chapter. In *Blockchain Scalability*; Chen, W., Zheng, Z., Huang, H., Eds.; Springer: Singapore. [CrossRef]
12. Liu, Y.; Lu, Q.; Zhu, L.; Paik, H.Y.; Staples, M. A Systematic Literature Review on Blockchain Governance. *arXiv* **2021**, arXiv:2105.05460. [CrossRef]
13. Barceló, E.; Dimić-Mišić, K.; Imani, M.; Brkić, V.S.; Hummel, M.; Gane, P. Regulatory Paradigm and Challenge for Blockchain Integration of Decentralized Systems: Example—Renewable Energy Grids. *Sustainability* **2023**, *15*, 2571. [CrossRef]
14. Zhuang, P.; Zamir, T.; Liang, H. Blockchain for Cyber Security in Smart Grid: A Comprehensive Survey. *IEEE Trans. Ind. Inform.* **2020**, *17*, 3–19. [CrossRef]
15. Chu, H.; Zhang, P.; Dong, H.; Xiao, Y.; Ji, S.; Li, W. A survey on smart contract vulnerabilities: Data sources, detection and repair. *Inf. Softw. Technol.* **2023**, *159*, 107221. [CrossRef]
16. Kolb, J.; AbdelBaky, M.; Katz, R.H.; Culler, D.E. Core Concepts, Challenges, and Future Directions in Blockchain. *ACM Comput. Surv. (CSUR)* **2020**, *53*, 1–39. [CrossRef]
17. Yap, K.Y.; Chin, H.H.; Klemeš, J.J. Blockchain technology for distributed generation: A review of current development, challenges and future prospect. *Renew. Sustain. Energy Rev.* **2023**, *175*, 113170. [CrossRef]
18. Nour, M.; Chaves-Avila, J.P.; Sanchez-Miralles, A. Review of Blockchain Potential Applications in the Electricity Sector and Challenges for Large Scale Adoption. *IEEE Access* **2022**, *10*, 47384–47418. [CrossRef]
19. Ji, Q.; Hu, C.; Duan, Q.; Huang, C.; Zhao, X. Decentralized power grid fault traceability system based on internet of things and blockchain technology. *Front. Energy Res.* **2023**, *10*, 861321. [CrossRef]
20. Piesciorovsky, E.C.; Hahn, G.; Borges Hink, R.; Werth, A.; Lee, A. Electrical substation grid testbed for DLT applications of electrical fault detection, power quality monitoring, DERs use cases and cyber-events. *Energy Rep.* **2023**, *10*, 1099–1115. [CrossRef]
21. He, H.; Luo, Z.; Wang, Q.; Chen, M.; He, H.; Gao, L. Joint Operation Mechanism of Distributed Photovoltaic Power Generation Market and Carbon Market Based on Cross-Chain Trading Technology. *IEEE Access* **2020**, *8*, 66116–66130. [CrossRef]
22. Bokkisam, H.R.; Singh, S.; Acharya, R.M.; Selvan, M.P. Blockchain-based peer-to-peer transactive energy system for community microgrid with demand response management. *CSEE J. Power Energy Syst.* **2022**, *8*, 198–211.
23. Liang, G.; Weller, S.R.; Luo, F.; Zhao, J.; Dong, Z.Y. Distributed Blockchain-Based Data Protection Framework for Modern Power Systems Against Cyber Attacks. *IEEE Trans. Smart Grid.* **2019**, *10*, 3162–3173. [CrossRef]
24. Mnatsakanyan, A.; Albeshr, H.; Al Marzooqi, A.; Bilbao, E. Blockchain-Integrated Virtual Power Plant Demonstration. In Proceedings of the 2020 2nd International Conference on Smart Power and Internet Energy Systems, Bangkok, Thailand, 15–18 September 2020; pp. 172–175.

25. Cioara, T.; Antal, M.; Mihailescu, V.T.; Antal, C.D.; Anghel, I.M.; Mitrea, D. Blockchain-Based Decentralized Virtual Power Plants of Small Prosumers. *IEEE Access* **2021**, *9*, 29490–29504. [CrossRef]
26. Schweitzer Engineering Laboratories Inc. SEL-700G Generator and Intertie Protection Relays Instruction Manual. Available online: https://selinc.com/products/700G/docs/ (accessed on 9 October 2024).
27. PowerLogic™ PM5500/PM5600/PM5700 Series User Manual. Schneider Electric, Document Number HRB1684301, Version 16, 31 December 2023. pp. 1–238. Available online: https://www.se.com/us/en/download/document/HRB1684301/ (accessed on 9 October 2024).
28. Piesciorovsky, E.C.; Borges Hink, R.; Werth, A.; Hahn, G.; Lee, A.; Richards, J.; Polsky, Y. *Assessment of the Electrical Substation-Grid Test Bed with Inside/Outside Devices and Distributed Ledger*; ORNL/TM-2022/1840; Oak Ridge National Laboratory: Oak Ridge, TN, USA, 2022; pp. 1–87.
29. Piesciorovsky, E.C.; Borges Hink, R.; Werth, A.; Hahn, G.; Lee, A.; Polsky, Y. Assessment and Commissioning of Electrical Substation Grid Testbed with a Real-Time Simulator and Protective Relays/Power Meters in the Loop. *Energies* **2023**, *16*, 4407. [CrossRef]
30. Borges Hink, R.; Hahn, G.; Werth, A.; Piesciorovsky, E.C.; Lee, A.; Monday, W.; Polsky, Y. *Oak Ridge National Laboratory Pilot Demonstration of an Attestation and Anomaly Detection Framework Using Distributed Ledger Technology for Power Grid Infrastructure*; ORNL/TM-2022/2527; Oak Ridge National Laboratory: Oak Ridge, TN, USA, 2022; pp. 1–56.
31. *IEEE Std 1459TM-2010*; Standard Definitions for the Measurement of Electric Power Quantities Under Sinusoidal, Nonsinusoidal, Balanced, or Unbalanced Conditions. IEEE Power & Energy Society: Piscataway, NJ, USA, 2010; pp. 1–40.
32. Introduction to Power Factor. Introduction to Power Factor. Wyandotte Municipal Services, Revision 15 September 2004. pp. 1–12. Available online: https://cms2.revize.com/revize/wyandottems/document_center/Electric/PowerFactor.pdf (accessed on 9 October 2024).
33. Zhao, C.; Han, D.; Li, C.; Wang, H. A Blockchain Consensus Mechanism to Optimize Reputation-Based Distributed Energy Trading in Urban Energy System. *IEEE Access* **2024**, *12*, 53698–53712. Available online: https://ieeexplore.ieee.org/document/10496579 (accessed on 13 October 2024). [CrossRef]
34. Yao, S.; Tian, X.; Chen, J.; Xiong, Y. Privacy preserving distributed smart grid system based on Hyperledger Fabric and Wireguard. *Int. J. Netw. Manag.* **2023**, *33*, e2193. [CrossRef]
35. Zhao, Z.; Guo, J.; Luo, X.; Xue, J.; Sing Lai, C.; Xu, Z.; Lei Lai, L. Energy Transaction for Multi-Microgrids and Internal Microgrid Based on Blockchain. *IEEE Access* **2020**, *8*, 144362–144372. [CrossRef]
36. Sikeridis, D.; Bidram, A.; Devetsikiotis, M.; Reno, M.J. A Blockchain-Based Mechanism for Secure Data Exchange in Smart Grid Protection Systems. In Proceedings of the 2020 IEEE 17th Annual Consumer Communications & Networking Conference (CCNC), Las Vegas, NV, USA, 10–13 January 2020; pp. 1–6.
37. Edvard, C. Six Common Bus Configurations in Substations up to 354 kV. Electrical Engineering Portal, 18 March 2019. Available online: https://electrical-engineering-portal.com/bus-configurations-substations-345-kv (accessed on 9 October 2024).
38. Schweitzer Engineering Laboratories Inc. SEL 735 Power Quality and Revenue Meter Instruction Manual. Available online: https://selinc.com/products/735/docs/ (accessed on 9 October 2024).
39. Schweitzer Engineering Laboratories Inc. SEL 421-4, -5, Protection, Automation, and Control System Instruction Manual. Available online: https://selinc.com/products/421/docs/ (accessed on 9 October 2024).
40. Schweitzer Engineering Laboratories Inc. SEL 451-5 Protection, Automation, and Bay Control System and SEL 400 Series Relays Instruction Manual. Available online: https://selinc.com/products/451/docs/ (accessed on 9 October 2024).
41. Schweitzer Engineering Laboratories Inc. SEL 351S Protection System Instruction Manual. Available online: https://selinc.com/products/351S/docs/ (accessed on 9 October 2024).
42. Du, B.; He, Y.; An, B.; Zhang, C. Remaining Useful Performance Estimation for Complex Analog Circuit Based on Maximal Information Coefficient and Bidirectional Gate Recurrent Unit. *IEEE Access* **2020**, *8*, 102449–102466. [CrossRef]
43. Piesciorovsky, E.C.; Smith, T.; Mukherjee, S.K.; Marshall, M.W. A Generic Method for Interfacing IEDs using Low Voltage Interfaces to Real-time Simulators with Hardware in the Loop. *Electr. Power Syst. Res. J.* **2021**, *199*, 107431. [CrossRef]
44. Piesciorovsky, E.C.; Schulz, N.N. Comparison of Programmable Logic and Setting Group Methods for Adaptive Overcurrent Protection in Microgrids. *Electr. Power Syst. Res. J.* **2017**, *151*, 273–282. [CrossRef]
45. Piesciorovsky, E.C.; Schulz, N.N. Fuse Relay Adaptive Overcurrent Protection Scheme for Microgrid with Distributed Generators. *IET Gener. Transm. Distrib.* **2016**, *11*, 540–549. [CrossRef]
46. S&C Electric Company. Total Clearing Time—Current Characteristic Curves Positrol® Fuse Links–S&C "T" Speed (TCC 170-6-2). Available online: https://www.sandc.com/en/products--services/products/positrol-fuse-links/ (accessed on 9 October 2024).
47. DTE Energy. Electric Choice Understanding Power Factor. Available online: https://www.dteenergy.com/content/dam/dteenergy/deg/website/business/service-request/electric/electric-choice/powerFactor.pdf (accessed on 9 October 2024).
48. Voloh, I.; Ernst, T. Review of Capacitor Bank Control Practices, GE Grid Solutions. In Proceedings of the 72nd Annual Conference for Protective Relay Engineers, Texas A&M University, College Station, TX, USA, 25–28 March 2019; pp. 1–13.
49. New, W.C. Load Shedding, Load Restoration and Generator Protection Using Solid-state and Electromechanical Underfrequency Relays. GET-6449, GE Power Management, 215 Anderson Avenue, Markham, Ontario, Canada. Available online: https://www.gevernova.com/grid-solutions/products/applications/get6449.pdf (accessed on 9 October 2024).

50. *ANSI/IEEE Standard C37.2*; Standard for Electrical Power System Device Function Numbers, Acronyms, and Contact Designations. IEEE Power and Energy Society: Piscataway, NJ, USA, 3 October 2008.
51. Hosseinzadeh, H. Distribution System, Protection ES586B: Power System Protection. University of Western Ontario, 5 May 2008. pp. 1–16. Available online: https://www.eng.uwo.ca/people/tsidhu/documents/es586b-hesam%20hosseinzadeh-250441131.pdf (accessed on 9 October 2024).
52. Dreyer, J.; Fischer, M.; Tönjes, R. Performance Analysis of Hyperledger Fabric 2.0 Blockchain Platform. In Proceedings of the Workshop on Cloud Continuum Services for Smart IoT Systems, New York, NY, USA, 16–19 November 2020; pp. 32–38. [CrossRef]
53. Zhou, Y.; Manea, A.N.; Hua, W.; Wu, J.; Zhou, W.; Yu, J. Application of Distributed Ledger Technology in Distribution Networks. *Proc. IEEE* **2022**, *110*, 1963–1975. [CrossRef]
54. Liu, Y.; He, J.; Li, X.; Chen, J.; Liu, X.; Peng, S.; Cao, H.; Wang, Y. An overview of blockchain smart contract execution mechanism. *J. Ind. Inf. Integr.* **2024**, *41*, 100674. [CrossRef]
55. Stefanescu, D.; Montalvillo, L.; Galán-García, P.; Unzilla, J.; Urbieta, A. A Systematic Literature Review of Lightweight Blockchain for IoT. *IEEE Access* **2022**, *10*, 123138–123159. [CrossRef]
56. Van Rensburg, M.; Dolezilek, D.; Dearien, J. Case Study: Using IEC 61850 Network Engineering Guideline Test Procedures to Diagnose and Analyze Ethernet Network Installations. In Proceedings of the PAC World Africa Conference, Johannesburg, South Africa, 12–13 November 2015; pp. 1–10.
57. Habib, A.A.; Hasan, M.K.; Alkhayyat, A.; Islam, S.; Sharma, R.; Alkwai, L.M. False data injection attack in smart grid cyber physical system: Issues, challenges, and future direction. *Comput. Electr. Eng.* **2023**, *107*, 108638. [CrossRef]
58. Daeli, A.; Mohagheghi, S. Power Grid Infrastructural Resilience against Extreme Events. *Energies* **2023**, *16*, 64. [CrossRef]
59. Immaniar, D.; Aryani, A.A.; Ula, S.Z. Challenges Smart Grid in Blockchain Applications. *Blockchain Front. Technol.* **2023**, *2*, 1–9. [CrossRef]

electronics

MDPI

Article

Mitigation Strategy of Neutral-Point DC for Transformer Caused by Metro Stray Currents

Aimin Wang [1], Sheng Lin [2,*], Guoxing Wu [3] and Xiaopeng Li [4]

1 School of Electrical Engineering and Electronic Information, Xihua University, Chengdu 610039, China; 1220210049@mail.xhu.edu.cn
2 School of Electrical Engineering, Southwest Jiaotong University, Chengdu 611756, China
3 Shenzhen Power Supply Bureau Co., Ltd., China Southern Power Grid, Shenzhen 518000, China; wuguoxing@sz.csg.cn
4 State Grid Sichuan Electric Power Research Institute, Chengdu 610044, China; lixp6053@sc.sgcc.com.cn
* Correspondence: slin@swjtu.edu.cn

Abstract: Metro stray currents flowing into neutral-point-grounded transformers can cause serious direct current (DC) bias. Affected by both metro train and urban power grid operations, transformer neutral-point DC caused by metro stray current is complex, dynamic, and time-varying, which changes the DC bias risk level of transformers. Thus, just installing blocking devices (BDs) at transformer neutral points with high instantaneous DC may make it difficult to comprehensively mitigate DC bias and lead to increased BD installation. To solve this, through optimizing BD installation placements, a mitigation method for transformer DC bias is proposed. In the proposed method, the DC bias risk level and BD installation number are included in the constraint and objective functions. To evaluate the risk level of transformer DC bias, four indicators are proposed considering the distribution characteristics of neutral DC. To optimize the BD installation placements, the effects of both the metro train dynamic operation and the operation topology of the urban power grid on neutral DCs are considered. The Monte Carlo method is used to sample the train operation conditions and a relation matrix among transformers is established. Applying the method to a certain power system, the result of BD installation placements shows that the transformers supplying the metro system must have BDs installed at their neutral points.

Keywords: metro stray current; transformer DC bias; mitigation strategy; blocking device; installation placement

1. Introduction

Since the direct current (DC) metro rail is not completely insulated from the soil, the DC current can leak from the rail and form metro stray currents [1–3]. The metro stray current can flow into the transformer's neutral points through the soil, ground wire and cable armoring, as shown in Figure 1, which can result in the transformer working in the DC bias state [4–6]. Because the metro network is widely distributed, stray currents may cause a large number of transformers to operate in the DC bias state [7]. DC bias not only damages the physical performance of transformers but also threatens the reliability of the urban power grid [8]. Therefore, to protect the transformer and system, it is urgent to comprehensively mitigate the transformer DC bias induced by DC metro stray currents.

To protect the transformer and urban power grid from transformer DC bias, one widely accepted existing method is to install a blocking device (BD) in series at the transformer neutral point [9–11]. Since the BD consists of capacitors, once installed, the BD truly and completely stops the flow of DC from the neutral point to the transformer windings [12]. Therefore, after the transformer neutral DC exceeds the threshold, the installation of BD can completely block neutral DC and mitigate the DC bias. However, when DCs simultaneously flow into a large number of transformers, the blocked DC by BD can flow to other neutral-point-grounded transformers, because transformers are interconnected by transmission lines and cables [6,7,13]. Thus, just installing BDs in series at the neutral points of the transformers that have high neutral DCs cannot eliminate the transformer DC bias risk of the whole urban power system.

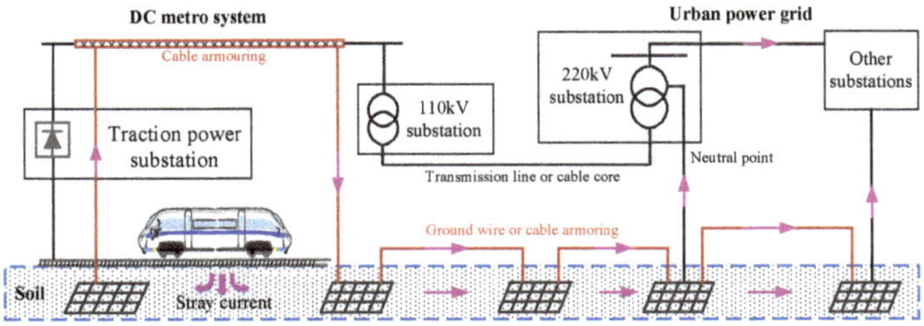

Figure 1. Paths of stray current flowing into transformer.

To comprehensively mitigate transformer DC bias, the BD installation placements have been optimized [11,14]. The neutral DC can be caused by geomagnetically induced current (GIC), grounding electrode direct current (GEDC) of high-voltage direct current systems, and metro stray current [4,15,16]. When the neutral DC is caused by GIC, the optimized methods of BD installation have considered equipment limits and operational constraints such as permissible voltages, power, and transformer heating [9,11]. The objective function is to minimize the cost function, which is affected by BD installation number [14]. When mitigating the transformer DC bias caused by GEDC, the fluctuation of neutral DCs is ignored and the maximum value is used to determine whether the transformer operates in the DC bias state [17,18]. The most significant constraint is that all neutral DC magnitudes are smaller than the threshold. And the objective function is to minimize the BD installation [10,12]. The research for mitigating neutral DCs caused by metro stray currents is light, and the maximum value is used to determine whether the transformer works in the DC bias state [13].

To this extent, when judging whether there is a DC bias risk, the magnitude of the neutral DC is the critical index. But, in the field, the transformer neutral DC caused by metro stray currents is dynamically affected by the traction and braking operations of metro trains [4]. For example, in a certain urban power grid, the neutral DC of a 220 kV transformer for 2 days is shown in Figure 2a. For 220 kV transformers, the neutral DC threshold is 5 A [17]. Thus, the transformer DC bias risk exists only during the peak periods (about 7:00–10:00 and 16:00–20:00) of train operations. If the number of times the neutral DC exceeds the threshold is very small, and the risk of DC bias is relatively very low, BD installation may be a waste.Moreover, the operation strategies of the urban power grid, such as loop-closed operation and isolated operation, also affect the neutral point DCs of transformers. For instance, the transformer neutral DCs of a certain 220 kV substation under different urban power grid operation strategies are shown in Figure 2b. The transformer DC bias risk exists only under loop-closed operation. Therefore, to mitigate transformer DC bias risk and minimize the installation of BDs, new evaluation indexes of DC bias risk should be proposed and the operations of the metro trains and urban power grid should be considered.

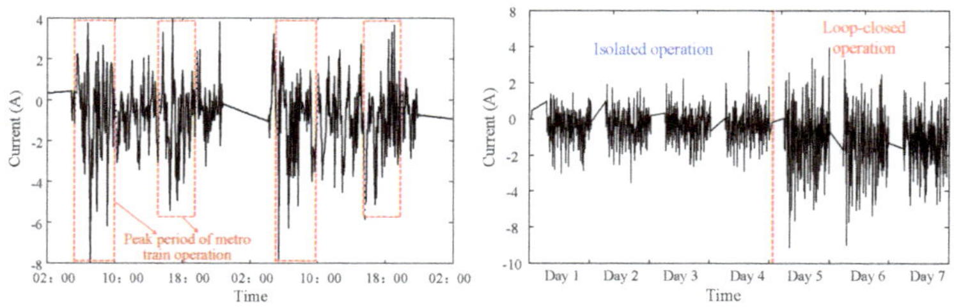

(a) **(b)**

Figure 2. Transformer neutral DCs. (**a**) Effect by dynamic metro train operations.(**b**) Effect by urban power system operations.

To address the above issues, a mitigation method for transformer DC bias risk caused by metro stray currents is proposed. The novelties are as follows:

(1) In the proposed method, considering the neutral DC magnitude and fluctuation characteristics, four indicators are proposed to evaluate the transformer DC bias risk. Then, the risk level of transformer DC bias is used as the mitigation constraint, and minimizing the number of BD installations is used as the optimizing objective.

(2) In the optimizing process, the effect of metro train and urban power system operations on the BD installation placements are both considered. The Monte Carlo method is used to sample the metro train operations and a relation matrix is proposed to describe the connection structures and network topology of AC power systems. The mitigation strategy of transformer DC bias risk is obtained by collecting the BD installation placements under each sampling condition.

By comparing it with the existing method, the proposed method can decrease the BD installations and keep the DC bias risk of transformers under an allowed level. The BD installation results show that the transformers supplying the metro system must have BDs installed at their neutral points.

The remainder of this paper is organized as follows. The evaluation indicators are presented in Section 2. Section 3 presents the optimized method of BD installation placements considering the operations of metro trains and the urban power system. Section 4 applies the proposed method and compares it with the existing method based on a certain urban power system. Section 5 concludes the whole paper.

2. Evaluation Indicators of Transformer DC Bias Risk

Affected by the operations of the metro trains and the urban power grid, the neutral DC of the transformer caused by the stray current is alternately positive and negative and randomly fluctuates [4].Considering the magnitude and fluctuation characteristics of transformer neutral DC, evaluation indicators are proposed to assess the transformer DC bias risk level.

2.1. Indicator P_i

The magnitude of the transformer neutral-point DC can reflect the severity of the transformer DC bias. The greater the neutral-point DC of the transformer, the higher the transformer DC bias risk level [7]. Thus, considering the amplitude of the transformer neutral-point DC, the first evaluation indicator is proposed. The evaluation indicator P_i is given as follows:

$$P_i = \begin{cases} \frac{|I_i| - I_{th}}{|I_i|} 100\%, & |I_i| > I_{th} \\ 0, & |I_i| \leq I_{th} \end{cases} \tag{1}$$

where I_i is the neutral DC value of the transformer; I_{th} is the magnitude threshold of the transformer neutral DC. The indicator is $0 \leq P_i < 1$. P_i is always less than 1. The greater the indicator P_i is, the larger the magnitude of neutral point DC is, then the higher the DC bias risk is. Thus, the indicator can be used to evaluate the DC bias risk for individual transformers.

2.2. Indicator P_T

Affected by the operations of metro trains, the higher the times of the neutral DC exceeding the threshold, the higher the risk of transformer DC bias. Therefore, considering the times of neutral DC exceeding the threshold, the second evaluation indicator P_T is given as follows:

$$P_T = \frac{T_{|I_i|>I_{th}}}{N} 100\% \tag{2}$$

where $T_{|I_i|>I_{th}}$ is the number of the exceeding threshold times of the neutral-point DC of the transformer; N is the sample number of transformer neutral DC in a certain period. The indicator is $0 \leq P_T \leq 1$. The larger the indicator P_T, the higher the DC bias risk level. When P_T equals 1, the transformer is always working under DC bias. The indicator can be used to evaluate the DC bias risk for individual transformers in a certain period.

2.3. Indicator P_X

In urban power grids, the number of DC bias transformers reflects the extent to which the system is affected by metro stray currents. As the number of DC bias transformers increasing, the DC bias risk level of the whole urban power grid gradually increases [7]. Therefore, considering the number of DC-biased transformers, the evaluation indicator P_X is

$$P_X = \frac{x}{K} 100\% \tag{3}$$

where x is the number of DC bias transformers in one sampling; K is the number of transformers in the urban power system. The indicator is $0 \leq P_X \leq 1$. When P_X is equal to 0, there is no transformer under DC bias. The greater the indicator P_X is, the higher the DC bias risk level of the whole urban power grid. The indicator is used to evaluate the transformer DC bias risk for the entire urban power grid.

2.4. Indicator P_E

Affected by train operations, neutral DCs of transformers are random. Considering the cumulative effect of random stray currents, the evaluation indicator P_E is proposed to evaluate the average risk probability of the urban power grid.

$$P_E = \sum_{t=1}^{N} \frac{x_t}{NK} 100\% \tag{4}$$

where N is the sample number of transformer neutral DC in a certain period; x_t is the number of the DC bias transformers in the sample of t_{th}; The indicator is $0 \leq P_E \leq 1$. The greater the indicator P_E is, the higher the cumulative effect of stray current on the urban power grid. The indicator can be used to evaluate the DC bias risk for the entire urban power grid in a period.

3. Mitigation Method of Transformer DC Bias

Using the proposed indicators, the DC bias risk is evaluated first. If the DC bias risk exists, the mitigation method is adopted. The BD is installed to mitigate transformer DC bias. To minimize the installation numbers and comprehensively mitigate the DC bias risk, a mitigation strategy method is proposed considering the operation of metro trains and the urban power grid.

3.1. Method Principle

The principle of the proposed mitigation method is shown in Figure 3. There are two parts, including establishing the optimization model of BD installation and calculating BD installation placements. In the optimization model of BD installation, the constraint is to control the risk level of transformer DC bias less than the setting. The objective is to minimize the BD installation. In the calculation process of BD installation, the operations of metro trains and the urban power system are sampled with the Monte Carlo method. In each sampling, the genetic algorithm (GA) is used to optimize the BD installation. Then, by merging the optimization results, the installation placements of BDs are obtained.

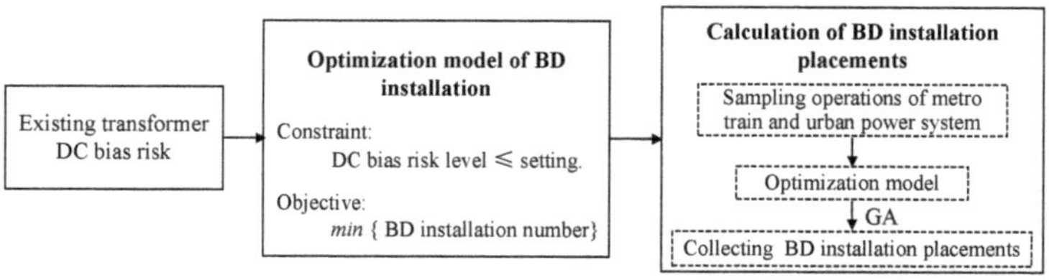

Figure 3. Principle of the proposed mitigation method.

3.2. Optimization Model of BD Installation

In the optimization model, the objective is to minimize the BD installation and mitigate the DC bias risk level of transformers. Assume that the total number of transformers is K. The transformer neutral point DCs are $I = \{I_1, I_2, \ldots, I_k\}$. If all neutral points have BDs installed, the risk of DC bias really can be reduced. But, the higher the number of BD installations, the higher the installation cost of BDs. Thus, all neutral points installing BDs is a costly and wasteful strategy. Under one sampling condition of metro train operation, the optimized objective is to minimize the BD installation as follows:

$$\min f_j = \min J_{i \cdot j} \tag{5}$$

where $J_{i \cdot j}$ is the installation number of the BDs. i is the i_{th} operation scenarios of urban power grid. j is the j_{th} sampling scenarios of train operations. Meanwhile, we use the proposed indicators to evaluate the transformer DC bias risk. We require that the indicators P_i and P_X are less than their corresponding thresholds of transformer DC bias. The constraint conditions are, respectively,

$$P_i \leq P_{i \cdot th} \tag{6}$$

$$P_X \leq P_{X \cdot th} \tag{7}$$

where $P_{i \cdot th}$ and $P_{X \cdot th}$ are thresholds of the indicators that are all greater than or equal to 0. If the optimization objective is to eliminate transformer DC bias risk, the two thresholds are both 0. If the objective is to reduce the transformer DC bias risk to an acceptable scope, thresholds can be set based on the field requirements. In the field, when the max magnitude of neutral point DC is less than 10 A, the average $P_{i \cdot th}$ is about 0.06 and the average $P_{X \cdot th}$ is about 0.01.

3.3. Calculation of BD Installation

3.3.1. Modeling of BD Installation

Considering the flow paths of stray current and the structures of the metro system and urban power grid shown in Figure 1, the DC resistance model for calculating the neutral DCs of transformers is established as shown in Figure 4. The metro system and urban power system are equivalent to the DC resistance network model considering the DC resistance parameters of devices [2,3,19]. In the urban power grid, the grounding grids of substations are connected by neutral-point-grounded transformers and grounding systems. The 500 kV and 220 kV transformers are grounded through neutral points. The 110 kV transformers are commonly ungrounded. Thus, only grounding wires or grounding cable armoring are modeled in 110 kV substations.

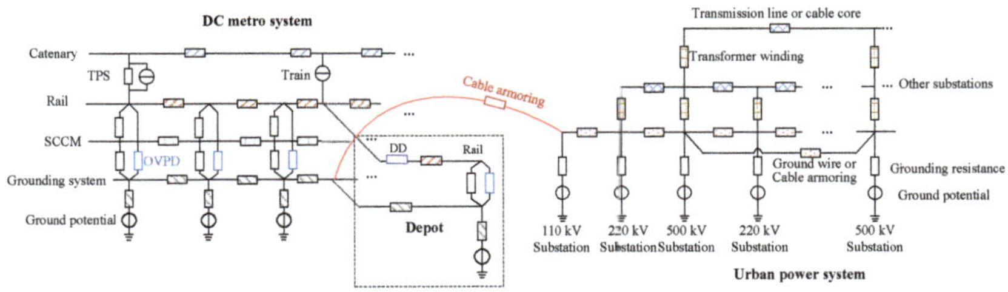

Figure 4. Schematic drawing of transformer neutral DC model.

After installing BDs at transformer neutral points, the BD installation models are shown in Figure 5. After installing BDs, the nodes i are disconnected to j in 500 kV substation. And the node $i+1$ and node $j+1$ are disconnected in the 220 kV substation. Therefore, when the BDs are installed at neutral points, the admittances $Y_{i+1,j+1}$, $Y_{j+1,i+1}$, Y_{ij}, and Y_{ji} change to 0. Using Kirchhoff's laws, the distribution of transformer neutral DCs after installing BDs can be obtained. Based on the model of BD installation, the operations of the metro train and urban power system are sampled first.

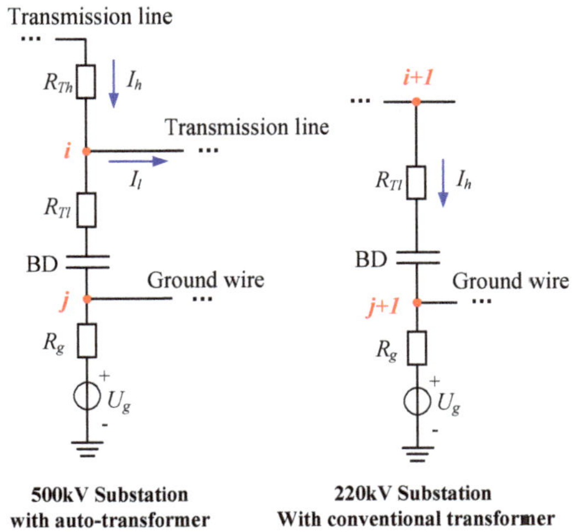

Figure 5. Model of BD installation.

3.3.2. Sampling Operation of Metro Trains

In the metro system, the train current and speed are obtained through the traction calculation method. The real train current and position in the field are tested and shown in Figure 6. There are two kinds of train operation periods, including dwell time and operation time. During the dwell time, the train stays in the station. During the operation time, the train operates between two stations at an average speed. The segmented linear relationship between train current, position, and time is obtained by a linearization process. The slope of the position linearization line during operation time corresponds to the train's average speed.

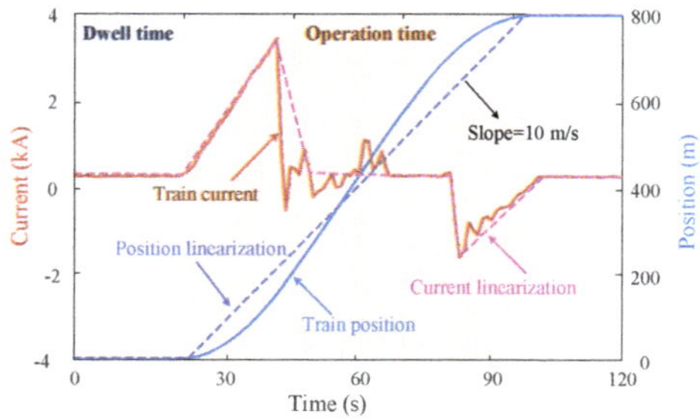

Figure 6. Train current and position data.

Based on the segmented linear relationship of train position and time, the probability distribution of train position between two stations can be obtained, as shown in Figure 7. *A* and *B* are the probability distributions of the train position at the stations (i.e., 0 and *L*). The two probabilities are calculated with the ratios of dwell time and total time. In the field, the dwell time and total time of each train are determined by the train timetable, which is pre-designed prior to the operation of the metro train. Thus, the probability distributions of *A* and *B* can be calculated based on the train timetable. In Figure 7, the *C* and *D* are the first and last samplings of train positions between the two stations. The two positions are determined by sampling time. For example, if the sampling time is 1 s and average speed is 10 m/s, *C* is 10 m, and *D* is *L*-10 m, considering the average speed. Using the Monte Carlo method, the train position can be sampled [7]. Then, based on the relationship of train current and position, the train current can also be sampled.

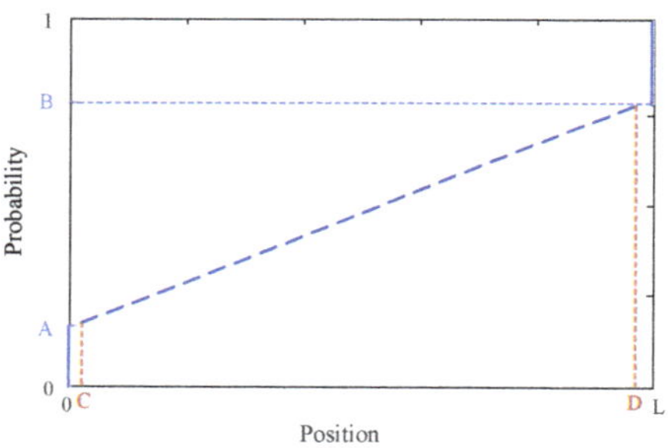

Figure 7. Probability distribution of train position between two stations.

3.3.3. Sampling Operation of Urban Power Grid

In the urban power grid, neutral-point-grounded transformers are connected by transmission lines. Thus, the connection relationship between transformers will change with the operation strategy of urban power grid switching. Thus, to characterize the operation strategy of the urban power grid, a connection relationship matrix is proposed. The admittance matrix $\boldsymbol{Y}_{power}$ of the urban power system is calculated

$$\mathbf{Y}_{power} = \mathbf{R}^T \cdot \mathbf{Y}_{sys} \tag{8}$$

where $\boldsymbol{Y}_{sys}$ is the entire admittance matrix of the urban power system, in which all possible transmission lines are modeled. $\boldsymbol{R}$ is the connection relationship matrix including 1 and 0. When a transmission line between two transformer windings is opened, the connection relationship between the two nodes of transformer windings is 0. Thus, by changing the relationship matrix, the operation of the urban power system can be sampled.

3.3.4. Calculation of BD Installation

To rapidly calculate the optimization model, the genetic algorithm (GA) is used. The optimization process with GA is shown in Figure 8. In the optimization process, the initial vector consists of K particles randomly populations $X = \{x_1, x_2, \ldots, x_K\}$. x_i is equal to 1 or 0. If $x_i = 1$, the BD is installed at transformer neutral point i. Using the populations, calculate the transformer neutral DCs and evaluate the DC bias risk. Consider that the genes replicate, cross, and mutate, forming new populations. If the number of iterations reaches the termination condition, the optimization is stopped. Finally, output the BD installation placements.

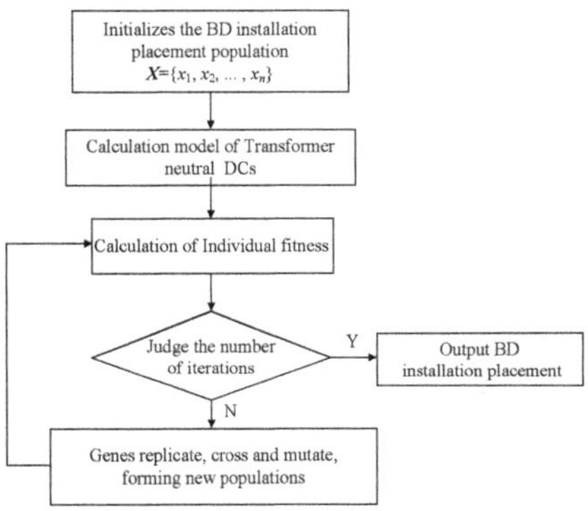

Figure 8. Optimization of BD installation placement with GA.

After calculating the optimization schemes of all the sampling scenarios of train operations, the optimization strategy is obtained by calculating the union of installation placements of BDs. The optimization strategy of BD installation placements J is calculated as follows:

$$\begin{aligned} J &= J_1 \cup J_2 \cup \ldots \cup J_i \cup \ldots \cup J_M \\ J_i &= J_{i\text{-}1} \cup J_{i\text{-}2} \cup \ldots \cup J_{i\text{-}j} \cup \ldots \cup J_{i\text{-}n} \end{aligned} \tag{9}$$

where J is the installation placement of BD under all operation scenarios of urban power grid. J_i is the installation placements of BD under the i_{th} operation scenarios of the urban power grid. M is the total number of operation strategies of the urban power grid. n is the Monte Carlo sampling number of the operation strategies of metro trains. Based on the results, the placements J should have BDs installed, while when the urban power grid works under the operation condition i, just the BDs with placement J_i are turned on.

In the field, the existing method is to install a BD at the transformer where the neutral DC is high above the threshold, which may be a waste of BD installation, and cannot mitigate the DC bias when the metro train operation changes. However, in the proposed method, the optimization objective is to minimize the BD installation. Moreover, because the mitigation strategy is obtained by collecting the BD installation placements under each sampling condition, the optimization results can be applied to any train operating condition.

4. Method Comparison and Application

4.1. Introduction of Urban Power System

The substations of the urban power grid and the line distributions of the metro system are shown in Figure 9. There are 73 substations and 11 metro lines. The urban power grids and metro lines are intertwined at high geographic densities. In Figure 9, substations 1–7 are power factories with conventional transformers, 8–16 are 500 kV substations with auto-transformers, and 17–73 are 220 kV substations with conventional transformers. And the transformers of 31–33, 35, 36, 38, 40, 42, 43, 45–47, 49–53, 59, 61, 62, and 72 supply power to metro lines.

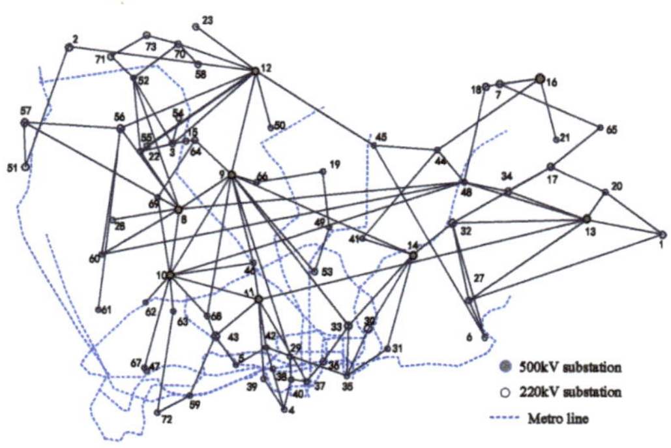

Figure 9. Topology structures of urban power grid and DC metro network.

Assume that the transformer works in the DC bias state when the transformer neutral DC exceeds 5 A. Choose a certain scenario of metro operation as the calculation condition. The DC calculation results of transformer neutral points are shown in Figure 10. There are 11 transformers that work in the DC bias state. The indicators P_i are larger than 0. The indicator P_E is 15.07%. The results show that the transformer DC bias risk exists in this urban power grid under this metro operation condition. The urban power system needs to take the measures to mitigate the transformer DC bias.

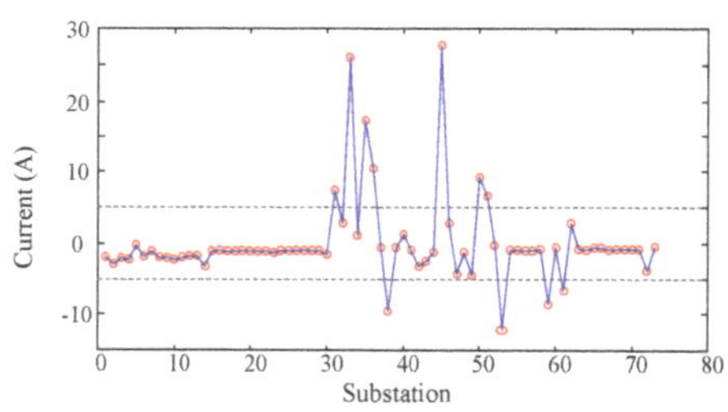

Figure 10. Transformer neutral DC under a certain operation of metro trains.

4.2. Method Comparison

In the field, the existing method used to mitigate transformer DC bias is to install a BD at the transformer neutral point, at which the neutral DC has exceeded the threshold. But, after installing the BD, the normal transformer without DC bias may experience DC bias.Thus, it is possible to install lots of BDs to mitigate transformers' DC bias. To verify that the proposed mitigation strategy can really lead to fewer BDs than the existing method, the numbers of BD installations of the different methods are compared.

4.3. Comparison in Static Scenario

In the static scenario, there are 419 metro trains operating on the metro lines, including 89 trains operating in the traction condition, 92 trains operating in the braking condition, and 238 trains operating in the coasting condition.

In the existing method, all transformers with neutral point DCs exceeding 5 A have BDs installed. After the neutral DC of the normal transformer exceeds the threshold value due to the BD installation, this new DC bias transformer also needs BDs to be installed. Until the neutral DCs of all transformers in the urban power grid are less than 5 A, the BD installations are the results of the existing method. In the proposed method, the GA parameters are set as follows: the initial number for GA population is 500; the number of iterations is 200; the parameter for gene replication is 0.7; and the thresholds of the indicators are all set to 0. The results of the existing and proposed methods are shown in Table 1.

Table 1. Installation placements of two methods.

Method	Number	Installation Placements
Existing method	66	3, 9–73
Proposed method	44	2, 5, 6, 12, 13, 22, 23, 30–59, 61–64, 67, 70, 72

The results show that using the existing method, 66 BDs are needed to mitigate transformer DC bias risk. But under the same train operations, 44 BDs can also mitigate transformer DC bias risk based on the proposed method. Compared to existing methods, the number of placements in the proposed method is reduced by 22 under the same train operating conditions, which is only 66.7% of the existing method. And the number of BD installations is reduced by 33.3% in this urban power grid. The comparison results show that the proposed method can optimize the installation placements of BD and reduce the number of installations.

4.4. Comparison in Dynamic Scenario

In the dynamic scenario, the metro train operations are sampled by the Monte Carlo method based on the metro train timetables. For each sample, the BD installations are calculated using two methods separately. A total of 100 iterations are sampled, and BD installation placements are obtained by collecting all installations. The BD installation placements of two methods are shown in Table 2.

Table 2. Installation placements of two methods.

Method	Number	Installation Placements
Existing method	67	3, 5, 9–73
Proposed method	46	2, 5, 6, 7, 12, 13, 22, 23, 26, 30–59, 61–64, 67, 70, 72

The results show that 67 BDs are needed to mitigate transformer DC bias risk using the existing method. But, 46 BDs can also mitigate transformer DC bias risk based on the proposed method. Compared to existing methods, the number of placements in the proposed method is reduced by 21 under the same sample conditions. The results show that the proposed method can optimize the BD installation placements and reduce the number of installations.

4.5. Method Application

Considering the metro train dynamic operations, the BD installation placements are optimized by the proposed method. The support threshold of the transformer neutral DC is 5 A. If the BDs are not installed, after 20,000 samplings, the maximum values of indicator P_i and indicator P_T of each transformer are shown in Table 3. The maximum value of indicatorP_X and indicator P_E are 31.10% and 10.74%. There are 36 substations with DC bias risk and 37 substations without DC bias risk.

Table 3. Risk indicators of transformer DC bias.

Substation	Indicator		Substation	Indicator	
	Max P_i(%)	P_T(%)		Max P_i(%)	P_T(%)
2	48.64	0.21	42	95.58	81.41
3	28.61	0.005	43	90.22	59.90
6	75.57	20.28	44	49.34	0.45
7	52.46	1.27	45	86.59	48.41
9	55.16	1.90	46	82.74	39.45
10	65.92	7.24	47	93.03	76.06
11	63.67	6.45	49	93.80	72.85
12	49.38	0.79	50	81.60	31.79
13	54.54	1.55	51	79.60	22.85
14	84.37	40.43	52	95.69	82.12
31	93.57	75.63	53	97.72	90.48
32	82.52	41.27	54	39.60	0.15
33	90.99	73.11	55	38.12	0.10
34	84.46	46.51	56	31.87	0.03
35	87.11	61.23	59	93.58	86.52
36	90.53	72.15	61	91.67	82.51
38	88.08	59.38	62	94.09	86.88
40	97.25	28.90	72	83.30	46.82

Through comparing the results of indicator $MaxP_i$, the maximum value is 97.72% in substation 53, 22 substations exceed 80%, and 13 substations exceed 90%. Comparing the indicator P_T, the maximum value is 90.48% in substation 53, 14 substations exceed 50%, and 6 substations exceeds 90%. The results show that there are 14 substations that have transformer DC bias risk during more than half the time of metro train operation. For some substations, the transformer DC bias risk is greater. The indicators $MaxP_X$ and P_E are 31.10% and 10.74%. The results show that the DC bias risk is very high. It is necessary to carry out a mitigation strategy in the urban power grid.

Using the proposed method, with thresholds of indicators set as 0, through 20,000 samplings of dynamic metro operation scenarios, the installation probabilities of optimized placement results are shown in Figure 11. Substations 2, 5, 6, 12, 13, 22, 23, 30–59, 61, 62, 64, 67, 70, and 71 need to have BDs installed, and the BD installation probability exceeds 90 % with the train operations. Among them, the probability percentage of substations 7, 26, 72, and 73 is less than 20%. The collection of BD installation placements in the power grid is shown in Table 4. Considering the dynamic train operation, a total of 48 substations need to have BDs installed. Moreover, all the 220 kV transformers supplying power to the metro system should have BDs installed.

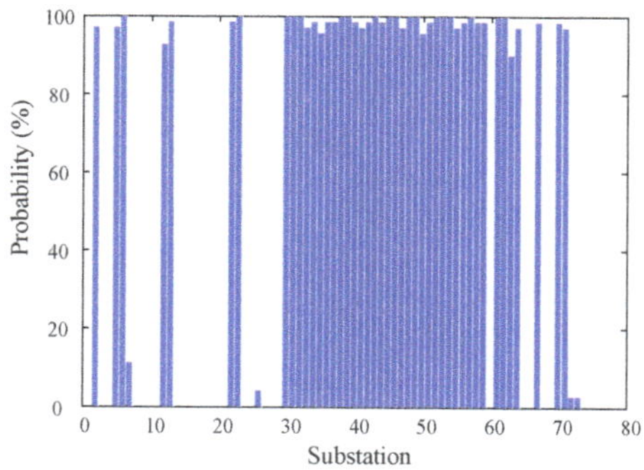

Figure 11. BD Installation probabilities of optimization results.

If the low transformer DC bias risk is allowed, the thresholds of indicators can be set to a small value. When both thresholds are 0.01, the BD installations are shown in Table 4. The results show that when the low-level DC bias is allowed, the BD installations can be reduced by the proposed method. Based on the method, BDs can be installed at transformers with high DC bias risk in the field, which can help control BD installation and the DC bias risk of transformers.

Table 4. Collection of BD installation placements.

Constraint Thresholds	Number	Installation Placement
0	48	2, 5, 6, 7, 12, 13, 22, 23, 26, 30–59, 61–64, 67, 70–73
0.01	44	2, 5, 6, 12, 13, 22, 23, 30–59, 61-64, 67, 70,71

5. Conclusions

To mitigate the transformer DC bias caused by metro stray current, an optimization method of BD installation has been proposed considering the effect of metro train and urban power system operations on the neutral DCs. Compared with the existing method, the BD installation numbers calculated by the proposed method were less than that of the existing method. Moreover, the BD installation probabilities at the transformers supporting power to the metro exceed 90%. Thus, when mitigating the transformer DC bias caused by metro stray current, the BD should be installed at the transformers supporting power to the metro first. However, the metro train operations are multitudinous and difficult to collect comprehensively. Thus, the effect of metro train operations should be analyzed deeply in the future.

Author Contributions: Conceptualization, investigation and methodology, A.W.; data curation, G.W. and X.L.; resources and supervision, S.L. All authors have read and agreed to the published version of the manuscript.

Funding: This work was supported in part by the National Natural Science Foundation of China (U2166212 and 52307139) and Science and Technology Project of CSG (090000KK52200145).

Data Availability Statement: No new data were created or analyzed in this study. Data sharing is not applicable to this article.

Conflicts of Interest: Author Guoxing Wu was employed by the company Shenzhen Power Supply Bureau Co., Ltd. The remaining authors declare that the research was conducted in the absence of any commercial or financial relationships that could be construed as a potential conflict of interest.

References

1. Lin, S.; Wang, A.; Liu, M.; Lin, X.; Zhou, Q.; Zhao, L. A Multiple Section Model of Stray Current of DC Metro Systems. *IEEE Trans. Power Deliv.* **2021**, *36*, 1582–1593. [CrossRef]
2. Du, G.; Wang, J.; Jiang, X.; Zhang, D.; Yang, L.; Hu, Y. Evaluation of Rail Potential and Stray Current With Dynamic Traction Networks in Multitrain Subway Systems. *IEEE Trans. Transp. Electrif.* **2020**, *6*, 784–796. [CrossRef]
3. Xu, S.Y.; Li, W.; Wang, Y. Effects of Vehicle Running Mode on Rail Potential and Stray Current in DC Mass Transit Systems. *IEEE Trans. Veh. Technol.* **2013**, *62*, 3569–3580.
4. Wang, A.; Lin, S.; Hu, Z.; Li, J.; Wang, F.; Wu, G.; He, Z. Evaluation Model of DC Current Distribution in AC Power Systems Caused by Stray Current of DC Metro Systems. *IEEE Trans. Power Deliv.* **2021**, *36*, 114–123. [CrossRef]
5. Wang, A.; Lin, S.; Wu, J.; Zhang, H.; Li, J.; Wu, G.; He, Z. Relationship Analysis Between Metro Rail Potential and Neutral Direct Current of Nearby Transformers. *IEEE Trans. Transp. Electrif.* **2021**, *7*, 1795–1804. [CrossRef]
6. Yu, K.; Ni, Y.; Zeng, X.; Peng, P.; Fan, X.; Leng, Y. Modeling and Analysis of Transformer DC Bias Current Caused by Metro Stray Current. *IEEJ Trans. Electr. Electron. Eng.* **2020**, *15*, 1507–1519. [CrossRef]
7. Wang, A.; Lin, S.; He, Z.; Jingzhuo, Z.; Wu, G. Probabilistic Evaluation Method of Transformer Neutral Direct Current Distribution in Urban Power Grid Caused by DC Metro Stray Current. *IEEE Trans. Power Deliv.* **2023**, *38*, 541–552. [CrossRef]
8. Rezaei-Zare, A. Behavior of Single-Phase Transformers Under Geomagnetically Induced Current Conditions. *IEEE Trans. Power Deliv.* **2014**, *29*, 916–925. [CrossRef]
9. Etemadi, A.H.; Rezaei-Zare, A. Optimal Placement of GIC Blocking Devices for Geomagnetic Disturbance Mitigation. *IEEE Trans. Power Syst.* **2014**, *29*, 2753–2762. [CrossRef]

10. Xie, Z.; Lin, X.; Zhang, Z.; Li, Z.; Xiong, W.; Hu, H.; Khalid, M.S.; Adio, O.S. Advanced DC Bias Suppression Strategy Based on Finite DC Blocking Devices. *IEEE Trans. Power Deliv.* **2017**, *32*, 2500–2509. [CrossRef]
11. Rezaei-Zare, A.; Etemadi, A.H. Optimal Placement of GIC Blocking Devices Considering Equipment Thermal Limits and Power System Operation Constraints. *IEEE Trans. Power Deliv.* **2018**, *33*, 200–208. [CrossRef]
12. Ma, S.; Lin, X.; Li, Z.; Jin, N.; Rong, Z.; Zhang, P.; Xu, H. A System-Level Suppression Method for DC Bias Based on Reverse Unbalanced Currents in the Same Transmission Section. *IEEE Access* **2021**, *9*, 126967–126975. [CrossRef]
13. Guo, Y.; Du, Q.; Liu, Y.; Yang, F.; Chen, L.; Zhang, X.; Xiao, S.; Li, C.; Wu, G. Systematic protective scheme for mega-city power systems against stray currents caused by metro systems. *High Volt.* **2023**, *8*, 943–953. [CrossRef]
14. Liang, Y.; He, D.; Zhu, H.; Chen, D. Optimal Blocking Device Placement for Geomagnetic Disturbance Mitigation. *IEEE Trans. Power Deliv.* **2019**, *34*, 2219–2231. [CrossRef]
15. Wu, F.; Yu, S.; Zhao, Z.; Quan, W. Calculation and control of DC bias current distribution in an AC power system around a typical ±800 kV DC grounding electrode. *J. Eng.* **2019**, *2019*, 3145–3149. [CrossRef]
16. Overbye, T.J.; Shetye, K.S.; Hutchins, T.R.; Qiu, Q.; Weber, J.D. Power Grid Sensitivity Analysis of Geomagnetically Induced Currents. *IEEE Trans. Power Syst.* **2013**, *28*, 4821–4828. [CrossRef]
17. Ma, S.; Rong, Z.; Lin, X.; Jin, N.; Wang, Z.; Xing, J.; Peifu, Z. Study on the Global Optimal Configuration of DC Bias Equipment Considering the Cooperation of Multiple Devices. *Proc. CSEE* **2020**, *40*, 4387–4399.
18. Ma, S.; Lin, X.; Li, Z.; Jin, N.; Rong, Z.; Zhang, P.; Xu, H. A Novel DC Bias Suppression Method Considering the Cooperation of Multiple Devices. *IEEE Access* **2021**, *9*, 130212–130220. [CrossRef]
19. Du, G.; Zhu, C.; Jiang, X.; Li, Q.; Huang, W.; Shi, J.; Zhu, Z. Multiobjective Optimization of Traction Substation Converter Characteristic and Train Timetable in Subway Systems. *IEEE Trans. Transp. Electrif.* **2023**, *9*, 2851–2864. [CrossRef]

electronics

MDPI

Article

A Prosumer Hydro Plant Network as a Sustainable Distributed Energy Depot

Michał Morawski and Przemysław Ignaciuk *

Faculty of Technical Physics, Information Technology and Applied Mathematics, Lodz University of Technology, 93-005 Lodz, Poland; michal.morawski@p.lodz.pl
* Correspondence: przemyslaw.ignaciuk@p.lodz.pl

Abstract: The shortage of efficient, low-cost storage depots inhibits the large-scale adoption of volatile-by-nature, renewable sources of energy (RSEs). In this paper, we outline how to utilize prosumer-owned hydro plants of a few to several kW as a distributed, short-term energy storage solution that is deployable with little investment and a low operational expenditure. The proposed solution is a system of interconnected hydro depots with an active water-flow control algorithm that reduces the grid's load variability and benefits prosumers. According to the tests conducted, prosumer revenue grows from several percent to over 30 percent, depending on weather conditions, in comparison to the free-flow case. In turn, the cushioning effect of the distributed energy buffer balances the fluctuations introduced by other RSEs, e.g., photovoltaic- or wind-based ones. Hence, while benefitting the involved parties, it also facilitates the inclusion of RSEs within the power distribution system.

Keywords: energy depot; hydro plants; distributed systems

1. Introduction

Electrical energy is considered the bloodstream of today's society. Electricity use in homes, industry, transportation, and building cooling and heating results in increased demand on a daily basis. Meanwhile, energy generation continues to be expensive, both from an environmental and an economic point of view. Global pollution, the growing cost of fossil fuels, and political pressure are just a few factors that encourage energy production from renewable sources (RSEs), e.g., the sun, wind, waves, geothermal energy, and water movement [1]. Unfortunately, the amount of energy retrieved from RSEs depends heavily on temporal weather conditions. RSEs provide inexpensive energy either at random hours (wind) or around noon (solar, mostly photovoltaic (PV)). In contrast, the demand remains at a similar level until the evening, as illustrated in Figure 1. Hence, costly, fossil-fuel-sourced plants need to be engaged in the evenings, but their activity should be constrained in favor of RSEs if the sun or wind is available. Moreover, energy generation from traditional thermal plants cannot be constrained too much due to technological factors. As a result of market dynamics, energy prices are set low (or even negative) at certain times to then reach a higher order of magnitude a few hours later, resembling the 'duck curve' depicted in Figure 2, which shows the initial days of July 2023 in the Polish energy market. In the interest of all stakeholders, this disparity should be reduced.

A self-imposed solution to the producer–consumer disparity is to store surplus energy when it is cheap for a later time when costly sources are incorporated. Unfortunately, today's technology does not permit executing this on a larger scale [2]. For instance, pumped hydro plants in Poland can deliver, at most, 5% of the overall power demand, and the battery depots only deliver 1% at most. Both of these traditional methods, as well as prospective hydrogen-based systems, are expensive to deploy and maintain. The low durability of electromechanical batteries further restricts a good return on one's investment. Hence, seeking a new, low-CAPEX (capital expenditures) and low-OPEX (operational expenditures) option for energy storage is desirable.

Recently, a substantial change in the electricity market has been observed. Traditional, large power plants are being supplemented by small-scale facilities installed by citizens on their own premises. These citizens are electricity consumers during certain hours of the day and producers at other times and are thus called prosumers. On the one hand, the popularity of prosumer installations advances the adoption of RSEs; on the other hand, it threatens the stability of the energy supply system by forcing extreme load changes on traditional plants.

This study contributes to the aforementioned challenges in the following ways:

- We propose engaging prosumer hydro plants with a peak power in the range of a few kW in order to form a distributed energy depot;
- We design an active control scheme, which allows for both increasing the revenue of the plant owners and reducing the load variability imposed on legacy thermal plants ("beheading the duck" [3]), thus achieving a win–win solution;
- Contrary to the typical approach, which focuses on a single hybrid generator, e.g., [4], we show how to co-ordinate a network of plants for a safe and efficient distributed system operation.

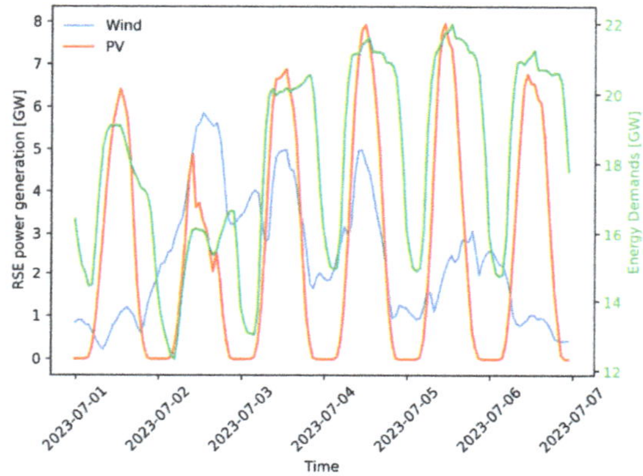

Figure 1. Energy generation from principal RSEs and predicted energy demands in Poland for 1–6 July 2023 (1 July was a Saturday [5]).

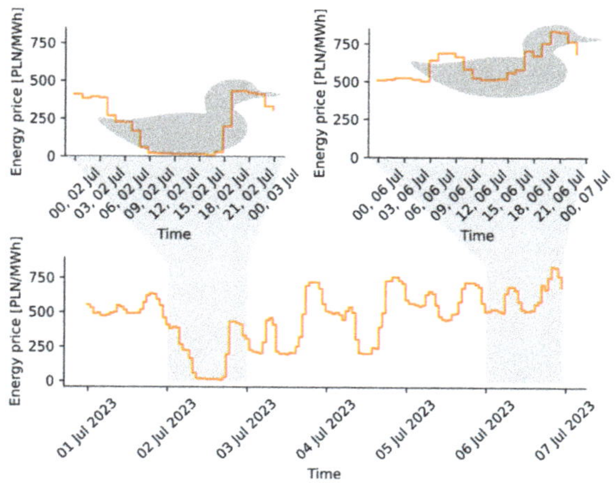

Figure 2. Fluctuation in next-day market energy prices [PLN/MWh] in the Polish market over 1–6 July 2023, following the 'duck curve' [5].

The remainder of the paper is organized as follows. In Section 2, the traditional methods of hydro plant control are discussed. In Section 3, we elaborate on the differences between conventional and prosumer plants. In Section 4, a discrete-time dynamic model of water flow in a multi-hydro plant system is constructed, and a procedure to optimize its operation is described. In Section 5, we evaluate the algorithm's performance for a broad spectrum of test data. Then, we discuss the implementation considerations in Section 6. Finally, conclusions are drawn in Section 7.

2. Control of Hydro Plants

In the literature, hydro plants (HPs) are classified as normal (>100 MW), medium (<100 MW), small (<15 MW), mini (<1 MW), micro (<100 kW), or pico (<5 kW) HPs, although the actual values vary by region [6,7]. Most of the research concentrates on small, medium, and normal HPs, which play a pronounced role in electricity supply. In this paper, owing to their significance for prosumers, micro- and pico-HPs are investigated. They encompass numerous formerly abandoned mills with small artificial ponds, swamps, and other natural formations, commonly known as run-of-river HPs [8,9], thus requiring insignificant capital expenditures for deployment. For example, currently, in Poland, less than 5% of possible installations are used for energy production [10]. Moreover, numerous online service publications show that pico-HPs can be easily deployed without specialized equipment [11]. Thus, the target area of the study conducted in this work is important. Of course, a single prosumer depot is not a game-changer. However, a revenue-oriented, optimally scheduled set of prosumer HPs may constitute an efficient energy storage system resistant to perturbances with low operational costs. In fact, similar circumstances led to the popularity of PV installations, which have become a common element of the present landscape.

However, there is a fundamental difference between PV and HP installations. The PV ones are mutually independent, whereas the proper operation of an HP depends on all the stations upstream being linked in a cascade. In the past, the scheduling of the operation of a cascade of HPs has been investigated as a case study [12] rather than as a networked system subjected to a control action, i.e., each plant was adjusted manually without enforcing a common framework. While a suitable scheduling plan can be defined in this way, it may not account for the optimal operating conditions or pose a risk in terms of drought/flood protection. Other approaches consider simplified models, reducing the cascade to a single-plant case [4], where the dams are at a close distance [8,13] or located on one river but not on its tributaries [9,14]. A solution conceptually closest to the one discussed here has been proposed in [13], albeit it targets large, concentrated installations and not the distributed HPs under prosumer management.

Technically, the steering of energy production at a hydro plant is not an intricate process. By raising the weirs, one constrains the water flow and, by lowering them, one increases energy generation. However, establishing the correct sequence of raising and lowering the weirs is far from obvious. Usually, the scheduling strategy is categorized as long-term (1–5 years), mid-term (3–18 months), and short-term (2–14 days) [15]. Nevertheless, in the case of prosumer HPs, such strategies are of limited usefulness due to the small volume of the ponds. In the considered case, a meaningful scheduling strategy should span the range of a few hours or, at most, one day. Moreover, it should focus on decreasing the generation–consumption imbalance in the evenings and mornings (Figure 2) or, equivalently, take advantage of the fluctuating market prices resulting from the disparity between forecasted and actual demands [16] (Figure 1).

The optimal scheduling of the HP water release problem has been investigated for a long time ([15,17] and the references therein). However, in those studies, HPs were examined only in the context of either flood/drought protection or power generation. Here, a third role of HPs is explored, i.e., the potential to provide an energy depot [2]. Instead of focusing just on the amount of generated electricity, HPs may constrain the river flow when the energy is cheap and increase it when the price grows, thus creating an efficient energy buffer. The overall amount of the generated energy will remain the same, but the HP owner revenue will grow by embracing the short-time changes in the price profile. Simultaneously, the fluctuation of energy generation from the traditional thermal plants decreases. A similar idea has been explored in [18], but it is destined for large installations, not the prosumer ones considered in this work.

3. Prosumer Hydro Plants

Not all RSEs are sensitive to the influence of weather fluctuations, e.g., geothermal, tide-based, or large, costly hydro HPs [19]. Here, the focus is on small, flexible installations, i.e., micro- [6] and pico-HPs [20]. They are typically deployed as run-on-river installations [8,21], i.e., without large dams and reservoirs, and thus they do not have significant expenditures or intrusion into the environment [21]. As opposed to larger HPs, which are not affected by local atmospheric conditions since the amount of water over the predictions is negligible compared to the reservoir volume, the limited storage space of micro- and pico-reservoirs does make them susceptible to weather variability. For instance, a local storm can promptly fill up the pond near one micro-HP while a nearby one simultaneously has sunny and dry conditions. Today, precipitation forecasting is not precise enough to deal with such situations. The water runoff models are not accurate either [22]. Therefore, when the weather is unstable, the plant schedule has to be recomputed frequently to adjust the system to the most recent circumstances or the price variability in a real-time, balancing market [16]. Contrary to the commonly used numerical optimization approaches, the algorithm presented in this manuscript takes a fraction of a second to execute, even on a low-end device.

A set of HPs uses the water from upstream plants and tributaries. Hence, improper control may breed a flood. Therefore, it is mandatory to synchronize the HP operation, explicitly taking into account diverse factors:

(1) Geographical factor—the distance between the plants and the ensuing delay of water reflow;
(2) Weather-related factor—the predicted precipitation level and uncertainty level;
(3) Application-oriented factor—the use of the HP cascade as a distributed energy depot combined with the common function of power generation.

Usually, the control schedule is established using numerical optimization procedures, incorporating artificial intelligence techniques to tackle the computational burden [15,17,21–29]. However, these procedures exhibit robustness issues in the considered application area, even in relatively simple cases [29]. Moreover, the large amount of time that one needs to arrive at a solution (using reasonable computing resources) contradicts the need for plan adaptability, i.e., introducing changes to the control schedule when the weather is unstable or a malfunction occurs.

To capture factors (1)–(3), information about different delays in the water flows on the links connecting the reservoirs is explicitly incorporated into the dynamical model constructed in this work. Using the system's dynamical representation, an optimization problem is stated. It may be solved analytically if saturation is absent [30]. In contrast, here, the natural constraints are embedded in the algorithmic solution. This proposal brings benefits not only to prosumers by increasing their economic gain but also to the power grid operators by reducing the load variation for the standard power plants.

4. System Model

Usually, HP modeling concentrates on the optimization of the work of generators, ignoring the supply system [23]. Here, the generator is considered a black box, and the focus is placed on the connected HPs as a water supply system. A key point to consider in a water reflow system is the non-negligible time between issuing the control action at one reservoir before it influences the water level at a downstream one. Therefore, as opposed to the earlier models of storage networks, e.g., [31], in the approach advocated here, the control principles from time-delay storage systems [32,33] will be applied. However, the models proposed in [32,33] assume continuous-time control adjustment, which is challenging to realize in a water control system owing to the specifics of the mechanical components steering the dam weirs. The model in this work explicitly covers the effects of a finite sample time and will be constructed directly in the discrete-time domain.

4.1. Single-Plant Depot

Let us consider the model of a single HP as a constituent element of the system illustrated in Figure 3. The water budget dynamics at the plant may be described via the recursive relation

$$s_j(k+1) = s_j(k) - f_j(k) + \sum_{\substack{i \in plants \\ upstream}} f_i(k - T_{ij}) + r_j(k), \tag{1}$$

where

- $k \in [1, m]$ is a time instant and m is a scheduling horizon;
- $s_j(k)$ is the amount of water stored in the reservoir near plant j (in Figure 3, plant $j = 3$ is shown in an enlarged view); $s_j(k)$ encompasses hydroengineering construction as well as nearby swamps and similar areas;
- $f_j(k)$ is the amount of water used to drive the power generators installed at dam j in the time between instants k and $k + 1$;
- $r_j(k)$ is the supply from external hydrological sources like rain (and its runoff), melting snow, uncontrolled tributaries, and vaporization. The $r_j(k)$ values can be obtained from the weather forecast and hydrological models within the planning horizon of m periods. $r_j(k)$ may be positive (prevalence of precipitation) or negative (prevalence of vaporization). $r_j(k)$ is assumed to be known in the model. When the true value of $r_j(k)$ differs from the forecast, the algorithm is run again, which does not involve complex computations, as discussed in Section 4.3.

The tributaries supply the reservoir with the water previously used by the plants upstream. The water from upstream plant i arrives at plant j with $T_{ij} > 0$ delay. The period length Δk, i.e., the time between instants k and $k + 1$, can be selected arbitrarily, but according to the pace of price changes, it is reasonable to choose 1 h down to 15 min, depending on local regulations. Similarly, the planning horizon m may cover a 24 h window of known energy prices (the next-day market). The initial flow $f_j(k \leq 0)$ and the initial water level $s_j(0)$ are assumed to be known. The terminal condition $s_j(m)$ can be selected arbitrarily.

The income from the plant obtained in one period may be calculated as

$$J_j(k) = \eta_j p_j(k) f_j(k) \Delta k, \tag{2}$$

where

- $p_j(k)$ is the energy price at instant k. The fact that all the HPs are connected to the same system does not induce the same price profile for each plant j. The owners may have a contract with different companies, use different tariffs, or have different consumption patterns themselves;
- η_j is the efficiency of power generators, incorporating the impact of the dam height. For the prosumer generators in the lowlands, the flow of 1 m^3/s corresponds to the power generation of 5–7 kW.

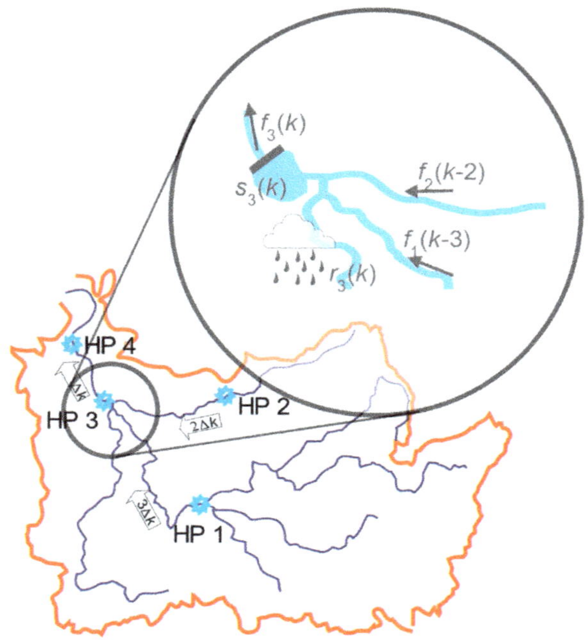

Figure 3. Model of connected hydro plants in a river basin. Plant 3 (magnified) is supplied by the inflows from reservoirs 1 and 2 with delays indicated by thick arrows (Δk—sampling period) and exogenous sources, e.g., rain or tributary in the south. The amount of water delivered to the plant from these sources within Δk at instant k is $f_1(k-3)$, $f_2(k-2)$, and $r_3(k)$, respectively. The outflow from the dam generators at plant 3 is $f_3(k)$. The input and output flow difference is buffered in the pond, whose current water volume is denoted by $s_3(k)$.

4.2. Multi-Plant Depot

Similar to PV, as the number of HPs grow, their impact on the grid increases. A collection of pico- and micro-HPs may constitute a large, distributed energy depot.

In the case of n power plants in the same river basin, the model variables are grouped into vectors:

$$\mathbf{s}(k) = \begin{bmatrix} s_1(k) \\ s_2(k) \\ \vdots \\ s_n(k) \end{bmatrix}, \ \mathbf{f}(k) = \begin{bmatrix} f_1(k) \\ f_2(k) \\ \vdots \\ f_n(k) \end{bmatrix}, \ \mathbf{r}(k) = \begin{bmatrix} r_1(k) \\ r_2(k) \\ \vdots \\ r_n(k) \end{bmatrix}, \tag{3}$$

where $\mathbf{s}(k)$ is the vector of reservoir water level, $\mathbf{f}(k)$ is the vector of water volumes in inter-reservoir flows, and $\mathbf{r}(k)$ is the vector of water volumes from exogenous sources, respectively.

Derived from (1), the networked system dynamics are described by

$$\mathbf{s}(k+1) = \mathbf{s}(k) + \sum_{t=0}^{T} \mathbf{\Theta}_t \mathbf{f}(k-t) + \mathbf{r}(k), \tag{4}$$

where T is the maximum delay, and matrices $\mathbf{\Theta}_t$ group the information about the geographical topology and flow delays. $\mathbf{\Theta}_t = \left[\theta_{ij}\right]_{n\times n}$, with $\theta_{ij} = 1$, if the flow from reservoir j reaches reservoir i with delay t, and otherwise, it is 0. $\mathbf{\Theta}_0$ describes the outflow from the system. Contrary to [13], here, the distance between plants is non-negligible. For the example from Figure 3, the longest delay $T = 3$ (the flow between reservoirs 1 and 3) and

$$\mathbf{\Theta}_0 = \begin{bmatrix} -1 & 0 & 0 & 0 \\ 0 & -1 & 0 & 0 \\ 0 & 0 & -1 & 0 \\ 0 & 0 & 0 & -1 \end{bmatrix}, \mathbf{\Theta}_1 = \begin{bmatrix} 0 & 0 & 0 & 0 \\ 0 & 0 & 0 & 0 \\ 0 & 0 & 0 & 0 \\ 0 & 0 & 1 & 0 \end{bmatrix}, \mathbf{\Theta}_2 = \begin{bmatrix} 0 & 0 & 0 & 0 \\ 0 & 0 & 0 & 0 \\ 0 & 1 & 0 & 0 \\ 0 & 0 & 0 & 0 \end{bmatrix}, \mathbf{\Theta}_3 = \begin{bmatrix} 0 & 0 & 0 & 0 \\ 0 & 0 & 0 & 0 \\ 1 & 0 & 0 & 0 \\ 0 & 0 & 0 & 0 \end{bmatrix}. \tag{5}$$

For the multi-plant system, the quality index (2) becomes

$$J = \sum_{k=1}^{n} J_j = \sum_{k=1}^{n} \sum_{j=1}^{m} J_j(k) = \sum_{k=1}^{n} \sum_{j=1}^{m} \eta_j p_j(k) f_j(k) \Delta k. \tag{6}$$

The stakeholders tend to have contrasting objectives. Prosumers try to maximize J_j, whereas the grid operators want to maximize J. A convenient way to reconcile their interests is a price system. The algorithm presented in the latter part of this paper satisfies the objectives of both plant and grid owners.

4.3. Control Algorithm

The main challenge to solving problem (6) relates to the complexity of system dynamics originating from the presence of non-negligible delay in (4). An additional difficulty stems from the physical restrictions imposed on the variables $\mathbf{f}(k)$ and $\mathbf{s}(k)$

$$\begin{aligned} &\forall_{j,k} f_j^{\min} \le f_j(k) \le f_j^{\max}, \\ &\forall_{j,k} s_j^{\min} \le s_j(k) \le s_j^{\max}, \\ &\forall_{j} s_j(m) = S_j, \end{aligned} \tag{7}$$

where S_j is an arbitrarily chosen terminal value.

As stated above, the standard solvers are not robust in the class of problems considered here [15,17,21–29]. Therefore, to address the computational aspects, a three-step min–max optimization procedure sketched in Figure 4 is applied. Each step of the algorithm is explained below. The communication among HPs conforms to the well-known industry standard Message Queuing Telemetry Transport (MQTT) [34], which provides both security and synchronization among the plant controllers.

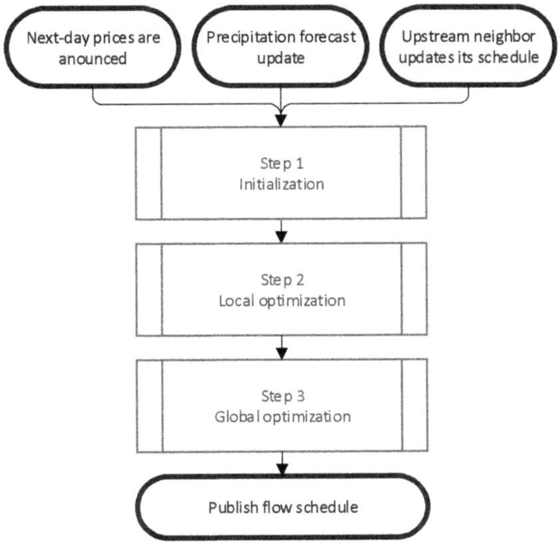

Figure 4. Algorithm flow chart.

Step 1, initialization:

Input: price profile.

Output: preliminary water flow schedule.

Objective: Determine the intervals of accumulation and earning within the considered time frame, as shown in Figure 5. When the energy prices are above the average, then it is time to earn money (earning interval). Otherwise, one should store the water (accumulation interval). In the first case, $\mathbf{f}(k) = \mathbf{f}^{\max}$, and in the latter case, $\mathbf{f}(k) = \mathbf{f}^{\min}$ for each k in the respective interval. As a result, for plant j, h_j periods with length m_{jp} are obtained. Since the HPs are managed independently owing to different tariffs or demands, the segmentation pattern may differ for each HP. Either way, this step does not involve complex computations. The approximation obtained in Step 1 may result in crossing the bounds of $\mathbf{s}(k)$ defined in (7).

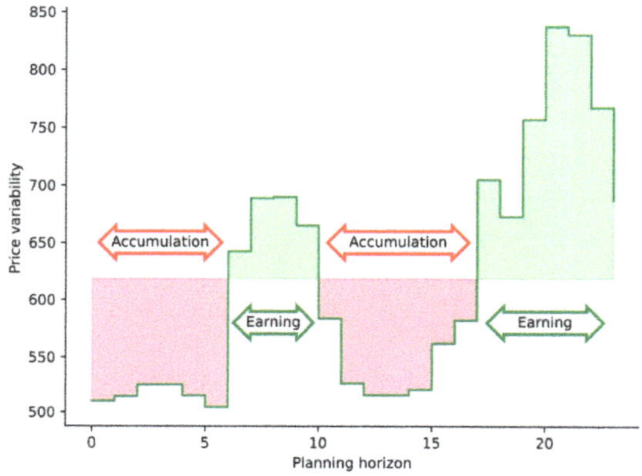

Figure 5. Illustration of Step 1: identifying accumulation and earning intervals for the price profile from 6 July. Marked periods impact the version of Step 2 of the presented algorithm.

Step 2, local optimization:

Input: price profile and preliminary water flow schedule computed at Step 1, as well as the inflow (i.e., predicted rain, tributaries inflow, the schedule of upstream plants) to the system.

Output: locally optimized flow schedule. The result does not cross the limits defined in (7).

Objective: Looking for the optimal flow within each period found by execution of Step 1.

Since two kinds of periods exist, it is reasonable to split Step 2, accordingly.

Step 2a (accumulation interval): If, for any k, $s_j(k) > s_j^{\max}$, it means that the water flow is constrained too much. Thus, the inflow of water floods the pond. Therefore, it is desirable to find instant k^* with the highest price within interval n_j and increase the flow by $s_j(k^*) - s_j^{\max}$, but not above $f_j^{\max}$. This step is repeated until the condition $s_j(k) \leq s_j^{\max}$ is satisfied for all k in the investigated interval.

Step 2b (earning interval): If, for any k, $s_j(k) < s_j^{\min}$, it means the flow is excessive. Thus, one should find an instant k^* with the lowest price in this interval and constrain the flow by $s_j^{\min} - s_j(k^*)$ but not below $f_j^{\min}$. This step is repeated until the condition $s_j(k) \geq s_j^{\min}$ is satisfied for all k in the investigated interval.

Step 2 is executed for each HP, starting from the upstream ones. The maximum number of iterations is $\sum_p m_{jp}^2$ for each HP j, which is low enough to perform in real-time, even for a large-scale system. Step 2 is illustrated in Figure 6. After computations, the maximum possible gain is obtained within each interval.

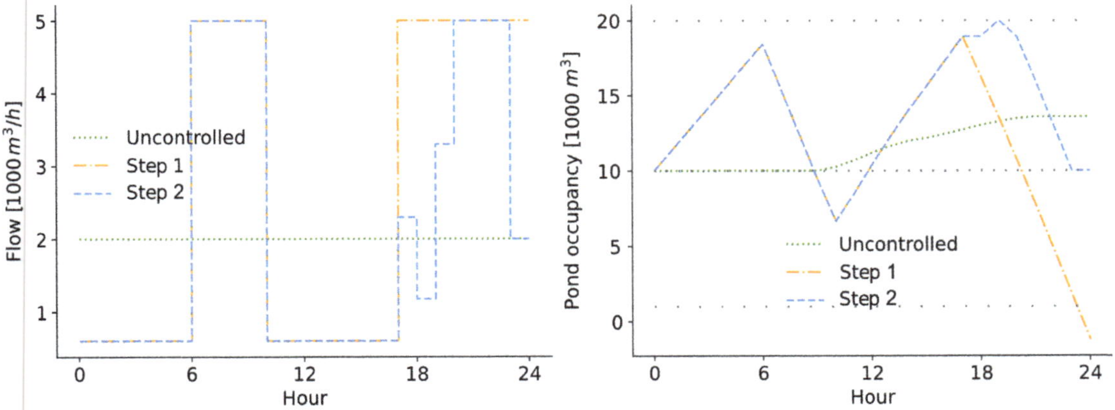

Figure 6. Flow intensity obtained in Step 2 for the pattern from Figure 5. No action is required for the first 3 periods. In the last period, terminal condition $\mathbf{s}(m)$ is violated, so the flow needs to be constrained.

Step 3, global optimization:

Input: price profile and locally optimized water flow schedule computed at Step 2.
Output: globally optimized flow schedule.
Objective: Rescheduling water flow within the whole optimization horizon.

Step 2 results in the optimal flow within each earning/accumulation interval and guarantees that condition (7) is satisfied for each interval n_j. Possibly, there is room for globally boosting the system's performance. By observing the events illustrated in Figure 6, one may note that in the hours 0–19, the pond occupancy grows nonmonotonically until the reservoir fills up. The filling and draining intervals are delineated by the instants when the pond saturates. The rescheduling inside these intervals does not influence the final volume of reservoirs, so it is possible to increase the flow when the energy is expensive and decrease it by the same value when the energy price is lower.

The maximum number of iterations in Step 3 is similar to those in Step 2. The result of an example Step 3 is shown in Figure 7.

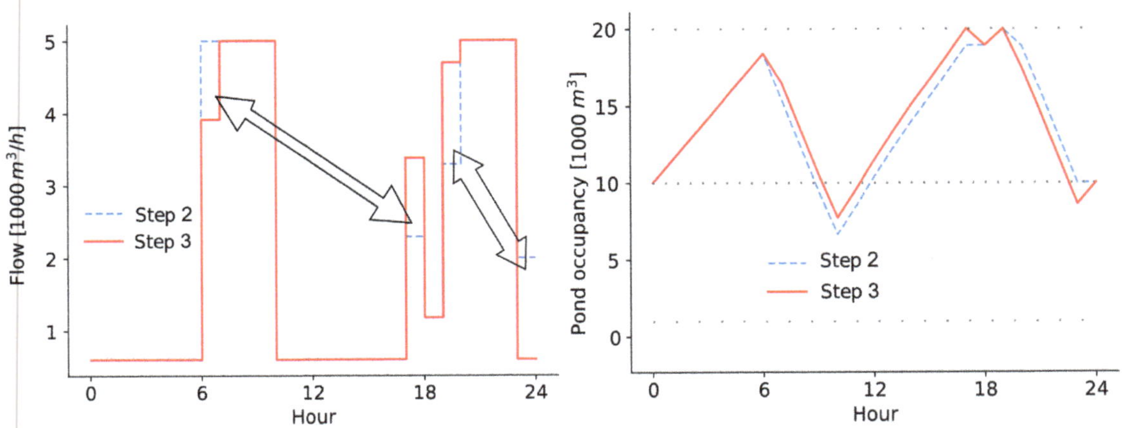

Figure 7. Outcome of Step 3 juxtaposed with the outcome of Step 2 (Figure 6). Arrows mark the rescheduled water volume.

Final considerations

The algorithm is executed whenever the input values change, e.g., when new energy prices are announced or a weather forecast update $r_j(k)$ is obtained. The results must be broadcast to all the concerned owners in the HP's network because Step 2 of the algorithm needs the output from upstream plants. The algorithm can be run locally using inexpensive hardware, e.g., a microcontroller.

5. Evaluation

The four-node topology in Figure 3 was used for the algorithm assessment. It is simple enough to illustrate the algorithm's properties while covering the major challenges of a prosumer HP network originating from arborescent topology and non-negligible delay among the plants. No common management is assumed—the prosumers make the control decisions at their plants independently from each other. Only broadcasting the control decisions is necessary. Formal regulations can easily enforce it.

The algorithm from Section 4 was tested using diverse price profiles taken from [5], including those illustrated in Figure 2, in the presence of a moving wave of intensive rain and the corresponding runoff depicted in Figure 8. The computations were executed for each profile and different volumes of ponds. In all the cases, an exogenous flow of constant intensity $\mathbf{f}(k) = [2, 3, 5, 5] \times 10^3$ [m^3/h] supplies the system. Other system parameters are taken as $\mathbf{f}^{\max} = [5, 6, 10, 10] \times 10^3$ [m^3/h], $\mathbf{f}^{\min} = [0.5, 0.6, 1, 1] \times 10^3$ [m^3/h], $\mathbf{s}(0) = \mathbf{s}(m)$, $\mathbf{s}^{\min} = 0.1 \times \mathbf{s}(0)$, $\mathbf{s}^{\max} = 2 \times \mathbf{s}(0)$, with the initial water level $\mathbf{s}(0)$ set as follows:

Scenario 1 (small-size reservoirs): $\mathbf{s}(0) = \mathbf{s}(m) = [0.5, 0.75, 1.25, 1.25] \times 10^4$ [m^3];

Scenario 2 (medium-size reservoirs): $\mathbf{s}(0) = \mathbf{s}(m) = [1, 1.5, 2.5, 2.5] \times 10^4$ [m^3];

Scenario 3 (large reservoirs): HPs are connected to larger artificial lakes with a negligible probability of saturation, $\mathbf{s}(0) = \mathbf{s}(m) = [2, 3, 5, 5] \times 10^4$ [m^3].

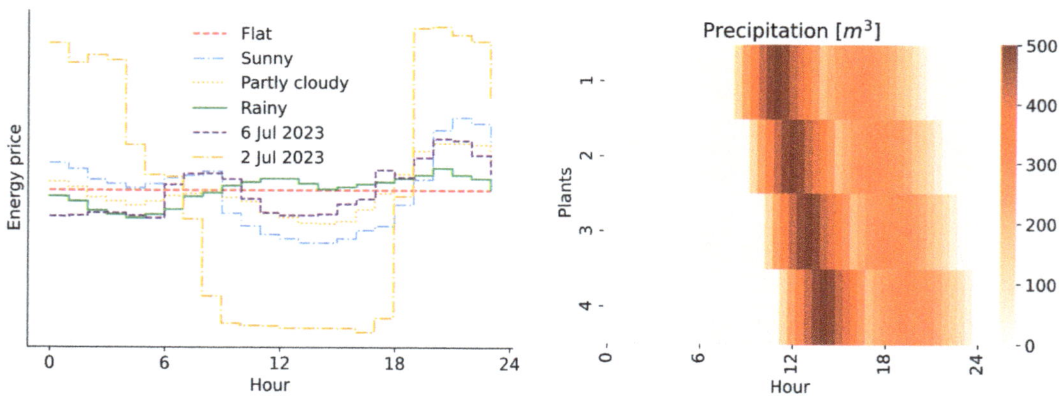

Figure 8. Example price profiles scaled to the same average and assumed precipitation pattern with runoff used to validate results.

While the algorithm does not expect any specific relation among the price profiles, for the sake of readability, all the HPs share the same data. For that reason, the turbine output is also taken as a linear function of the flow intensity $\eta_j(f_j(k)) = const$.

Table 1 summarizes the prosumer's gain and the grid operator's overall payoff. The gain is established with respect to the case when the HPs are not controlled. The discrepancy in the final state under free flow is refilled or released at the average price.

Table 1. Profit gain [%] with respect to free flow [m^3] for the price profiles from Figure 7 and different reservoir sizes.

Profile	Plant 1	Plant 2	Plant 3	Plant 4	All Plants	Remark
Small-volume ponds (Scenario 1)						
Sunny	10.85	11.80	10.98	6.74	9.71	
Partly cloudy	4.52	4.80	6.95	1.69	4.44	
Rainy	2.54	2.41	6.62	198.52	34.55	When Plant 4 is left uncontrolled, then the production drops below the minimum for 5 h.
6 July 2023	10.13	10.35	13.39	28.16	17.30	
2 July 2023	29.25	33.81	59.53	46.64	47.30	
Medium-volume ponds (Scenario 2)						
Sunny	20.13	19.38	15.91	11.44	15.65	
Partly cloudy	11.98	12.07	13.86	38.17	21.29	When Plant 4 is left uncontrolled, then the production drops below the minimum for 3 h.
Rainy	9.31	11.37	17.44	65.61	33.05	When Plants 3 and 4 are left uncontrolled, then the production drops below the minimum for 5 h (both), and Plant 3 experiences a flood during 2 h.
6 July 2023	14.47	13.77	14.25	49.92	25.15	When Plant 4 is left uncontrolled, then the production drops below the minimum for 4 h.
2 July 2023	68.50	68.34	69.79	82.83	74.03	When Plant 4 is left uncontrolled, then the production drops below the minimum for 2 h.
Large-volume ponds (Scenario 3)						
Sunny	21.04	20.82	19.52	16.15	18.86	
Partly cloudy	14.73	14.72	14.43	19.43	16.76	When Plant 4 is left uncontrolled, then the production drops below the minimum for 1 h.
Rainy	6.93	6.79	6.61	5.78	6.41	
6 July 2023	15.39	15.18	13.83	12.36	13.82	
2 July 2023	59.52	56.83	80.13	65.47	67.83	

The data shown in Table 1 reveal that the larger the pond volume, the more profit obtained. However, with the proposed algorithm applied, even for small reservoirs, the relative monetary benefit reaches several dozen and grows with price variability, albeit not linearly. The data also show that small changes in $r_j(k)$ or $p_j(k)$ may lead to a substantial discrepancy in the outcome. The gain grows when the distance between the HPs increases, which is attributed to a larger difference between $p_j(k)$ and $p_j(k - T_{ij})$. It is worth noting that the algorithm tends to fill up the ponds at low prices around midnight yet does not allow for floods. This behavior is advantageous as the ponds contribute to the soil moisture, thus alleviating the problem of persistent droughts in various regions.

The algorithm opts for the maximum or minimum flow intensity $\mathbf{f}(k)$ rather than intermediate ones. This property facilitates turbine construction and maintenance by increasing efficiency in the nominal conditions.

The results obtained using the proposed algorithm, summarized in Table 1, are much better than those obtained by standard optimization algorithms [29]. Moreover, all the downstream HPs are protected from otherwise common floods and droughts (Scenario 2).

The balancing market may earn additional profit for both the prosumers and grid operators [16]. When the grid operator to which the prosumer network is connected detects a lack of energy, it may trigger an immediate water release or, conversely, constrain the flow in the case of surplus.

6. Implementation Aspects

Prior to deployment, a few more aspects should be considered:

1. Legal issues. System deployment requires consent from the regulator of water resources and corresponding agreements from the grid operators.
2. Information exchange. The designed algorithm requires a reliable communication platform among prosumer plants. We propose MQTT [35] for this purpose.

3. Weather forecast. Next-day forecasting is currently not accurate enough to predict local storms several hours ahead. However, radar-based imaging provides thorough precipitation estimates in a timeframe of a few hours.
4. Depots as a real-time system. Each HP's controller operates independently from others. Hence, the system is potentially prone to unfavorable phenomena from distributed systems like rushes or deadlocks. The MQTT protocol solves this issue without increasing computational load as the schedule calculation takes only a fraction of a second, and the forecasting updates are infrequent.
5. Depots as a control system. The system operates in an open loop. The updates to the schedule only impact the downstream HPs. The system is bounded-input bounded-output (BIBO) stable.
6. Security. The proposed communication platform conforms to high industry security standards [35].
7. Computational resources. The proposed algorithm does not require high-end computing resources to execute. It operates and communicates efficiently on low-end, inexpensive devices.

7. Conclusions

This paper presents an optimal strategy to control a system of connected HPs, forming a sustainable, distributed energy depot. Such a system benefits both the power grid operators by reducing fluctuations in the grid load and the plant owners by increasing their income. It can be deployed with low capital and operational expenditures. In addition, by slowing down the precipitation runoff, the system elevates the resilience to floods and droughts in the area, which is of paramount importance in the face of climate change. The control algorithm does not assume the same price pattern or the scheduling horizon used by different stakeholders. The HP owners are not forced to co-operate—they only need to broadcast their control decisions to neighbors. As opposed to the adverse PV "inverter combat" [35], the co-operation among prosumers in the proposed scheme boosts their profits. Also, contrary to pumped HPs, energy accumulation does not incur extra costs, so even small price variations allow for increasing the revenue.

The presented algorithm performs better than classical optimization procedures in terms of numerical stability and robustness [29]. It can be implemented with negligible computing resources and triggered asynchronously at the HPs. While the final income depends on the weather, forecast accuracy, variability of energy prices, and other external conditions, the increase compared to a non-controlled case falls in the 20–70% range and should grow with the impact of RSEs, especially PV and wind RSEs.

Author Contributions: Conceptualization, M.M. and P.I.; methodology, M.M.; software, M.M.; validation, M.M. and P.I.; formal analysis, M.M. and P.I.; investigation, M.M.; resources, M.M. and P.I.; data curation, M.M. and P.I.; writing—original draft preparation, M.M.; writing—review and editing, P.I.; visualization, M.M. and P.I.; supervision, P.I.; project administration, P.I.; funding acquisition, P.I. All authors have read and agreed to the published version of the manuscript.

Funding: This work was performed in the framework of project no. 2021/41/B/ST7/00108: "Robust control solutions for multi-channel networked flows", financed by the National Science Centre, Poland.

Data Availability Statement: All the data are available publicly [5].

Conflicts of Interest: The authors declare no conflicts of interest. The funders had no role in the design of the study; in the collection, analyses, or interpretation of data; in the writing of the manuscript; or in the decision to publish the results.

References

1. Kolosok, S.; Bilan; Vasylieva, T.; Wojciechowski, A.; Morawski, M. A scoping review of renewable energy, sustainability and the environment. *Energies* **2021**, *14*, 4490. [CrossRef]

2. Rahman, M.M.; Oni, A.O.; Gemechu, E.; Kumar, A. Assessment of energy storage technologies: A review. *Energy Convers. Manag.* **2020**, *223*, 113295. [CrossRef]
3. Wirfs-Brock, J. IE Questions: Why Is California Trying to Behead the Duck? 2014. Available online: https://insideenergy.org/2014/10/02/ie-questions-why-is-california-trying-to-behead-the-duck/ (accessed on 12 May 2023).
4. Borkowski, D.; Cholewa, D.; Korzeń, A. A run-of-the-river Hydro-PV battery hybrid system as an energy supplier for local loads. *Energies* **2021**, *14*, 5160. [CrossRef]
5. PSE. Market Energy Prices. Available online: https://www.pse.pl/dane-systemowe/funkcjonowanie-rb/raporty-dobowe-z-funkcjonowania-rb/podstawowe-wskazniki-cenowe-i-kosztowe/rynkowa-cena-energii-elektrycznej-rce (accessed on 1 August 2023).
6. Anaza, S.O.; Abdulazeez, M.S.; Yisah, Y.A.; Yusuf, Y.O.; Salawu, B.U.; Momoh, S.U. Micro Hydro-Electric Energy Generation—An Overview. *Am. J. Eng. Res.* **2017**, *6*, 5–12.
7. Sachdev, H.; Akella, A.K.; Kumar, N. Analysis and evaluation of small hydropower plants: A bibliographical survey. *Renew. Sustain. Energy Rev.* **2015**, *51*, 1013–1022. [CrossRef]
8. Yiwei, Q.; Jin, L.; Liu, F.; Dai, N.; Song, Y.; Chen, G.; Ding, L. Tertiary Regulation of Cascaded Run-of-the-River Hydropower in the Islanded Renewable Power System Considering Multi-Timescale Dynamics. *IET Renew. Power Gener.* **2021**, *15*, 1778–1795.
9. Jamali, S.; Jamali, B. Cascade hydropower systems optimal operation: Implications for Iran's Great Karun hydropower systems. *Appl. Water Sci.* **2019**, *9*, 66. [CrossRef]
10. Wyszkowski, K.; Piwowarek, Z.; Pałejko, Z. Małe Elektrowne Wodne w Polsce. Technical Report, UN Global. 2022. Available online: https://ungc.org.pl/wp-content/uploads/2022/03/Raport_Male_elektrownie_wodne_w_Polsce.pdf (accessed on 1 June 2024). (In Polish).
11. YouTube. Constriction General Channel. 2024. Available online: https://www.youtube.com/watch?v=MBDV_Xj6Dj8 (accessed on 1 June 2024).
12. IEA Hydropower. Management Models for Hydropower Cascade Reservoirs, Compilation of Cases. Tech Report. 2021. Available online: https://www.ieahydro.org/media/02e6ec8e/EAHydro_AnnexXIV_Management%20Models%20for%20Hydropower%20Cascade%20Reservoirs_CASE%20COMPILATION.pdf (accessed on 1 June 2024).
13. Yu, Y.; Wu, Y.; Sheng, Q. Optimal scheduling strategy of cascade hydropower plants under the joint market of day-ahead energy and frequency regulation. *IEEE Access* **2021**, *9*, 87749–87762. [CrossRef]
14. Zhang, H.; Chang, J.; Gao, C.; Wu, H.; Wang, Y.; Lei, K.; Long, R.; Zhang, L. Cascade hydropower plants operation considering comprehensive ecological water demands. *Energy Convers. Manag.* **2019**, *180*, 119–133. [CrossRef]
15. Thaeer Hammid, A.; Awad, O.I.; Sulaiman, M.H.; Gunasekaran, S.S.; Mostafa, S.A.; Manoj Kumar, N.; Khalaf, B.A.; Al-Jawhar, Y.A.; Abdulhasan, R.A. A Review of Optimization Algorithms in Solving Hydro Generation Scheduling Problems. *Energies* **2020**, *13*, 2787. [CrossRef]
16. Mielczarski, W. *Energy Systems & Markets*; Association of Polish Electrical Engineers: Warszawa, Poland, 2018.
17. Singh, V.K.; Singal, S. Operation of hydro power plants—A review. *Renew. Sustain. Energy Rev.* **2017**, *69*, 610–619. [CrossRef]
18. Li, J.; Zhao, Z.; Li, P.; Mahmud, M.A.; Liu, Y.; Chen, D.; Han, W. Comprehensive benefit evaluations for integrating off-river pumped hydro storage and floating photovoltaic. *Energy Convers. Manag.* **2023**, *296*, 117651. [CrossRef]
19. Wijesinghe, A.; Lai, L. Small hydro power plant analysis and development. In Proceedings of the 2011 4th International Conference on Electric Utility Deregulation and Restructuring and Power Technologies (DRPT), Weihai, China, 6–9 July 2011; pp. 25–30.
20. Khomsah, A.; Laksono, A.S. Pico-hydro as A Renewable Energy: Local Natural Resources and Equipment Availability in Efforts to Generate Electricity. *IOP Conf. Ser. Mater. Sci. Eng.* **2018**, *462*, 012047. [CrossRef]
21. Anderson, D.; Moggridge, H.; Warren, P.; Shucksmith, J. The impacts of 'run-of-river' hydropower on the physical and ecological condition of rivers. *Water Environ. J.* **2015**, *29*, 268–276. [CrossRef]
22. Nguyen, A.; Vu, V.; Hoang, D.; Nguyen, T.; Nguyen, K.; Nguyen, P.; Ji, Y. Attentional ensemble model for accurate discharge and water level prediction with training data enhancement. *Eng. Appl. Artif. Intell.* **2023**, *126*, 107073. [CrossRef]
23. Alvarez, G. An optimization model for operations of large scale hydro power plants. *IEEE Lat. Am. Trans.* **2020**, *18*, 1631–1638. [CrossRef]
24. Shaw, A.R.; Sawyer, H.; LeBoeuf, E.; McDonald, M.; Hadjerioua, B. Hydropower optimization using artificial neural network surrogate models of a high-fidelity hydrodynamics and water quality model. *Water Resour. Res.* **2017**, *53*, 9444–9461. [CrossRef]
25. Ahmadianfar, I.; Samadi-Koucheksaraee, A.; Asadzadeh, M. Extract nonlinear operating rules of multi-reservoir systems using an efficient optimization method. *Sci. Rep.* **2022**, *12*, 18880. [CrossRef]
26. Bernardes, J., Jr.; Santos, M.; Abreu, T.; Prado, L., Jr.; Miranda, D.; Julio, R.; Viana, P.; Fonseca, M.; Bortoni, E.; Bastos, G.S. Hydropower operation optimization using machine learning: A systematic review. *AI* **2022**, *3*, 78–99. [CrossRef]
27. Marcelino, C.G.; Leite, G.M.C.; Delgado, C.A.D.M.; de Oliveira, L.B.; Wanner, E.F.; Jiménez-Fernández, S.; Salcedo-Sanz, S. An efficient multi-objective evolutionary approach for solving the operation of multi-reservoir system scheduling in hydro-power plants. *Expert Syst. Appl.* **2021**, *185*, 115638. [CrossRef]
28. Linke, H. A model-predictive controller for optimal hydro-power utilization of river reservoirs. In Proceedings of the 2010 IEEE International Conference on Control Applications, Yokohama, Japan, 8–10 September 2010; pp. 1868–1873.
29. Morawski, M.; Ignaciuk, P. Balancing energy budget in prosumer water plant installations with explicit consideration of flow delay. *IFAC-PapersOnLine* **2023**, *56*, 11748–11753. [CrossRef]

30. Ignaciuk, P.; Morawski, M. Optimal Control of Cascade Hydro Plants as a Prosumer-Oriented Distributed Energy Depot. *Energies* **2024**, *17*, 469. [CrossRef]
31. Danielson, C.; Borrellia, F.; Oliver, D.; Anderson, D.; Phillips, T. Constrained flow control in storage networks: Capacity maximization and balancing. *Automatica* **2013**, *49*, 2612–2621. [CrossRef]
32. Basin, M.V.; Guerra-Avellaneda, F.; Shtessel, Y.B. Stock management problem: Adaptive fixed-time convergent continuous controller design. *IEEE Trans. Syst. Man Cybern. Syst.* **2020**, *50*, 4974–4983. [CrossRef]
33. Ignaciuk, P. Dead-time compensation in continuous-review perishable inventory systems with multiple supply alternatives. *J. Process Control* **2012**, *22*, 915–924. [CrossRef]
34. Sengul, C. Message Queuing Telemetry Transport (MQTT) and Transport Layer Security (TLS) Profile of Authentication and Authorization for Constrained Environments (ACE) Framework; RFC 9431. 2023. Available online: https://datatracker.ietf.org/doc/rfc9431/ (accessed on 1 June 2024).
35. Gu, Y.; Green, T.C. Power System Stability With a High Penetration of Inverter-Based Resources. *Proc. IEEE* **2023**, *111*, 832–853. [CrossRef]

electronics

MDPI

Article

Modular Multilevel Converter Control Strategy for AC Fault Current Maximization and Grid Code Compliance

Ricardo Vidal-Albalate *, Enrique Belenguer and Francisco Magraner

Department of Industrial Systems Engineering and Design, Universitat Jaume I, 12071 Castelló, Spain; efbeleng@uji.es (E.B.); al431019@uji.es (F.M.)
* Correspondence: rvidal@uji.es

Abstract: This paper proposes a dynamic current limit for modular multilevel converters (MMCs) that maximizes the injection of current during grid faults in order to mitigate the voltage dip, reduce voltage imbalances in case of an asymmetrical fault, and ensure the proper operation of protective relays. The reduced short-circuit capacity of MMCs, and power converters in general, is one of their main limitations. In the event of a fault, the converter's current is significantly lower than that of the synchronous generators, which may impact both the performance of power system protective relays and the mitigation of voltage drops during faults. Usually, to protect the MMCs themselves, their output current is limited by their control. However, the current flowing through the power semiconductors is the arm current, not the output current, and this consists of an AC and a DC component. A new current saturation strategy aiming at maximizing fault current injection, in compliance with the most recent grid codes, is proposed. This strategy limits the arm currents by dynamically adjusting the output current limit while injecting reactive currents (both positive- and negative-sequence) and active current according to the grid codes, the fault type, and voltage sag level. A theoretical analysis is carried out to determine the maximum current injection that will not exceed the arm limits, and this is then validated through detailed PSCAD simulations. With the proposed strategy, the supplied current can be increased by approximately 40%.

Keywords: MMC; fault currents; grid codes; protective relays; current saturation; sequence components

1. Introduction

Power systems are experiencing an increase in the number of high-voltage direct current (HVDC) links and the renewable power plants they contain, which are interfaced with the AC grid through power electronic converters (PECs), which are based on voltage source converters (VSCs). However, PECs present some challenges for the protective relays of AC power systems because their response differs significantly from that of synchronous generators (SGs). While SGs can provide fault currents of up to 6 times their rated current, PECs have a very limited overcurrent capability of about 1.2 times their rated current. Moreover, in cases of asymmetrical faults, SGs naturally behave as a voltage source, injecting both positive- and negative-sequence currents into the system. Conversely, PECs behave during faults like a controlled current source; hence, the injection of negative-sequence currents depends on the PEC's control [1].

A declining short-circuit level (SCL) may negatively impact power systems, as identified by National Grid ESO, the British system operator [2]. As the SCL declines, the

 https://doi.org/10.3390/electronics14091763

voltage waveform will become more distorted during faults and the retained voltage will be lower [3]. In the case of grid-following converters, which rely on a Phase-Locked Loop (PLL) to determine the voltage angle, low SCLs can cause the PLL to malfunction, potentially leading to the disconnection of a converter from the network. Thus, the source of the risk is twofold: the declining SCLs and the growing number of PLL-based converters. High SCLs, therefore, are required to maintain a stable voltage during a fault and limit the extent of voltage sags.

On the other hand, with low SCLs—and, consequently, low fault currents—protective relays could take longer to operate or not operate as designed, which could result in longer faults in the system, posing a safety risk to people, damaging equipment, and leading to instability [2]. Moreover, although an increasing number of distributed generation facilities are being connected to distribution networks, the contribution of fault current from the distribution networks towards transmission grid faults is expected to increase only slightly [4].

Several countermeasures have been proposed to mitigate the impact of fault level shortfall, such as increasing the k-factor, increasing the converter's power rating, and installing synchronous condensers [5]. Increasing the k-factor (i.e., the gain that connects the reactive current injection to the voltage deviation during a fault) allows for an increase in fault currents. However, its effectiveness is limited due to the current constraints of the power converters. Thus, for deep voltage sags, the effect of a high k-factor may be negligible [6]. Increasing the converter's rated power results in higher costs that are roughly proportional to the power increment [7]. Finally, synchronous condensers, which provide a boost to system inertia and an increase in the system fault level, could facilitate the operation of the protection system in future scenarios [8].

In response to an expected drop in fault levels in Scotland as a result of the closure of large SGs, National Grid ESO launched the Stability Pathfinder (SPf) Phase 2 project to seek new sources of fault current, inertia provision and dynamic voltage regulation that could be added to the system services [9]. During the Expression of Interest (EOI) phase, more than 50% of the proposals from participants included synchronous condensers [10]. Ultimately, several synchronous condensers were selected for use in the Stability Pathfinder Project [11,12]. Similarly, the Spanish TSO Red Eléctrica de España (REE) will install three synchronous condensers in the Canary Islands and another one in the Balearic Islands [13].

Many protective relays, such as directional relays, fault-type-identification algorithms, and distance relays, use negative-sequence currents for fault detection [14]. Consequently, some transmission system operators (TSOs) require PECs to inject both positive- and negative-sequence reactive currents based on the magnitudes of the positive and negative voltages in the system, respectively, to emulate the behavior of SGs [15,16]. However, the limited overload capability of PECs can hinder fault detection, as the values of the fault currents are similar to the PECs' rated current. For this reason, in addition to injecting reactive positive- and negative-sequence currents, TSOs also require the maximization of the injected current to within the limits of the power converter [17].

In recent years, differential protection has become the most secure and reliable method used by TSOs [18], with distance protection widely used as the back-up protection on new lines and the primary protection on legacy transmission lines. It is, in fact, in distance protection where the impact of PECs can pose a risk to the behavior of the protection algorithms, leading to under- or overreach events [19]. Moreover, voltage drops during short-circuits in networks with a high penetration of PECs are significantly larger than those in systems dominated by synchronous generation. This can introduce voltage measurement errors when using a self-polarized mho algorithm for distance calculations. Additionally, the impact of the pre-fault load flow and fault resistance can further distort impedance calculations, potentially leading to false trips for out-of-zone faults and no tripping for in-zone faults [20]. This effect becomes even more pronounced as the SCLs' difference at both ends of the line increases, which is common in lines ending at HVDC links connected to offshore wind power plants.

Modular Multilevel Converters (MMCs), which are usually connected to transmission grids, have become the preferred converter topology for high-power and high-voltage applications. Moreover, a growing number of this kind of converter is expected in transmission grids. In the North Sea, several offshore wind power plants are already connected to the AC grid through MMC-HVDC links [21]. Germany plans to build 35 HVDC lines to connect 70 GW of offshore wind power plants to the onshore grid by 2045, according to the Network Development Plan 2023–2037/2045 published by its Federal Network Agency (Bundesnetzagentur) [22]. MMC-HVDC links are also utilized in places with technical or environmental constraints, such as the INELFE interconnection between Spain and France [23]. Therefore, their use is clearly increasing in power systems.

To overcome the limited capability of power converters to inject fault currents, ref. [24] proposes controlling the phase of the injected currents to a specific value that avoids the incorrect operation of differential relays. An algorithm for the calculation of the optimal current reference is proposed in [25]; however, it is based on an optimization algorithm that uses 72 system variables and 72 linear and non-linear equations, which may be challenging to apply in real time. For power systems with MMC-HVDC links, ref. [26] analyses a new distance relay scheme for zone 1 that facilitates an accurate calculation of the distance to the fault point.

To protect the MMCs, their output current is typically limited by their control. However, the current that flows through the power semiconductors is the arm current, not the output current, which consists of an AC and a DC component. In the event of a fault, the active power that is exchanged with the AC grid is significantly reduced. This fact, in turn, decreases the DC current that flows through the arms and allows for an increase in the AC current. This paper proposes a new current saturation strategy that is aimed at maximizing fault current injection in compliance with the most recent grid codes. This strategy helps to mitigate the voltage dip, reduce voltage imbalances in cases of asymmetrical faults, and ensure the proper operation of protective relays. This strategy saturates the arm currents by dynamically adjusting the output current limit while injecting reactive currents (both positive- and negative-sequence) and active current according to the grid codes, the fault type, and voltage sag level. A theoretical analysis is carried out to determine the maximum current injection that will not exceed the arm limits, and this is then validated through detailed PSCAD simulations.

The paper is structured as follows. Section 2 describes the main requirements imposed by some European grid codes about fast fault current injection and fault-ride-through capability of HVDC systems and power park modules. The structure of the MMC, its parameters, and its control are detailed in Section 3. Section 4 describes the proposed saturation algorithm that allows for an increase in fault current injection. The new saturation strategy is validated through detailed PSCAD simulations in Section 5. Finally, Section 6 encapsulates the concluding remarks.

2. Grid Codes

According to ENTSO-E [27], an HVDC converter must meet several requirements to provide voltage support and fault ride-through:

- Capability of staying connected to the network within given ranges of the network voltage.
- Capability to provide fast fault current injection at the connection point in case of a symmetrical fault (in coordination with the relevant TSO).

- Asymmetrical current injection in case of an asymmetrical fault (in coordination with the relevant TSO).
- Fault-ride-through capability.

These requirements may be further detailed in national grid codes. For example, in Spain [15], fast fault current requirements for power park modules are defined as follows:

- Power park modules must be capable of managing the rapid injection of fault current through a continuous control system.
- They shall have the capability to inject apparent current per phase that is at least equal to their nominal current.
- For the positive-sequence current, power park modules must inject or absorb, depending on the positive-sequence voltage error ΔU_1, a reactive current ΔI_{1r} (pu) through a continuous proportional control (see Figure 1). The default value for the limit of this current is ± 0.9 pu.
- In addition to the total reactive current component I_{1r}, power park modules must inject an active current component.
- For the negative sequence current, power park modules must inject or absorb, depending on the negative sequence voltage error ΔU_2, a negative sequence current ΔI_{2r} through a continuous proportional control (see Figure 1).

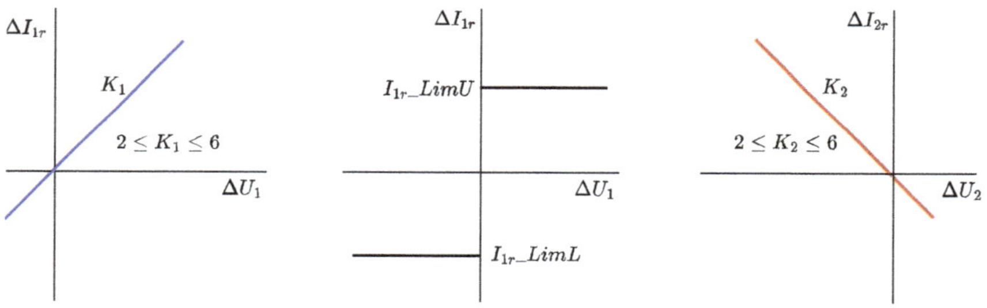

Figure 1. Spanish grid code requirements for the injection of positive- and negative-sequence reactive currents as a function of the sequence voltages at the point of connection.

The IEEE Std. 2800-2022 [28] also establishes the requirements for the injection of fault currents by inverter-based resources (IBRs) during the fault-ride-through mode:

- For balanced faults, IBR units shall inject reactive current dependent on the terminal voltage of the IBR unit. The difference between the reactive current during a fault and the pre-fault reactive current is the incremental positive-sequence reactive current (ΔI_{1r}). During a fault condition, priority shall be given to reactive current injection with any residual capacity being supplied as active current unless the IBR unit is specified to operate in active current priority mode by the TSO.
- For unbalanced faults, in addition to increasing positive-sequence reactive current, the IBR unit shall inject negative-sequence current (ΔI_{2r}), dependent on IBR unit terminal negative-sequence voltage. If the IBR unit's total current limit is reached, either ΔI_{1r} or ΔI_{2r}, or both may be reduced with a preference of equal reduction in both currents.
- This standard intentionally does not specify magnitude of incremental positive- and negative-sequence reactive current injection during a fault condition. The TSO should consider specifying the required magnitudes.

These network codes aim at minimizing the voltage drop during faults and enhancing the fault-ride-through capability of the converters connected to the grid. As established, the grid codes prioritize the injection of reactive current over active current to mitigate the voltage drop in the power system. In the case of an unbalanced fault, the grid codes also require the injection of negative-sequence current to emulate the behavior of synchronous generators and improve the performance of the protective relays [15,16,28]. Maximizing the AC-side current during faults improves these capabilities. The key question is how to determine the value of each current component during the fault that will not exceed the converter's capacity while simultaneously optimizing the performance of the protection system.

3. System and Control Description

Figure 2 shows the structure of a modular multilevel converter. This consists of three phases and each phase has an upper and a lower arm. Each arm is formed by N cascaded submodules. The parameters of the MMC are presented in Table 1.

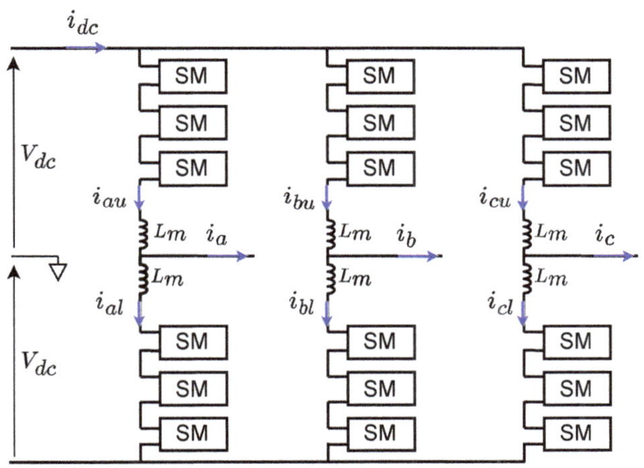

Figure 2. Modular multilevel converter.

Table 1. Parameters of the MMC.

Description	Parameter	Value
Rated apparent power	S_{MMC}	435 MVA
Rated active power	P_{MMC}	400 MW
Arm inductance	L_m	76.16 mH
Cells per arm	N	315 (HB-SMs)
Submodule capacitor	C_{SM}	6000 µF
Line-to-line grid voltage	V_g	260 kV
Angular frequency	ω	$2\pi 50$ rad/s
Pole-to-ground dc voltage	V_{dc}	±250 kV

The positive and negative sequences of the currents that are injected by the MMC into the grid are independently controlled in two synchronous reference frames (SRFs), one for each sequence, as shown in Figure 3. A Second Order Generalized Integrator (SOGI) is used to extract the positive and negative sequences of the grid voltage (v_{1abc} and v_{2abc}, respectively). Next, two Phase Locked Loops (PLLs) determine the voltage angles of the positive and negative sequences (θ_1 and θ_2, respectively), aligning the d-axis of each frame with the grid voltage [29]. This ensures that the voltages V_{1q} and V_{2q} are zero. Using the SOGI and the Park transform with the voltage angles of each sequence, the d- and q-axis components of the positive- and negative-sequence voltages (V_{1dq} and V_{2dq}, respectively) and currents (I_{1dq} and I_{2dq}, respectively) are computed in their respective reference frames [30]. Finally, two independent control loops with two PI regulators each along with the feedforward terms determine the voltage to be generated by the MMC. The energy control of the MMC consists of three parts [31]: (i) a loop for the inner positive-sequence circulating current that is used to transfer energy from the top to the bottom arms, (ii) a loop for the inner negative-sequence circulating current that transfers energy among the three legs, (iii) a loop for the inner DC circulating current that controls the total energy of each leg. Further details can be found in Appendix A.

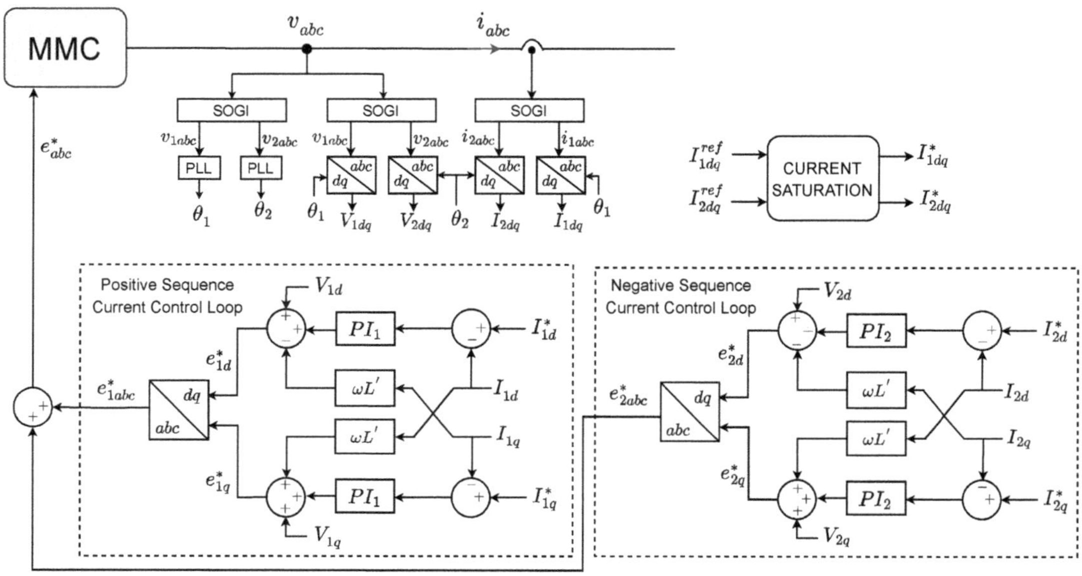

Figure 3. Positive- and negative-sequence current control.

To protect the MMC from overcurrents that could damage their power semiconductor devices, their output currents I_x (where x stands for phase x = a, b, c) are limited to the maximum allowed current (I_{lim}). The current through each phase is:

$$I_a = \sqrt{I_1^2 + I_2^2 + 2I_1I_2cos(\theta)} \tag{1}$$

$$I_b = \sqrt{I_1^2 + I_2^2 + 2I_1I_2cos\left(\theta - \frac{2\pi}{3}\right)} \tag{2}$$

$$I_c = \sqrt{I_1^2 + I_2^2 + 2I_1I_2cos\left(\theta + \frac{2\pi}{3}\right)} \tag{3}$$

where I_1 ($I_1 = \sqrt{I_{1d}^2 + I_{1q}^2}$) and I_2 ($I_2 = \sqrt{I_{2d}^2 + I_{2q}^2}$) are the magnitudes of the positive- and negative-sequence currents, respectively, and $\theta = \theta_2 + \theta_{2i} - \theta_1 - \theta_{1i}$. I_{1d} and I_{1q} are the positive-sequence active and reactive currents, respectively. I_{2d} and I_{2q} are the negative-sequence active and reactive currents, respectively. θ_1 and θ_2 are the space vector angles of the positive and negative SRFs, respectively. θ_{1i} and θ_{2i} are the angles of the positive- and negative-sequence currents, respectively, in their own reference frames (see Figure 4).

In case of a grid unbalance, the current through each phase is different. The phase with the highest current is:

$$I_{max} = max(I_a, I_b, I_c) \tag{4}$$

Expression (4) can also be written as:

$$I_{max} = \sqrt{I_1^2 + I_2^2 + 2I_1 I_2 cos(\theta)_{max}} \quad (5)$$

where $cos(\theta)_{max} = max(cos(\theta), cos(\theta - 2\pi/3), cos(\theta + 2\pi/3))$.

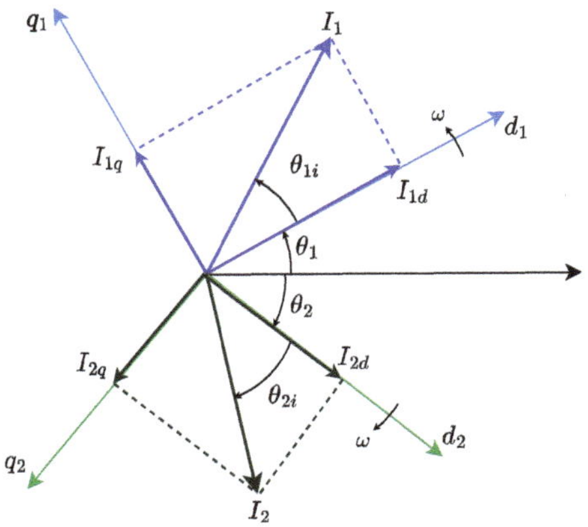

Figure 4. Synchronous reference frames for the positive and negative sequences.

In case of an overcurrent, different limits can be applied depending on the sequence that is prioritized.

$$\text{L1:} \begin{cases} I_1^* = min\left(I_1^{ref}, I_{1lim}\right) & (6) \\ I_2^* = min\left(I_2^{ref}, \dfrac{-2I_1^* cos(\theta)_{max} + \sqrt{(2I_1^* cos(\theta)_{max})^2 - 4(I_1^{*2} - I_{lim}^2)}}{2}\right) & (7) \end{cases}$$

$$\text{L2:} \begin{cases} I_2^* = min\left(I_2^{ref}, I_{2lim}\right) & (8) \\ I_1^* = min\left(I_1^{ref}, \dfrac{-2I_2^* cos(\theta)_{max} + \sqrt{(2I_2^* cos(\theta)_{max})^2 - 4(I_2^{*2} - I_{lim}^2)}}{2}\right) & (9) \end{cases}$$

$$\text{L3:} \begin{cases} I_1^* = min\left(I_1^{ref}, \dfrac{I_{lim}}{\sqrt{I_1^2 + I_2^2 + 2I_1 I_2 cos(\theta)_{max}}} I_1^{ref}\right) & (10) \\ I_2^* = min\left(I_2^{ref}, \dfrac{I_{lim}}{\sqrt{I_1^2 + I_2^2 + 2I_1 I_2 cos(\theta)_{max}}} I_2^{ref}\right) & (11) \end{cases}$$

where I_1^{ref} and I_2^{ref} are the current references before applying the saturation (hereinafter, they are referred to as non-saturated current references). I_1^* and I_2^* are the current references after applying the saturation if needed (hereinafter, they are referred to as saturated current references). I_{1lim} and I_{2lim} are the limits for the positive- and negative-sequence currents, respectively, and I_{lim} is the limit for the total current. These limits may be equal or different depending on whether both sequences are required to be injected or not (subject to $I_{1lim} \leq I_{lim}$ and $I_{2lim} \leq I_{lim}$). For instance, for the limit L1, if $I_{1lim} = I_{lim}$, when the positive-sequence current reaches the saturation value, $I_1 = I_{lim}$ according to Equation (6) and no negative-sequence current can be injected as per Equation (7). However, if $I_{1lim} < I_{lim}$, even if I_1 saturates, there is some margin for injecting negative-sequence current.

Note that in cases of unbalanced currents, these limits must be applied to the phase with the highest current, which is determined through the term $\cos(\theta)_{max}$. L1 prioritizes the injection of positive-sequence current, L2 prioritizes the injection of negative-sequence current, and L3 limits both sequences in the same proportion.

The d- and q-axis current references of the positive sequence are obtained as:

$$\text{L4:}\begin{cases} I_{1d}^{*} = min\left(I_{1d}^{ref}, I_{1dlim}\right) & (12) \\ I_{1q}^{*} = min\left(I_{1q}^{ref}, \sqrt{I_{1lim}^{2} - I_{1d}^{*2}}\right) & (13) \end{cases}$$

$$\text{L5:}\begin{cases} I_{1q}^{*} = min\left(I_{1q}^{ref}, I_{1qlim}\right) & (14) \\ I_{1d}^{*} = min\left(I_{1d}^{ref}, \sqrt{I_{1lim}^{2} - I_{1q}^{*2}}\right) & (15) \end{cases}$$

$$\text{L6:}\begin{cases} I_{1d}^{*} = min\left(I_{1d}^{ref}, \dfrac{I_{1lim}}{\sqrt{\left(I_{1d}^{ref}\right)^{2} + \left(I_{1q}^{ref}\right)^{2}}} I_{1d}^{ref}\right) & (16) \\ I_{1q}^{*} = min\left(I_{1q}^{ref}, \dfrac{I_{1lim}}{\sqrt{\left(I_{1d}^{ref}\right)^{2} + \left(I_{1q}^{ref}\right)^{2}}} I_{1q}^{ref}\right) & (17) \end{cases}$$

where I_{1d}^{ref} and I_{1q}^{ref} are the non-saturated current references for the d- and q-axis, respectively. I_{1d}^{*} and I_{1q}^{*} are the saturated current references for the d- and q-axis, respectively. I_{1dlim} and I_{1qlim} are the limits for the positive-sequence active and reactive currents, respectively, subject to $I_{1dlim} \leq I_{1lim}$ and $I_{1qlim} \leq I_{1lim}$. I_{1lim} is the limit for the total current. L4 prioritizes the injection of active power (positive-sequence), L5 prioritizes the injection of reactive power (positive-sequence), and L6 limits both components in the same proportion.

Similar expressions are used to compute the d- and q-axis current references of the negative-sequence current:

$$\text{L7:}\begin{cases} I_{2d}^{*} = min\left(I_{2d}^{ref}, I_{2dlim}\right) & (18) \\ I_{2q}^{*} = min\left(I_{2q}^{ref}, \sqrt{I_{2lim}^{2} - I_{2d}^{*2}}\right) & (19) \end{cases}$$

$$\text{L8:}\begin{cases} I_{2q}^{*} = min\left(I_{2q}^{ref}, I_{2qlim}\right) & (20) \\ I_{2d}^{*} = min\left(I_{2d}^{ref}, \sqrt{I_{2lim}^{2} - I_{2q}^{*2}}\right) & (21) \end{cases}$$

$$\text{L9:}\begin{cases} I_{2d}^{*} = min\left(I_{2d}^{ref}, \dfrac{I_{2lim}}{\sqrt{\left(I_{2d}^{ref}\right)^{2} + \left(I_{2q}^{ref}\right)^{2}}} I_{2d}^{ref}\right) & (22) \\ I_{2q}^{*} = min\left(I_{2q}^{ref}, \dfrac{I_{2lim}}{\sqrt{\left(I_{2d}^{ref}\right)^{2} + \left(I_{2q}^{ref}\right)^{2}}} I_{2q}^{ref}\right) & (23) \end{cases}$$

where I_{2d}^{ref} and I_{2q}^{ref} are the non-saturated current references for the d- and q- axis, respectively. I_{1d}^{*} and I_{1q}^{*} are the saturated current references for the d- and q-axis, respectively. I_{2dlim} and I_{2qlim} are the limits for the negative-sequence active and reactive currents, respectively, subject to $I_{2dlim} \leq I_{2lim}$ and $I_{2qlim} \leq I_{2lim}$. I_{2lim} is the limit for total current. L7 prioritizes the injection of active power (negative-sequence), L8 prioritizes the injection of reactive power (negative-sequence), and L9 limits both components in the same proportion.

In normal operation and considering balanced grid voltages, the current references for the positive sequence (I_{1d}^{ref} and I_{1q}^{ref}) are computed from the active and reactive power references (P_1^{ref} and Q_1^{ref}, respectively) while the negative-sequence currents (I_{2d}^{ref} and I_{2q}^{ref}) are set to zero.

$$I_{1d}^{ref} = \frac{P_1^{ref}}{3V_{1d}} \tag{24}$$

$$I_{1q}^{ref} = -\frac{Q_1^{ref}}{3V_{1d}} \tag{25}$$

$$I_{2d}^{ref} = 0 \tag{26}$$

$$I_{2q}^{ref} = 0 \tag{27}$$

In case of a fault, the reactive current references are computed according to the Spanish Grid Code (see Figure 1, [15]):

$$I_{1q}^{ref} = I_{1qi} - K_1(V_{1i} - V_1) \tag{28}$$

$$I_{2q}^{ref} = I_{2qi} - K_2(V_{2i} - V_2) \tag{29}$$

where I_{1qi} and I_{2qi} are the pre-fault positive and negative q-axis (reactive) currents, respectively, in per-unit terms. Similarly, V_{1i} and V_{2i} are the pre-fault positive- and negative-sequence voltages in per-unit terms. K_1 and K_2 are the proportional gains (also called k-factor in the literature), which can be set in the range $2 \leq K_{1,2} \leq 6$. If the TSO does not specify any value, they are set to 3.5.

The Spanish Grid Code also states that the sign of the reactive current must be such that the MMC behaves like a synchronous generator In the event of a fault, synchronous generators inject positive-sequence reactive current ($Q_1 > 0$), which helps to minimize voltage drops, and absorb negative-sequence reactive current ($Q_2 < 0$), which helps to reduce voltage imbalances in case of an asymmetrical fault. For the MMC, with the SRFs aligned with the d-axis, the reactive power can be computed as $Q = -3V_d I_q$. To inject positive-sequence reactive power, the value of I_{1q} must be negative. Conversely, to absorb negative-sequence reactive power, the value of I_{2q} must be positive. Thus, considering positive values for the gains K_1 and K_2, the sign before these constants must be negative.

The grid code, therefore, defines the reactive current increments in order to minimize voltage sags (positive-sequence reactive current) and voltage imbalances (negative-sequence reactive current). It prioritizes the injection of positive-sequence reactive current up to a value of 0.9 pu, while the remaining current capability is used for the injection of negative-sequence reactive current and positive-sequence active current.

4. Current Saturation Strategies

4.1. Output Current Saturation with Fixed Limits: Saturation Algorithm 1 (sat1)

Fixed saturation limits for the output currents are usually employed; however, these do not guarantee that the maximum current is injected. For instance, consider the following saturation logic. The positive-sequence reactive current is prioritized up to the limit $I_{1qlim} = 0.9$ pu as per Equation (14). Then, positive-sequence active current is injected according to Equation (15) up to the limit $I_{1lim} = 0.92$ pu. Finally, the remaining current capacity is used for injecting negative-sequence reactive current according to Equation (7).

Consider that the power generated by the MMC is P = 1 pu ($I_{1ai} = 0.92$ pu) and Q = 0 pu ($I_{1ri} = 0$ pu). If a three-phase voltage sag of 0.5 pu takes place, the non-saturated references for the positive- and negative-sequence reactive currents, according to Equations (28) and (29), are $I_{1q}^{ref} = 0 - 3.5(1 - 0.5) = -1.75$ pu and

$I_{2q}^{ref} = 0 - 3.5(0 - 0) = 0$ pu. The non-saturated reference of the positive-sequence active current will double because the voltage is halved ($I_{1a} = 1.84$ pu).

The positive-sequence reactive current will be limited to 0.9 pu and the active current to 0.19 pu ($\sqrt{0.92^2 - 0.9^2}$). The remaining current capability of the converter is kept for the negative-sequence current. However, in this case, the negative-sequence current is zero; hence, the current injection is not maximized.

4.2. Output Current Saturation with Dynamic Limits: Saturation Algorithm 2 (sat2)

To overcome the aforementioned problem, if any of the currents I_{1r}, I_{1a}, or I_{2r} are saturated while the output current I is not, their limits can be increased until either the output current limit is reached or the currents are no longer saturated. The currents that saturate depend on the fault type, the voltage drop and the pre-fault active power. Therefore, a saturation scheme that dynamically adjusts the limits is proposed.

Figure 5 shows the algorithm that is used to maximize the current that is injected by the MMC when the saturation is applied to the output AC current. This consists of the following steps:

- Step 1: Define the limits of the output current (I_{lim}) and the positive-sequence current (I_{1lim}). In the figure, these limits are set to $I_{lim} = 1.2$ pu (assuming a 20% overload capability) and $I_{1lim} = 0.92r$ pu (with $r = 1$), but other values could be used. To maximize the injected current, the positive sequence current limit is adjusted through an iterative process that modifies the variable r. Initially, this variable is set to 1. For subsequent iterations, an updated value of r is computed using its value in the two previous iterations, which are denoted as r_1 and r_2, respectively. Since there are no prior values of r for the second iteration, the initial values of 0.8 and 1.35 are assigned to r_1 and r_2, respectively.
- Step 2: Compute the positive- and negative-sequence voltages (V_{1d} and V_{2d}) and their angles (θ_1 and θ_2) from the grid measurements by using the SOGI and the PLL (see Figure 3).
- Step 3: Compute the non-saturated positive-sequence reactive current reference (I_{1q}^{ref}) according to the grid code (Equation (28)) and the non-saturated positive-sequence active current reference (I_{1d}^{ref}) from the active power reference (Equation (24)).
- Step 4: Obtain the saturated current references I_{1d}^* and I_{1q}^* of the positive-sequence according to Equations (15) and (14), respectively. Note that the injection of the reactive current is prioritized. The limit I_{1qlim} is set at 0.9 according to the Spanish Grid Code.
- Step 5: Compute the magnitude (I_1^*) and the angle (θ_{1i}) of the saturated positive-sequence current.
- Step 6: Compute the non-saturated reference of the negative-sequence reactive current (I_{2q}^{ref}) according to the grid code (Equation (29)).
- Step 7: Compute the magnitude (I_2^{ref}) and the angle (θ_{2i}) of the negative-sequence current.
- Step 8: Compute $\cos(\theta)_{max}$ to determine the phase with the highest current. Note that it is necessary to know the angles θ_1, θ_2, θ_{1i} and θ_{2i}. The first three angles were already calculated in steps 2 and 5. θ_{2i} depends on the values of the active and reactive currents. However, given that the negative-sequence active current is always set to zero, θ_{2i} is always 90°, regardless of the saturation.
- Step 9: The maximum negative-sequence current is calculated using I_1^* and the output current limit according to Equation (5). Using Equation (7), the saturated negative-sequence current reference (I_{2q}^*) is calculated (note that $I_2^* = I_{2q}^*$ given that $I_{2d}^{ref} = 0$).

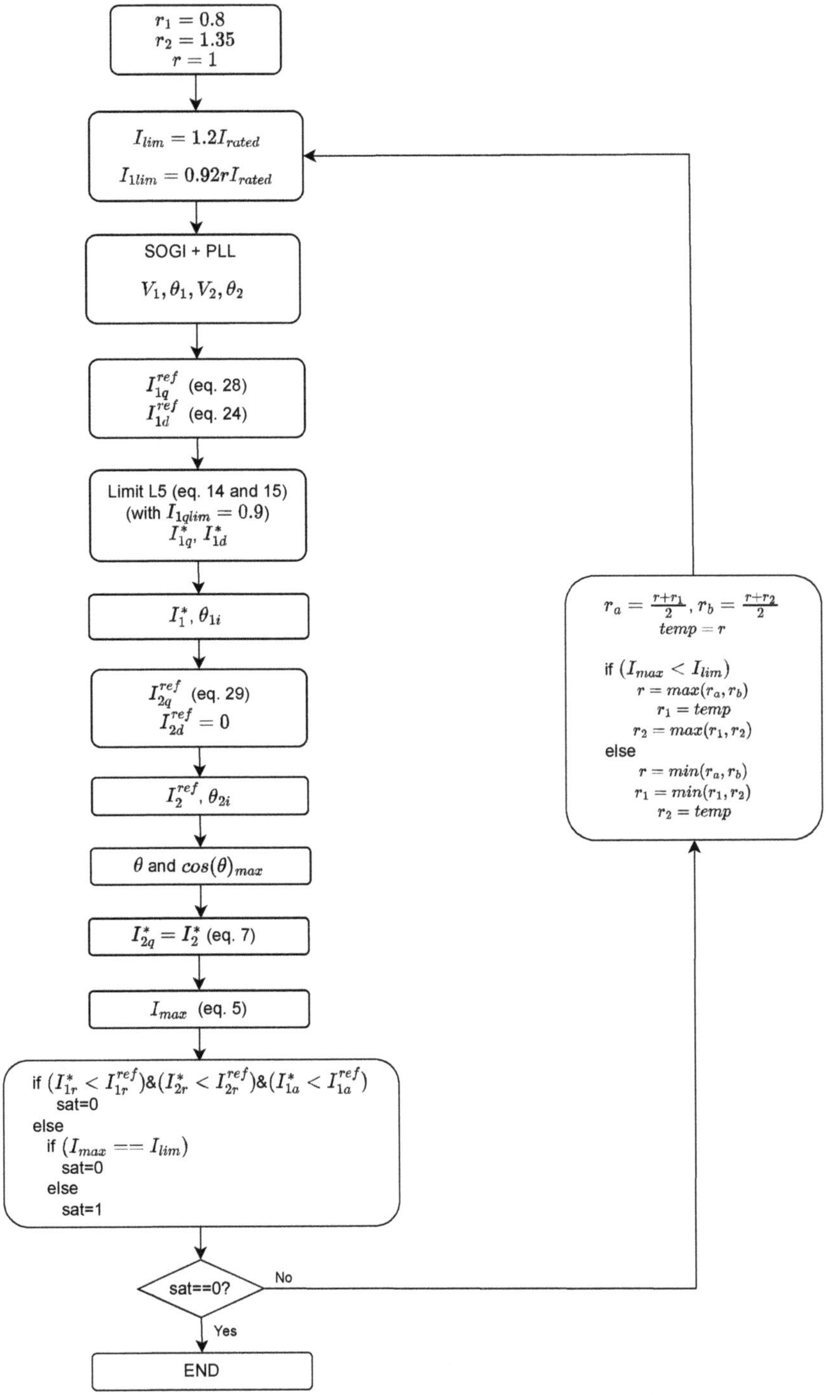

Figure 5. Flow chart for the computation of the active and reactive current references of the positive and negative sequences. The dynamic current limit is applied to the output current of the MMC.

- Step 10: Using the saturated currents I_1^* and I_2^*, and $cos(\theta)_{max}$, the output current of the phase with the highest value is computed using Equation (5).
- Step 11: Four cases may occur:
  - $I_1^* < I_{1lim}$ and $I_{max} < I_{lim}$: None of the currents saturate; hence, the MMC tracks its current references and the loop ends.
  - $I_1^* = I_{1lim}$ and $I_{max} = I_{lim}$: Both currents are saturated; hence, the MMC injects its maximum current according to the grid code. The loop ends.
  - $I_1^* = I_{1lim}$ and $I_{max} < I_{lim}$: The positive-sequence current reaches its limit but the output current does not (for instance, for a three-phase short-circuit that does not require injecting negative-sequence current). In this case, the loop returns to the first step and increases the limit of the positive-sequence current. This is achieved by means of the variable r, which is modified in each iteration to search for the current limits.
  - $I_1^* < I_{1lim}$ and $I_{max} = I_{lim}$: The MMC tracks the positive-current reference because I_1^* is not saturated. The remaining margin is assigned to the negative-sequence current up to the output current limit. The loop ends because the output current reaches its limit.

4.3. Arm Current Saturation with Dynamic Limits: Saturation Algorithm 3 (sat3)

The current flowing through the upper (i_u) and lower (i_l) arms is:

$$i_u = \frac{I_{dc}}{3} + \frac{1}{2}i \tag{30}$$

$$i_l = \frac{I_{dc}}{3} - \frac{1}{2}i \tag{31}$$

where I_{dc} is the pole DC current and i is the output AC current of the MMC.

In case of a fault, the active power injected into (or absorbed from) the grid is reduced because of the voltage drop and the prioritization of the reactive current according to the grid codes. Consequently, the term $I_{dc}/3$ is small. On the other hand, applying the limits presented in Sections 4.1 and 4.2 to the output AC current, the maximum value of $i/2$ remains unchanged regardless of the active power that is transmitted. Thus, since the current through the MMC arms consists of both a DC and an AC component, during a fault, the DC component decreases while the maximum value of the AC component remains unchanged. As a result, the current through the arms of the MMC is below its limits.

The rated output current of the MMC is:

$$I_{1,rated} = \frac{S_{MMC}}{3V_{1,rated}} = \frac{P_{MMC}}{3V_{1,rated}cos(\phi_{1,rated})} \tag{32}$$

The relation between the DC and the AC currents is:

$$I_{dc} = \frac{3I_1V_1cos(\phi_1)}{2V_{dc}} \tag{33}$$

where V_1 is the positive-sequence rms value of the output AC voltage (line-to-ground voltage), I_1 is the positive-sequence rms value of the output AC current, I_{dc} and V_{dc} are the DC current and the pole-to-ground DC voltage, respectively, and $cos(\phi_1)$ is the power factor of the positive sequence.

The rated peak current of the upper and the lower arms of the MMC is:

$$\hat{I}_{u,l,rated} = \frac{I_{1,rated}V_{1,rated}cos(\phi_{1,rated})}{2V_{dc}} + \frac{\sqrt{2}}{2}I_{1,rated} \tag{34}$$

If the power factor decreases because the active power is reduced; the first term on the right-hand side of Equation (34) decreases and the second term can be increased without exceeding the arm limits.

$$\frac{I_1V_{1,rated}cos(\phi)}{2V_{dc}} + \frac{\sqrt{2}}{2}I_1 = \frac{I_{1,rated}V_{1,rated}cos(\phi_{1,rated})}{2V_{dc}} + \frac{\sqrt{2}}{2}I_{1,rated} \tag{35}$$

From Equation (35), the maximum rms output current is:

$$I_1 = \frac{\frac{V_{1,rated}cos(\phi_{1,rated})}{V_{dc}} + \sqrt{2}}{\frac{V_{1,rated}cos(\phi_1)}{V_{dc}} + \sqrt{2}}I_{1,rated} \tag{36}$$

The AC and DC voltages are related by the modulation factor:

$$V_1 = m\frac{V_{dc}}{\sqrt{2}} \tag{37}$$

Replacing Equation (37) in Equation (36):

$$I_1 = \frac{m \cdot cos(\phi_{1,rated}) + 2}{m \cdot cos(\phi_1) + 2}I_{1,rated} \tag{38}$$

Note that as the value of $cos(\phi_1)$ decreases, the current I_1 increases. In case of a fault, the grid codes prioritize the injection of reactive current; hence, the term $cos(\phi_1)$ is small. The maximum value of the output current that does not exceed the arm limit occurs when the DC current is zero, that is, when $\cos(\phi) = 0$.

$$I_{max} = \left(1 + \frac{m}{2}cos(\phi_{1,rated})\right)I_{1,rated} \tag{39}$$

With the parameters provided in Table 1, the term inside the parentheses in Equation (39) is 1.39. Hence, the output current can be increased by 39% with respect to the rated output current without exceeding the arm current limit. This is a theoretical value that is only achieved when the active power drops to zero. However, in the event of a voltage sag, the injection of reactive power should be prioritized to ensure the voltage stability. Moreover, a reduced AC voltage also limits the active power that can be exchanged with the ac grid. Thus, despite the injection of active power, the current increase does not change considerably. On the other hand, in power converters, the most critical components are the semiconductor devices. Other components, such as transformers, disconnectors, and cables can typically withstand much higher currents than IGBTs for short periods. All these components are also used in conventional power plants that are based on synchronous generators, which can inject fault currents of up to approximately 6 times their rated current. Therefore, increasing the output current by 39% should not pose significant issues.

To maximize the current injected by the MMC, the current limit has to be applied to the arm current according to the flow chart shown in Figure 6. The algorithm is the same as that explained in Section 4.2 except for the last two steps, which are as follows:

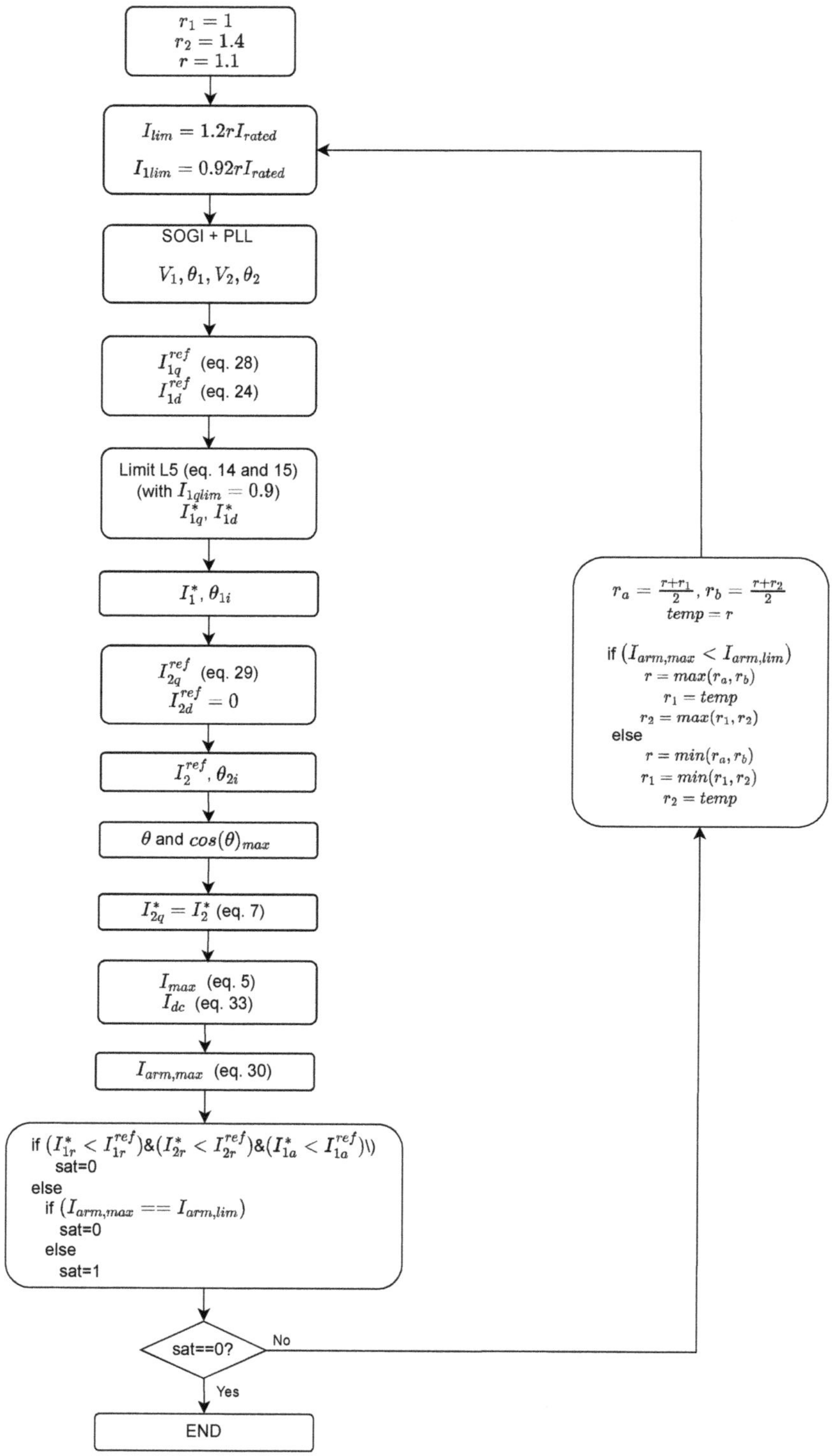

Figure 6. Flow chart for the computation of the active and reactive current references of the positive and negative sequences. The dynamic current limit is applied to the arm current of the MMC.

- Step 10: Using the saturated currents I_1^* and I_2^*, and $cos(\theta)_{max}$, the output current of the phase with the highest value (I_{max}) is computed using Equation (5). Additionally, the DC current I_{dc} is calculated using Equation (33).
- Step 11: Using the phase with the highest output current and the DC current, the arm with the highest current is computed (Equation (34)). Note that the grid voltage V_{1d}, obtained in step 2, the output current I_{max}, and the DC current I_{dc}, calculated in step 10, are used instead of the rated values in Equation (34).
- Step 12: Three cases may occur:
  - $I_{1q}^* < I_{1r}^{ref}$ and $I_{1d}^* < I_{1d}^{ref}$ and $I_{2q}^* < I_{2q}^{ref}$: Since none of the currents reach saturation, the MMC is able to track the current references, and the control loop terminates.
  - $I_{1q}^* = I_{1q}^{ref}$ or $I_{1d}^* = I_{1d}^{ref}$: The positive-sequence current saturates.
    - $I_{arm}^* < I_{arm,lim}$: The arm current does not saturate; hence, the negative-sequence component does not saturate either. Therefore, there is still some margin to increase the positive-sequence current. In this case, the loop returns to the initial step and increases both the output current limit and the positive-sequence current limit. This is achieved through the variable r, which is modified in each iteration to search for the current limits.
    - $I_{arm}^* = I_{arm,lim}$: The arm current saturates and the loop ends.

5. Results

The saturation algorithms presented in Sections 4.2 and 4.3 are compared for seven types of voltage dips (see Figure 7). In this figure, E is the pre-fault voltage, which is considered to be 1 pu, and V is the retained voltage in the phase or phases affected by the fault. Although all types of fault (three-phase, two-phase, phase-to-ground, and phase-to-phase-to-ground faults) may happen, the most common fault is, by far, the phase-to-ground fault (70–80% of the cases). Three-phase faults represent only 2–3% of the cases while phase-to-phase-to-ground faults account for 10–17% and phase-to-phase faults account for 8–10% of the cases [32]. Apart from the voltage (magnitude) imbalance caused by asymmetrical faults, the transformer connection may also cause phase imbalances.

Figure 8 shows the theoretical results for voltage dips ranging from 0 pu to 1 pu:

- Top-left graph: It shows the output AC current considering the saturation algorithms described in Section 4.2 (red line) and Section 4.3 (blue line). In both cases, the phase with the highest current is plotted.
- Bottom-left graph: The ratio of the output current with saturation strategy 3 to the output current with saturation strategy 2 is plotted.
- Top-center graph: It shows the output positive-sequence reactive current ($I_{1r} = I_{1q}$), the positive-sequence active current ($I_{1a} = I_{1d}$), and the negative-sequence reactive current ($I_{2r} = I_{2q}$). Again, the saturation algorithm described in Section 4.2 (dashed lines) is compared with the saturation algorithm described in Section 4.3 (solid lines). In both cases, the phase with the highest current is plotted.
- Bottom-center graph: The ratio of the sequence currents with saturation strategy 3 to the sequence currents with saturation strategy 2 is plotted.
- Top-right graph: It shows the peak arm current of the MMC considering the saturation algorithms presented in Section 4.2 (red line) and that in Section 4.3 (blue line). In both cases, the arm with the highest current is plotted.
- Bottom-left graph: The ratio of the arm current with saturation strategy 3 to the arm current with saturation strategy 2 is plotted.

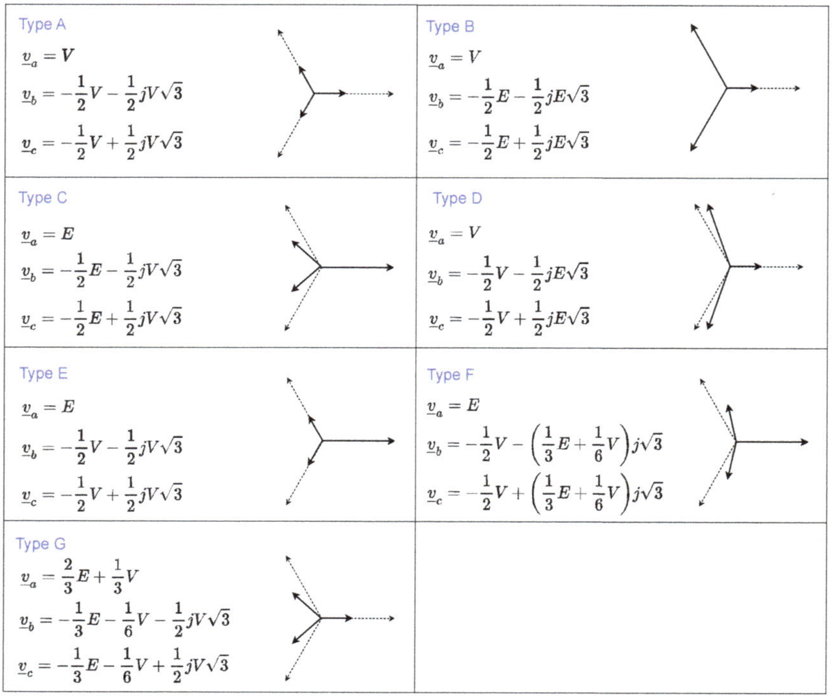

Figure 7. Types of voltage dips caused by symmetrical and asymmetrical faults [33].

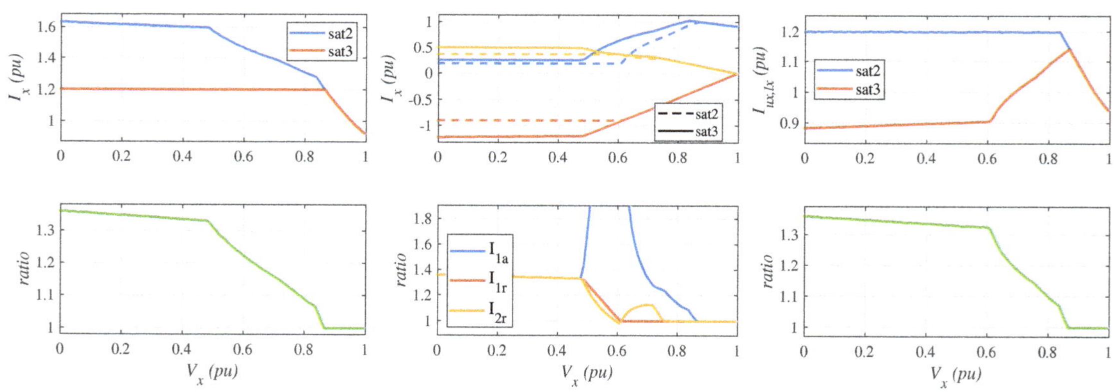

(**a**) Output AC current. (**b**) Output sequence currents. (**c**) Arm current of the MMC.

Figure 8. Type E fault (two-phase voltage dip).

Figure 8 shows the results for a type E voltage dip when the MMC delivers P = 400 MW (0.92 pu) and Q = 0 MVAr (0 pu). For this type of asymmetrical voltage dip, the grid voltage consists of both positive and negative sequences. As a result, according to the aforementioned grid codes, positive- and negative-sequence reactive currents have to be injected into the grid according to Equations (28) and (29). Moreover, given that the Spanish Grid Code requires the power converter to behave like a synchronous generator, a positive-sequence reactive power must be injected (thus, a negative value of I_{1q} is required) and a negative-sequence reactive power must be absorbed (hence, a positive value of I_{2q} is needed).

The theoretical results have been obtained considering that the phasor values of the phase-to-ground voltages at the MMC connection point are those shown in Figure 7. Using these values, the positive- and negative-sequence voltages are obtained by means of Fortescue's theorem. Next, the saturation algorithms are applied to the reactive current references according to Equations (28) and (29). The current references are then saturated according to the algorithms presented in Section 4. Once the output active and reactive positive- and negative-sequence currents are known, the current of the phase with the highest value is computed using Equation (4). Finally, the current through the arm with the highest value is calculated using Equation (30).

Focusing on the top-center graph and the dashed lines (saturation algorithm 2), it can be observed that, as the voltage drops below 1 pu, both the positive- and the negative-sequence reactive currents increase. Simultaneously, to keep the same active power injection, the active current must increase. For small voltage dips, the negative-sequence current is low; hence, the positive-sequence limit (I_{1lim}) is increased beyond its initial value of 0.92 pu to accommodate the increasing active current and leverage the overall current capability of the converter. For $V_x = 0.87$ pu, I_{1q} is -0.303 pu and I_{1a} is 1.006 pu, resulting in $I_1 = 1.051$ pu. This implies that the positive-sequence limit has been increased from its initial value of 0.92 pu to 1.051 pu due to the low negative-sequence current demand, thereby maximizing the injection of fault current. However, at this point, the output current saturates for sat2 (see the top-left graph), indicating that the converter can no longer track all the current references below this voltage level. Consequently, I_{1d} begins to decrease.

For $V_x = 0.76$ pu, I_{1q} is 0.56 pu and I_{1a} is 0.733 pu, resulting in $I_1 = 0.92$ pu. At this point, the positive-sequence current limit is 0,92 and cannot be reduced any further. Consequently, the negative-sequence current saturates.

For $V_x = 0.61$ pu, I_{1q} is 0.9 pu and I_{1a} is 0.191 pu, resulting in $I_1 = 0.92$ pu. The positive-sequence reactive current saturates to 0.9 pu, and the remaining current capability for the positive-sequence current (up to 0.92 pu) is used for the active current. The current margin up to 1.2 pu is allocated to the negative-sequence current. At this point, all the currents are saturated; therefore, they remain constant for voltages below 0.61 pu.

The same analysis can be carried out for the solid lines, which correspond to saturation strategy 3. In this case, the arm current saturates for a voltage dip with a lower retained voltage ($V_x = 0842$ pu), which indicates that all the currents are higher at the instant they begin to saturate. This can be seen in Figure 8, where the solid lines are above the dashed lines.

The bottom-center graph shows the ratio between the currents injected into the grid using both saturation algorithms. For small voltage dips, the currents do not saturate so the same currents are injected into the grid with both saturation algorithms, resulting in a ratio of one.

For $V_x = 0.84$, the active current saturates and the ratio of the positive-sequence current is greater than one, indicating that higher active currents can be injected into the grid with saturation algorithm 3. Conversely, since neither the positive- nor negative-sequence currents are saturated, the same reactive currents are injected with both algorithms, keeping the ratio at one.

For $V_x = 0.72$, the negative-sequence current reaches its limit for saturation algorithm 2, while it does not saturate for saturation algorithm 3, which allows higher negative-sequence currents to be injected into the grid. In this case, the ratio is greater than 1. At this point, the positive-sequence current is not saturated so this current is the same for both saturation algorithms. For $V_x = 0.61$ pu, I_{1q} reaches its limit when saturation strategy 2 is used, while it does not reach its limit when saturation strategy 3 is utilized; hence, the ratio increases.

Finally, for $V_x = 0.48$ pu, all the currents saturate for algorithm 3. Below this voltage, the ratio for all the currents is around 1.36; hence, they increase by 36%. All the currents increase in the same proportion as defined in Figure 6, although other strategies could also be utilized. Therefore, the currents injected into the grid when strategy 3 is used are always higher than those injected using strategy 2, except for low-voltage dips in which the currents do not saturate.

The top-right graph shows the arm current. For saturation strategy 2, the output current saturates when $V_x = 0.87$ pu; therefore, at this point, the arm current reaches its maximum value. For lower retained voltages, the arm current does not increase; therefore, the AC current through the arms remains constant. However, as the voltage drops, the active power decreases and the DC current through the arms also decreases. As a result, the arm current remains below 0.9 pu for voltages lower than 0.4 pu. Conversely, when saturation strategy 3 is applied, the arm current saturates at 1.2 pu. The bottom-right graph shows the ratio of the arm currents between both saturation techniques.

The top-left graph shows the output current. For saturation strategy 2, the output current is limited to 1.2 pu; however, it keeps increasing as the voltage drops for saturation strategy 3. This is due to the reduction in the arm DC current, which allows for higher AC currents. The bottom-left graph shows the output current ratio. Note that the output current is increased by 36% when saturation strategy 3 is used. Theoretically, the maximum increase is 39%, as computed in Section 4 (see Equation (39)). In this case, such a value is not reached because some active power is still injected into the grid; thus, some DC current flows through the arms. If the limits I_{1qlim} and I_{1lim} were the same, all the positive-sequence current would be reactive and the active current would be zero. In this scenario, no DC circulating current would flow through the arms and an increase of 39% would be reached.

The same analysis can be carried out for the other types of voltage sags, leading to similar conclusions. Therefore, the detailed analysis is omitted here but the results are presented in Appendix B (Figures A2–A7). Table 2 provides a summary of the current increase for saturation strategy 3 under different voltage sags.

Table 2. Output current increase in % for strategy 3 and five voltage sag dips.

Voltage Sag Type	Voltage Dip				
	0 pu	**0.20 pu**	**0.40 pu**	**0.60 pu**	**0.80 pu**
A	38%	36%	34%	26%	8%
B	31%	23%	16%	11%	3%
C	35%	34%	27%	16%	9%
D	35%	34%	27%	16%	9%
E	36%	35%	34%	23%	9%
F	36%	35%	34%	23%	9%
G	36%	35%	34%	23%	9%

Simulation Results

To verify the proposed saturation algorithms, detailed PSCAD simulations have been carried out. The MMC presented in Section 3 is connected to an AC grid and several voltage sags are applied. The voltage sag is simulated by changing the voltage at the connection point of the MMC.

Figures 9a and 9b show the PSCAD results for saturation strategies 2 and 3, respectively. The following data are plotted from top to bottom: (1) phase-to-ground voltages at the point of connection of the MMC; (2) output current of the MMC; (3) d- and q-components of the voltage at the point of connection for the positive and negative sequences; (4) d- and q-components of the output current for the positive and negative sequences, that is, the active and the reactive currents of each sequence; (5) upper and lower arm currents of the MMC for the phase with the highest current; (6) average upper and lower submodule capacitor voltages of the phase with the highest current.

The MMC is initially injecting into the grid 400 MW (0.92 pu) and 0 MVAr (0 pu). At t = 0.02 s, a two-phase voltage dip occurs (first graph) with a retained voltage of 0.3 pu. This causes a rapid increase in the output current (second graph). The SOGI detects the asymmetrical conditions and separates the positive- and negative-sequence voltages. Then, the PLLs align the d-axis of the positive and the negative reference frames with the corresponding grid voltages. This can be seen in the third graph where the q-components of both sequence voltages are zero, while V_{1d} is 0.535 pu and V_{2d} is 0.235 pu. Once the voltage dip is detected, the current references of the positive- and the negative-sequence reactive currents are determined based on the grid codes. According to Equations (28) and (29), the values of the non-saturated current references are $I_{1q}^{ref} = 0 - 3.5(1 - 0.535) = 1.628$ pu and $I_{2q}^{ref} = -3.5(0 - 0.235) = 0.825$ pu, respectively. The non-saturated reference of the positive-sequence active current is $I_{1a}^{ref} = 0.92/0.535 = 1.720$ pu. The previous values would cause a high overcurrent in the MMC; therefore, they must be limited.

For saturation strategy 2 (Figure 9a), the current I_{1q} is limited to 0.9 pu. The value of I_{1d} is 0.19 pu, which corresponds to the positive-sequence current limit of 0.92 pu ($\sqrt{0.92^2 - 0.9^2} = 0.19$ pu). The remaining current capacity is used to inject negative-sequence reactive current, resulting in a value of 0.363 pu. These values match with the theoretical ones shown with dashed lines in the top-center graph of Figure 8. The output current of the phase with the highest value is 1.2 pu (second graph). Since the saturation is applied to the output current, the value of the arm current is only 0.91 pu, which matches with the value shown in the top-right graph of Figure 8. The voltage dip causes a disturbance in the capacitor voltages of the submodules; however, the inner energy control effectively stabilizes them. The capacitor voltages are restored to their reference values within approximately 150 ms (sixth graph).

For saturation strategy 3 (Figure 9b), the limits of the output current and the positive-sequence currents are increased until the arm current reaches its limit. In this case, the current I_{1q} is limited to 1.164 pu and the current I_{1d} to 0.248 pu, which corresponds to a positive-sequence current limit of 1.19 pu. The remaining current capacity is used to inject negative-sequence reactive current, resulting in a value of 0.489 pu. These values match with the theoretical ones shown with solid lines in the top-center graph of Figure 8. The output current for the phase with the highest value is 1.60 pu (second graph), which matches with the value shown in the top-left graph of Figure 8. Since the saturation is now applied to the arm current, its value is 1.2 pu. Again, the inner energy control effectively controls the capacitor voltages of the submodules to 1.6 kV within approximately 150 ms (sixth graph).

One-phase and three-phase voltage dips have also been simulated in PSCAD. The results are similar to those presented in Figure 9 for a two-phase voltage dip; therefore, the analysis is not repeated here, although the results are included in Appendix B (Figures A8 and A9).

Finally, the system shown in Figure 10 has been simulated. It consists of a 400 MW offshore wind power plant connected to the inland grid through an HVDC link. The onshore MMC is connected to an AC grid with an SCR of 5 through two parallel 60 km long AC transmission lines. Given that the objective is to analyze the response of MMC2 in the event of an AC fault, the wind power plant and the offshore MMC have been replaced by a DC source. The system parameters are listed in Table 3. The parameters of the MMC are those included in Table 1.

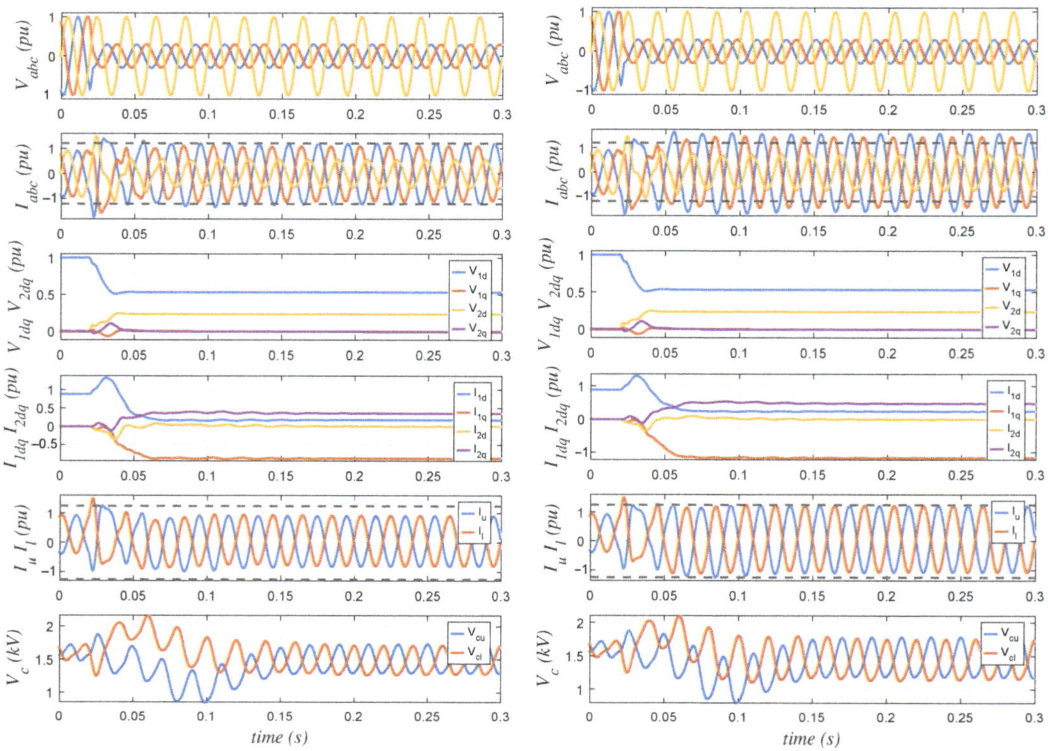

(**a**) Output current saturation (sat2). (**b**) Arm current saturation (sat3).

Figure 9. Type E voltage dip (two-phase voltage dip).

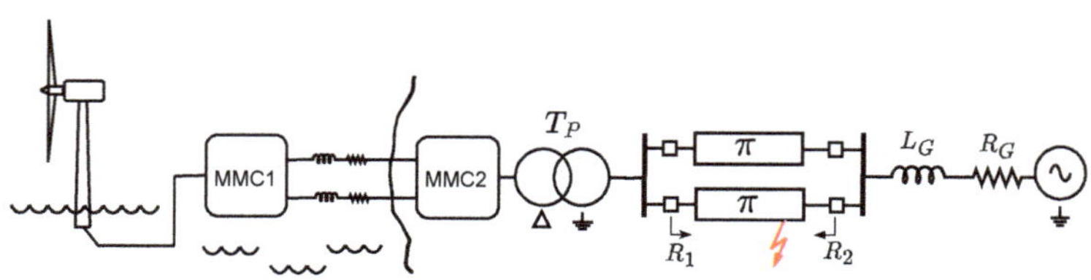

Figure 10. Offshore wind power plant connected through an HVDC link.

Table 3. System parameters.

Description	Parameter	Value
Nominal voltage (line-to-line)	V_G	220 kV
Resistance of the grid	R_G	2.41 Ω
Inductance of the grid	L_G	76.65 mH
Resistance of the line	r_l	7×10^{-5} Ω/m
Inductive reactance of the line	x_l	4×10^{-4} Ω/m
Capacitive reactance of the line	x_c	385 MΩm
Length of the line	l_l	60 km
Voltage ratio of the transformer	r_T	260/220 kV
Rated power of the transformer	S_T	450 MVA
Leakage inductance of the transformer	L_T	47.82 mH (220 kV side)

The wind power plant is delivering 400 MW when a phase-to-phase fault occurs at t = 0.02 s, with zero ohms fault resistance, located at 90% of the length of one of the AC transmission lines. The protective relays at both ends of the line (R1 and R2, respectively) should detect the fault and isolate it by tripping. Considering distance relays [34], R1 would identify the fault in zone 2 because it is beyond 80% of the line length. Unless teleprotection schemes with communication signals are used, the trip of the relay for a fault in zone 2 is intentionally delayed about 300–500 ms. Figure 11 shows the results for saturation algorithms 2 and 3. For saturation strategy 2, the low injected currents provoke the malfunction of the SOGI-PLL, which is not able to obtain the sequence voltages and align the d-xis of the reference frames with the grid voltage. As a result, the MMC loses synchronism, which causes overcurrents that would trip the converter.

On the other hand, with saturation strategy 3, the SOGI-PLL keeps working properly and the MMC is able to remain connected to the grid and inject active and reactive power. Saturation strategy 3 allows for larger currents that increase the retained voltage and reduce the unbalance. Therefore, the MMC can ride through the fault and resume normal operation as soon as the fault is isolated by the relays and their breakers.

Note that during the first cycle after the fault onset, the control of the MMC is not able to effectively limit the fault currents to 1.2 pu due to the system and control dynamics, as well as the time required to compute the sequence components. To avoid high overcurrents that could damage the power semiconductor devices, the MMC is temporarily blocked and immediately deblocked. This process is repeated several times if required, which allows the converter to remain connected to the grid. The limit for blocking the converter is set at 1.35 pu, which provides some margin and coordination over the saturation limit of 1.2 pu. If the current approaches the maximum repetitive peak forward current of the IGBTs (typically around 2 pu), the MMC would be permanently blocked and the AC breaker would trip [35].

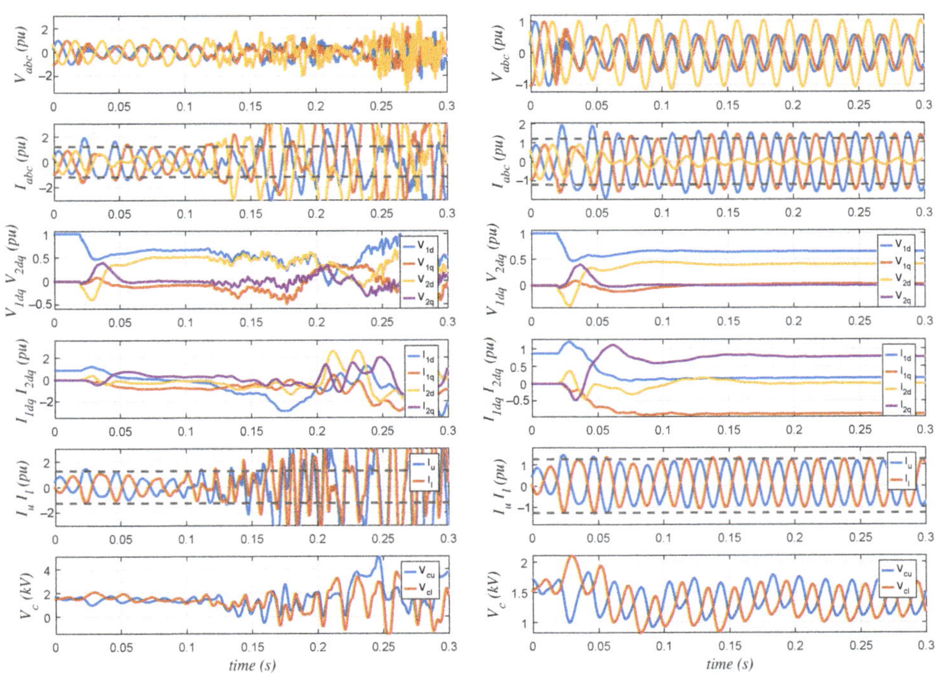

(a) Output current saturation (sat2). (b) Arm current saturation (sat3).

Figure 11. MMC fault ride-through for a two-phase fault at the end of the output ac line.

6. Conclusions

With the increasing number of renewable power plants, future power systems will be dominated by power electronic converters. In this context, MMCs are the preferred topology for high-voltage and high-power applications. However, power electronic converters present some challenges. In case of a fault, due to their limited overcurrent capability, retained voltages will be lower and voltage dips will spread further than in the case of power systems that are based on synchronous generation. This may lead to a malfunction of some controls like the PLL. Additionally, low voltages and currents may complicate the correct operation of the protective relays. To address this issue, several TSOs have developed new grid codes with updated requirements for power electronic converters. These include the need for fast current injection of both positive- and negative-sequence currents and fault current maximization.

The control of MMCs includes saturation limits in the output AC currents to avoid overcurrents. However, using fixed current limits for the positive-sequence and total output currents does not necessarily maximize the fault current, for instance, in case of a three-phase fault that does not require the injection of negative-sequence current. To address this, as a first improvement, this paper proposes dynamic limits that adjust the positive-sequence current limit as a function of the requirements for the negative-sequence current.

However, the current flowing through the power semiconductor devices is the arm current, not the output AC current. The arm current consists of half the AC current plus one-third of the DC current. In the event of a fault, the active power is significantly reduced, which leads to a reduction in the DC current. As a result, the output AC current can be increased without exceeding the arm current limits.

Therefore, as a second improvement, the saturation limit is applied to the arm current. The proposed algorithm for such a saturation strategy is detailed in the paper. The values of the output current with both saturation strategies are compared for several fault types and voltage dips and it is concluded that the output fault current can be increased by up to 40% when the arm current saturation is used. The theoretical analysis is validated with detailed PSCAD simulations. Finally, to illustrate the superiority of the arm saturation strategy, an example of a wind power plant connected to the AC grid through an MMC-based HVDC link is presented. This example shows that the higher fault currents injected into the grid when the arm saturation strategy is used allow for a more resilient fault ride-through.

Author Contributions: Conceptualization, R.V.-A. and E.B.; methodology, R.V.-A. and F.M.; software, R.V.-A. and F.M.; validation, E.B. and F.M.; formal analysis, R.V.-A., E.B. and F.M.; investigation, R.V.-A., E.B. and F.M.; writing, R.V.-A., E.B. and F.M.; supervision, R.V.-A. and E.B. All authors have read and agreed to the published version of the manuscript.

Funding: The authors would would like to acknowledge the support of the Spanish Research Agency through grant PID2020-112943RBI00 funded by MCIN/AEI/10.13039/501100011033; grant PID2021-125634OB-I00 funded by MCIN/AEI/10.13039/501100011033 and ERDF a way of making Europe; and grants TED2021-130120BB-C21 and TED2021-130120B-C22 funded by MCIN/AEI/10.13039/501100011033 and by the European Union NextGenerationEU/PRTR.

Data Availability Statement: The data are included in the article. Additional data are available from the corresponding author upon request.

Conflicts of Interest: The authors declare no conflicts of interest.

Abbreviations

The following abbreviations are used in this manuscript:

GC	Grid Code
HVDC	High-Voltage Direct Current

MMC	Modular Multilevel Converter
PEC	Power Electronic Converter
PLL	Phase-Locked Loop
SC	Synchronous Condenser
SCL	Short-Circuit Level
SG	Synchronous Generator
SOGI	Second Order Generalized Integrator
TSO	Transmission System Operator
VSC	Voltage Source Converter

Appendix A

Appendix A.1. Inner Control of the MMC

Figure A1 shows the inner control of the MMC when this regulates the active and reactive power injected into the AC grid. An upper-level control regulates the total energy (E_{tot}) stored in all thee submodule capacitors of the MMC, which determines the DC power reference (P_{dc}^{*}) through a PI controller and the feedforward term of the measured AC power, as shown in Figure A1a. Additionally, another PI controller determines the DC voltage (V_{dc}^{*}) to be generated by the MMC in order to obtain the desired DC current (I_{dc}^{*}), which is computed from the DC power reference and the DC voltage.

The total energy of each arm (the sum of the upper and the lower arm energies of each phase, E_x^{Σ}) is controlled using the circulating DC current ($I_{x,dc}$), as shown in Figure A1b. This current reference is obtained through a PI controller and the feedforward term equal to one-third of the AC power. While this control regulates the total leg energy, it does not ensure a proper energy distribution between the upper and the lower arms (E_{com}^{Δ}). For this purpose, a positive-sequence circulating current ($i_{x,pos}$) transfers energy from the upper arms to the lower arms, or vice versa (Figure A1c). Finally, a negative-sequence circulating current ($i_{x,neg}$) transfers energy among the three legs, as shown in Figure A1d.

The DC and AC circulating currents are controlled by means of a Proportional–Integral–Resonant (PIR) controller, as shown in Figure A1e. A detailed description of the equations governing the dynamics of the MMC can be found in [36]. Finally, a phase-shifted carrier-based PWM along with a short and select algorithm determines the submodules to be connected in each arm. Further details can be found in [37].

Appendix A.2. Control Parameters

The values of the controllers and filters are presented in Table A1.

Table A1. Control parameters.

Controllers/Filters	Values
PI_1 [1]: positive sequence output current control	$k_p = 45.8$ V/A, $k_i = 7539.5$ V/As
PI_2: negative sequence output current control	$k_p = 45.8$ V/A, $k_i = 7539.5$ V/As
PI_3: total MMC energy control	$k_p = 40.0$ W/J, $k_i = 877.92$ W/Js
PI_4: dc current control	$k_p = 56$ V/A, $k_i = 5600$ V/As
PI_5: leg energy control	$k_p = 40.0$ W/J, $k_i = 175.6$ W/Js
PI_6: upper and lower arms common energy difference control	$k_p = 40.0$ W/J, $k_i = 175.6$ W/Js
PI_7: upper and lower arms differential energy control	$k_p = 40.0$ W/J, $k_i = 175.6$ W/Js
PIR [2]: MMC circulating current control	$k_p = 15.2$ V/A, $k_i = 1687.7$ V/As $k_{pr} = 222.1$ V/A, $\omega_r = 314.16$ rad/s
Band reject filter [3]	$a_2 = 1$ Vs^2/A, $a_1 = 0$ Vs/A $a_0 = 628.3^2$ V/A, $b_2 = 1$ s^2 $b_1 = 890$ s, $b_0 = 628.3^2$

[1] $PI(s) = k_p + \frac{k_i}{s}$. [2] $PIR(s) = k_p + \frac{k_i}{s} + \frac{k_{pr} s}{s^2 + \omega_r^2}$. [3] $BRF(s) = \frac{a_2 s^2 + a_1 s + a_0}{b_2 s^2 + b_1 s + b_0}$.

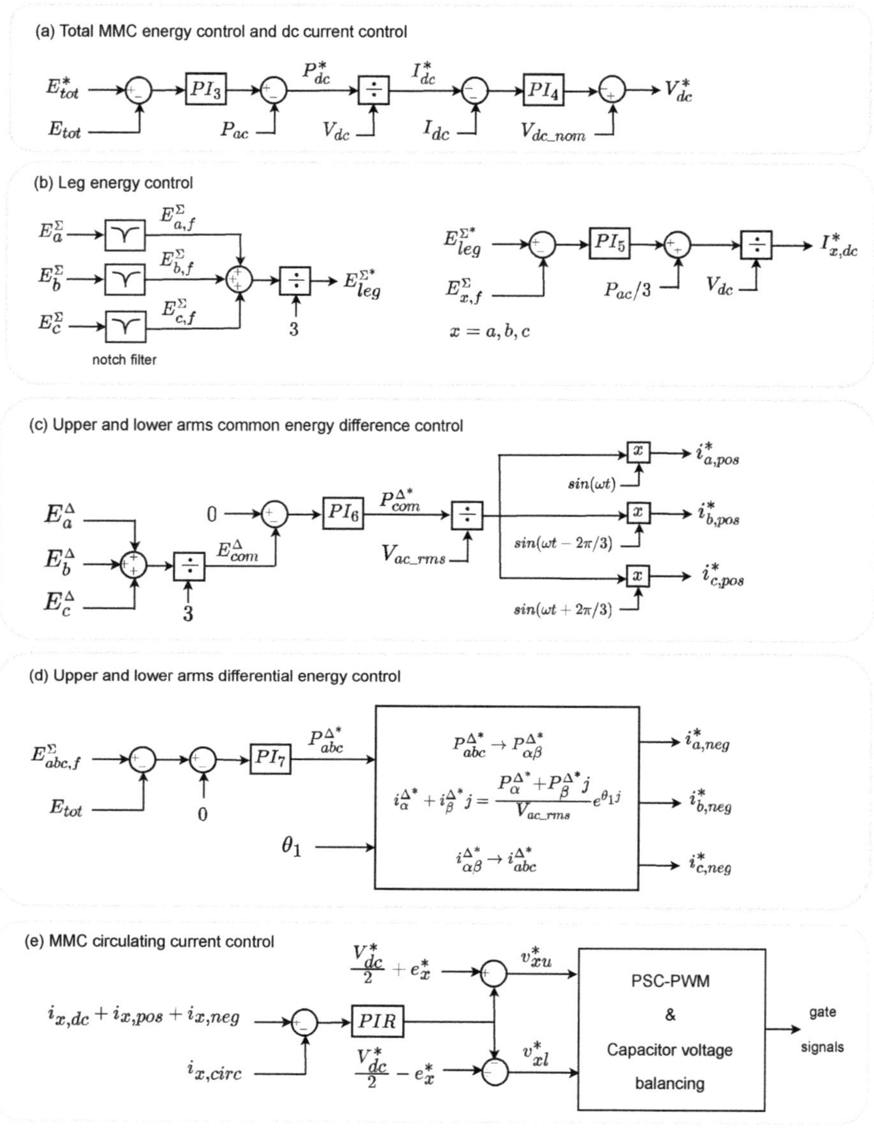

Figure A1. Inner control of the MMC.

Appendix B

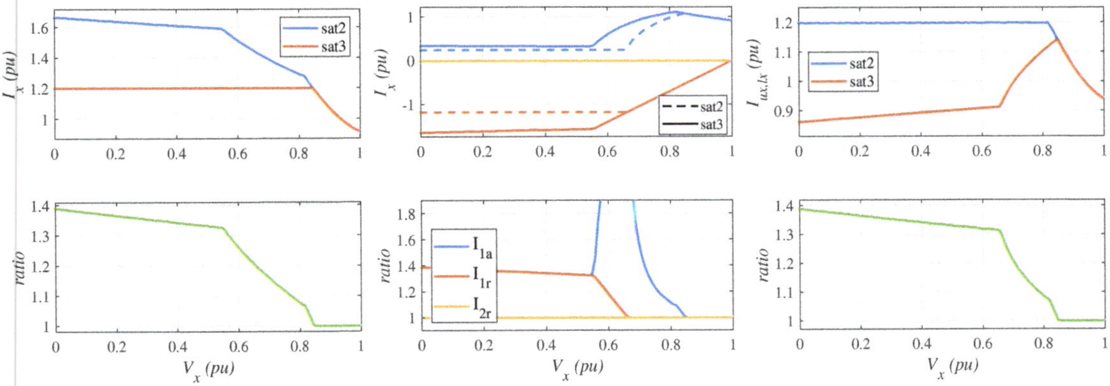

(**a**) Output ac current. (**b**) Output sequence currents. (**c**) Arm current of the MMC.

Figure A2. Type A voltage dip.

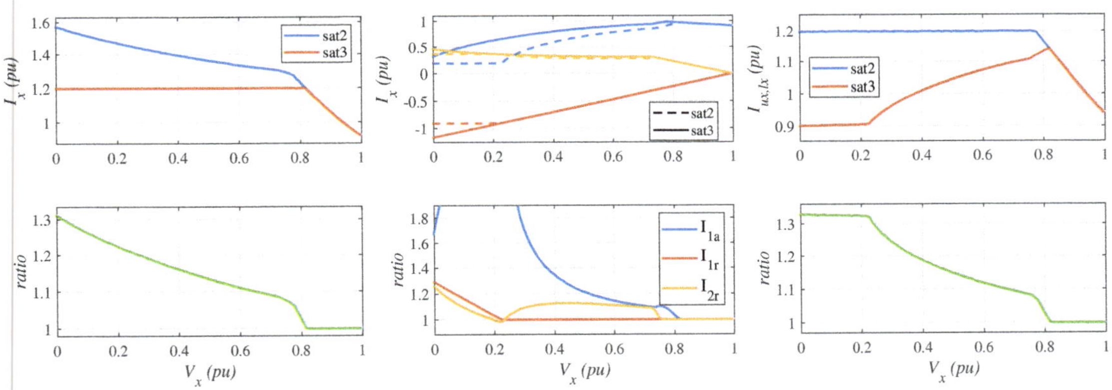

(**a**) Output ac current. (**b**) Output sequence currents. (**c**) Arm current of the MMC.

Figure A3. Type B voltage dip.

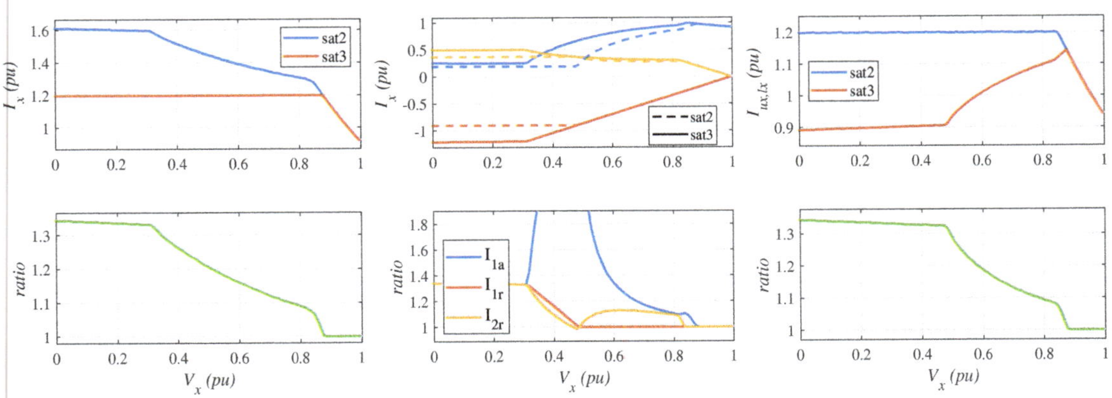

(**a**) Output ac current. (**b**) Output sequence currents. (**c**) Arm current of the MMC.

Figure A4. Type C voltage dip.

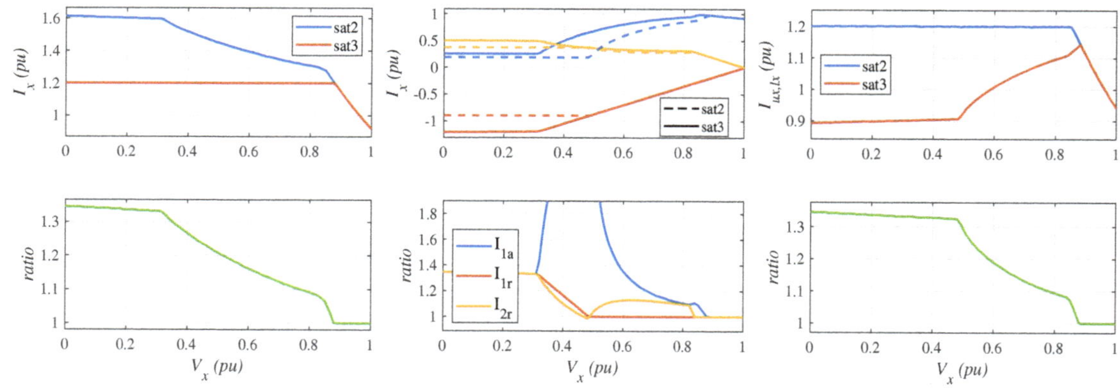

(**a**) Output ac current. (**b**) Output sequence currents. (**c**) Arm current of the MMC.

Figure A5. Type D voltage dip.

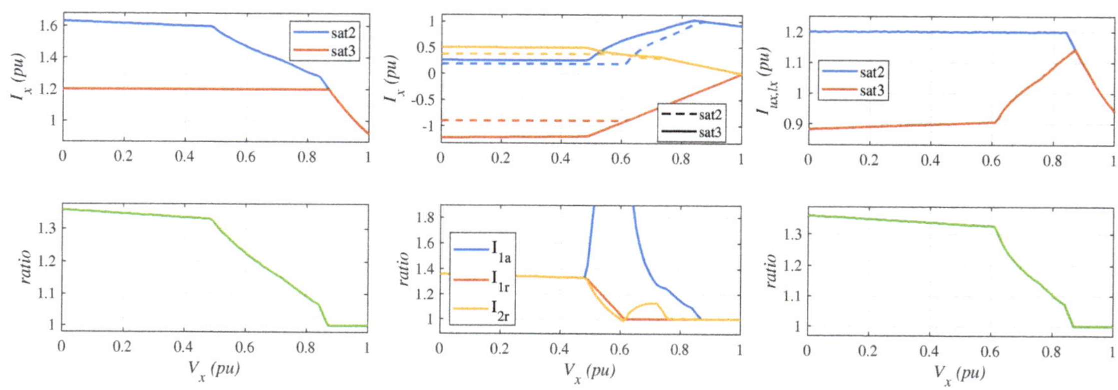

(**a**) Output ac current. (**b**) Output sequence currents. (**c**) Arm current of the MMC.

Figure A6. Type F voltage dip.

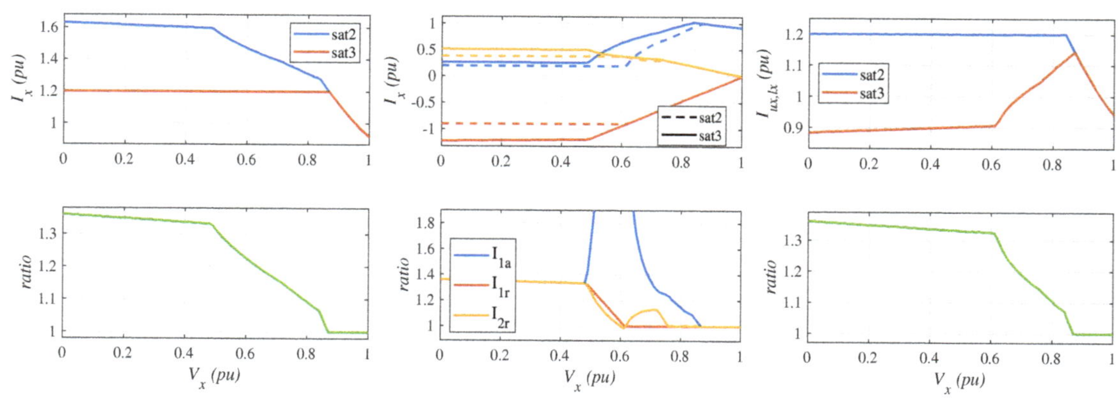

(**a**) Output ac current. (**b**) Output sequence currents. (**c**) Arm current of the MMC.

Figure A7. Type G voltage dip.

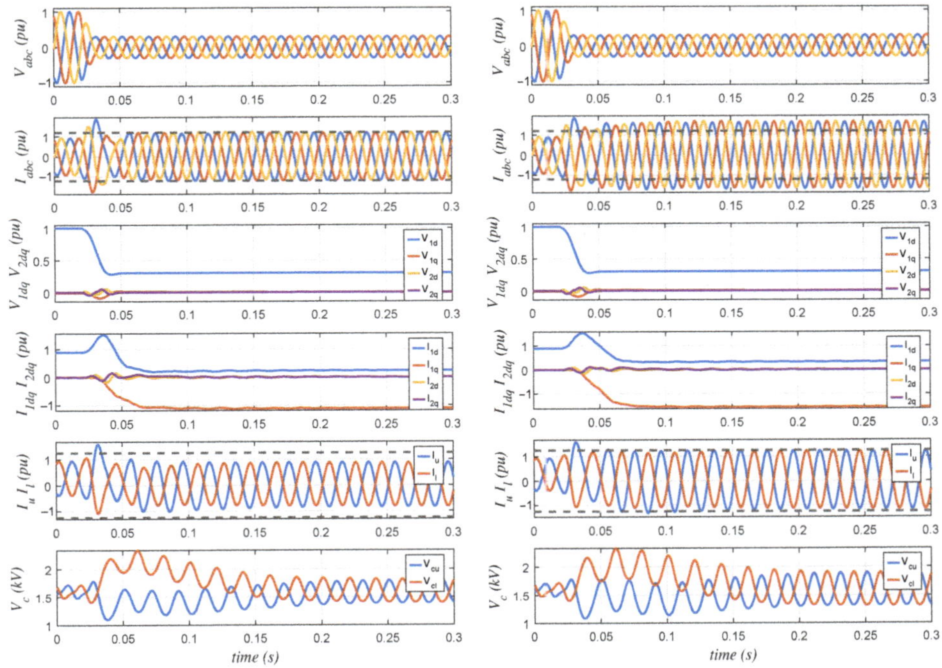

(**a**) Output current saturation. (**b**) Arm current saturation.

Figure A8. Type A voltage dip (three-phase dip).

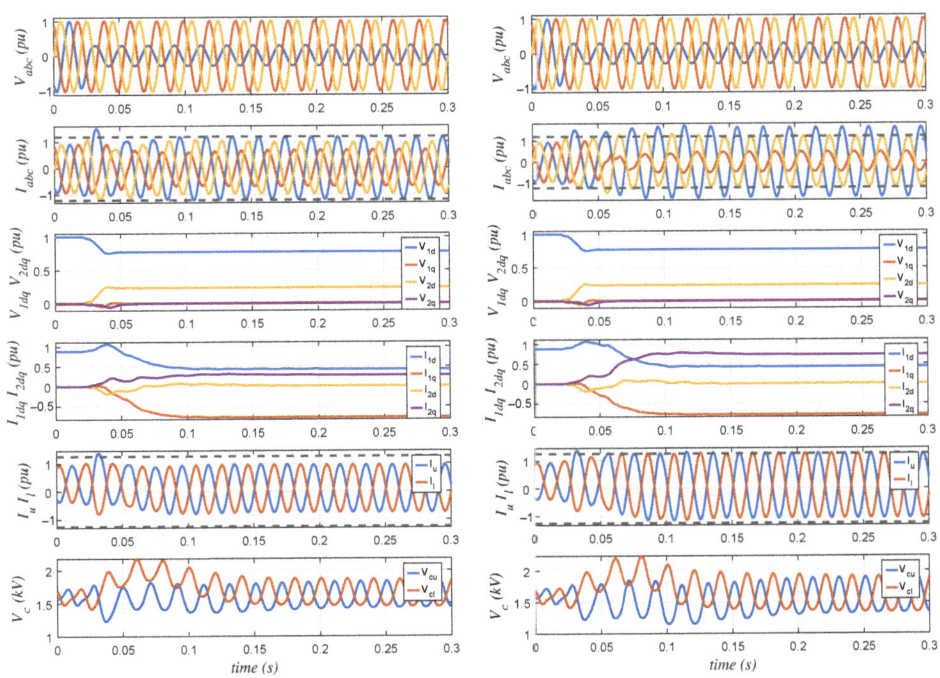

(**a**) Output current saturation. (**b**) Arm current saturation.

Figure A9. Type B voltage dip (one-phase dip).

References

1. IEEE Power & Energy Society. Impact of Inverter Based Generation on Bulk Power System Dynamics and Short-Circuit Performance. Technical Report. 2018. Available online: https://resourcecenter.ieee.org/publications/technical-reports/pes_tr_7-18_0068 (accessed on 14 March 2025).
2. National Grid ESO. *System Operability Framework Lmpact of Declining Short Circuit Levels*; Technical Report; National Grid ESO: Warwick, UK, 2018.
3. Belenguer, E.; Vidal-Albalate, R.; Magraner, F. Fault analysis of power systems dominated by Power Electronic Interfaced Devices. In Proceedings of the ELECTRIMACS 2024—International Conference on Modeling and Simulation of Electric Machines, Converters and Systems, Castelló, Spain, 27–30 May 2024.
4. National Grid ESO. *System Operability Framework Whole System Short Circuit Levels*; Technical Report; National Grid ESO: Warwick, UK, 2018.
5. Aljarrah, R.; Marzooghi, H.; Terzija, V. Mitigating the impact of fault level shortfall in future power systems with high penetration of converter-interfaced renewable energy sources. *Int. J. Electr. Power Energy Syst.* **2023**, *149*, 109058. [CrossRef]
6. Weise, B. Impact of K-factor and active current reduction during fault-ride-through of generating units connected via voltage-sourced converters on power system stability. *IET Renew. Power Gener.* **2015**, *9*, 25–36. [CrossRef]
7. Wang, S.; Egea-Alvarez, A. Operating a Zero Carbon GB Power System in 2025: Frequency and Fault Current. Power Electronic Devices and Fault Current. Technical Report. 2020. Available online: https://strathprints.strath.ac.uk/74793/ (accessed on 14 March 2025).
8. Nedd, M.; Hong, Q.; Bell, K.; Booth, C.; Mohapatra, P. Application of Synchronous Compensators in the GB Transmission Network to Address Protection Challenges from Increasing Renewable Generation. In Proceedings of the CIGRE (Study Committee B5 Colloquium), Auckland, New Zealand, 11–15 September 2017.
9. Gordon, S.; Hong, Q.; Bell, K. Implications of reduced fault level and its relationship to system strength: A Scotland case study. In Proceedings of the CIGRE, Paris, France, 28 August–2 September 2022.
10. National Grid ESO. NOA Stability Pathfinder Phase 2. Expression of Interest Summary. Available online: https://www.neso.energy/document/187371/download (accessed on 18 February 2025).
11. pse2consulting. The National Grid Pathfinder 2 Project. Available online: https://pse2consulting.com/national-grid-pathfinder (accessed on 18 February 2025).
12. Elecnor. Elecnor to Participate in the NOA Stability Pathfinder Programme in Scotland. Available online: https://www.grupoelecnor.com/storage/media/files/shares/noticias/en/elecnorukwpthurso-and-neilstonnpen.pdf (accessed on 18 February 2025).
13. ABB. ABB's Integrated Technology Will Stabilize the Power Grid as Spanish Islands Transition to Green Energy. Available online: https://new.abb.com/news/detail/116318/abbs-integrated-technology-will-stabilize-the-power-grid-as-spanish-islands-transition-to-green-energy (accessed on 18 February 2025).
14. Vidal-Albalate, R.; Belenguer, E.; Magraner, F.; El Ghoufairi, Y. Influence of converter control on directional and fault identification algorithms of protective relays. In Proceedings of the 2024 ONCON—The 3rd IEEE Industrial Electronics Society Annual Online Conference, Beijing, China, 8–10 December 2024.
15. Ministry for the Ecological Transition and the Demographic Challenge. Orden TED/749/2020, de 16 de Julio, por la que se Establecen los Requisitos téCnicos Para la Conexión a la Red Necesarios Para la Implementación de los Códigos de Red de Conexión (Technical Requirements for Grid Connection Necessary for the Implementation of the Network Connection Codes). Available online: https://www.boe.es/buscar/act.php?id=BOE-A-2020-8965 (accessed on 15 February 2025).
16. Technical Connection Rules for High-Voltage (VDE-AR-N 4120). Available online: https://www.vde.com/en/fnn/topics/technical-connection-rules/tar-for-high-voltage (accessed on 25 February 2025).
17. Torresan, G.; Saad, H.; Gomes, V.; Gartmann, P. Study on the impact of fault current injection in a wind power plant using EMT-type tool. In Proceedings of the 20th WInd Integration Workshop, Berlin, Germany, 29–30 September 2021; Volume 2021, pp. 449–456. [CrossRef]
18. REE (Spanish TSO). Criterios Generales de Protección de los Sistemas Eléctricos Insulares y Extrapeninsulares (General Protection Criteria for Insular and Extra-Peninsular Electrical Systems). Technical Report. Available online: https://www.ree.es/sites/default/files/14_OPERACION/Documentos/criterios_proteccion_sistema_2005_v2.pdf (accessed on 5 March 2025).
19. Fentie, D.D. Understanding the dynamic mho distance characteristic. In Proceedings of the 2016 69th Annual Conference for Protective Relay Engineers (CPRE), College Station, TX, USA, 4–7 April 2016. [CrossRef]
20. C37.113-2015—IEEE Guide for Protective Relay Applications to Transmission Lines. IEEE. Available online: https://ieeexplore.ieee.org/document/7502047 (accessed on 3 March 2025).
21. Ryndzionek, R.; Sienkiewicz, L. Evolution of the HVDC Link Connecting Offshore Wind Farms to Onshore Power Systems. *Energies* **2020**, *13*, 1914. [CrossRef]
22. The Federal Network Agency. Network Development Plan 2037/2045. 2023. Available online: https://www.netzentwicklungsplan.de/en/nep-aktuell/netzentwicklungsplan-20372045-2023 (accessed on 21 February 2025).

23. Francos, P.; Verdugo, S.; Alvarez, H.; Guyomarch, S.; Loncle, J. INELFE—Europe's first integrated onshore HVDC interconnection. In Proceedings of the 2012 IEEE Power and Energy Society General Meeting, San Diego, CA, USA, 22–26 July 2012; pp. 1–8. [CrossRef]
24. Gao, G.; Wu, H.; Blaabjerg, F.; Wang, X. Fault current control of MMC in HVDC-connected offshore wind farm: A coordinated perspective with current differential protection. *Int. J. Electr. Power Energy Syst.* **2023**, *148*, 108952. [CrossRef]
25. Westerman Spier, D.; Prieto-Araujo, E.; Lopez-Mestre, J.; Gomis-Bellmunt, O. Optimal Current Reference Calculation for MMCs Considering Converter Limitations. *IEEE Trans. Power Deliv.* **2021**, *36*, 2097–2108. [CrossRef]
26. Liang, Y.; Li, W.; Huo, Y. Zone I Distance Relaying Scheme of Lines Connected to MMC-HVDC Stations During Asymmetrical Faults: Problems, Challenges, and Solutions. *IEEE Trans. Power Deliv.* **2021**, *36*, 2929–2941. [CrossRef]
27. ENTSO-E. Network Code on Requirements for Grid Connection of High Voltage Direct Current Systems and Direct Current-Connected Power Park Modules. Available online: https://www.entsoe.eu/network_codes/hvdc/ (accessed on 11 February 2025).
28. *IEEE Std 2800–2022*; IEEE Standard for Interconnection and Interoperability of Inverter-Based Resources (IBRs) Interconnecting with Associated Transmission Electric Power Systems. IEEE: Piscataway, NJ, USA, 2022; pp. 1–180. [CrossRef]
29. Rodríguez, P.; Teodorescu, R.; Candela, I.; Timbus, A.V.; Liserre, M.; Blaabjerg, F. New positive-sequence voltage detector for grid synchronization of power converters under faulty grid conditions. In Proceedings of the 2006 37th IEEE Power Electronics Specialists Conference, Jeju, Republic of Korea, 18–22 June 2006; pp. 1–7. [CrossRef]
30. O'Rourke, C.J.; Qasim, M.M.; Overlin, M.R.; Kirtley, J.L. A Geometric Interpretation of Reference Frames and Transformations: Dq0, Clarke, and Park. *IEEE Trans. Energy Convers.* **2019**, *34*, 2070–2083. [CrossRef]
31. Vidal-Albalate, R.; Forner, J. Modeling and Enhanced Control of Hybrid Full Bridge–Half Bridge MMCs for HVDC Grid Studies. *Energies* **2020**, *13*, 180. [CrossRef]
32. Blackburn, J.L.; Domin, T.J. *Protective Relaying*; CRC Press: Boca Raton, FL, USA; Taylor & Francis Group: London, UK, 2014.
33. Bollen, M.; Zhang, L. Different methods for classification of three-phase unbalanced voltage dips due to faults. *Electr. Power Syst. Res.* **2003**, *66*, 59–69. [CrossRef]
34. Kasztenny, B.; Finney, D. Fundamentals of Distance Protection. In Proceedings of the 61st Annual Conference for Protective Relay Engineers, College Station, TX, USA, 1–3 April 2008; pp. 1–34. [CrossRef]
35. Tu, Q.; Li, Y.; Liu, W.; Huang, M.; Zeng, G.; Du, B.; Wu, Z. Arm overcurrent protection and coordination in MMC-HVDC. In Proceedings of the 2018 IEEE Power & Energy Society General Meeting (PESGM), Portland, OR, USA, 5–10 August 2018. [CrossRef]
36. Cui, S.; Kim, S.; Jung, J.J.; Sul, S.K. A comprehensive cell capacitor energy control strategy of a modular multilevel converter (MMC) without a stiff DC bus voltage source. In Proceedings of the Applied Power Electronics Conference and Exposition (APEC), 2014 Twenty-Ninth Annual IEEE, Fort Worth, TX, USA, 16–20 March 2014; IEEE: Piscataway, NJ, USA, 2014; pp. 602–609.
37. Tu, Q.; Xu, Z.; Xu, L. Reduced Switching-Frequency Modulation and Circulating Current Suppression for Modular Multilevel Converters. *IEEE Trans. Power Deliv.* **2011**, *26*, 2009–2017. [CrossRef]

MDPI AG
Grosspeteranlage 5
4052 Basel
Switzerland
Tel.: +41 61 683 77 34

Electronics Editorial Office
E-mail: electronics@mdpi.com
www.mdpi.com/journal/electronics

www.ingramcontent.com/pod-product-compliance
Lightning Source LLC
Chambersburg PA
CBHW061805210926
51850CB00006B/615
9783725862702